RADAR
DETECTION

THE ARTECH RADAR LIBRARY
David K. Barton, Editor

Laser Radar Systems and Techniques
Christian G. Bachman

MTI Radar
D. Curtis Schleher

Radar Anti-Jamming Techniques:
A Translation from the Russian
of Zashchita Ot Radiopomekh
M.V. Maksimov *et al.*

Phased Array Antennas
Arthur A. Oliner
George H. Knittel

Radar Detection and Tracking Systems
Shahen A. Hovanessian

Radar Signal Simulation
Richard L. Mitchell

Radar System Analysis
David K. Barton

Radar Technology
Eli Brookner

RADARS — in seven volumes
I.   Monopulse Radar
II. The Radar Equation
III. Pulse Compression
IV. Radar Resolution and Multipath Effects
V.   Radar Clutter
VI. Frequency Agility and Diversity
VII. CW and Doppler Radar

David K. Barton

Significant Phased Array Papers
Robert C. Hansen

Synthetic Aperture Radar
John J. Kovaly

Introduction to Synthetic Array and Imaging Radars
Shahen A. Hovanessian

Radar Electronic Counter-Countermeasures
Stephen L. Johnston

RADAR
DETECTION

J. V. DiFranco

Research Department Head for Advanced Radar Studies,
Sperry Gyroscope Division of Sperry Rand Corporation

and

W. L. Rubin

Director of Research,
Sperry Gyroscope Division of Sperry Rand Corporation
and Adjunct Professor of Electrical Engineering, New York University

ARTECH HOUSE, INC. ● DEDHAM, MA

To our wives and families

PREFACE

The purpose of this book is to present a comprehensive tutorial exposition of radar detection using the methods and techniques of mathematical statistics. The application of the statistical approach to radar has resulted not only in a better understanding of the performance of existing radars but also in the optimization of new sophisticated radar systems.

The book is intended as a text for a graduate course in radar detection theory, and as a reference for the practicing radar engineer who seeks optimal radar system designs. These dual objectives are reflected in the organization of the book's subject matter, in the inclusion of problems and exercises following each chapter, and in the presentation of important practical results on large full-page graphs.

The book is divided into six parts; of these the first two are introductory. The first part is included for those unfamiliar with radar. The second part includes the mathematical tools that are necessary for a study of detection theory. Part III contains comprehensive tutorial expositions in a radar context of the classical signal-to-noise and a posteriori theories; both have played important roles in the evolution of modern radar. The unifying theme of the book, however, is provided by statistical decision theory, which is introduced in the last chapter of Part III; this discussion provides the framework for the chapters that follow.

The first three chapters of Part IV contain a unified tutorial treatment of single-hit and multiple-hit detection. Detailed results for frequently used radar target models are presented in approximately sixty full-page graphs for the calculation of expected radar detection performance. The final two chapters of Part IV are devoted to the application of detection theory; they include a chapter on the use of the radar equation and a discussion of cumulative detection probability, including a procedure for minimizing the power-aperture product of a search radar.

In Part V the performance of important near-optimum multiple-hit detection techniques are considered from a tutorial viewpoint. These include binary detection and delay-line weighted integrators. Part VI is devoted to two advanced topics in detection theory. First, sequential detection theory is applied to the radar detection problem. This discussion includes the Marcus and Swerling test strategy and a two-step approximation to sequential detection. The book concludes with the development of Bayes decision rules and Bayes receivers for optimizing the detection of multiple targets with unknown parameters such as range, velocity, angle, and so forth.

This book evolved from a one year after-hours course in modern radar theory prepared by both authors and presented to engineers at the Sperry Gyroscope Division of the Sperry Rand Corporation. The authors wish to acknowledge the numerous suggestions and ideas offered by their students at Sperry during the meetings of the radar course and by their colleagues, who reviewed the book. In particular, discussions with C. Kaiteris, P. Develet, P. Kahrilas, C. Cook, and M. Bernfeld resulted in many insights from which the book has benefited. Many thanks are due to the management of the Sperry Rand Corporation for permission to publish the book and for enrichment of the text provided by the many radar programs and studies with which the authors were associated at Sperry.

Special thanks are due to Mrs. Eloise Dodd, who typed most of the preliminary and final material. In addition, we wish to thank Mr. Angelo Cincotta for his preparation of many of the graphs and figures used in the text and for his general assistance in the overall preparation of the manuscript. Mr. Edward Cheatham is deserving of our gratitude for his continuous assistance in the preparation of the manuscript and figures.

J. V. DiFranco
W. L. Rubin

CONTENTS

PART I: INTRODUCTION, 1

1 INTRODUCTION TO RADAR, 3

 1.1. Introduction, 3
 1.2. Recent Developments, 5
 1.3. Radar Classifications, 7
 1.4. The Search Problem, 8
 1.5. The Track Problem, 16
 1.6. Nature of the Radar Statistical Problem, 24
 1.7. The Uses of Radar and Special-Purpose Systems, 29
 1.8. Integration of Signal and Data Processing, 30

 References, 38

PART II: MATHEMATICAL DESCRIPTION OF SIGNAL AND
NOISE WAVEFORMS, 41

2 MATHEMATICAL REPRESENTATION OF WAVEFORMS, 43

2.1. Introduction, 43
2.2. Fourier Integral Representation, 44
2.3. Fourier Series Representation, 49
2.4. Low-Frequency (Video) Sampling Theorem, 51
2.5. Real and Complex (Analog) Representations of Narrowband Waveforms, 57
2.6. High-Frequency Sampling Theorem, 61
2.7. Representation in Terms of Orthogonal Functions, 66
2.8. Some Useful Relations, 68

Problems, 71
References, 73

3 PROBABILITY THEORY, 74

3.1. Introduction, 74
3.2. Definition of Probability, 74
3.3. Sample Space, 76
3.4. Independent Events, 78
3.5. Conditional Probability, 79
3.6. Random Variables, 79
3.7. Probability Distribution and Probability Density Functions, 81
3.8. Joint Probability Distribution and Density Functions, 84
3.9. Sums of Random Variables, 86
3.10. Transformations of Random Variables, 88
3.11. Statistical Averages, 89
3.12. Characteristic Functions, 94
3.13. The Gaussian (Normal) Distribution and the Central Limit Theorem, 96
3.14. The Multidimensional Gaussian (Normal) Distribution, 98
3.15. Summary, 101

Problems, 104
Reference, 107
Recommended Bibliography, 107

4 RANDOM PROCESSES, 108

4.1. Introduction, 108
4.2. Characterization of a Random Process, 109

4.3. Statistically Independent Random Processes, 112

4.4. Stationarity and Ergodicity, 113

4.5. Power Spectral Density, 115

4.6. Wiener-Khintchine Theorem, 117

4.7. Shape of Power Spectral Density, 121

4.8. The Gaussian Random Process, 122

4.9. Fourier Series Representation of a Gaussian Random Process, 124

4.10. (Karhunen-Loéve) Orthonormal Expansion of a
Gaussian Random Process, 127

4.11. The Narrowband Gaussian Random Process, 129

Problems, 136
Recommended Bibliography, 138

PART III: OPTIMUM RECEPTION OF SIGNALS
IN NOISE, 139

5 MAXIMIZATION OF SIGNAL-TO-NOISE RATIO AND
THE MATCHED FILTER, 143

5.1. Introduction, 143

5.2. Maximizing Signal-to-Noise Ratio, 144

5.3. The Matched Filter, 146

5.4. Example I: Matched Filter Approximation for a Rectangular rf Pulse, 149

5.5. Example II: Matched Filter Approximation for a Linear fm Pulse with
Rectangular Envelope, 155

5.6. Example III: Matched Filter Approximation for Finite Pulse Train
(Comb Filter), 161

5.7. Example IV: Matched Filter for Digital-Coded Pulse with Rectangular
Envelope, 167

5.8. The Stationary Clutter Filter (Urkowitz Filter), 170

5.9. Parameter Estimation and Matched Filtering, 177

5.10. Remarks, 181

Problems, 181
References, 184

6 OPTIMUM FILTER FOR COLORED NOISE, 185

6.1. Introduction, 185

6.2. Colored *RC* Noise, 186

6.3. Solution of the Integral Equation for the Rational Kernel, 191

6.4. A Mixed Kernel (White plus Colored Noise), 193
6.5. Colored Noise with Infinite Observation Interval, 196
6.6. Maximum Signal-to-Noise Ratio, 198
    Problems, 199
    References, 200

7   THE A POSTERIORI THEORY OF RECEPTION, 202

7.1. Introduction, 202
7.2. Information Measure—Discrete Case, 203
7.3. Information Measure—Continuous Case, 208
7.4. Ideal Receiver, 209
7.5. The A Posteriori Distribution in the Presence of White Gaussian Noise, 210
7.6. Sufficient Receiver (Correlation Receiver), 214
7.7. Estimating Radar Range of a Stationary Target with Known Cross Section, 219
7.8. A Posteriori Detection, 229
7.9. Band-Limited Signals in Colored Noise, 233
7.10. Arbitrary Signals in Colored Noise, 239
7.11. Signals Perturbed by a Random Channel, 243
7.12. Summary, 249
    Problems, 249
    References, 251

8   STATISTICAL DECISION THEORY, 252

8.1. Introduction, 252
8.2. Elements of a Statistical Decision Problem, 253
8.3. Reception as a Decision Problem, 257
8.4. Definition of Loss Function, 261
8.5. Binary Detection, 263
8.6. Bayes Decision Rule, 265
8.7. Examples of Bayes Detectors, 267
8.8. Error Probabilities, 270
8.9. The Ideal Observer, 271
8.10. The Neyman-Pearson Criterion, 272
8.11. The Minimax Approach, 274
8.12. Review, 277
8.13. Bayes Solutions for Complex Cost Functions, 278
8.14. *Preferred* Neyman-Pearson Strategy, 280
8.15. *Intuitive* Substitute, 281
8.16. Fixed and Sequential Testing, 282
8.17. Concluding Remarks, 283
    Problems, 284
    References, 285

PART IV: OPTIMUM RADAR DETECTION, 287

9 DETECTION BASED ON A SINGLE OBSERVATION, 289

9.1. Introduction, 289
9.2. Optimum Detection of an Exactly Known Signal, 291
9.3. Optimum Detection of a Signal of Unknown Phase, 298
9.4. Optimum Detection of a Rayleigh Fluctuating Signal, 309
9.5. Optimum Detection of a One-Dominant-Plus-Rayleigh Fluctuating Signal; Structure and Performance, 313
9.6. Analytic Approximations, 315
9.7. Likelihood Ratio for Colored Noise, 319
9.8. Optimum Detection of an Exactly Known Signal in Colored Noise, 324
9.9. Optimum Detection of a Signal of Unknown Phase in Colored Noise, 328
9.10. Optimum Detection of a Signal of Unknown Phase and Amplitude in Colored Noise, 332
9.11. Summary, 332

Problems, 333
References, 335

10 DETECTION BASED ON MULTIPLE OBSERVATIONS: NONFLUCTUATING MODEL, 336

10.1. Introduction, 336
10.2. Nonfluctuating Coherent Pulse Train, 338
10.3. Optimum Receiver Structure for a Nonfluctuating Incoherent Pulse Train, 339
10.4. Small-Signal Optimum Receiver Performance, 344
10.5. Integration Loss for Nonfluctuating Incoherent Detector, 359
10.6. Gram-Charlier Series, 359
10.7. Analytic Approximations (Nonfluctuating Incoherent Detector), 364
10.8. Linear versus Square-Law Detection, 370

Problems, 371
References, 372

11 DETECTION BASED ON MULTIPLE OBSERVATIONS: SWERLING FLUCTUATING MODELS, 373

11.1. Introduction, 373
11.2. Scan-to-Scan Rayleigh Fluctuating Incoherent Pulse Train (Swerling I), 374
11.3. Pulse-to-Pulse Rayleigh Fluctuating Incoherent Pulse Train (Swerling II), 392

11.4. Scan-to-Scan One-Dominant-Plus-Rayleigh Fluctuating Incoherent Pulse Train (Swerling III), 407

11.5. Pulse-to-Pulse One-Dominant-Plus-Rayleigh Fluctuating Incoherent Pulse Train (Swerling IV), 424

11.6. Performance of Partially Correlated Fluctuating Pulse Trains, 439

11.7. Summary, 440

Problems, 444

References, 445

12   THE RADAR EQUATION, 446

12.1. Introduction, 446

12.2. Classical Radar Equation, 446

12.3. Radar Equation and Signal-to-Noise Ratio, 448

12.4. Antenna Gain and Effective Aperture, 450

12.5. Radar Cross Section, 455

12.6. Noise Factor and Noise Temperature, 456

12.7. System Loss Factor, 461

12.8. Example of Radar Equation Calculation, 467

12.9. Search Radar Equation, 468

12.10. Summary of Search Radar Status, 470

12.11. The Beacon and Jamming Radar Equations, 472

Problems, 473

References, 475

13   CUMULATIVE DETECTION PROBABILITY OF STATIONARY AND MOVING TARGETS, 476

13.1. Introduction, 476

13.2. Cumulative Detection Probability, 477

13.3. Cumulative Detection Probability of Stationary Targets, 477

13.4. Cumulative Detection Probability of Moving Targets with a Uniformly Scanning Radar, 481

13.5. Minimization of Power-Aperture Product, 484

Problems, 493

References, 494

PART V: SUBOPTIMUM DETECTION TECHNIQUES, 495

14   THE BINARY DETECTOR, 497

14.1. Introduction, 497

14.2. Description of Binary Detector, 498

14.3. Performance of Binary Integrator, 499

14.4. Extension to Fluctuating Targets, 514

14.5. Coincidence Detection, 515

Problems, 518

References, 519

15 WEIGHTED INTEGRATORS, 521

15.1. Introduction, 521

15.2. Implementation, 524

15.3. Parameter Optimization, 528

15.4. Performance Evaluation, 531

15.5. Detection Probability for Uniform Integration of Constant Amplitude Pulses, 536

15.6. Detection Probability for a Single Loop-Integrator Assuming a Gaussian Pulse Train Envelope, 538

15.7. Detection Probability for a Double-Loop Integrator Assuming a Gaussian Pulse Train Envelope, 540

15.8. Comparison of Single- and Double-Loop Integrators, 542

Problems, 542

References, 543

PART VI: SPECIAL TOPICS IN DETECTION, 545

16 SEQUENTIAL DETECTION, 547

16.1. Introduction, 547

16.2. Notion of a Sequential Test, 548

16.3. The Operating Characteristic Function (OCF), 550

16.4. The Average Sample Number (ASN), 553

16.5. Coherent Detection in White Gaussian Noise, 556

16.6. Coherent Detection in Colored Gaussian Noise, 559

16.7. Wald's Fundamental Identity, 562

16.8. The Characteristic Function of n, 564

16.9. The Distribution of n for the Coherent Detector, 566

16.10. Some Remarks on the PDF of n, 571

16.11. ASN for Incoherent Detection of Signals of Unknown Phase, 572

16.12. The Need for Truncation, 575

16.13. Application of Sequential Detection to a Multiple-Resolution-Bin Radar, 575

16.14. Analogue Marcus and Swerling Test, 582

16.15. Binomial Marcus and Swerling Test, 588

16.16. Average Loss Formulation, 591

16.17. Two-Step Sequential Detection, 592

Problems, 595

References, 597

17   MULTIPLE-TARGET DETECTION, 598

    17.1. Introduction, 598
    17.2. Approach to the Problem, 599
    17.3. Detection of Two Known Targets, 599
    17.4. Interpretation of Two-Target Detection Strategy for Small Signal-to-Noise Ratios, 603
    17.5. Interpretation of the Two-Target Detection Strategy for Large Signal-to-Noise Ratios, 605
    17.6. Bayes Strategy for N Exactly Known Signals, 606
    17.7. Bayes Detection Strategy for N Targets with Unknown Times of Arrival, 609
    17.8. Existence of a Single Target with Unknown Time of Arrival, 611
    17.9. Approximation of Bayes Receiver, 613
    17.10. Extension of Bayes Receiver to Angle and Doppler, 615
    17.11. Difficulties with the Theory, 618

       Problems, 620
       References, 620

APPENDIX **A**   NARROWBAND REPRESENTATION OF ECHOES FROM MOVING TARGETS, 623

APPENDIX **B**   FALSE-ALARM RATE OF NARROWBAND NOISE, 626

GLOSSARY OF PRINCIPAL SYMBOLS, 630

AUTHOR INDEX, 645

SUBJECT INDEX, 647

PART I

INTRODUCTION

1

INTRODUCTION
TO RADAR

1.1. Introduction

Radar has evolved from a very simple device for the detection of aircraft in the 1930's to the present almost wholly automatic systems in which the radar sensor is an integrated part of a computer complex with completely programmed functions, decision-making operations, and self-check features. The major impetus for the rapid development of radar was the urgent air-defense requirements of the Second World War; this stimulus led to many advances in radar technology in a relatively short period of time. Radar was raised to a level of sophistication comparable to older technologies in a period of less than ten years.†

The postwar era saw an accelerated development of new high-performance weapons and devices, including missiles and satellites. The perfor-

†The Radiation Laboratory Series of 28 volumes prepared by members of the MIT Radiation Laboratory and published by McGraw-Hill between 1946 and 1950 provides a detailed record of this advance.

mance capabilities of these weapons created in turn a need for sensor systems capable of early detection, multiple target sorting, rapid and accurate tracking, and weapon guidance. This need accelerated the development of radar into a new phase in which radar search and track functions have been integrated into computer-directed automatic systems. The prophetic words in Volume I [1] of the Radiation Laboratory Series have indeed come to pass (p. 214): "The major improvements to be looked for in the use of radar over the next few years will lie, for the most part, not in the category of technical radar design, but in the field of fitting the entire radar system, including its operational organization, to the detailed needs of the use and the user." System integration and complex functional duties are the most distinguishing features of modern radar systems. The significant development in technology that has occurred in this area is exemplified by the evolution from the celebrated "fruit machine" employed by the British during the Second World War to the modern sophisticated high-speed digital computer. The "fruit machine" was a simple slot-machine type of computer, consisting of selector switches and relays, which corrected radar data in the British Home Chain (CH) systems.†

The ability of a radar to measure target range—its greatest asset—has been extended to include target direction, velocity, and size. Further, these capabilities are available under conditions that normally impair optical vision, such as night, rain, fog, smoke, snow, and clouds. Not all of these capabilities are available at low frequencies (less than 100 MHz), where the measurement of direction becomes severely hampered by restrictions on antenna aperture size, or at extremely high frequencies (greater than 15,000 MHz), where the attenuation of electromagnetic energy due to weather conditions becomes excessive, and hence radar has achieved its greatest growth in the microwave spectrum. When weather constraints are absent, such as in outer space, the principal advantage of microwave radar over optical systems, for example, is diminished, and sensors located in these regions may employ higher and higher frequencies as new power-generation techniques become available.

Although radar has found maximum utilization in earthbound functions, its earthbound arms are being extended thousands of miles out into space for the detection, inspection, tracking, and control of space vehicles; radar astronomy is also a rapidly developing science. The earthbound uses of radar include mapping, weather prediction, guidance, identification, navigation, air traffic control, aircraft landing, altitude measurement, collision avoidance, and many others, and are continually increasing. It may be surmised that the applications of radar will grow with time to aid in man's domination of his environment.

†See reference [1], p. 226.

1.2. Recent Developments

Since the Second World War radar has advanced in sophistication in both technology and theory. These developments, spurred primarily by military needs, have taken place in universities, industrial concerns, and U.S. government laboratories.

Microwave tubes have been developed that are capable of high power, high gain, and large bandwidth, such as klystrons and traveling-wave tubes. These tubes permit the amplification of complex coded radar waveforms to extremely high power levels and have all but replaced the magnetron, which dominated the transmitter design of early radars.

Low-noise preamplifiers, such as parametric amplifiers, masers, and low-noise traveling-wave tubes, have been developed that reduce internally generated noise to a level below that received from the external environment. These receiver developments, together with substantial increases in transmitted power levels, have improved detection ranges by several orders of magnitude.

Radar has been significantly affected by developments in solid state technology. One result has been the development of highly reliable, lightweight radars for portable use. Another is the introduction of digital computers and digital logic into small battlefield surveillance radars as well as giant search and tracking radars.

Early search and track radars employed large dishes, or heavy mechanical antennas, which were subject to rotational problems caused by large inertia. With the advent of electronic scan radars employing phased array types of antennas, extremely rapid changes in beam position are possible with minimal inertial effects. This development has also facilitated the generation of extremely high radiated-power levels by the use of a multiplicity of low-power transmitting devices connected to one or more array elements. In addition, array steering lends itself to computer control and rapid data handling and has facilitated the union of radar and computer technology.

Advances in radar techniques have been paralleled by significant advances in radar theory. Radar theory has been raised to a high level of sophistication and has become an important branch of statistical communication theory. Too much space would be required to name all the significant contributors in this field; hence we shall mention only a few of the outstanding innovators.

The application of information theory to radar was stimulated by the studies of Woodward [2]. He introduced the important concept of the "ideal receiver," which is discussed in detail in Chapter 7 of Part III. Woodward also introduced a signal function called the *ambiguity function*, which provides a useful measure of radar resolution and ambiguity. This function, which depends upon the radar waveform, has stimulated studies into signal design and optimization techniques. These studies in turn have yielded insight and understanding in such related areas as target detection in noise and clutter, antenna theory, parameter estimation, and matched filter theory. Woodward's results acted as a catalyst to other early investigators; Siebert, for example, extended some of Woodward's results in a well-known paper [3] in 1956.

Matched filter theory was introduced by D. O. North [4] in a classic report in 1943, which has since been republished as a special paper in the *Proceedings of the IEEE* (1963). Considerable work has been done in refining the matched filter concept and particularizing it to different signals and special problems. This approach, discussed in detail in Chapters 5 and 6, is based on the concept of maximizing signal-to-noise ratio for a known signal. The matched filter solution can also be derived from consideration of either statistical decision theory or information theory. The matched filter concept facilitates an understanding of optimum filter theory for signals immersed in white- and colored-noise backgrounds. In addition, signal-coding techniques such as pulse compression [5] can be interpreted and better understood with this approach.

A comprehensive analysis of the statistics of target detection is contained in the classic works of Marcum [6] and Swerling [7]. Marcum considered the problem of detecting an incoherent train of nonfluctuating pulsed signals in a random white Gaussian noise background. He showed that postdetection integration is optimum for processing an incoherent train and he related detection performance to that obtained using coherent or predetection integration. These results were extended by Swerling to include fluctuating targets whose radar cross section could be approximated either by a large number of small scatterers or by a large scatterer plus many small scatterers. These two target models have been extensively documented by experimental data and are widely used by engineers. The contributions of Marcum and Swerling are considered in Chapters 10 and 11, respectively.

In many radars decisions must be made at speeds beyond the capability of an operator, and so the operator must be replaced by automatic signal-processing and data-processing equipment. A design philosphy for such equipment is provided by statistical decision theory, which is introduced in Chapter 8.† Statistical decision theory provides insight into both the signal-detection problem and the signal-extraction problem. In addition, it satisfies the need for a general theory into whose framework much of radar theory fits. It seems likely that decision theory will be increasingly used to solve new and difficult radar problems, such as the resolution of closely spaced multiple targets.

---

†Decision theory is extensively treated in the exhaustive work of Middleton [8]. Related books on this subject are by Helstrom [9] and Wainstein and Zubakov [10].

This section has reviewed the principal developments in radar technology and theory. The direction these developments have taken has been profoundly influenced by the trend toward radar system automatization and the marriage of radar sensors with computer technology. Because radar is still in an active stage of development, some of the problems discussed in this book have not been completely resolved and discussions pertaining to them may be somewhat controversial.

1.3. Radar Classifications

The search or surveillance function of a radar is primarily concerned with the elementary question: *Is a target present?* In radar theory this simple question is referred to as the binary detection problem—that is, *yes* or *no* is the final expected result. Search radars are planned and designed with major emphasis on the solution of the detection problem. For this purpose the entire search volume, which consists of some prescribed sector of space, must be filled with sufficient average power to provide reliable detection within the given region and within the allotted time permitted by a specific surveillance problem.

Another major radar classification is the tracking radar. It is discussed in greater detail in Section 1.5 of this chapter. This type of radar is intimately related to the signal-extraction problem or, as it is often called, parameter estimation; that is, once target existence has been established, we ask: *What are the significant target parameters such as range, angle, and so on?* This is, of course, an oversimplified statement of the problem, but it serves to distinguish a tracking radar from a search radar.

No radar is either a pure search or a pure tracking radar; most radars perform both functions in varying degrees. However, the major distinguishing features of a radar usually permit its classification as search or track. Signal-processing and data-processing differ considerably between systems. Indeed, it will become clear in subsequent discussion that the optimum solution for a search radar is not the same as that for a tracking radar.

A notable exception is the electronic scan array radar system, which cannot easily be classified in the elementary fashion above since it can perform both functions with nearly equal facility. Generally, an electronic scan array radar is programmed on a time-shared basis to perform both surveillance and tracking functions, as well as many others. As a result, the equipment design represents an integration of the optimum structures for both radar types.

Nevertheless, the separation of radar systems into search or track permits a better understanding of the concepts of target detection and signal extraction, and this organization will be retained in most of our discussions.

There are, however, several radar problems in which the two theoretical approaches cannot be easily separated. For example, the multitarget resolution problem cannot be classified simply as a question of detection or extraction, for the question that must be answered is: *How many targets are present and where are they located?* The solution involves the joint application of both approaches to the question via some common organizing principle. This joint detection and extraction problem is treated within the framework of statistical decision theory and is discussed at greater length in a companion volume, which is in preparation.

1.4. The Search Problem

The simplest type of search problem consists of detecting moving targets in a region surrounding the radar site as shown in Fig. 1.4-1. A simple vertical antenna is employed which provides almost complete volumetric coverage surrounding the radar site (except for a hole directly above the antenna). In order to detect any object entering this region a radar can employ a high-power oscillator operating in a cw mode. Detection is accomplished by means of a bank of doppler filters which are matched to all possible radial target velocities. Stationary targets (or very slow-moving targets) cannot be detected, since their reflected signals are practically at the same frequency as the transmitted signal and are orders of magnitude smaller, and therefore they are masked by the transmitted signal. It is also usually necessary in a cw radar to suppress the transmitted carrier-frequency signal in the remaining filters in the doppler bank, since all received signals are many orders of magnitude less than the transmitted signal.

A pulse system, as shown in Fig. 1.4-2, avoids this isolation problem and permits detection of stationary (and very slow-moving) targets. In this radar the receiver is turned off during transmission and turned on again after the transmitted pulse has been radiated. Further, the use of a pulse transmission introduces a capability for measurement of target range. Each

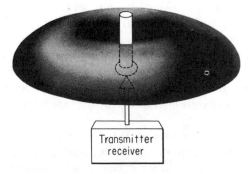

Fig. 1.4-1. Elementary search system.

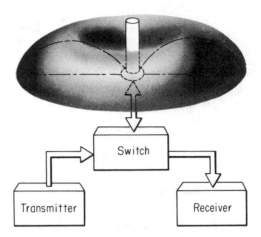

Fig. 1.4-2. Elementary search system.

microsecond of delay of the reflected pulse corresponds to an increment in target range of 164 yards.

Owing to transmitter power limitations and background noise in the receiver the maximum range of a radar is limited to detection of targets whose radar cross section is greater than some minimum value. Assume, for example, a detection range of 100 statute miles, which corresponds to a maximum echo delay of roughly 1080 microseconds. At the end of this period the system can theoretically be turned off. In practice the radar interrogation may be repeated every 1080 microseconds; this corresponds to a prf (pulse repetition frequency) of approximately 925 cps. In many applications it is unreasonable to interrogate the entire region almost 1000 times a second, since there are few objects that could pentetrate deeply into the search volume before being detected. Suppose a maximum penetration of 1 per cent into the search volume or approximately one mile in depth (see Fig. 1.4-3) before detection is permitted. Then, an aircraft flying at a speed of 900 miles per hour, for example, will take four seconds to penetrate this outer guard ring. If, instead of the volumetric search pattern previously described, the azimuth beamwidth is set to one degree, as shown in Fig. 1.4-4, each successive region in the entire volume can be scanned sequentially

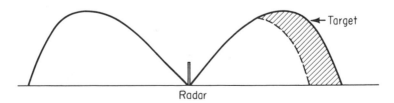

Fig. 1.4-3. Permissible target penetration.

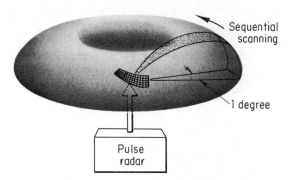

Fig. 1.4-4. Elementary scan surveillance.

in roughly 0.4 second. This will provide ten opportunities to detect a target before it has penetrated the outer one-mile zone. This system appears more attractive, since the antenna aperture may be increased by approximately 400, permitting a reduction in the transmitted pulse energy by roughly a factor of 52 db.† In addition, it is now possible to determine the angular position of a detected target as well as its range.

From the illustrative example it can be seen that a detection radar not only answers the question of target existence but also provides information as to gross target position. Figure 1.4-5 shows pictorially how detection is closely related to the existence of a target in a given resolution cell whose dimensions depend on antenna beamwidth and the transmitted pulse width (assuming a simple uncoded pulse transmission). A search radar effectively interrogates every system resolution cell in the surveillance region, that is, every range element in each angular beamwidth is tested successively for target existence.‡ The information on gross target position provided by

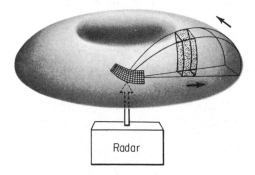

Fig. 1.4-5. Resolution element of simple search radar.

†See Chapter 12.
‡This approach is developed at length in Chapter 17.

detection in a resolution cell is seldom sufficient and some parameter estimation is almost always required in search functions. We shall see later that parameter estimation is a more sophisticated problem than the location of an object in a resolution cell. Parameter estimates are usually sought that are an order of magnitude more precise than the size of the resolution cell in which an object is located.

The radar described above is an important type of surveillance radar, distinguished by a narrow azimuth beamwidth and a broad elevation beam (sometimes called a fan beam). A radar of this class is shown in Fig. 1.4-6. This is a long-range sophisticated search radar for both military and non-

Fig. 1.4-6. FPS-35 long-range search radar (*courtesy of Sperry Rand Corporation*).

military uses. An airborne radar of the fan-beam type employed for ground or surface search is shown conceptually in Fig. 1.4-7.

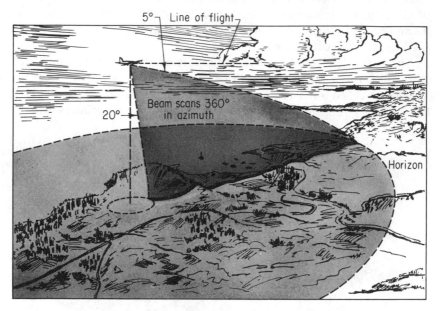

Fig. 1.4-7. A fan-beam airborne radar (*after Ridenour* [1], *courtesy of McGraw-Hill Book Company*).

A modification of the fan-beam search radar that also provides elevation information is the V-beam radar (Fig. 1.4-8). Two fan beams are generated at different angles, and the length of time required for an object to penetrate both beams, at a particular antenna rotation rate, is a measure of the target elevation angle. An early radar of this type is shown in Fig. 1.4-9 and a more sophisticated one in Fig. 1.4-10. The radar shown in Fig. 1.4-10 employs two separate antenna feeds to form the two beams with a dual reflecting surface. The beams are separated by polarization techniques. A radar with the addition of height-measuring capability is commonly called a 3-D radar.

We have considered a simple example above in which there were roughly ten target-detection opportunities in a four-second surveillance period. If elevation-angle information is also desired, the elevation search angle can be divided into ten parts as shown in Fig. 1.4-11. This beam shape is commonly called a pencil beam. In order to search the ten vertical beam positions at each azimuth angle, the antenna rotation rate must be slowed by a factor of ten. In this case every target in the original search volume is interrogated at least once in four seconds. As a result, no target can penetrate the outer guard ring more than one mile before a detection opportunity has occurred. Note that the antenna aperture has been increased by an addi-

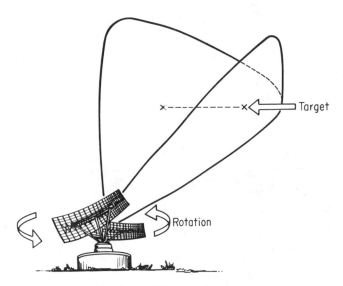

Fig. 1.4-8. V-beam radar concept.

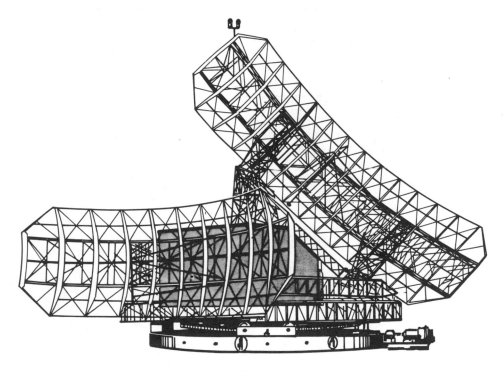

Fig. 1.4-9. Early V-beam radar (*after Ridenour* [1], *courtesy of McGraw-Hill Book Company*).

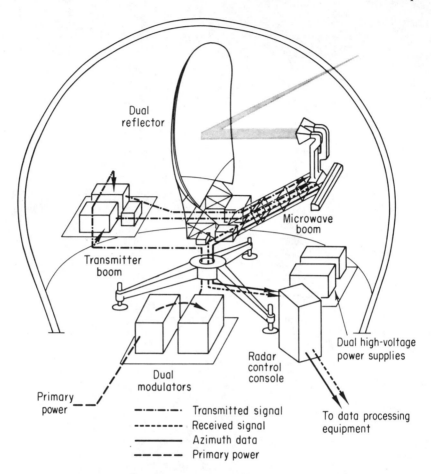

Dual
reflector

Microwave
boom

Transmitter
boom

Dual high-voltage
power supplies

Radar
control
console

Dual
modulators

Primary
power

To data processing
equipment

-·-·-·- Transmitted signal
-------- Received signal
———— Azimuth data
- - - - - Primary power

Fig. 1.4-10. TPS-34 tactical V-beam radar (*courtesy of Sperry Rand Corporation*).

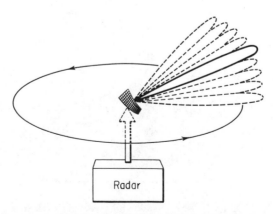

Radar

Fig. 1.4-11. Pencil beam search.

tional factor of ten so that transmitter pulse energy can be reduced further by 20 db. Elevation-angle information is now available by virtue of the narrow elevation beam.

The volumetric search may also be accomplished in a fencelike manner by scanning a ring at a constant elevation, then scanning a ring at the next elevation, and so on. A ring search pattern may be employed in a stacked beam system, in which simultaneous vertical receiver beams search a hemispherical volume, or in a limited sector scan radar employing electronic beam steering. An electronic scan radar is shown in Fig. 1.4-12. Radars of this

Fig. 1.4-12. HAPDAR, electronically steered array radar (*courtesy of Sperry Rand Corporation*).

type can also be arranged to provide almost complete hemispheric coverage by proper pointing of three or four array antenna surfaces, as shown in Fig. 1.4-13.

A resolution cell in a surveillance radar employing a pencil beam is shown in Fig. 1.4-14. The resolution cell size is approximately defined by pulse width, azimuth, and elevation beamwidths.† This elementary definition of resolution will be retained through most of the following discussions (see Chapter 17). Parameter estimation permits the subdivision of each resolution cell to achieve greater accuracy than the dimensions of the resolution cell.

†In more complex search radars, target doppler is frequently an additional dimension of a radar resolution cell (in addition to range and angles).

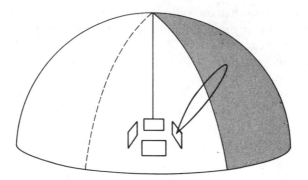

Fig. 1.4-13. Hemispheric search with array antennas.

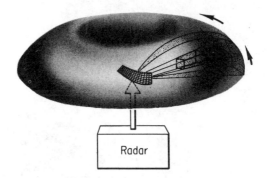

Fig. 1.4-14. Resolution cell in space.

1.5. The Track Problem

The tracking problem begins after a target is detected in a resolution cell, as shown in Fig. 1.4-14, and attempts to answer the question: *Where is the target within this cell?*† This problem is related to the extraction of the target echo out of background receiver noise, which ultimately limits the accuracy achieved by a tracking radar. The precise location of a target within a resolution cell is useful for a variety of purposes, such as prediction, communications, destruction, guidance, and navigation.

The location of a target in a resolution cell is especially significant when the dimensions of the cells are large. For example, consider a radar employing a 33-microsecond pulse and a one-degree rectangular beam. The

---

†Presumably the question of the existence of one and only one target in the cell has been answered. This complex question will be treated in greater detail in Chapter 17 and in the companion volume.

resolution cell, at a range of 200 miles, is roughly a 3-mile cube with a volume of 27 cubic miles. Better range accuracy is obtained by using the leading edge of the radar echo as a measure of target range; better angle information is obtained by simultaneously generating two adjacent antenna beams, commonly referred to as *monopulse tracking.* These techniques are more fully discussed in connection with parameter estimation in the companion volume.

Target parameters of frequent interest in radar are target cross section, radial distance (range), elevation and azimuth angle, radial velocity (doppler), radial acceleration, angular velocity, and angular acceleration. The number and type of parameters to be estimated are indications of the complexity of a radar system. In most cases tracking radars are more complex in structure and implementation than search radars, although the theory is simpler. Methods for estimating target parameters using a single hit (that is, a single target echo) have evolved in recent years. Early tracking radars employed multiple-hit information.

Let us select target azimuth angle as a representative parameter to be measured to give insight into the parameter-estimation problem. Using a single-beam antenna pattern, the target cannot be located more accurately than an angular beamwidth (with a single echo). With monopulse angle tracking an additional antenna beam is added to the search pattern as shown in Fig. 1.5-1. Let us designate the magnitude of the coherent sum of the two rf antenna outputs by Σ and their difference by Δ. Plots of Σ and Δ as a function of target angle are shown in Fig. 1.5-2. The difference Δ is a measure of target location off the antenna boresight (center of the two beams). However, since target echo magnitude also influences the amplitude of Δ, this quantity is not by itself a correct indication of target angular deviation from boresight; by dividing the difference Δ by the sum Σ, the amplitude of the target echo is eliminated from consideration. Thus,

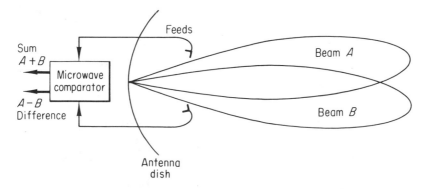

Fig. 1.5-1. Use of two beams for amplitude comparison.

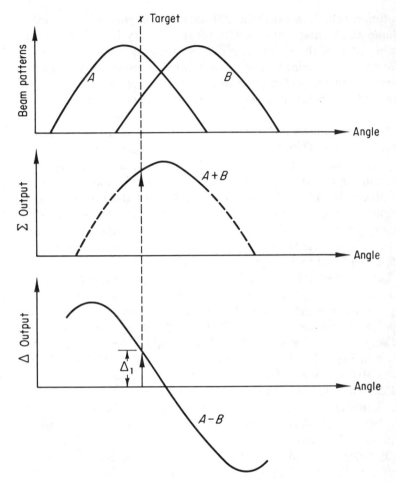

Fig. 1.5-2. Representative sum and difference patterns.

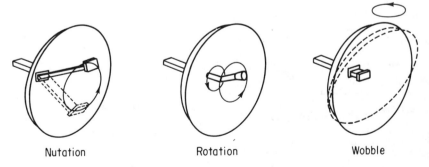

Nutation   Rotation   Wobble

Fig. 1.5-3. Three methods of obtaining conical scan (*after Brockner and Price* [12], *courtesy of Sperry Rand Corp.*).

Fig. 1.5-4. MIT Lincoln Laboratory Millstone Hill Radar
(*courtesy of MIT Lincoln Laboratory*).

Δ/Σ may be used as a measure of the angular deviation of the target from boresight to yield fine-grain angle estimation. The noise in the system will perturb the value of Δ/Σ and limit the absolute accuracy that can be achieved. However, at signal-to-noise levels normally required for target detection, beam splitting of an order of magnitude or more can be achieved with this technique.

The technique just described is called monopulse tracking, since angle information is obtained with a single pulse transmission. Other techniques employed for angle measurement that require more than one hit include *sequential lobing* (sequential beam switching) and *conical scan*. Many early trackers employed some form of conical scan by means of a rotary nutation of the feed horn with respect to the dish antenna, or a wobble of the dish, or other variants on this theme (see reference [12] and Fig. 1.5-3). A typical tracking radar employing conical scan is shown in Fig. 1.5-4.

Modern tracking radars generally employ some form of monopulse processing for angle estimation. To obtain angle estimates in both azimuth and elevation, a four-horn monopulse feed may be employed, as shown in Fig. 1.5-5. This feed develops a cluster of adjacent beams, which is re-

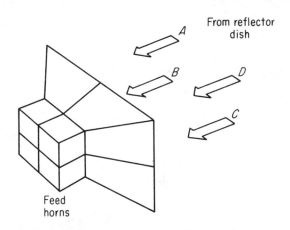

Fig. 1.5-5. Four-horn monopulse-feed configuration.

quired for an amplitude monopulse system. A typical monopulse radar is shown in Figure 1.5-6.

Range and doppler estimation are closely related to angle estimation. For example: the optimum structure of a range estimator, subject to certain constraints, involves signal normalization with respect to total pulse energy and a differencing operation in the form of an "early-late" gate. This is discussed in greater detail in the companion volume. Further, doppler estimation is similar to angle estimation, since doppler channels are analogous to angle beams.

When insufficient estimation accuracy is achieved on a single observation —say in angle—the accuracy can be improved by means of multiple-hit

Fig. 1.5-6. AN/FPQ-6 tracking radar (*courtesy of Radio Corporation of America*).

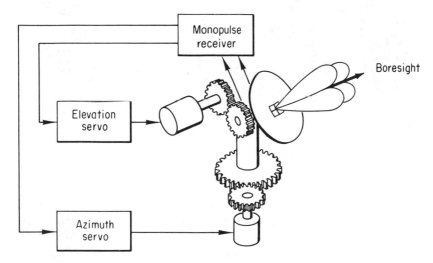

Fig. 1.5-7. Closed-loop null tracking system.

integration (smoothing). The early tracking radars all used multiple-hit integration accomplished through closed servo loops as shown in Fig. 1.5-7. A system of this type is called a *null tracking system*. Angular deviation errors were used to drive servos, which trained the antenna in a direction that reduced the target deviation from boresight. The angle deviation errors are usually obtained either by monopulse, sequential lobing or conical scan. Smoothing of the data by a properly designed narrowband servo tracking loop provides considerable integration and high accuracy. Some typical tracking radars are shown in Fig. 1.5-8.

Fig. 1.5-8. AN/SPG-49 Talos missile-tracking radars (*official U.S. Navy Photograph*).

In a null tracking system the angle estimate, or more generally the target angle trajectory estimate as a function of time, is provided by the positions traversed by the antenna boresight, since the antenna boresight continuously points at the smoothed (or best) estimate of the angular target position. Analog servos can be replaced by a digital computer that stores N observations over a period of time T. The computer loop is shown in Fig. 1.5-9. The computer, using a predetermined estimation technique, obtains

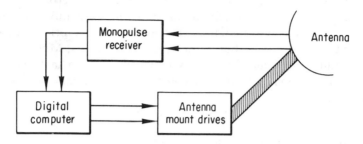

Fig. 1.5-9. Computer null tracking.

the curve or trajectory best fitting the data (for example, in the least-squares sense). If the computer data-storage time T is commensurate with the analog servo integration period, then the smoothed data in both cases are essentially equivalent (provided the order of prediction in both servo and computer are equivalent).

The use of the computer in parameter estimation is of great importance in modern radar technology. In addition to null track, a computer can provide interpolative tracking, which is highly desirable in many situations. For example, a search pattern may be conducted as shown in Fig. 1.5-10,

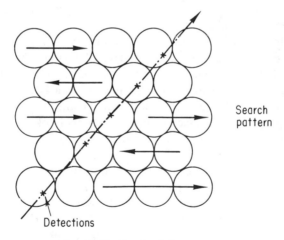

Fig. 1.5-10. Track-while-scan.

where each circle represents a monopulse cluster. Detections that occur off the boresight or center of a monopulse cluster result in angle-deviated monopulse signals, which the computer can curve-fit and smooth over N observations. A null tracking system is more accurate, since the sum beam signal-to-noise ratio is greatest at boresight; however, if interpolative data are weighted by the received sum beam signal-to-noise ratio off boresight, accuracy degradation is minimized. This technique is a form of track-while-scan and has wide application in electronic scan radars. Multiple functions are generally programmed by the system computer, so that search, track, and other functions are simultaneously conducted and adapted to the environment.

1.6. Nature of the Radar Statistical Problem

There are two aspects to the radar statistical problem. The first is concerned with the background noise, which is random in character. In the absence of background noise detection poses no difficulty; that is, however small the reflected signal from a target, in theory it may be detected with sufficient gain in the receiver. Background-noise interference, however, imposes a limit on the minimum detectable signal. The question of target existence is, in fact, a choice of deciding between noise alone or signal plus noise.

The noise background includes both random-noise interference, which is present in all systems, and often another type of noise of entirely different character called *clutter*. Random-noise interference exists in all physical devices at temperatures above absolute zero; clutter generally refers to unwanted back-scattered radar signals from ground objects, sea waves, rain, and chaff (metalized strips).

Random-noise interference arises from many sources, including radiation from the external environment, internal Johnson (thermal) noise, shot (vacuum-tube) noise, semiconductor (transistor) noise, and so on. This noise generally is wideband with a white—or nearly flat—spectral density. The noise is generally Gaussian in character and can be described by the statistical properties discussed in Part II of this book.

The noise performance of early radar receivers was described by means of a *noise-figure* criterion. This figure of merit, which related the increase in noise added by the receiver to the input noise, was particularly useful when the equivalent receiver temperature, referenced to the antenna input terminals, was of the order of a few thousand degrees Kelvin. However, with the development of low-noise preamplifiers such as masers and parametric amplifiers, receiver contributions have been reduced to the order of magnitude of external noise (a few hundred degrees Kelvin). As a result, receiver noise-figure has generally been replaced by *noise temperature* as a figure of merit.

Present systems are limited mainly by the black-body radiation of the antenna environment and the thermal noise generated in the microwave "plumbing." External noise temperature depends on both the antenna beam shape and pointing direction. For example, the noise level is much greater for an antenna pointing at the sun, or some other large radiating body, than when pointing at a cool region of the sky. These considerations complicate the radar problem, since noise levels may be dependent on radar function. For analytic purposes, it is convenient to assume that thermal noise is drawn from a stationary process.

The other major background-noise source has been referred to as clutter. This type of noise represents the aggregate radar return from a collection of many small scatterers. For example, ground return, sea return, reflections from rain, chaff, and decoy clouds represent unwanted target echoes when the desired target is an aircraft or ship. (Clutter must be carefully defined for each situation since clutter in one context may be a desired target in another. For example, while rain represents clutter to an aircraft-search radar, it is the desired information in a weather radar.) Large background clutter can mask a target and prevent reliable detection and signal-parameter estimation. Detection and estimation in a clutter environment is a major problem in modern radar.

A distinguishing feature between clutter and noise interference is that the clutter spectral density is inherently related to the transmitted signal spectrum, since it represents a composite return of the transmitted signal from a group of random scatterers. Background noise—in contrast—has a spectrum that is independent of the transmitted signal. In addition, random-noise waveforms are statistically independent on successive radar scans while clutter waveforms may be highly correlated. Further, while noise power is independent of range, clutter return is range-dependent.

The problems of detection and estimation yield different optimum solutions that depend on the nature of the noise background with which the target echo competes. Background-noise interference has been thoroughly studied and an adequate theory is available for solving problems. Solutions to the clutter problem in many cases are inadequate, owing to complexities in defining mathematically tractable clutter models. In general, a real radar system must be capable of operating against both types of background interference. When clutter is present, it is usually the limiting factor on system performance. There are intermediate cases where performance against both types of interference must be considered. Some of these will be studied in the companion volume.

The second statistical aspect of the radar problem stems from the reflective properties of radar targets. If the radar cross section of aircraft, or other complex target structures, is observed as a function of aspect angle, patterns are obtained similar to those shown in Figs. 1.6-1 and 1.6-2. The patterns are characterized by rapid fluctuations in amplitude with minute changes in aspect angle; for example, a 15-db change in cross section can occur for only a 0.3-degree change in angle. The behavior shown results

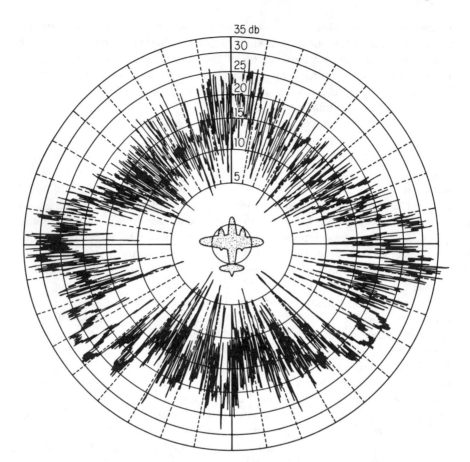

Fig. 1.6-1. Return power from an aircraft at 10-cm wavelength as a function of azimuth angle (*after Ridenour* [1], *courtesy of McGraw-Hill Book Company*).

from the phase-sensitive rf addition of returns from many point and/or surface scatterers making up a composite target echo. Slight changes in the relative positions of reflecting surfaces and scatterers can create a transition from phase addition to phase cancellation and change the cross section drastically. This shift need be only of the order of a few centimeters for a 10-cm wavelength radar, as shown in the examples in Figs. 1.6-1 and 1.6-2. A simple model that demonstrates this phenomenon is shown in Fig. 1.6-3, in which two scatterers are observed at different aspect angles for various separations.

In a typical radar situation the target is observed many times. The aspect angle at a particular time will govern its observed radar cross section. Since many targets have relative motion with respect to the radar because

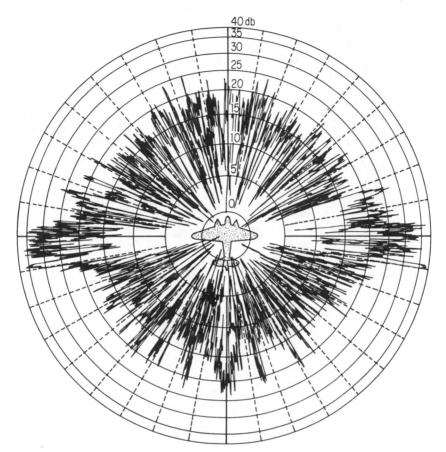

Fig. 1.6-2. Return power from a small aircraft at 10-cm wavelength as a function of azimuth angle (*after Ridenour* [1], *courtesy of McGraw-Hill Book Company*).

of roll, yaw, pitch, spin, and so on, aspect-angle changes on successive observations alter the rf phase relationships, thereby modifying the radar cross section. This change may be a slow variation and occur on a scan-to-scan basis (on successive antenna scans across a target) or it may be on a pulse-to-pulse basis (on successive sweeps). Because the exact nature of the change is difficult to predict, a statistical description is often adopted to characterize the target radar cross section.

Interesting work has been done on the scatter cross section of complex targets by K. Siegel [11] and others at the University of Michigan in which average cross section has been related to a variety of shapes and structures. However, the statistical nature of cross-section fluctuations—which are also a function of the radar and its environment—has been principally based

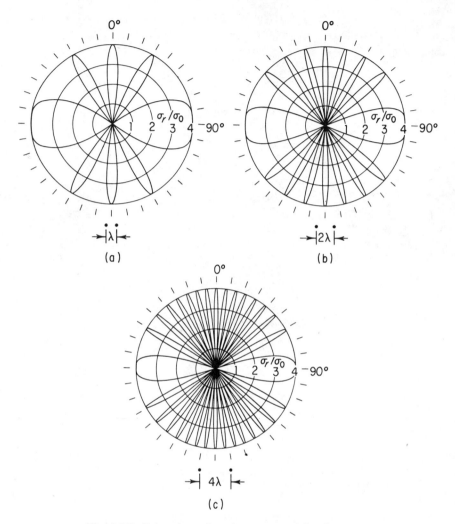

Fig. 1.6-3. Polar plots of resultant cross section for a two-scatterer complex target with (σ_0) the cross section of each and spacing shown on figure (*after Skolnik* [15], *courtesy of McGraw-Hill Book Company*).

on experimental observations. Two statistical models have been treated with respect to radar detection in the classic work of Swerling [7]. These two models are fairly representative of actual experimental radar data and are used in this text. The first model assumes a Rayleigh fluctuating target; it has a signal echo amplitude characteristic that is a random variable defined by a Rayleigh probability density function. This model is representative of a target complex composed of many scatterers of comparable size and cross section. The second target model is called a one-dominant-plus-Rayleigh fluctuating target and has a random amplitude characteristic that is defined by a probability density function given in Chapter 9. This model is characteristic of a target composed of one large reflector plus many small scatterers of comparable size.

In summary, detection is complicated by the fact that the target cross section is a random variable fluctuating with time and must be detected in the presence of interference, including both noise and clutter. Extraction, or parameter estimation, is likewise complicated by random fluctuations of the target echo.

Most radar theory is concerned with the detection and extraction of signal returns from *point* targets. This approach is reasonably valid when radar range resolution is significantly greater than target extent. However, there is a strong trend in modern technology to much higher range resolution. When the effective range resolution is less than the target extent, a more complex optimization theory is required. With a priori knowledge of target structures, and hence radar echo signatures, conceivably one could construct a system for the separation of these known targets from higher background levels of noise, clutter, and other target shapes.

1.7. The Uses of Radar and Special-Purpose Systems

In the preceding discussion it was assumed that radars are either *search* or *track*. As previously noted, this distinction permits simple radar classifications. Real radar systems usually combine both functions. In the multitude of uses to which radar has been put, there are applications in which it is difficult to distinguish clearly between the two types.

The search, or existence, concept is generally fundamental to radars employed for shipboard and aircraft navigation. Many military radars used for surveillance are, of course, also of this type; these include the DEW radars (Distant Early Warning), BMEWS (Ballistic Missile Early Warning System), the search radars of SAGE (Semiautomatic Ground Environment System), AEW (Airborne Early Warning) radars, and the Nike missile defense system acquisition radars. Weather and mapping radars are also generally configured with emphasis on the search function. In addition, space-exploration radars and communication-link radars have a strong search emphasis.

The track family of radars include fire-control radars employed in military applications, weapon homing radars, and AI (Airborne Intercept) radars which guide fighter craft to targets. The radar altimeter is of this type as are the police speed-control radars. In addition, there are many other tracking radar types for space guidance, satellite tracking, radar astronomy, and for communication links via satellite backscatter and beacons.

The systems providing both functions with almost equal emphasis are generally of the central station type, such as those for air traffic control, collision avoidance, and space surveillance and detection.

1.8. Integration of Signal and Data Processing

One of the outstanding trends in radar technology is the integration of equipments into large system configurations and the automatization of many of the functions previously performed by radar operators. Emphasis is being placed on fitting the radar to its use and user by means of sophisticated signal and data processing. Digital computers are being incorporated, in one fashion or another, in almost all modern radar systems.

One of the early systems incorporating well-developed techniques for utilizing radar data was the U.S. Tactical Air Command Forward Director Post in the Second World War. In this system a radar station located at a strategic point would control aircraft operations directly from the site. This included guiding aircraft intercept operations and supplying enemy dispositions and other information. The layout of a typical radar control center is shown in Fig. 1.8-1 and the operations center in Fig. 1.8-2. The equipment was designed for portable use.

The heart of the Forward Director Post (FDP) was the operations shelter shown in Fig. 1.8-2. On the dais at the rear sat the chief controller, who was responsible for the general operation of the station, for the officers identifying tracks and aircraft, for liaison functions with other arms of the service, and for sundry other functions. The data were displayed on large plotting boards, behind which a number of plotters designated aircraft tracks. The plotters were instructed by phone from information obtained on a B-Scope in the reporting shelter. This system saw little combat use owing to its development late in the war. However, it represents a fairly high level of radar integration into a complex network, although all functions except detection were conducted by men rather than machine.

A more recent example of a highly developed integrated system is the ARIS (Advance Range Instrumentation Ship) mobile Atlantic range station (MARS). An entire ship has been integrated for the specific purpose of providing a mobile site for the study of ICBM terminal characteristics and re-entry phenomena, much as performed at large land sites such as Cape Kennedy. The ship gathers trajectory, signature, and telemetry data from re-entry vehicles and can track targets at ranges far exceeding previous shipboard tracking radar capabilities. The ship contains meteorological equipment, data-conversion equipment, and communications facilities. Its instrumentation is diagrammed in Fig. 1.8-3.

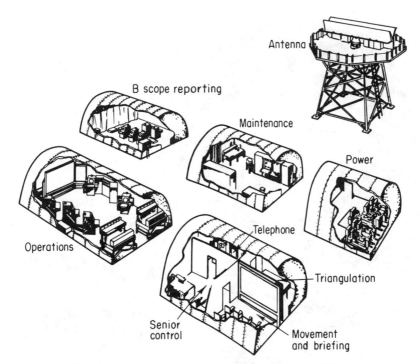

Fig. 1.8-1. Layout of radar control center (*after Ridenour* [1], *courtesy of McGraw-Hill Book Company*).

At the heart of the ship equipment complex is a well-organized data-processing system, shown in Fig. 1.8-4, which performs a multitude of functions including data processing, data recording, and data conversion. The data processor includes a digital computer and peripheral equipment such as paper-tape perforators and readers, magnetic-tape recorders and reproducers, and teletype units. Emphasis in the computer is on large random-access internal storage and rapid communication with external devices. Its general-purpose nature allows for parallel processing of both "batch" and real-time data. All mathematical manipulations necessary to accomplish a mission are performed by this unit. The computer also makes logical decisions and integrates and controls other portions of the system. Functions aided or controlled by the machine include system checkout, target acquisition, target tracking, data format, data transmission, and navigational fixing and updating. Additional storage is provided by the CDCE (Central Data Conversion Equipment), shown in Fig. 1.8-3, which serves as interface and main storage of all trajectory data collected during a mission.

The command post of the ship is manned by personnel in the OCC (Operations Control Center), shown in Fig. 1.8-5. This centralized control facility provides smooth internal coordination and assures proper integration

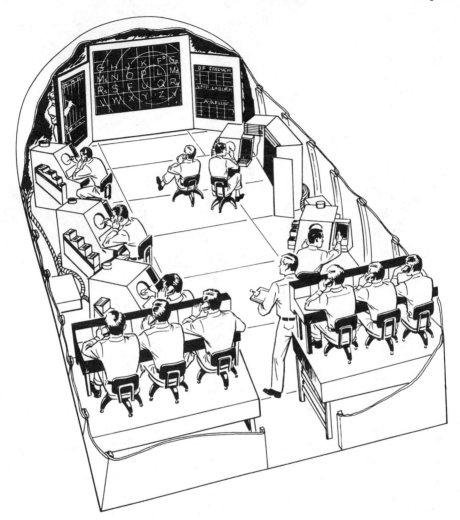

Fig. 1.8-2. Interior of operations shelter (*after Ridenour* [1],
courtesy of McGraw-Hill Book Company).

of the tracking station into the complex AMR (Atlantic Missile Range)
operation. The OCC personnel perform the following essential management
activities during a mission: (1) coordinate with other AMR stations, aircraft,
and ships; (2) review system and subsystem equipment status; (3) assure
correctness of ship's position, speed, and heading; (4) monitor and, if necessary, countermand subordinate decisions; (5) select the best sensor for a

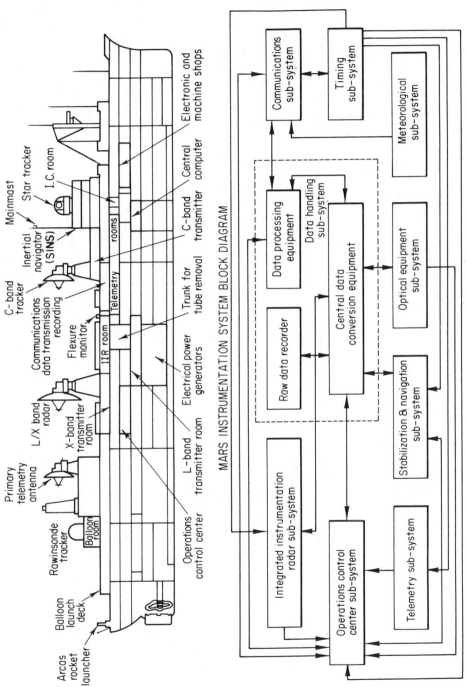

MARS INSTRUMENTATION SYSTEM BLOCK DIAGRAM

Fig. 1.8-3. The MARS instrumentation system block diagram and ship profile (*courtesy of Sperry Rand Corporation*).

33

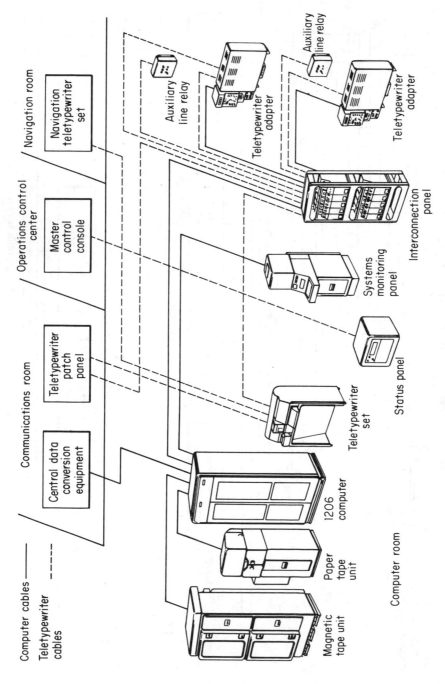

Fig. 1.8-4. MARS data-processing system with UNIVAC® 1206 computer *(courtesy of Sperry Rand Corporation).*

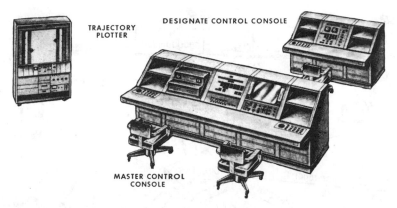

Fig. 1.8-5. MARS operations control center (*courtesy of Sperry Rand Corporation*).

mission; and (6) synchronize the countdown clock at liftoff. It may be noted that—in addition to the multitude of tasks performed and organized by the computer—there is still considerable activity required of personnel.

One of the largest system integrations attempted since the Second World War is the SAGE (Semiautomatic Ground Environment System) complex for the continental defense of the United States (see references [13] and [14]). In SAGE the United States is divided into about thirty sectors; each sector contains a SAGE direction center, which conducts all air defense in the sector. A typical direction center is shown in Fig. 1.8-6. The sector that contains Boston, for example, has a direction center at the Stewart Air Force Base in New York. The area of responsibility of this center extends from Maine on the north to Connecticut on the south; from New York on the west to a point hundreds of miles off the seacoast on the east.

Fig. 1.8-6. A SAGE direction center (*after Everett et al.* [13], *courtesy of IEEE*).

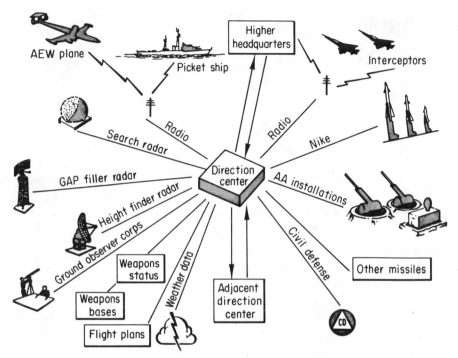

Fig. 1.8-7. A direction-center information and communication network (*after Everett et al.* [13], *courtesy of IEEE*).

The direction center is the nucleus of a sector and communicates with over one hundred adjacent installations, as shown in Fig. 1.8-7. Air-surveillance data are received from long-range search and gap-filler radars, from picket ships, early warning (AEW) aircraft, and formerly from Texas Towers; height finders supply altitude data. The direction center includes equipment for the conversion of these data to a single positional frame of reference and the generation of a complete picture of the air situation. Other inputs are supplied also, such as missile and other weapon status, weather, and flight plans of friendly aircraft. Such data are automatically processed by the computer and used by operational personnel to identify aircraft, employ and assign weapons, and select strategy and tactics to be employed.

The SAGE computer occupies an entire floor of the direction center. The computer is shown in block form in Fig. 1.8-8. It consists of a central computer, air-defense computer programs, and system status data-storage facilities on magnetic drums. The central computer is buffered from almost all sector and console equipment by magnetic drums. A real-time clock and four magnetic-tape units complete the computing system.

The general-purpose binary computer is a single-address machine with

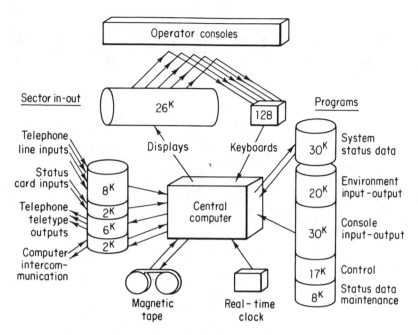

Fig. 1.8-8. Each of two identical computers consists of the following major components: a central computer, which performs all calculations, a 75,000-instruction air-defense program, and a million bits of system status data. Both of the latter are stored on auxiliary magnetic drums (*after Everett et al.* [13], *courtesy of IEEE*).

32-bit word length and a core memory of 8192 words. The memory-cycle time is 6 microseconds. A short *sequence-control* program in the core memory transfers appropriate programs into the core memory, initiates processing, and then returns appropriate table blocks (but no programs) back to the drum. The operation of each air-defense routine is coordinated with the sequence-control program so that data and programs are transferred during data processing. By time-sharing the computer in this manner, each of the air-defense routines is operated at least once every minute; many routines are operated every few seconds. An interesting feature of the machine is the synchronization of the machine with real time, which simplifies many of the control and input-output functions.

Although most of the tasks of SAGE are performed by machine, many depend on the versatility of man. A large group of operations personnel are required in the system for proper functioning. They are located in areas for each of the distinct major air-defense functions, as shown in Fig. 1.8-9: radar inputs, status inputs, air surveillance, identification, weapons control,

Fig. 1.8-9. The fourth floor of the direction center contains separate operational rooms for surveillance, identification, status input, weapons assignment and control, and command functions. Up to 50 operators are required in one room to man the consoles, which are directly connected to the computer (*after Everett et al.* [13], *courtesy of IEEE*).

operations analysis, training simulation, and sector command. The operators perform a variety of tasks, which include phone and radio communications, pattern recognition, tactical judgments required for aircraft identification, and weapons deployment and commitment. It appears that automation and computer technology tend to lag behind the complex tasks that man creates for control of his environment.

REFERENCES

1. Ridenour, L. N.: "Radar System Engineering," MIT Radiation Laboratory Series, Vol. I, McGraw-Hill, New York, 1947.

2. Woodward, P. M.: "Probability and Information Theory with Applications to Radar," McGraw-Hill, New York, 1953.

3. Siebert, W. M.: A Radar Detection Philosophy, *IRE Trans.*, **IT-2**: 204–221 (September, 1956).

4. North, D.O.: Analysis of the Factors which Determine Signal/Noise Discrimination in Pulsed-Carrier Systems, *RCA Laboratories Rept.*, PTR-6C (June, 1943); reprinted in *Proc. IEEE*, **51**: (7), 1015–1027 (July, 1963).

5. Cook, C. E.: Pulse Compression—Key to More Efficient Radar Transmission, *Proc. IRE*, **48**: (3), 310–316 (March, 1960).

6. Marcum, J. I.: A Statistical Theory of Target Detection by Pulsed Radar, *Rand Research Memo.*, RM-754 (December, 1947); reprinted in *IRE Trans.*, **IT-6:** (*2*), 59–144 (April, 1960).

7. Swerling, P.: Probability of Detection for Fluctuating Targets, *Rand Report*, RM-1217 (March 5, 1954); reprinted in *IRE Trans.*, **IT-6:** (*2*), 269–308 (April, 1960).

8. Middleton, D.: "An Introduction to Statistical Communication Theory," McGraw-Hill, New York, 1960.

9. Helstrom, C. W.: "Statistical Theory of Signal Detection," Pergamon, Oxford, England, 1960.

10. Wainstein, L. A., and Zubakov, V. P.: "Extraction of Signals from Noise," Prentice-Hall, Englewood Cliffs, N.J., 1962.

11. Siegel, K. M.: Far Field Scattering from Bodies of Revolution, *Appl. Sci. Research*, Sec. B, **7**: 293–328 (1958).

12. Brockner, C. E., and Price, R. C.: Fire Control Radar Techniques, *Sperry Engineering Review*, **2**: (*1*), 2–10 (March, 1958).

13. Everett, R. R., Zraket, C. A., and Bennington, H. D.: SAGE—A Data-Processing System for Air Defense, *Proc. Eastern Joint Computer Conference*, Washington, D. C., 148–155 (December, 1957).

14. Vance, P. R., Dooley, L. G., and Diss, C. E.: Operation of the SAGE Duplex Computers, *Proc. Eastern Joint Computer Conference*, Washington, D. C., 160–163 (December, 1957).

15. Skolnik, M. I.: "Radar Systems," McGraw-Hill, New York, 1962.

PART II

MATHEMATICAL DESCRIPTION OF SIGNAL AND NOISE WAVEFORMS

Background interfering noise is encountered in every branch of radio engineering. A study of noise is of fundamental importance to radar-system understanding, since it imposes definite limitations on performance. Noise limits the detectability of small radar echos and the accuracy with which the physical parameters of a radar target can be estimated from a reflected echo.

There are many types of noise. In fact, any unpredictable interfering signal is often loosely referred to as noise. Included in this category are impulse noise, thermal noise, channel cross-talk, cw interference, and so on. The mathematical theory of noise described in the following chapters is not valid for all such interfering signals. It is confined to the study of noise signals that originate in a truly random background. In practice many types of noise come close to meeting the mathematical requirements of the theory. Some of these are thermal or Johnson noise, which results from the random motion of electrons within a resistor; shot noise, which is generated by random fluctuations in electron flow through a vacuum tube; and semiconductor noise, which occurs in transistors and crystals.

41

In treating such problems as the detection of a radar signal immersed in noise and the estimation of the signal parameters in the presence of noise, it is convenient to be able to choose from among a variety of mathematical descriptions in order to facilitate solution of the problem. For that reason a number of different mathematical representations of both signals and noise are presented in the following chapters. The mathematical representations in Chapter 2 apply to deterministic and random waveforms. A random waveform, however, must be further constrained for completion of its mathematical description. For this purpose, a knowledge of probability theory and random processes is required; these subjects are discussed in Chapters 3 and 4.

2

MATHEMATICAL REPRESENTATION
OF WAVEFORMS

2.1. Introduction

Waveforms normally encountered in radar are essentially time-limited and band-limited. (With proper interpretation, this characterization also applies to cw radar systems.) Those familiar with Fourier transform theory will realize, of course, that a waveform cannot be simultaneously time- and band-limited. But in many practical cases this approximation permits a better understanding of radar performance and simplifies calculations. Hence we shall be interested in approximate mathematical representations based on this assumption.

Other radar problems, however, require greater mathematical rigor for their solution. In this chapter, therefore, several approximate and exact waveform representations are described that are used in the remainder of the book. Many of these mathematical representations will be familiar to the reader as a result of previous contact.

2.2. Fourier Integral Representation

One of the most important and widely used representations of a waveform $f(t)$ is its *Fourier transform $F(\omega)$* which is defined by

$$F(\omega) = \int_{-\infty}^{\infty} f(t)e^{-j\omega t}\, dt \qquad (2.2\text{-}1)$$

The function $f(t)$ is related to $F(\omega)$ through the *Fourier inversion formula:*

$$f(t) = \frac{1}{2\pi}\int_{-\infty}^{\infty} F(\omega)e^{j\omega t}\, d\omega \qquad (2.2\text{-}2)$$

Equation (2.2-2) describes $f(t)$ as the superposition of an infinite number of complex amplitude-weighted exponential functions $e^{j\omega t}$. When $f(t)$ has points of discontinuity, Eq. (2.2-2) is still valid if it is assumed that

$$f(t) = \frac{f(t^{+}) + f(t^{-})}{2} \qquad (2.2\text{-}3)$$

at discontinuity points.

Not all functions have Fourier transforms. A sufficient condition on $f(t)$ for the existence of its Fourier transform $F(\omega)$ is that $f(t)$ be absolutely integrable in the interval $(-\infty, \infty)$—that is,

$$\int_{-\infty}^{\infty} |f(t)|\, dt < \infty \qquad (2.2\text{-}4)$$

In such cases $F(\omega)$ can be evaluated by means of (2.2-1). If the transform $F(\omega)$ obtained in this manner is also absolutely integrable in the interval $(-\infty, \infty)$, then $f(t)$ can be expressed in terms of $F(\omega)$ by means of (2.2-2). An example of an absolutely integrable function $f(t)$ is a pulse-modulated rf signal,

$$f(t) = \begin{cases} \cos \omega_0 t, & -\dfrac{T}{2} \le t \le \dfrac{T}{2} \\ 0, & \text{otherwise} \end{cases} \qquad (2.2\text{-}5)$$

Substituting (2.2-5) into (2.2-1) and carrying out the integration results in

$$F(\omega) = \frac{\sin \frac{1}{2}(\omega - \omega_0)T}{(\omega - \omega_0)} + \frac{\sin \frac{1}{2}(\omega + \omega_0)T}{(\omega + \omega_0)} \qquad (2.2\text{-}6)$$

The Fourier transform pair described in (2.2-5) and (2.2-6) is shown in Fig. 2.2-1, assuming $\omega_0 T \gg 1$.

Many functions [for example, $(\sin \omega_0 t)/t$] do not meet the absolute integrability condition (2.2-4), for which, nevertheless, the Fourier representation exists in a special sense. In general, the transforms of these functions cannot be obtained by means of Eqs. (2.2-1) and (2.2-2). The Fourier integral theorem, as stated by Plancherel, gives sufficient conditions for which such transforms exist: If $f(t)$ is of integrable square over the interval $-\infty < t < \infty$—that is,

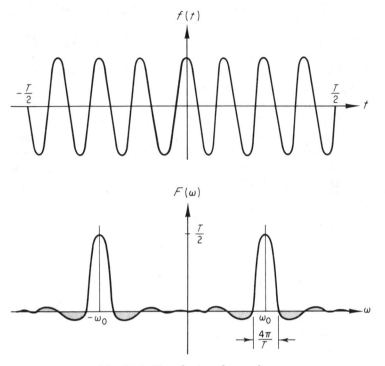

Fig. 2.2-1. Fourier transform pair.

$$\int_{-\infty}^{\infty} |f(t)|^2 \, dt < \infty \qquad (2.2\text{-}7)$$

there exists a function $F(\omega)$ that is also of integrable square which is related to $f(t)$ by

$$F(\omega) = \text{l.i.m.}_{a \to \infty} \int_{-a}^{a} f(t)e^{-j\omega t} \, dt \qquad (2.2\text{-}8)$$

and

$$f(t) = \text{l.i.m.}_{a \to \infty} \frac{1}{2\pi} \int_{-a}^{a} F(\omega)e^{j\omega t} \, d\omega \qquad (2.2\text{-}9)$$

where l.i.m. denotes *limit in the mean*. If we define $F(\omega, a)$ by

$$F(\omega, a) = \int_{-a}^{a} f(t)e^{-j\omega t} \, dt$$

the limit in the mean in Eq. (2.2-8) implies that

$$\lim_{a \to \infty} \int_{-a}^{a} |F(\omega) - F(\omega, a)|^2 \, d\omega \to 0 \qquad (2.2\text{-}10)$$

—that is, the mean-square difference between $F(\omega)$ and $F(\omega, a)$ approaches zero as a approaches infinity. A similar statement can be made for Eq. (2.2-9). In practice, limit-in-the-mean Fourier transforms evaluated by Eqs. (2.2-8) and (2.2-9) are usually not distinguished from those obtained from (2.2-1) and (2.2-2). This custom will be followed hereafter.

In Fig. 2.2-2 additional examples of Fourier transform pairs are shown for several types of waveforms. The first two waveforms illustrate the change in $F(\omega)$ that results when a pulse waveform is modulated onto a radio-frequency carrier. The last two waveforms demonstrate the phenomenon of reciprocal spreading, in which the stretching of a time waveform has the opposite effect on its Fourier spectrum and vice versa. These examples also illustrate the self-transform characteristic of a Gaussian waveform.

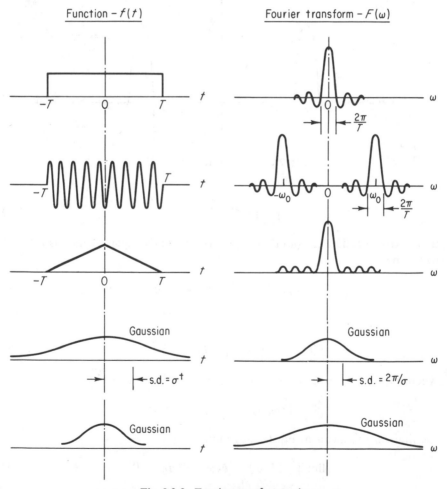

Fig. 2.2-2. Fourier transform pairs.

A number of important results related to Fourier transform theory are used in the text. These are presented here without proof. First, if $f(t)$ is a real function, its Fourier transform $F(\omega)$ satisfies the relation

$$F(\omega) = F^*(-\omega) \tag{2.2-11}$$

If $f(t)$ results from the *convolution* of $f_1(t)$ and $f_2(t)$—that is,

$$f(t) = \int_{-\infty}^{\infty} f_1(\tau) \cdot f_2(t - \tau) \, d\tau \tag{2.2-12a}$$

$$= f_1(t) * f_2(t) \tag{2.2-12b}$$

—then $F(\omega)$, the Fourier transform of $f(t)$, is given by the product of $F_1(\omega)$ and $F_2(\omega)$, the transforms of $f_1(t)$ and $f_2(t)$, respectively. That is,

$$F(\omega) = F_1(\omega) \cdot F_2(\omega) \tag{2.2-13}$$

There is an analogous result for the convolution of two spectra. Thus if $F(\omega)$ is equal to the convolution of $F_1(\omega)$ and $F_2(\omega)$ divided by 2π,

$$F(\omega) = \frac{1}{2\pi} \int_{-\infty}^{\infty} F_1(y) F_2(\omega - y) \, dy \tag{2.2-14}$$

then its inverse Fourier transform $f(t)$ is given by

$$f(t) = f_1(t) \cdot f_2(t) \tag{2.2-15}$$

where $f_1(t)$ and $f_2(t)$, as previously stated, are the inverse transforms of $F_1(\omega)$ and $F_2(\omega)$.

The convolution theorem is very useful in evaluating the performance of a linear time-invariant system. Figure 2.2-3 illustrates such a system,

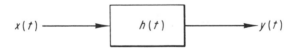

Fig. 2.2-3. Linear system representation.

with impulse response $h(t)$, that is excited by an input voltage $x(t)$ and with system output voltage $y(t)$. If input $x(t)$ is approximated by a series of delayed impulses of width $d\tau$ and height $x(\tau)$, as shown in Fig. 2.2-4, then for $d\tau$ vanishingly small, an impulse at time τ results in an output signal $x(\tau) \, d\tau \, h(t - \tau)$. The system output $y(t)$ is the sum or superposition of the system reponse due to each input impulse up to time t. Thus,

$$y(t) = \int_{-\infty}^{t} x(\tau) h(t - \tau) \, d\tau \tag{2.2-16}$$

The integral in (2.2-16) is called the *superposition integral*. For a passive realizable system, the impulse response $h(t)$ is zero for negative values of its argument—that is, there is no output until an input is applied. Consequently, $h(t - \tau)$ is zero for $\tau > t$; then the upper limit of superposition

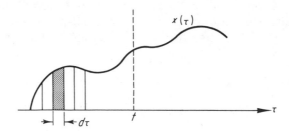

Fig. 2.2-4. Excitation signal $x(t)$.

integral (2.2-16) may be replaced by infinity without affecting its value. When this is done, superposition integral (2.2-16) and convolution integral (2.2-12) are identical in form. It follows from (2.2-13) that $Y(\omega)$, the Fourier transform of $y(t)$, is equal to the product of $X(\omega)$ and $H(\omega)$, the Fourier transforms of $x(t)$ and $h(t)$, respectively. The output signal $y(t)$ can be obtained from $Y(\omega)$ by means of Eq. (2.2-2). Owing to the existence of extensive tables of Fourier transforms, this procedure is often simpler to carry out than a direct evaluation of $y(t)$ by the superposition or convolution integral.

The following basic result is known as Parseval's formula. Let $F(\omega)$ be the Fourier transform of $[f_1(t) \cdot f_2(t)]$. Then, from (2.2-1),

$$F(\omega) = \int_{-\infty}^{\infty} f_1(t) f_2(t) e^{-j\omega t} \, dt \qquad (2.2\text{-}17)$$

For the special case $\omega = 0$, $F(0)$ is equal to

$$F(0) = \int_{-\infty}^{\infty} f_1(t) f_2(t) \, dt \qquad (2.2\text{-}18)$$

and from (2.2-14)

$$F(0) = \frac{1}{2\pi} \int_{-\infty}^{\infty} F_1(y) F_2(-y) \, dy \qquad (2.2\text{-}19)$$

Combining (2.2-18) and (2.2-19) yields a general form of Parseval's formula:

$$\int_{-\infty}^{\infty} f_1(t) f_2(t) \, dt = \frac{1}{2\pi} \int_{-\infty}^{\infty} F_1(\omega) F_2(-\omega) \, d\omega \qquad (2.2\text{-}20)$$

If $f_1^*(t) = f_2(t)$, (2.2-20) reduces to a widely used form of Parseval's formula:

$$\int_{-\infty}^{\infty} |f_1(t)|^2 \, dt = \frac{1}{2\pi} \int_{-\infty}^{\infty} |F(\omega)|^2 \, d\omega \qquad (2.2\text{-}21)$$

Some useful Fourier transforms are provided in Table 2.2-1 for convenience.

TABLE 2.2-1

Waveform, $f(t)$	Fourier transform, $F(\omega)$
$bf(t)$	$bF(\omega)$
$f(at)$	$\lvert 1/a \rvert F(\omega/a)$
$(d^n/dt^n)f(t)$	$(j\omega)^n F(\omega)$
$f(t + t_0)$	$F(\omega)e^{j\omega t_0}$
$f(t)e^{-j\omega_0 t}$	$F(\omega + \omega_0)$
$f(-t)$	$F(-\omega)$
$F(t)$	$f(-\omega)$
$\delta(t)$ (impulse)	1

2.3. Fourier Series Representation

The validity of Fourier integral relations (2.2-1) and (2.2-2) is often established as a related result of the theory of *Fourier series*. Alternatively, it is possible and perhaps more logical mathematically to consider Fourier series as a special case of the Fourier integral relations when $f(t)$ is restricted to *periodic* functions. It can be shown by means of Fourier integral relations that a periodic function $f(t)$ of period T,

$$f(t + T) = f(t) \qquad (2.3\text{-}1)$$

can be written as a sum of exponentials,

$$f(t) = \sum_{k=-\infty}^{\infty} c_k e^{jk\omega_0 t}, \qquad \omega_0 = \frac{2\pi}{T} \qquad (2.3\text{-}2)$$

where k is a positive integer and the constant coefficients c_k are given by

$$c_k = \frac{1}{T} \int_{-T/2}^{T/2} f(t)e^{-jk\omega_0 t}\, dt \qquad (2.3\text{-}3)$$

In order to demonstrate (2.3-2) and (2.3-3), it is convenient first to introduce the following definition of an *impulse function:* If $g(t)$ is an arbitrary function that is continuous at a point t_0, then the impulse function $\delta(t)$ is defined by the relation†

$$\int_{-\infty}^{\infty} \delta(t - t_0)g(t)\, dt = g(t_0) \qquad (2.3\text{-}4)$$

By means of (2.3-4), a periodic function $f(t)$ satisfying (2.3-1) can be described in a different manner. Let

$$f_0(t) = \begin{cases} f(t), & \lvert t \rvert < \dfrac{T}{2} \\ 0, & \text{otherwise} \end{cases} \qquad (2.3\text{-}5)$$

†See Papoulis [1], p. 273, for definition of an impulse function as a *distribution*.

in which case $f(t)$ can be written as a sum:

$$f(t) = \sum_{k=-\infty}^{\infty} f_0(t + kT) \qquad (2.3\text{-}6)$$

It follows from (2.3-4) and the definition of the convolution integral, Eq. (2.2-12), that (2.3-6) may be rewritten as

$$f(t) = f_0(t) * \sum_{k=-\infty}^{\infty} \delta(t - kT) \qquad (2.3\text{-}7)$$

From (2.2-12) and (2.2-13), the Fourier transform of $f(t)$ is the product of the transforms of $f_0(t)$ and $\sum_k \delta(t - kT)$. The Fourier transform of an infinite sum of equally spaced impulses in the time domain is another infinite sum of equally spaced impulses in the frequency domain; specifically, the Fourier transform pair is given by [1]

$$\sum_{k=-\infty}^{\infty} \delta(t - kT) \longleftrightarrow \frac{2\pi}{T} \sum_{k=-\infty}^{\infty} \delta\left(\omega - \frac{2\pi k}{T}\right) \qquad (2.3\text{-}8)$$

Denoting the Fourier transform of $f(t)$ and $f_0(t)$ by $F(\omega)$ and $F_0(\omega)$, respectively, we find the Fourier transform of (2.3-7), employing (2.3-8), to be

$$F(\omega) = F_0(\omega) \cdot \frac{2\pi}{T} \sum_{k=-\infty}^{\infty} \delta\left(\omega - \frac{2\pi k}{T}\right) \qquad (2.3\text{-}9)$$

The inverse transform of $F(\omega)$ is found by substituting (2.3-9) into (2.2-2), as follows:

$$f(t) = \frac{1}{2\pi} \int_{-\infty}^{\infty} F_0(\omega) \cdot \frac{2\pi}{T} \sum_{k=-\infty}^{\infty} \delta\left(\omega - \frac{2\pi k}{T}\right) e^{j\omega t} \, d\omega \qquad (2.3\text{-}10)$$

Interchanging the order of summation and integration in (2.3-10), and using (2.3-4) once again, we obtain finally

$$f(t) = \frac{1}{T} \sum_{k=-\infty}^{\infty} F_0\left(\frac{2\pi k}{T}\right) e^{j(2\pi k t/T)} \qquad (2.3\text{-}11)$$

Equation (2.3-11) is identical to (2.3-2) if $2\pi/T$ is replaced by ω_0 and

$$c_k = \frac{1}{T} F_0\left(\frac{2\pi k}{T}\right) \qquad (2.3\text{-}12)$$

Figure 2.3-1 illustrates the relationships between the Fourier transform of a pulse, the transform of an infinite train of identical pulses, and the transform of a finite train of identical pulses. In the case of an infinite train, the Fourier transform consists of equally spaced impulses with amplitudes governed by the transform of a single pulse. In the case of a finite pulse train, the transform-domain impulses are replaced by nonzero-width pulses with a shape given by the Fourier transform of the envelope of the finite pulse train.

An arbitrary function $f(t)$ defined only over an interval T (and hence not periodic) can also be represented by a Fourier series. Assume a new function

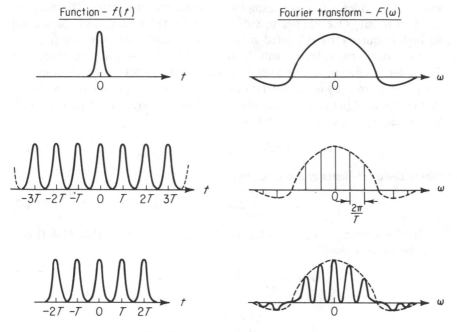

Fig. 2.3-1. Fourier transform pairs.

that is periodic in T and that coincides with the original function $f(t)$ within the interval T. This function clearly has a Fourier series representation that describes $f(t)$ within the region of definition (but not elsewhere).

We shall also have occasion to use the sine-cosine form of Fourier series in which the exponentials in (2.3-2) are replaced by their sine-cosine trigonometric expansions. Thus,

$$f(t) = \frac{a_0}{2} + \sum_{k=1}^{\infty} a_k \cos k\omega_0 t + b_k \sin k\omega_0 t \qquad (2.3\text{-}13)$$

where $\omega_0 = 2\pi/T$ as before and

$$a_k = \frac{2}{T} \int_{-T/2}^{T/2} f(t) \cos k\omega_0 t \, dt \qquad (2.3\text{-}14)$$

$$b_k = \frac{2}{T} \int_{-T/2}^{T/2} f(t) \sin k\omega_0 t \, dt \qquad (2.3\text{-}15)$$

2.4. Low-Frequency (Video) Sampling Theorem

When the Fourier transform $S(\omega)$ of a waveform $s(t)$ is essentially band-limited, the *sampling theorem*, also known as the *Shannon-Kotelnikov theorem*, provides a discrete "sampled" representation of $s(t)$. The theorem has two forms. One applies to video band-limited signals and the second to high-frequency band-limited signals. Both results are very useful.

Consider first video signals band-limited to $(-\omega_c, \omega_c)$ as shown in Fig. 2.4-1. Define next a periodic Fourier transform $S_p(\omega)$, which coincides with transform $S(\omega)$ in the interval $(-\omega_c, \omega_c)$ and has period $2\omega_c$ as shown in Fig. 2.4-2. $S_p(\omega)$ can be expressed in a Fourier series as in (2.3-11) with T replaced by $2\omega_c$ and t replaced by ω:

$$S_p(\omega) = \sum_{k=-\infty}^{\infty} c_k e^{j(k\pi\omega/\omega_c)} \tag{2.4-1}$$

where the coefficients c_k are given by

$$c_k = \frac{1}{2\omega_c} \int_{-\omega_c}^{\omega_c} S_p(\omega) e^{-j(k\pi\omega/\omega_c)} \, d\omega \tag{2.4-2}$$

In the interval $(-\omega_c, \omega_c)$ $S_p(\omega)$ is identical to $S(\omega)$, so that (2.4-2) can also be written as

$$c_k = \frac{1}{2\omega_c} \int_{-\omega_c}^{\omega_c} S(\omega) e^{-j(k\pi\omega/\omega_c)} \, d\omega \tag{2.4-3}$$

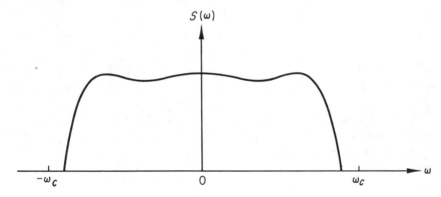

Fig. 2.4-1. Transform $S(\omega)$ of band-limited video signal.

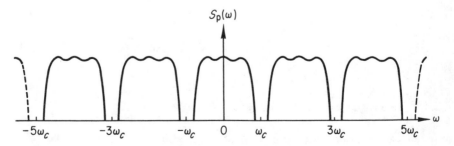

Fig. 2.4-2. Transform $S_p(\omega)$.

Let $\tau = -k\pi/\omega_c$; then (2.4-3) can be rewritten as

$$c_k = \frac{1}{2\omega_c} \cdot 2\pi \left[\frac{1}{2\pi} \int_{-\omega_c}^{\omega_c} S(\omega)e^{j\omega\tau} \, d\omega \right] \qquad (2.4\text{-}4)$$

Since $S(\omega)$ is band-limited to $(-\omega_c, \omega_c)$, the bracketed quantity in (2.4-4) is actually $s(\tau)$, the inverse Fourier transform of $S(\omega)$. Thus c_k is given by

$$c_k = \frac{\pi}{\omega_c} s\left(-\frac{k\pi}{\omega_c} \right) \qquad (2.4\text{-}5)$$

Substituting (2.4-5) into (2.4-1) results in

$$S_p(\omega) = \sum_{k=-\infty}^{\infty} \frac{\pi}{\omega_c} s\left(-\frac{k\pi}{\omega_c} \right) e^{j(k\pi\omega/\omega_c)} \qquad (2.4\text{-}6)$$

or, with $-k$ replaced by k,

$$S_p(\omega) = \frac{\pi}{\omega_c} \sum_{k=-\infty}^{\infty} s\left(\frac{k\pi}{\omega_c} \right) e^{-j(k\pi\omega/\omega_c)} \qquad (2.4\text{-}7)$$

In the interval $(-\omega_c, \omega_c)$,

$$S(\omega) = S_p(\omega) = \frac{\pi}{\omega_c} \sum_{k=-\infty}^{\infty} s\left(\frac{k\pi}{\omega_c} \right) e^{-j(k\pi\omega/\omega_c)}, \qquad -\omega_c < \omega < \omega_c \qquad (2.4\text{-}8)$$

By substituting (2.4-8) into Fourier relation (2.2-1), we find the inverse transform $s(t)$ to be

$$s(t) = \frac{1}{2\pi} \int_{-\omega_c}^{\omega_c} \left[\frac{\pi}{\omega_c} \sum_{k=-\infty}^{\infty} s\left(\frac{k\pi}{\omega_c} \right) e^{-j(k\pi\omega/\omega_c)} \right] e^{j\omega t} \, d\omega$$

$$= \frac{\pi}{\omega_c} \sum_{k=-\infty}^{\infty} s\left(\frac{k\pi}{\omega_c} \right) \int_{-\omega_c}^{\omega_c} e^{j\omega\{t-(k\pi/\omega_c)\}} \frac{d\omega}{2\pi}$$

$$s(t) = \frac{\pi}{\omega_c} \sum_{k=-\infty}^{\infty} s\left(\frac{k\pi}{\omega_c} \right) \frac{\sin \omega_c\left(t - \dfrac{k\pi}{\omega_c} \right)}{\pi \left(t - \dfrac{k\pi}{\omega_c} \right)} \qquad (2.4\text{-}9)\dagger$$

With ω_c replaced by $2\pi f_c$, (2.4-9) may be rewritten as

$$s(t) = \sum_{k=-\infty}^{\infty} s\left(\frac{k}{2f_c} \right) \frac{\sin 2\pi f_c\left(t - \dfrac{k}{2f_c} \right)}{2\pi f_c\left(t - \dfrac{k}{2f_c} \right)} \qquad (2.4\text{-}10)$$

Equation (2.4-10) is the desired result. It states that a band-limited function $s(t)$ is uniquely characterized over all time by samples of $s(t)$ at intervals $1/2f_c$ apart, provided the sampling frequency f_c is equal to or greater than the highest positive frequency present in the spectrum of $s(t)$.

---

†To obtain (2.4-9) we have used the relation

$$\frac{1}{2\pi} \int_{-\omega_c}^{\omega_c} e^{j\omega x} \, d\omega = \frac{\sin \omega_c x}{\pi x}$$

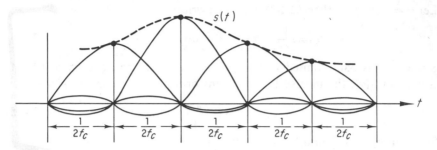

Fig. 2.4-3. Reconstruction of $s(t)$ from its sampled value.

Consider the example in Fig. 2.4-3. Suppose the values of $s(t)$ are available at the sampling instants and it is desired to reconstruct $s(t)$. In accordance with (2.4-10) $s(t)$ is described by an infinite sum of $(\sin x)/x$ functions, each $(\sin x)/x$ function centered at a sampling instant and weighted by its corresponding sample value, as shown in Fig. 2.4-3. Observe that precisely at the sampling instants only one $(\sin x)/x$ function, centered at that instant in time, contributes to the value of the sum, all other $(\sin x)/x$ functions being instantaneously zero. However, between sampling instants, every $(\sin x)/x$ function in the infinite sum is nonzero and contributes to the value of $s(t)$. Hence an exact reconstruction of $s(t)$, even over a finite interval, requires knowledge of the sampled values of $s(t)$ over all time.

Suppose it is desired to reconstruct $s(t)$ from its sampled values within a finite interval $(0, T)$. An excellent approximation of $s(t)$ in this interval—acceptable in many applications where $f_c T$ is much greater than one—is obtained by summing only $(\sin x)/x$ functions centered at sampled values of $s(t)$ within the interval $(0, T)$. This results in a small error, since $(\sin x)/x$ functions that are centered outside the interval $(0, T)$ contribute relatively little to the reconstruction of $s(t)$ within the interval; in particular, they contribute nothing at the sampling instants, so that errors in approximation can occur only between sampling instants. With this approximation, a video band-limited waveform, with greatest (positive) frequency f_c, is specified by $2f_c T$ samples within the interval $(0, T)$. See Fig. 2.4-4a. Such waveforms are said to have $2f_c T$ degrees of freedom. Conceptually a waveform defined by $2f_c T$ samples may be represented by a vector **s** in a waveform or signal space with $2f_c T$ dimensions as shown in Fig. 2.4-4b. The components of vector **s** are notationally given by

$$\mathbf{s} = (s_1, s_2, s_3, \ldots, s_{2f_c T}) \qquad (2.4\text{-}11)$$

where

$$s_k = s\left(\frac{k}{2f_c}\right) \qquad (2.4\text{-}12)$$

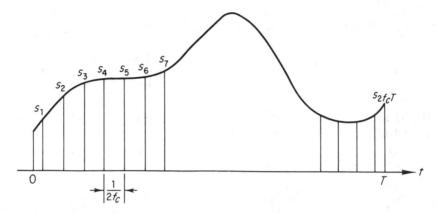

Fig. 2.4-4(a). Sampled band-limited waveform $s(t)$.

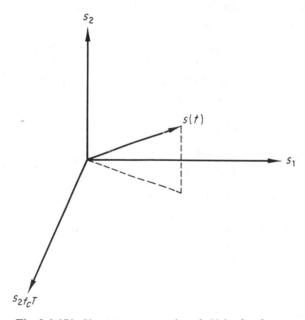

Fig. 2.4-4(b). Vector representation of $s(t)$ in signal space.

The vector description in (2.4-11), for band-limited signals, is very convenient for solving noise problems and is used extensively in later chapters.

The discussion of the low-frequency sampling theorem is concluded with the derivation of several useful relations. Assume that $f(t)$ and $g(t)$ are each band-limited to $(-f_c, f_c)$. Direct substitution for $f(t)$ and $g(t)$ in the following integral, using (2.4-10), results in

$$\int_{-\infty}^{\infty} f(t)g(t)\,dt = \int_{-\infty}^{\infty} \sum_{i=-\infty}^{\infty} \sum_{k=-\infty}^{\infty} f\left(\frac{i}{2f_c}\right) g\left(\frac{k}{2f_c}\right)$$

$$\cdot \frac{\sin 2\pi f_c\left(t - \dfrac{i}{2f_c}\right)}{2\pi f_c\left(t - \dfrac{i}{2f_c}\right)} \cdot \frac{\sin 2\pi f_c\left(t - \dfrac{k}{2f_c}\right)}{2\pi f_c\left(t - \dfrac{k}{2f_c}\right)}\,dt \qquad (2.4\text{-}13)$$

If $f(t)$ and $g(t)$ are well-behaved functions,† integration and summation in (2.4-13) can be interchanged as follows:

$$\int_{-\infty}^{\infty} f(t)g(t)\,dt = \sum_{i=-\infty}^{\infty} \sum_{k=-\infty}^{\infty} f\left(\frac{i}{2f_c}\right) g\left(\frac{k}{2f_c}\right)$$

$$\cdot \int_{-\infty}^{\infty} \frac{\sin 2\pi f_c\left(t - \dfrac{i}{2f_c}\right)}{2\pi f_c\left(t - \dfrac{i}{2f_c}\right)} \cdot \frac{\sin 2\pi f_c\left(t - \dfrac{k}{2f_c}\right)}{2\pi f_c\left(t - \dfrac{k}{2f_c}\right)}\,dt \qquad (2.4\text{-}14)$$

The integral on the right side of (2.4-14) is zero unless $i = k$, in which case it is equal to $1/2f_c$. Thus,

$$\int_{-\infty}^{\infty} f(t)g(t)\,dt = \frac{1}{2f_c} \sum_{k=-\infty}^{\infty} f\left(\frac{k}{2f_c}\right) g\left(\frac{k}{2f_c}\right) \qquad (2.4\text{-}15)$$

Equation (2.4-15) is an exact relation. However, if segments of $f(t)$ and $g(t)$, each T seconds long, are considered, the approximate relation

$$\int_{0}^{T} f(t)g(t)\,dt \cong \frac{1}{2f_c} \sum_{k=1}^{2f_c T} f\left(\frac{k}{2f_c}\right) g\left(\frac{k}{2f_c}\right) \qquad (2.4\text{-}16)$$

is obtained. Defining vectors: $\mathbf{f} = (f_1, f_2, \ldots, f_{2f_c T})$ and $\mathbf{g} = (g_1, g_2, \ldots, g_{2f_c T})$, the right side of (2.4-16) is the dot product of $\mathbf{f}$ and $\mathbf{g}$—that is,

$$\int_{0}^{T} f(t)g(t)\,dt \cong \frac{1}{2f_c} (\mathbf{f} \cdot \mathbf{g}) \qquad (2.4\text{-}17)$$

For the special case in which $f(t) = g(t)$, (2.4-16) and (2.4-17) reduce to

$$\int_{0}^{T} f^2(t)\,dt \cong \frac{1}{2f_c} \sum_{k=1}^{2f_c T} f^2\left(\frac{k}{2f_c}\right) \qquad (2.4\text{-}18a)$$

$$\cong \frac{1}{2f_c} (\mathbf{f} \cdot \mathbf{f}) \qquad (2.4\text{-}18b)$$

The integral on the left of (2.4-18) describes the energy in waveform $f(t)$ within the interval $(0, T)$. The right sides of (2.4-18a) and (2.4-18b) conform to the vector calculation of energy.

In a strict sense, the sampling theorem and related results have been proven only for deterministic waveforms, since the voltage spectrum (Fourier transform) of $s(t)$ was used in the proof. For stationary random processes, which are discussed in Chapter 4, the power spectrum rather

†A sufficient condition is the absolute integrability of each function.

than the voltage spectrum is normally specified. It can be shown [2], however, that a band-limited waveform of a stationary random process can also
be described by the sampling-theorem representation in a mean-square sense.

2.5. Real and Complex (Analog) Representations of Narrowband Waveforms

A waveform is said to be *narrowband* if its principal Fourier components
are confined to a band whose width is small compared to the band center
frequency. A typical narrowband Fourier transform of width Δ_ω and center
frequency ω_0 is shown in Fig. 2.5-1. The waveform corresponding to the
narrowband transform shown in Fig. 2.5-1 might well resemble that in Fig.

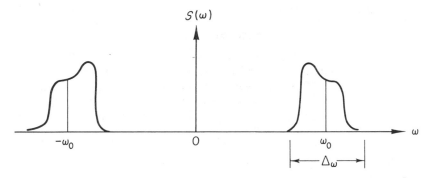

Fig. 2.5-1. A narrowband Fourier transform.

2.5-2. Note that the waveform has the appearance of a sine wave whose
amplitude and phase are both varying slowly with respect to the sine-wave
fluctuations.

A useful representation for such a signal is given by

$$s(t) = a(t) \cos [\omega_0 t + \theta(t)] \qquad (2.5\text{-}1)$$

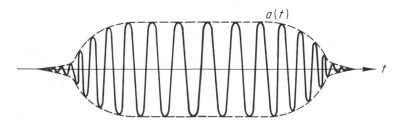

Fig. 2.5-2. A narrowband waveform.

where $a(t)$ describes the amplitude (or envelope) modulation and $\theta(t)$ describes the phase modulation of a monochromatic cosine wave of angular frequency ω_0. Waveform representation (2.5-1) is actually not restricted to narrowband signals; however, the concepts of envelope and phase modulations lose their physical significance unless $a(t)$ and $\theta(t)$ vary slowly in comparison to $\cos \omega_0 t$.

A simplification in mathematical notation results in both theoretical and practical problems when a narrowband waveform $s(t)$ is expressed as the real part of a complex exponential function. Inspection of (2.5-1) immediately leads to one such representation; thus

$$s(t) = \text{Re}\,[a(t)e^{j[\omega_0 t + \theta(t)]}] \tag{2.5-2}$$

Equation (2.5-2) can be modified as follows:

$$s(t) = \text{Re}\,[a(t)e^{j\theta(t)}e^{j\omega_0 t}] \tag{2.5-3}$$

or, alternatively:

$$s(t) = \text{Re}\,[\tilde{s}(t)e^{j\omega_0 t}] \tag{2.5-4}$$

where $a(t)e^{j\theta(t)}$ in (2.5-3) is replaced by $\tilde{s}(t)$ in (2.5-4). $\tilde{s}(t)$ is a complex time function containing both the amplitude and phase modulation of waveform $s(t)$. It is often referred to as the *complex envelope* of narrowband signal $s(t)$.

Another related complex representation is extensively used in the literature in which $s(t)$ is defined as the real part of a complex function $\Psi(t)$,

$$\Psi(t) = s(t) + j\hat{s}(t) \tag{2.5-5}$$

where $\hat{s}(t)$ denotes the Hilbert transform of $s(t)$ and is a real function defined by

$$\hat{s}(t) = \frac{1}{\pi} \int_{-\infty}^{\infty} \frac{s(\tau)}{t - \tau}\,d\tau \tag{2.5-6}$$

$\Psi(t)$ is called the *analytic signal*, since it arises from the analytic continuation of real signal $s(t)$ with t considered as a complex variable. Gabor [3] and Ville [4] have shown that if $S(\omega)$ is the Fourier transform of $s(t)$, then $\Psi(\omega)$, the Fourier transform of $\Psi(t)$, can be expressed in terms of $S(\omega)$ as

$$\Psi(\omega) = \begin{cases} 2S(\omega), & \omega > 0 \\ S(\omega), & \omega = 0 \\ 0, & \omega < 0 \end{cases} \tag{2.5-7}$$

Thus, the Fourier transform of an analytic signal $\Psi(t)$ (see Fig. 2.5-3) is found by suppressing the negative Fourier spectrum of $s(t)$ and doubling its positive Fourier spectrum, except at $\omega = 0$ where $\Psi(0) = S(0)$. Note that from (2.5-5)

$$s(t) = \text{Re}\,(|\Psi(t)|\,e^{j\,\arg\,\Psi(t)}) \tag{2.5-8}$$

$$|\Psi(t)| = [s^2(t) + \hat{s}^2(t)]^{1/2} \tag{2.5-9}$$

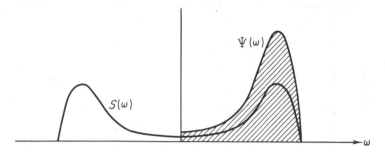

Fig. 2.5-3. Transform of analytic signal.

and

$$\arg \Psi(t) = \tan^{-1}\left[\frac{\hat{s}(t)}{s(t)}\right] \tag{2.5-10}$$

When $s(t)$ is a narrowband signal, $\Psi(t)$ approximates closely the amplitude modulation $a(t)$, and $\arg \Psi(t)$ approximates $\omega_0 t + \theta(t)$. It can be shown [5] that they are identical when $s(t)$ is not only narrowband but also band-limited about ω_0. For this case, if

$$s(t) = a(t) \cos\left[\omega_0 t + \theta(t)\right] \tag{2.5-11}$$

then direct calculation of the Hilbert transform of band-limited signal $s(t)$ leads to

$$\hat{s}(t) = a(t) \sin\left[\omega_0 t + \theta(t)\right] \tag{2.5-12}$$

Equations (2.5-4), (2.5-5), (2.5-11), and (2.5-12) lead to the result

$$\Psi(t) = a(t)e^{j[\omega_0 t + \theta(t)]} = \tilde{s}(t)e^{j\omega_0 t} \tag{2.5-13}$$

Further, it can be shown [5] that for $s(t)$ not band-limited, the modulus and phase of $\Psi(t)$ approximate very closely (in an rms sense) the envelope and phase functions of $s(t)$, even for fractional bandwidths approaching one-half. This is shown for two typical cases [5] in Fig. 2.5-4. For these reasons complex representations (2.5-4) and (2.5-8) are used interchangeably in the remainder of this book.

We have described a narrowband waveform in terms of envelope and phase modulations $a(t)$ and $\theta(t)$. A related representation, which is more suitable for narrowband noise waveforms, can be derived by expanding (2.5-1) trigonometrically as

$$s(t) = a(t) \cos \theta(t) \cos \omega_0 t - a(t) \sin \theta(t) \sin \omega_0 t \tag{2.5-14}$$

Let

$$s_I(t) = a(t) \cos \theta(t) \tag{2.5-15a}$$

$$s_Q(t) = a(t) \sin \theta(t) \tag{2.5-15b}$$

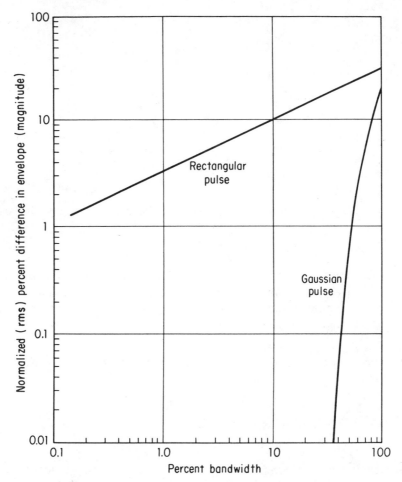

Fig. 2.5-4. Normalized rms difference between $|\Psi(t)|$ and $a(t)$ for narrowband rf pulses with rectangular and Gaussian envelopes (*after Rubin and DiFranco*[5], *courtesy of Journal of the Franklin Institute*).

Then (2.5-14) can be rewritten as

$$s(t) = s_I(t) \cos \omega_0 t - s_Q(t) \sin \omega_0 t \qquad (2.5\text{-}16)$$

$s_I(t)$ and $s_Q(t)$ are commonly called the *quadrature components* of narrowband signal $s(t)$. It follows from the definition of complex envelope $\tilde{s}(t)$, and from Eqs. (2.5-15a) and (2.5-15b) that

$$\tilde{s}(t) = s_I(t) + js_Q(t) \qquad (2.5\text{-}17)$$

—that is, the quadrature components of $s(t)$ are the real and imaginary components of complex envelope $\tilde{s}(t)$. If $s_I(t)$ and $s_Q(t)$ can be described within a time interval by vectors, each having N dimensions in signal space, then $\tilde{s}(t)$ is described over the same interval by a vector $\tilde{\mathbf{s}}$ with $2N$ dimensions.

We can relate the quadrature components of a narrowband waveform defined on an interval $(0, T)$ to its Fourier series expansion. For a narrrow-band waveform with transform centered at ω_0, the Fourier series representation in the interval $(0, T)$ can be written in the following modified form (with dc component equal to zero):

$$s(t) = \sum_{k=1}^{\infty} \alpha_k \cos \omega_k t + \beta_k \sin \omega_k t \qquad (2.5\text{-}18a)$$

$$= \sum_{k=1}^{\infty} \alpha_k \cos [(\omega_k - \omega_0) + \omega_0]t + \beta_k \sin [(\omega_k - \omega_0) + \omega_0]t \qquad (2.5\text{-}18b)$$

where $\omega_k = 2\pi k/T$, and α_k and β_k are given by

$$\alpha_k = \frac{2}{T} \int_0^T s(t) \cos \omega_k t \, dt \qquad (2.5\text{-}19)$$

$$\beta_k = \frac{2}{T} \int_0^T s(t) \sin \omega_k t \, dt \qquad (2.5\text{-}20)$$

Equation (2.5-18b) can be expanded trigonometrically as

$$\begin{aligned} s(t) = \sum_{k=1}^{\infty} &\alpha_k \cos (\omega_k - \omega_0)t \cos \omega_0 t - \alpha_k \sin (\omega_k - \omega_0)t \sin \omega_0 t \\ &+ \beta_k \sin (\omega_k - \omega_0)t \cos \omega_0 t + \beta_k \cos (\omega_k - \omega_0)t \sin \omega_0 t \end{aligned} \qquad (2.5\text{-}21)$$

Rearranging (2.5-21) yields

$$\begin{aligned} s(t) = \sum_{k=1}^{\infty} &[\alpha_k \cos (\omega_k - \omega_0)t + \beta_k \sin (\omega_k - \omega_0)t] \cos \omega_0 t \\ &- [\alpha_k \sin (\omega_k - \omega_0)t - \beta_k \cos (\omega_k - \omega_0)t] \sin \omega_0 t \end{aligned} \qquad (2.5\text{-}22)$$

Comparing (2.5-22) with (2.5-16) leads to the identification of the quadrature components

$$s_I(t) = \sum_{k=1}^{\infty} \alpha_k \cos (\omega_k - \omega_0)t + \beta_k \sin (\omega_k - \omega_0)t \qquad (2.5\text{-}23)$$

$$s_Q(t) = \sum_{k=1}^{\infty} \alpha_k \sin (\omega_k - \omega_0)t - \beta_k \cos (\omega_k - \omega_0)t \qquad (2.5\text{-}24)$$

These auxiliary relations will be useful when we consider random processes in Chapter 4.

2.6. High-Frequency Sampling Theorem

With the relations of Section 2.5, the low-frequency sampling theorem can be extended to high-frequency band-limited waveforms. Consider the Fourier transform of $s(t)$, band-limited as shown in Fig. 2.6-1, with angular spectral width Δ_ω about ω_0. Waveform $s(t)$ can be expressed in terms of its complex envelope function $\tilde{s}(t)$ as follows:

$$s(t) = \frac{\tilde{s}(t)e^{j\omega_0 t} + \tilde{s}^*(t)e^{-j\omega_0 t}}{2} \tag{2.6-1}$$

By substituting (2.6-1) into Fourier integral relation (2.2-1), we can relate $S(\omega)$, the transform of $s(t)$, to $\tilde{S}(\omega)$, the transform of $\tilde{s}(t)$. Thus

$$S(\omega) = \frac{1}{2}\int_{-\infty}^{\infty} \tilde{s}(t)e^{-j(\omega-\omega_0)t}\,dt + \frac{1}{2}\int_{-\infty}^{\infty} \tilde{s}^*(t)e^{-j(\omega+\omega_0)t}\,dt \tag{2.6-2}$$

The first integral in (2.6-2) is the Fourier transform of $\tilde{s}(t)$ with ω replaced by $\omega - \omega_0$. Similarly, the second integral is the Fourier transform of $\tilde{s}^*(t)$ with ω replaced by $\omega + \omega_0$. Thus (2.6-2) may be rewritten as

$$S(\omega) = \frac{1}{2}[\tilde{S}(\omega - \omega_0) + \tilde{S}^*(-\omega - \omega_0)] \tag{2.6-3}$$

$\tilde{S}(\omega - \omega_0)$ is the Fourier transform of (complex) video signal $\tilde{s}(t)$ shifted to the right from the region of zero frequency to the vicinity of ω_0; and $\tilde{S}^*(-\omega - \omega_0)$ is the transform of $\tilde{s}^*(t)$ shifted to the left from the neighborhood of zero frequency to the vicinity of $-\omega_0$. From (2.6-3) and the hypothesis that $s(t)$ is band-limited, it follows that $\frac{1}{2}\tilde{S}(\omega - \omega_0)$ is identical to the positive band-limited portion of spectrum $S(\omega)$ (see Fig. 2.6-1), and

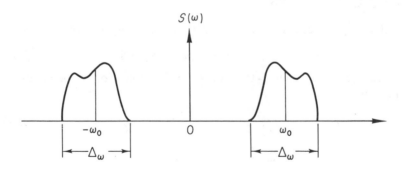

Fig. 2.6-1. High-frequency band-limited transform

$\frac{1}{2}\tilde{S}^*(-\omega - \omega_0)$ describes the negative band-limited portion. Since $\tilde{S}(\omega - \omega_0)$ is band-limited to $(\omega_0 - \Delta_\omega/2, \omega_0 + \Delta_\omega/2)$, $\tilde{S}(\omega)$ is band-limited to $(-\Delta_\omega/2, \Delta_\omega/2)$, as shown in Fig. 2.6-2. It follows that $\tilde{s}(t)$ is a band-limited (complex) video signal that satisfies the requirements of the low-frequency sampling theorem derived in Section 2.4. (Note that the low-frequency sampling theorem applies to real and complex time functions.) Therefore $\tilde{s}(t)$ can be written as

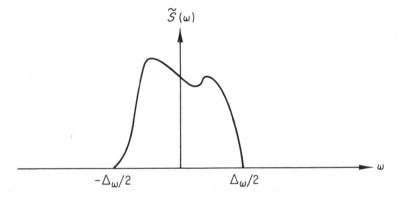

Fig. 2.6-2. Transform of complex envelope $\tilde{s}(t)$.

$$\tilde{s}(t) = \sum_{k=-\infty}^{\infty} \tilde{s}\left(\frac{k}{\Delta_f}\right) \frac{\sin \pi\Delta_f \left(t - \frac{k}{\Delta_f}\right)}{\pi\Delta_f \left(t - \frac{k}{\Delta_f}\right)} \qquad (2.6\text{-}4)$$

where $\Delta_f = \Delta_\omega/2\pi$. Substituting (2.6-4) into (2.5-4) gives

$$s(t) = \text{Re}\left[\sum_{k=-\infty}^{\infty} \tilde{s}\left(\frac{k}{\Delta_f}\right) e^{j\omega_0 t} \frac{\sin \pi\Delta_f \left(t - \frac{k}{\Delta_f}\right)}{\pi\Delta_f \left(t - \frac{k}{\Delta_f}\right)} \right] \qquad (2.6\text{-}5)$$

With $\tilde{s}(t)$ replaced by $a(t)e^{j\theta(t)}$ in (2.6-5),

$$s(t) = \text{Re}\left[\sum_{k=-\infty}^{\infty} a\left(\frac{k}{\Delta_f}\right) e^{j[\omega_0 t + \theta(k/\Delta_f)]} \frac{\sin \pi\Delta_f \left(t - \frac{k}{\Delta_f}\right)}{\pi\Delta_f \left(t - \frac{k}{\Delta_f}\right)} \right] \qquad (2.6\text{-}6)$$

The real part of the bracketed quantity in (2.6-6) yields one form of the high-frequency sampling theorem:

$$s(t) = \sum_{k=-\infty}^{\infty} a\left(\frac{k}{\Delta_f}\right) \cos\left[\omega_0 t + \theta\left(\frac{k}{\Delta_f}\right)\right] \frac{\sin \pi\Delta_f \left(t - \frac{k}{\Delta_f}\right)}{\pi\Delta_f \left(t - \frac{k}{\Delta_f}\right)} \qquad (2.6\text{-}7)$$

A second version of the theorem, in terms of the quadrature components of $s(t)$, is given by

$$s(t) = \sum_{k=-\infty}^{\infty} \left[s_I\left(\frac{k}{\Delta_f}\right) \cos \omega_0 t - s_Q\left(\frac{k}{\Delta_f}\right) \sin \omega_0 t \right] \frac{\sin \pi\Delta_f \left(t - \frac{k}{\Delta_f}\right)}{\pi\Delta_f \left(t - \frac{k}{\Delta_f}\right)} \qquad (2.6\text{-}8)$$

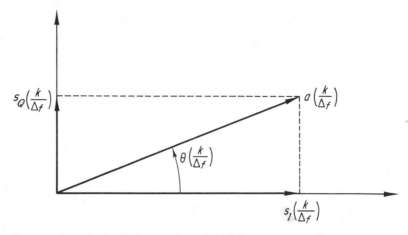

Fig. 2.6-3. Representation of high-frequency sampling at one instant of time in Cartesian and polar coordinates.

To completely specify $s(t)$ over a finite interval T seconds long, sampled values of $s(t)$ over all time are required, as in the case of the low-frequency sampling theorem. From (2.6-7) and (2.6-8) it is evident that both $a(t)$ and $\theta(t)$, or $s_I(t)$ and $s_Q(t)$, must be sampled every $1/\Delta_f$ seconds. The relationship between these quantities is illustrated in Fig. 2.6-3. In an interval T seconds long, $2T\Delta_f$ samples within the interval provide a good approximation of a high-frequency waveform. Formally this result is similar to that obtained for the low-frequency sampling theorem.

We conclude our discussion of the high-frequency sampling theorem with the derivation of integral relations similar to (2.4-16) and (2.4-17). Let waveforms $f(t)$ and $g(t)$ each have a (positive frequency) transform bandwidth Δ_f located about carrier frequency $f_0 = \omega_0/2\pi$ and denote the complex envelopes of $f(t)$ and $g(t)$ by $\tilde{f}(t)$ and $\tilde{g}(t)$, respectively. Then $f(t)$ and $g(t)$ may be described in sampled form by

$$f(t) = \mathrm{Re}\left[\sum_{i=-\infty}^{\infty} \tilde{f}\left(\frac{i}{\Delta_f}\right) e^{j\omega_0 t} \frac{\sin \pi\Delta_f\left(t - \dfrac{i}{\Delta_f}\right)}{\pi\Delta_f\left(t - \dfrac{i}{\Delta_f}\right)}\right] \qquad (2.6\text{-}9)$$

$$g(t) = \mathrm{Re}\left[\sum_{k=-\infty}^{\infty} \tilde{g}\left(\frac{k}{\Delta_f}\right) e^{j\omega_0 t} \frac{\sin \pi\Delta_f\left(t - \dfrac{k}{\Delta_f}\right)}{\pi\Delta_f\left(t - \dfrac{k}{\Delta_f}\right)}\right] \qquad (2.6\text{-}10)$$

Equations (2.6-9) and (2.6-10) can be rewritten as

$$f(t) = \frac{1}{2}\sum_{i=-\infty}^{\infty}\left[\tilde{f}\left(\frac{i}{\Delta_f}\right) e^{j\omega_0 t} + \tilde{f}^*\left(\frac{i}{\Delta_f}\right) e^{-j\omega_0 t}\right] \frac{\sin \pi\Delta_f\left(t - \dfrac{i}{\Delta_f}\right)}{\pi\Delta_f\left(t - \dfrac{i}{\Delta_f}\right)} \qquad (2.6\text{-}11)$$

and

$$g(t) = \frac{1}{2} \sum_{k=-\infty}^{\infty} \left[\tilde{g}\left(\frac{k}{\Delta_f}\right) e^{j\omega_0 t} + \tilde{g}^*\left(\frac{k}{\Delta_f}\right) e^{-j\omega_0 t} \right] \frac{\sin \pi\Delta_f \left(t - \dfrac{k}{\Delta_f}\right)}{\pi\Delta_f \left(t - \dfrac{k}{\Delta_f}\right)} \qquad (2.6\text{-}12)$$

Substitution of (2.6-11) and (2.6-12) into the following product integral gives

$$\int_{-\infty}^{\infty} f(t)g(t)\, dt$$

$$= \int_{-\infty}^{\infty} \frac{1}{4} \sum_{i=-\infty}^{\infty} \sum_{k=-\infty}^{\infty} \left[\tilde{f}\left(\frac{i}{\Delta_f}\right) \tilde{g}\left(\frac{k}{\Delta_f}\right) e^{j2\omega_0 t} + \tilde{f}^*\left(\frac{i}{\Delta_f}\right) \tilde{g}^*\left(\frac{k}{\Delta_f}\right) e^{-j2\omega_0 t} \right.$$

$$\left. + \tilde{f}\left(\frac{i}{\Delta_f}\right) \tilde{g}^*\left(\frac{k}{\Delta_f}\right) + \tilde{f}^*\left(\frac{i}{\Delta_f}\right) \tilde{g}\left(\frac{k}{\Delta_f}\right) \right]$$

$$\cdot \frac{\sin \pi\Delta_f \left(t - \dfrac{i}{\Delta_f}\right)}{\pi\Delta_f \left(t - \dfrac{i}{\Delta_f}\right)} \cdot \frac{\sin \pi\Delta_f \left(t - \dfrac{k}{\Delta_f}\right)}{\pi\Delta_f \left(t - \dfrac{k}{\Delta_f}\right)}\, dt \qquad (2.6\text{-}13)$$

Combining terms and interchanging the operations of integration and summation results in

$$\int_{-\infty}^{\infty} f(t)g(t)\, dt$$

$$= \frac{1}{2} \operatorname{Re} \sum_{i=-\infty}^{\infty} \sum_{k=-\infty}^{\infty} \int_{-\infty}^{\infty} \left[\tilde{f}\left(\frac{i}{\Delta_f}\right) \tilde{g}\left(\frac{k}{\Delta_f}\right) e^{j2\omega_0 t} + \tilde{f}\left(\frac{i}{\Delta_f}\right) \tilde{g}^*\left(\frac{k}{\Delta_f}\right) \right]$$

$$\cdot \frac{\sin \pi\Delta_f \left(t - \dfrac{i}{\Delta_f}\right)}{\pi\Delta_f \left(t - \dfrac{i}{\Delta_f}\right)} \cdot \frac{\sin \pi\Delta_f \left(t - \dfrac{k}{\Delta_f}\right)}{\pi\Delta_f \left(t - \dfrac{k}{\Delta_f}\right)}\, dt \qquad (2.6\text{-}14)$$

The first term in the bracket in (2.6-14) contains a rapidly oscillating term, $e^{j2\omega_0 t}$. The value of the integral of this term will be small compared to the remaining integral and can be neglected. The second integral was evaluated earlier in connection with the low-frequency sampling theorem (see Section 2.4), so that (2.6-14) reduces to

$$\int_{-\infty}^{\infty} f(t)g(t)\, dt = \frac{1}{2\Delta_f} \operatorname{Re} \sum_{k=-\infty}^{\infty} \tilde{f}\left(\frac{k}{\Delta_f}\right) \tilde{g}^*\left(\frac{k}{\Delta_f}\right) \qquad (2.6\text{-}15)$$

From Eq. (2.5-17), the complex envelopes $\tilde{f}(t)$ and $\tilde{g}(t)$ are related to their respective quadrature components $f_I(t)$, $f_Q(t)$ and $g_I(t)$, $g_Q(t)$ by

$$\tilde{f}(t) = f_I(t) + jf_Q(t) \qquad (2.6\text{-}16)$$

$$\tilde{g}(t) = g_I(t) + jg_Q(t) \qquad (2.6\text{-}17)$$

Substituting (2.6-16) and (2.6-17) into (2.6-15) gives the following relation:

$$\int_{-\infty}^{\infty} f(t)g(t)\, dt = \frac{1}{2\Delta_f} \sum_{k=-\infty}^{\infty} f_I\left(\frac{k}{\Delta_f}\right) g_I\left(\frac{k}{\Delta_f}\right) + f_Q\left(\frac{k}{\Delta_f}\right) g_Q\left(\frac{k}{\Delta_f}\right) \qquad (2.6\text{-}18)$$

If $f(t)$ and $g(t)$ are now restricted to an interval $(0, T)$, the following approximate relation results:

$$\int_0^T f(t)g(t)\, dt \cong \frac{1}{2\Delta_f} \sum_{k=1}^{T\Delta_f} f_I\left(\frac{k}{\Delta_f}\right) g_I\left(\frac{k}{\Delta_f}\right) + f_Q\left(\frac{k}{\Delta_f}\right) g_Q\left(\frac{k}{\Delta_f}\right) \qquad (2.6\text{-}19)$$

Defining vectors $\mathbf{f}$ and $\mathbf{g}$ as having components $f_I(k/\Delta_f)$, $f_Q(k/\Delta_f)$ and $g_I(k/\Delta_f)$, $g_Q(k/\Delta_f)$, respectively, we may replace the right side of (2.6-19) by a vector dot product:

$$\int_0^T f(t)g(t)\, dt \cong \frac{1}{2\Delta_f} (\mathbf{f} \cdot \mathbf{g}) \qquad (2.6\text{-}20)$$

Vectors $\mathbf{f}$ and $\mathbf{g}$ can be represented as points in a waveform space with approximately $2T\Delta_f$ dimensions.

2.7. Representation in Terms of Orthogonal Functions

A real or complex waveform $g(t)$ of finite energy may be described by an orthogonal expansion. Mathematically, a finite-energy waveform is classified as *integrable square*, or a *member of Lebesgue class L^2*, and satisfies the constraint

$$\int_a^b |g(t)|^2\, dt < \infty \qquad (2.7\text{-}1)$$

Functions $f_k(t)$ are defined to be *orthogonal* over an interval $a \leqslant t \leqslant b$ if

$$\int_a^b f_i^*(t) f_k(t)\, dt = \begin{cases} \lambda_k, & i = k \\ 0, & i \neq k \end{cases} \qquad (2.7\text{-}2)$$

for every pair of functions in the set. If $\lambda_k = 1$ for all k, the set is *orthonormal*. The set $\{f_k(t)\}$ is said to be *complete* in the class of integrable square functions defined over the interval $a \leqslant t \leqslant b$ if every $g(t)$ satisfying (2.7-1) can be approximated arbitrarily closely, in a mean-square sense, by a linear combination of orthogonal functions $f_k(t)$. Thus, if there exist constants c_k such that

$$\lim_{N \to \infty} \int_a^b \left| g(t) - \sum_{k=1}^N c_k f_k(t) \right|^2 dt = 0 \qquad (2.7\text{-}3)$$

for all $g(t)$ of integrable square defined over the interval $a \leqslant t \leqslant b$, then the set $f_k(t)$ is complete and $g(t)$ can be represented by the series

$$g(t) = \mathop{\text{l.i.m.}}_{N \to \infty} \sum_{k=1}^N c_k f_k(t) \qquad (2.7\text{-}4)$$

where, as previously noted, l.i.m. denotes "limit in the mean." Fourier series description (2.3-2) and sampling-theorem representation (2.4-10) are examples of complete orthogonal sets.

A set of orthogonal functions may be thought of as the components of a finite- or infinite-dimensional vector space. Just as there are many sets of axes that describe a point in vector space, there are many possible complete sets of orthogonal functions that describe (in a mean-square sense) a waveform in waveform space. Depending on the application, one set may be more convenient to use than another.

In the case of random processes, it is particularly convenient to choose an orthogonal set composed of solutions to the following integral equation, mathematically known as the *homogeneous Fredholm equation of the second kind:*

$$\int_a^b K(t, s)f(s)\, ds = \lambda f(t) \qquad a \leqslant t \leqslant b \tag{2.7-5}$$

When the *kernel* of the integral equation $K(t, s)$ satisfies

$$K(t, s) = K^*(s, t) \tag{2.7-6}$$

the kernel is said to be *hermitian symmetric*. If $K(t, s)$ is real, Eq. (2.7-6) becomes

$$K(t, s) = K(s, t) \tag{2.7-7}$$

and the kernel is called *real symmetric*. Depending on the kernel $K(t, s)$ and the region over which the integral is defined, solutions of (2.7-5) exist only for certain values of λ. These values of λ are called *eigenvalues*. Corresponding to each eigenvalue λ_k, there is at least one function $f_k(t)$ that satisfies the integral equation. These functions are termed *eigenfunctions*.

Under certain conditions the eigenfunctions of the homogeneous integral equation constitute a complete and orthogonal set over the interval $a \leqslant t \leqslant b$. These conditions are now briefly reviewed. First we show that if the kernel $K(t, s)$ is hermitian symmetric, the eigenfunctions are orthogonal. Let $f_i(t)$ and $f_k(t)$ be eigenfunctions corresponding to two different eigenvalues λ_i and λ_k of Eq. (2.7-5). Then

$$\int_a^b K(t, s)f_i(s)\, ds = \lambda_i f_i(t), \qquad a \leqslant t \leqslant b \tag{2.7-8}$$

and

$$\int_a^b K(t, s)f_k(s)\, ds = \lambda_k f_k(t) \qquad a \leqslant t \leqslant b \tag{2.7-9}$$

First both sides of (2.7-8) are multiplied by $f_k^*(t)$ and the result is integrated with respect to t over (a, b), in which case

$$\int_a^b f_k^*(t) \left[\int_a^b K(t, s)f_i(s)\, ds \right] dt = \lambda_i \int_a^b f_i(t)f_k^*(t)\, dt \tag{2.7-10}$$

Next the order of the integration on the left-hand side of (2.7-10) is reversed, which yields

$$\int_a^b f_i(s) \left[\int_a^b K(t, s) f_k^*(t)\, dt \right] ds = \lambda_i \int_a^b f_i(t) f_k^*(t)\, dt \qquad (2.7\text{-}11)$$

By use of (2.7-6), the bracketed integral in (2.7-11) can be rewritten as

$$\int_a^b K^*(s, t) f_k^*(t)\, dt$$

but this integral is simply the conjugated left member of (2.7-9) (except for a trivial interchange of variables s and t). Hence the bracketed integral in (2.7-11) can be replaced by $\lambda_k^* f_k^*(s)$ as follows:

$$\lambda_k^* \int_a^b f_i(s) f_k^*(s)\, ds = \lambda_i \int_a^b f_i(t) f_k^*(t)\, dt \qquad (2.7\text{-}12)$$

Except for the variable of integration the two integrals in (2.7-12) are identical, so that (2.7-12) simplifies to

$$(\lambda_k^* - \lambda_i) \int_a^b f_i(s) f_k^*(s)\, ds = 0 \qquad (2.7\text{-}13)$$

Since it was assumed that $\lambda_k \neq \lambda_i$, (2.7-13) requires that

$$\int_a^b f_i(s) f_k^*(s)\, ds = 0 \qquad (2.7\text{-}14)$$

Equation (2.7-14) verifies orthogonality condition (2.7-2) for eigenfunctions that correspond to different eigenvalues. When more than one linearly independent eigenfunction corresponds to a single eigenvalue, such eigenfunctions can be transformed into orthogonal functions by the Gram-Schmidt orthogonalization procedure.

Another interesting result follows from (2.7-13). For $i = k$, (2.7-13) becomes

$$(\lambda_k^* - \lambda_k) \int_a^b |f_k(s)|^2\, ds = 0 \qquad (2.7\text{-}15)$$

Since the integral in (2.7-15) is always positive, $(\lambda_k^* - \lambda_k)$ must be zero; hence λ_k is a real number. Real eigenvalues result from the hermitian symmetry of the kernel $K(t, s)$.

There are a variety of sufficient conditions on a hermitian symmetric kernel $K(t, s)$ that insure that the set of solutions to the Fredholm homogeneous integral equation is complete. One condition is that $K(t, s)$ can be written in the form $K(t - s)$, or $K(\tau)$, where $K(\tau)$ is the inverse Fourier transform of a power spectral density function [6]. This result will be very useful in the study of stationary random processes.

2.8. Some Useful Relations

An elegant Fourier relation between a (complex) time function $f(t)$ and its Fourier transform $F(\omega)$ is obtained by repeated application of the Fourier transform of the derivative of a time function (see Table 2.2-1) in Parseval's formula (2.2-20); the result is

$$\int_{-\infty}^{\infty} f^*(t) \frac{d^n f(t)}{dt^n} dt = \frac{(j)^n}{2\pi} \int_{-\infty}^{\infty} \omega^n |F(\omega)|^2 d\omega \qquad (2.8\text{-}1)$$

Starting with (2.8-1), Gabor [3] developed a number of useful formulas for the moments of $\Psi(\omega)$, the Fourier transform of analytic signal $\Psi(t)$; these functions are related to real signal $s(t)$, its complex envelope $\tilde{s}(t)$, and its Fourier transform $S(\omega)$ by Eqs. (2.5-5) through (2.5-13).

For $n = 0$ and $f(t) = \Psi(t)$, Eq. (2.8-1) becomes

$$\int_{-\infty}^{\infty} |\Psi(t)|^2 dt = \frac{1}{2\pi} \int_{-\infty}^{\infty} |\Psi(\omega)|^2 d\omega \qquad (2.8\text{-}2)$$

which is Parseval's relation (2.2-21) for analytic signals. Since $\Psi(\omega)$ is equal to $2S(\omega)$ for positive frequencies and zero for negative frequencies [see Eq. (2.5-7)], (2.8-2) can be rewritten as

$$\int_{-\infty}^{\infty} |\Psi(t)|^2 dt = 4 \left[\frac{1}{2\pi} \int_{0}^{\infty} |S(\omega)|^2 d\omega \right] \qquad (2.8\text{-}3)$$

The spectral symmetry of $|S(\omega)|$ and Parseval's formula (2.2-20) permits (2.8-3) to be successively modified as follows:

$$\int_{-\infty}^{\infty} |\Psi(t)|^2 dt = 2 \left[\frac{1}{2\pi} \int_{-\infty}^{\infty} |S(\omega)|^2 d\omega \right] \qquad (2.8\text{-}4)$$

$$\int_{-\infty}^{\infty} |\Psi(t)|^2 dt = 2 \left[\int_{-\infty}^{\infty} s^2(t) dt \right] \qquad (2.8\text{-}5)$$

$$\int_{-\infty}^{\infty} |\Psi(t)|^2 dt = 2E \qquad (2.8\text{-}6)$$

where E is defined to be the energy in $s(t)$. Since $|\Psi(t)|$ is approximately equal to $|\tilde{s}(t)|$ for narrowband signals, Eq. (2.8-6) can also be written as

$$\int_{-\infty}^{\infty} |\tilde{s}(t)|^2 dt \cong 2E \qquad (2.8\text{-}7)$$

For $n = 1$, Eq. (2.8-1) becomes

$$\int_{-\infty}^{\infty} \Psi^*(t) \frac{d\Psi(t)}{dt} dt = \frac{j}{2\pi} \left[\int_{-\infty}^{\infty} \omega |\Psi(\omega)|^2 d\omega \right] \qquad (2.8\text{-}8)$$

The bracketed term in (2.8-8) can be interpreted as the first moment of $|\Psi(\omega)|^2$. It is convenient to define a *normalized* first moment $\bar{\omega}$ by

$$\bar{\omega} = \frac{\dfrac{1}{2\pi} \displaystyle\int_{-\infty}^{\infty} \omega |\Psi(\omega)|^2 d\omega}{\dfrac{1}{2\pi} \displaystyle\int_{-\infty}^{\infty} |\Psi(\omega)|^2 d\omega} \qquad (2.8\text{-}9)$$

From (2.8-2) and (2.8-6), $\bar{\omega}$ can also be expressed as

$$\bar{\omega} = \frac{\dfrac{1}{2\pi} \displaystyle\int_{-\infty}^{\infty} \omega |\Psi(\omega)|^2 d\omega}{2E} \qquad (2.8\text{-}10)$$

Substituting definition (2.8-10) into (2.8-8) yields

$$\int_{-\infty}^{\infty} \Psi^*(t)\frac{d\Psi(t)}{dt}\,dt = 2jE\bar{\omega} \tag{2.8-11a}$$

In narrowband applications, $\Psi(t)$ is approximately equal to $\tilde{s}(t)e^{j\omega_0 t}$; with this substitution Eq. (2.8-11a) becomes

$$\int_{-\infty}^{\infty} \tilde{s}^*(t)\frac{d\tilde{s}(t)}{dt}\,dt + j\omega_0 \int_{-\infty}^{\infty} |\tilde{s}(t)|^2\,dt = 2jE\bar{\omega} \tag{2.8-11b}$$

By means of (2.8-5), Eq. (2.8-11b) can be written more simply as

$$\int_{-\infty}^{\infty} \tilde{s}^*(t)\frac{d\tilde{s}(t)}{dt}\,dt = 2jE(\bar{\omega} - \omega_0) \tag{2.8-12}$$

If $\omega_0 = \bar{\omega}$ (a frequent narrowband assumption), Eq. (2.8-12) finally reduces to

$$\int_{-\infty}^{\infty} \tilde{s}^*(t)\frac{d\tilde{s}(t)}{dt}\,dt = 0 \tag{2.8-13}$$

For $n = 2$, Eq. (2.8-1) becomes

$$\int_{-\infty}^{\infty} \Psi^*(t)\frac{d^2\Psi(t)}{dt^2}\,dt = -\frac{1}{2\pi}\left[\int_{-\infty}^{\infty} \omega^2 |\Psi(\omega)|^2\,d\omega\right] \tag{2.8-14}$$

The bracketed quantity in (2.8-14) can be interpreted as the (unnormalized) second moment of $|\Psi(\omega)|^2$. If a normalized second moment $\overline{\omega^2}$ is defined by

$$\overline{\omega^2} = \frac{\dfrac{1}{2\pi}\displaystyle\int_{-\infty}^{\infty} \omega^2 |\Psi(\omega)|^2\,d\omega}{2E} \tag{2.8-15}$$

then Eq. (2.8-14) can be rewritten as

$$\int_{-\infty}^{\infty} \Psi^*(t)\frac{d^2\Psi(t)}{dt^2}\,dt = -2E\overline{\omega^2} \tag{2.8-16}$$

For a narrowband signal, $\Psi(t)$ can be replaced by $\tilde{s}(t)e^{j\omega_0 t}$, which gives

$$\int_{-\infty}^{\infty} \tilde{s}^*(t)\frac{d^2\tilde{s}(t)}{dt^2}\,dt + 2j\omega_0 \int_{-\infty}^{\infty} \tilde{s}^*(t)\frac{d\tilde{s}(t)}{dt}\,dt$$
$$- \omega_0^2 \int_{-\infty}^{\infty} |\tilde{s}(t)|^2\,dt = -2E\overline{\omega^2} \tag{2.8-17}$$

The second integral in (2.8-17) is zero by (2.8-13) and the third is equal to $2E$ by (2.8-7), so that (2.8-17) simplifies to

$$\int_{-\infty}^{\infty} \tilde{s}^*(t)\frac{d^2\tilde{s}(t)}{dt^2}\,dt = 2E(\omega_0^2 - \overline{\omega^2}) \tag{2.8-18}$$

By means of the identity

$$\overline{(\omega - \bar{\omega})^2} = \overline{\omega^2} - \bar{\omega}^2 \tag{2.8-19}$$

we may rewrite Eq. (2.8-18) as

$$\int_{-\infty}^{\infty} \tilde{s}^*(t)\frac{d^2\tilde{s}(t)}{dt^2}\,dt = 2E[\omega_0^2 - \bar{\omega}^2 - \overline{(\omega - \bar{\omega})^2}] \tag{2.8-20}$$

If $\bar{\omega}$ is once again set equal to ω_0, (2.8-20) reduces to the Gabor relation,

$$\overline{(\omega - \bar{\omega})^2} = -\frac{1}{2E} \int_{-\infty}^{\infty} \bar{s}^*(t) \frac{d^2\bar{s}(t)}{dt^2}\, dt \qquad (2.8\text{-}21)$$

The quantity $\overline{(\omega - \bar{\omega})^2}$ is often interpreted as the mean-square bandwidth of narrowband signal $s(t)$.

PROBLEMS

2.1. If $f(t) = 0$ for $t < 0$, show that

$$f(t) = \frac{2}{\pi} \int_0^{\infty} R(\omega) \cos \omega t\, d\omega, \qquad t > 0$$

$$= \frac{2}{\pi} \int_0^{\infty} X(\omega) \sin \omega t\, d\omega, \qquad t > 0$$

$$f(0) = \frac{1}{\pi} \int_0^{\infty} R(\omega)\, d\omega$$

when the Fourier transform of $f(t)$ is given by

$$F(\omega) = R(\omega) + jX(\omega)$$

2.2. If $f(t)$ is a real function, show that its Fourier transform $F(\omega)$ obeys the constraint

$$F(-\omega) = F^*(\omega)$$

2.3. If a is a real constant, show that the Fourier transform of $f(at)$ is given by

$$\frac{1}{|a|} F\left(\frac{\omega}{a}\right)$$

where $F(\omega)$ is the Fourier transform of $f(t)$.

2.4. Show that the Fourier transform of $f^*(t)$ is given by $F^*(-\omega)$.

2.5. Show that the Fourier transform of the convolution of two functions, $f_1(t)$ and $f_2(t)$, equals the product of their respective Fourier transforms, $F_1(\omega)$ and $F_2(\omega)$.

2.6. Let $f(t)$ be a time function whose transform $F(\omega)$ is band-limited to $(-\omega_c, \omega_c)$. Time function $f_s(t)$ is obtained by sampling $f(t)$ at regular intervals spaced τ seconds apart:

$$f_s(t) = f(t) \sum_{k=-\infty}^{\infty} \delta(t - k\tau) = \sum_{k=-\infty}^{\infty} f(k\tau)\, \delta(t - k\tau)$$

Show first that the Fourier transform $F_s(\omega)$ of $f_s(t)$ is given by

$$F_s(\omega) = \sum_{k=-\infty}^{\infty} \frac{1}{\tau} F\left(\omega - \frac{2\pi k}{\tau}\right)$$

Apart from the factor $1/\tau$, $F_s(\omega)$ consists of replicas of $F(\omega)$ centered on the spectral lines $\delta(\omega - 2\pi k/\tau)$ as shown in Fig. 2P-6. For what range of

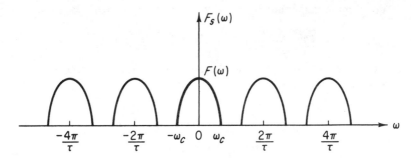

Fig. 2P-6. Transform of sampled low-pass function.

values of τ can the original transform be recovered? Next assume that sampling rate $\tau = 1/2f_c$, where $f_c = \omega_c/2\pi$. The original transform can be recovered by multiplying $F_s(\omega)$ by a spectral gating function $G(\omega)$ whose value is unity over $(-\omega_c, \omega_c)$. Derive the low-frequency sampling representation by taking the inverse transform of the product $G(\omega)F_s(\omega)$.

2.7. Show that

$$\int_{-\infty}^{\infty} \frac{\sin 2\pi f_c\left(t - \dfrac{i}{2f_c}\right)}{2\pi f_c\left(t - \dfrac{i}{2f_c}\right)} \cdot \frac{\sin 2\pi f_c\left(t - \dfrac{k}{2f_c}\right)}{2\pi f_c\left(t - \dfrac{k}{2f_c}\right)}\, dt = \begin{cases} \dfrac{1}{2f_c} & i = k \\ 0, & i \neq k \end{cases}$$

2.8. Show that $e^{j\omega_k t}$ is an eigenfunction of the integral equation

$$\int_{-\infty}^{\infty} K(t - s)f(s)\, ds = \lambda f(t), \qquad -\infty < t < +\infty$$

where the Fourier transform of $K(\tau)$ is given by $S(\omega)$. (Substitute and solve.) What eigenvalue corresponds to eigenfunction $e^{j\omega_k t}$?

2.9. Prove the Fourier transform shift theorems:

$$s(t + \tau) \longleftrightarrow S(\omega)e^{j\omega\tau}$$
$$s(t)e^{-j\omega_0 t} \longleftrightarrow S(\omega + \omega_0)$$

2.10. Express the average power in a periodic function, described by Fourier series (2.3-13), in terms of Fourier coefficients a_k and b_k, where the average power is defined by

$$P_{AV} = \lim_{T \to \infty} \frac{1}{T} \int_{-T/2}^{T/2} f^2(t)\, dt$$

What power is associated with radian frequency ω_k?

2.11. Show that the Fourier transform of $ds(t)/dt$ is given by $j\omega S(\omega)$ (if it exists).

2.12. Show that the function

$$f_1(x) = \frac{k}{\sqrt{\pi}} e^{-k^2 x^2}$$

has the properties of a delta function as $k \to \infty$. Repeat for the function

$$f_2(x) = \begin{cases} ke^{-kx}, & x > 0 \\ 0, & x < 0 \end{cases}$$

2.13. Show that the inverse Fourier transform of $(2/j\omega)$ is given by sgn t, defined as shown in Fig. 2P-13.

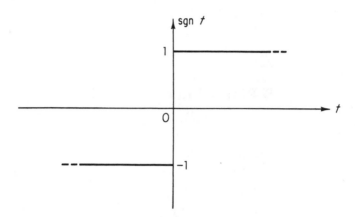

Fig. 2P-13. Function: sgn t.

REFERENCES

1. Papoulis, A.: "The Fourier Integral and its Applications," McGraw-Hill, New York, 1962.
2. Wainstein, L. A., and Zubakov, V. P.: "Extraction of Signals from Noise," Chap. 3, Prentice-Hall, Englewood Cliffs, N.J., 1962.
3. Gabor, D.: Theory of Communications, *J. IEE* (*London*), **93**: (*Pt. III*), 429–457 (1946).
4. Ville, J.: Théorie et applications de la notion de signal analytique, *Cables et transmissions*, **2**: (*1*), 61–74 (1948).
5. Rubin, W. L., and DiFranco, J. V.: Analytic Representation of Wide Band Radio Frequency Signals, *J. Franklin Inst.*, **275**: (*3*), 197–204 (March, 1963).
6. Youla, D. C.: The Use of the Method of Maximum Likelihood in Estimating Continuous Modulated Intelligence Which Has Been Corrupted by Noise, *Trans. IRE Prof. Group on Information Theory*, **PGIT-3**: 90–106 (March, 1954).

3

PROBABILITY THEORY

3.1. Introduction

The specific value of a noisy waveform at any instant of time is inherently unpredictable. Some values, however, tend to occur more often than others. The frequency with which each value occurs can be characterized by a *probability density function*. This concept and other related probabilistic concepts are presented in this chapter to provide a theoretical framework for subsequent discussions. Readers desiring a more detailed and rigorous exposition should consult the references recommended at the end of the chapter.

3.2. Definition of Probability

Suppose a particular experiment is repeated a great number of times under similar circumstances. For example, consider a coin-tossing experiment in which it is desired to predict heads or tails on successive trials. If the physical phenomena underlying the experiment were sufficiently understood and susceptible to quantitative evaluation, the result of each trial could be predicted exactly. It is well known, however, that in a long sequence of apparently identical tossings of a coin, approximately one-half of the outcomes will be heads and one-half tails. This is a typical example of *statistical regularity*, which constitutes the empirical basis of probability theory. This result is illustrated in Fig. 3.2-1, in which the ratio of the number of heads occurring

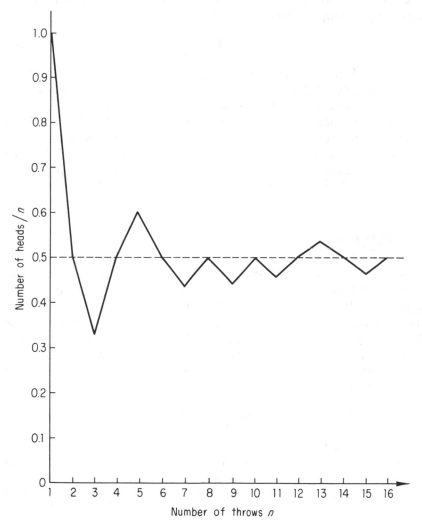

Fig. 3.2-1. Relative frequency of number of heads in a sequence of coin tossings.

in a particular series of n trials is plotted against n. The apparent reason for such behavior is that the outcome of each trial is affected by many small factors which cannot be evaluated simply. Probability theory has been invented, in part, to provide an alternative mathematical description of such experiments, in which the outcomes are predicted on a probabilistic basis.

The probability of a *random event* may be defined as follows. Suppose in a sequence of n trials or experiments, an event A occurs $\nu_A(n)$ times. The notation $\nu_A(n)$ is used to indicate that ν_A is a function of n. The *relative frequency* of occurrence of event A is $\nu_A(n)/n$. When the mathematical limit exists, the probability of event A is defined as

$$P(A) = \lim_{n \to \infty} \frac{\nu_A(n)}{n} \qquad (3.2\text{-}1)$$

This definition is often referred to as the frequency interpretation of the probability $P(A)$. It follows immediately from (3.2-1) that the mathematical probability of an event is a number between zero and one.

Because of difficulties associated with definition (3.2-1), probability theory has been recast in recent years into a more rigorous mathematical framework; a discussion of this approach is given in several of the texts listed in the bibliography. However, the frequency definition of probability (3.2-1) suffices for our purposes.

3.3. Sample Space

Let the results of *trials, experiments*, or *observations* be called *outcomes* or *events*. All the possible outcomes of an experiment form a *sample space*. Figure 3.3-1 illustrates the sample space for a die-tossing experiment. If the experiment has a finite or countably infinite number of outcomes, the sample space is called *discrete*. With each element (event) in a discrete sample space we can associate a probability whose value is between zero and one.

If a particular event A has zero probability, we mean that in a long sequence of trials

$$\lim_{n \to \infty} \frac{\nu_A(n)}{n} \longrightarrow 0$$

that is, the frequency of occurrence of A is vanishingly small. Note that event A is not an *impossible* event. However, an impossible event—one that can never occur—also has zero probability since $\nu_A(n) = 0$ for all n. In a similar way, if the probability of A is unity, we mean that in a long sequence of n trials

$$\lim_{n \to \infty} \frac{\nu_A(n)}{n} \longrightarrow 1$$

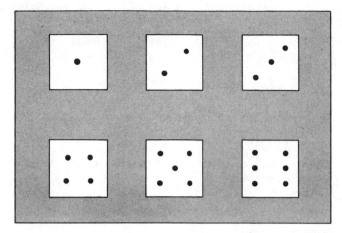

Fig. 3.3-1. Sample space for a die.

Note again that *A* is not a *certain* event, since it need not occur on every trial. It also follows that an event that is certain to occur has unity probability, since in this case $v_A(n) = n$ for all *n*.

Consider an experiment in which there are three *mutually exclusive* outcomes *A, B,* and *C.* Mutually exclusive events are defined to be events such that only one can occur on any trial. The sample space for this experiment contains three sample points (see Fig. 3.3-2). What is the probability that the outcome of an experiment is *A or B*? In this case the outcome is said to be a *compound event*, since it is a subset of simple events in the sample space. The probability of the compound event (*A or B*) may be found in

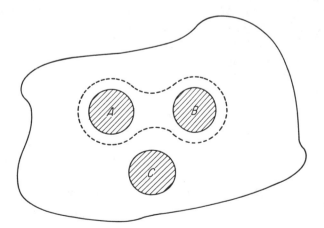

Fig. 3.3-2. Mutually exclusive (nonoverlapping) events with compound event (*A* or *B*) shown enclosed.

terms of the probabilities of the simple events A and B as follows. If in a long sequence of n trials A occurs $\nu_A(n)$ times and B occurs $\nu_B(n)$ times, then, since A and B are mutually exclusive, the outcome $(A$ or $B)$ occurs $\nu_A(n) + \nu_B(n)$ times. It follows from Eq. (3.2-1) that

$$P(A \text{ or } B) = \lim_{n \to \infty} \frac{\nu_A(n) + \nu_B(n)}{n} = P(A) + P(B) \qquad (3.3\text{-}1)$$

By a similar argument we can show that

$$P(A \text{ or } B \text{ or } C) = P(A) + P(B) + P(C) = 1 \qquad (3.3\text{-}2)$$

The reader can easily extend the above results to an experiment with k mutually exclusive outcomes.

3.4. Independent Events

Consider next two events A and B, with given probabilities $P(A)$ and $P(B)$. What is the probability that both A and B occur simultaneously? This probability is expressed by the notation $P(A, B)$ and is conventionally called the *joint probability of A and B*. Clearly, if A and B are mutually exclusive outcomes of the same experiment, the joint probability $P(A, B)$ must be zero, since both cannot occur at the same time. Suppose, however, that the two events A and B are *independent*. By independence we mean that both events can occur simultaneously but the occurrence of one does not influence or affect the occurrence of the other. In a large number n of trials of a given experiment let $\nu_A(n)$ and $\nu_B(n)$ be the respective number of occurrences of A and B. From Eq. (3.2-1), it is expected that $P(A)$ will be approximated closely by $\nu_A(n)/n$ and $P(B)$ will very nearly equal $\nu_B(n)/n$. Consider next only those trials in which the event A has occurred. If $\nu_A(n)$ is a large number, we should also expect that the event B will occur approximately $\nu_A P(B)$ times in the ν_A trials in which A has occurred. It follows that $\nu_A P(B)$ is also equal to the number of events in which both A and B occur simultaneously. Hence, we obtain the simple relationship

$$nP(A, B) = \nu_A P(B) \qquad (3.4\text{-}1)$$

When both sides of (3.4-1) are divided by n, and n is allowed to become indefinitely large, then by Eq. (3.2-1) the joint probability of two independent events is

$$P(A, B) = P(A \text{ and } B) = P(A) \cdot P(B) \qquad (3.4\text{-}2)$$

Result (3.4-2) can be generalized to the joint probability of any number of independent events as follows:

$$P(A, B, C, \ldots) = P(A) \cdot P(B) \cdot P(C) \cdot \ldots \qquad (3.4\text{-}3)$$

3.5. Conditional Probability

Suppose that two events, A and B, are not independent. That is, factors which influence the occurrence of one influence the occurrence of the other. The sample space might be represented as shown in Fig. 3.5-1. What is the probability of occurrence of A *and* B in this case? To deal with this situation, define the *conditional probability* of event A given that B has occurred by $P(A \mid B)$ and the conditional probability of event B assuming A has occurred by $P(B \mid A)$. By an argument analogous to the one that led to Eq. (3.4-2), the joint probability $P(A, B)$ when A and B are not independent events can be shown to be given by either of the following expressions:

$$P(A, B) = P(A) \cdot P(B \mid A) \tag{3.5-1}$$

$$P(A, B) = P(B) \cdot P(A \mid B) \tag{3.5-2}$$

Note that when A and B are independent, $P(A \mid B) = P(A)$ and $P(B \mid A) = P(B)$, so that Eqs. (3.5-1) and (3.5-2) reduce to Eq. (3.4-2).

To illustrate, let us find the probability that two white balls will be selected (without replacement) from an urn containing 6 white and 4 black balls. The probability $P(A)$ of initially selecting a white ball is $\frac{6}{10}$. The probability of selecting a second white ball, given that the first was white [designated by conditional probability $P(B \mid A)$] is given by $\frac{5}{9}$, since only 5 out of 9 balls are white at the start of this selection. By Eq. (3.5-1), the probability of successively selecting the two white balls is $(\frac{6}{10})(\frac{5}{9}) = \frac{1}{3}$.

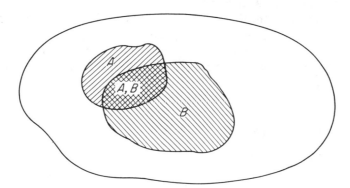

Fig. 3.5-1. Sample space of overlapping events.

3.6. Random Variables

Up to this point the possible outcomes of a random experiment have been identified by letters of the alphabet, either singly or in combinations. This representation may be clumsy when the number of outcomes of an experiment is very large or infinite. Suppose instead we ascribe to each outcome or event a real number. Thus to a sample space composed of n discrete events, n distinct real numbers, $x_i, i = 1, 2, \ldots, n$, are assigned. Define now a *random variable* X that can assume only values in the discrete set $\{x_i\}$. Clearly there is a one-to-one correspondence between the set of values assumed by the random variable X and the sample space of events.

In general, any function that assigns a real number to each point in the sample space of events is defined as a random variable. (Although it would appear from the definition that random function is a more appropriate designation, the term random variable is universally used.) In many cases the outcomes of a particular experiment may already be associated with a set of real numbers. If convenient, such numbers can also be used to define the set of values assumed by the random variable. For example, in an experiment consisting of the roll of a die it is natural to let the random variable take on the values of the integers one through six, coinciding with the number of dots on each of the six faces of the die.

From this point on, we designate the probability of an event in a discrete sample space (with associated real numbers x_i) by the probability that the random variable X assumes the value x_i, written as $\Pr (X = x_i)$. In a continuous sample space it is more meaningful to speak of the probability that the random variable X lies in a subset S_i of values assumed by the random variable—that is, $\Pr (X \text{ in } S_i)$. To unify notation, a discrete sample space is considered a special case of a continuous sample space for which each subset S_i contains only a single point x_i.

Suppose an experiment is repeated n times. Then random variables $X_1, X_2, \ldots, X_n$ can be assigned, respectively, to designate the possible outcomes of each repetition of the experiment. Each random variable X_i is defined over the same sample space. If each repetition of the experiment is independent of all the others, the random variables $X_1, X_2, \ldots, X_n$ are statistically independent. This is stated formally as follows. The random variables $X_1, X_2, \ldots, X_n$ will be statistically independent if, and only if, for all sets of real numbers S_i,

$$\Pr (X_i \text{ in } S_i, X_j \text{ in } S_j) = \Pr (X_i \text{ in } S_i) \cdot \Pr (X_j \text{ in } S_j)$$
$$i, j = 1, 2, \ldots, n, \qquad i \neq j \qquad (3.6\text{-}1a)$$

$$\Pr (X_i \text{ in } S_i, X_j \text{ in } S_j, X_k \text{ in } S_k) = \Pr (X_i \text{ in } S_i) \cdot \Pr (X_j \text{ in } S_j) \cdot \Pr (X_k \text{ in } S_k)$$
$$i, j, k = 1, 2, \ldots, n, \qquad i \neq j \neq k \qquad (3.6\text{-}1b)$$

. . . .

$$\Pr (X_1 \text{ in } S_1, X_2 \text{ in } S_2, \ldots, X_n \text{ in } S_n)$$
$$= \Pr (X_1 \text{ in } S_1) \cdot \Pr (X_2 \text{ in } S_2) \cdot \ldots \cdot \Pr (X_n \text{ in } S_n) \qquad (3.6\text{-}1c)$$

Thus, in order that random variables $X_1, X_2, \ldots, X_n$ be statistically independent, they must be independent in pairs, triples, and so forth.

3.7. Probability Distribution and Probability Density Functions

By introducing the concept of a random variable, we have mapped the space of events onto a set of real numbers. It is convenient to map this set of numbers onto a real line that extends from minus infinity to plus infinity. A subset of events in sample space will then be a collection of points or intervals on the real line.

A probability distribution function $P(x)$ can now be defined as

$$P(x) = \Pr(X \leqslant x) \tag{3.7-1}$$

where x can assume all values on the real line, $-\infty < x < \infty$. $P(x)$ is simply the probability that an event occurs in a subset of the sample space for which the corresponding value of the random variable X is less than or equal to x. The probability distribution function provides a method of ordering events in the sample space so that a physical picture of the distribution of probability over the space can be inferred.

The function $P(x)$ has the following properties. From definition (3.7-1) $P(x)$ is a monotonically nondecreasing function of x. Further, since the probability of any set of events is always a positive number between zero and one, $P(x)$ lies between zero and one for all x. In particular, $P(-\infty) = 0$ and $P(+\infty) = 1$. Figure 3.7-1a is an example of a Gaussian probability distribution function. The probability that X lies in the interval (a, b) is simply given by

$$\Pr(a < X \leqslant b) = P(b) - P(a) \tag{3.7-2}$$

If $P(x)$ is a continuous† differentiable function, we can define a new function

$$p(x) = \frac{dP(x)}{dx} \tag{3.7-3}$$

which is called a *probability density function*. Figure 3.7-1b illustrates a Gaussian probability density function. To obtain the probability that the random variable X lies in an infinitesimal interval dx centered at x, $p(x)$ is multiplied by dx. To find the probability that X lies in a larger specified interval (a, b), $p(x)$ is integrated over this interval:

$$\Pr(a < X \leqslant b) = \int_a^b p(x)\, dx \tag{3.7-4}$$

The probability distribution $P(x)$ can be expressed in terms of the probability density function by

$$P(x) = \int_{-\infty}^{x} p(u)\, du \tag{3.7-5}$$

---

†A function is *absolutely continuous* if the number of points at which it is not differentiable is countable.

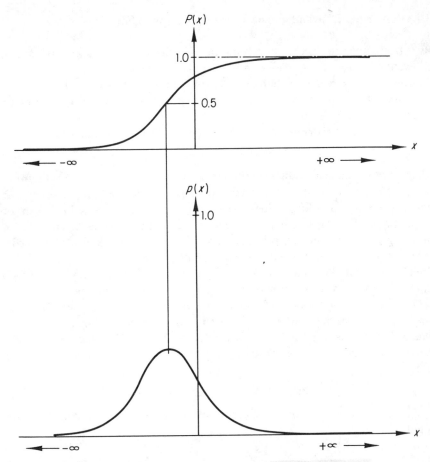

Fig. 3.7-1(a). *Upper curve:* Gaussian probability distribution function. **(b).** *Lower curve:* Gaussian probability density function.

When $x = +\infty$, we obtain the following important constraint on $p(x)$ from Eq. (3.7-5):

$$\int_{-\infty}^{\infty} p(x)\,dx = P(+\infty) = +1 \qquad (3.7\text{-}6)$$

Because of properties (3.7-4) and (3.7-6), the probability density function can be interpreted as the *mass density* of a unit of mass distributed over the entire real line $-\infty < x < \infty$ (see Fig. 3.7-1b). The mass allocated to any set of points on the real line is a measure of the probability that the random variable will assume a value in that set. The probability distribution function $P(x)$ can be interpreted as the portion of mass lying to the left of, and including, the point x.

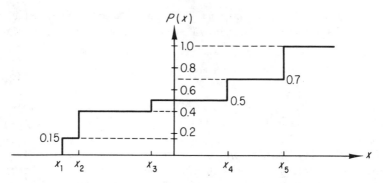

Fig. 3.7-2(a). Example of discrete probability distribution function.

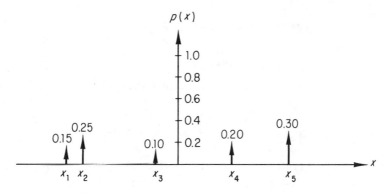

Fig. 3.7-2(b). Example of discrete probability density function.

Suppose the random variable X takes on only discrete values x_1, x_2, x_3, ..., such that $\Pr(x = x_1) = p_1$, $\Pr(x = x_2) = p_2$, $\Pr(x = x_3) = p_3$, and so on. The probability distribution of X is not continuous in this case but has a staircase appearance as shown in Fig. 3.7-2a. The distribution function $P(x)$ is given by

$$P(x) = \sum_{k=1}^{n} p_k U(x - x_k) \tag{3.7-7}$$

where $U(x)$ is the unit step function

$$U(x) = \begin{cases} 1, & x \geqslant 0 \\ 0, & x < 0 \end{cases} \tag{3.7-8}$$

and where p_k is the magnitude of the discontinuity at $x = x_k$, that is,

$$p_k = P(x_k^+) - P(x_k^-) \tag{3.7-9}$$

From the definition of the impulse or delta function $\delta(x)$ given by Eq. (2.3-4) in Chapter 2, it can be shown† that the derivative of the unit step function is

$$\frac{dU(x)}{dx} = \delta(x) \qquad (3.7\text{-}10)$$

The probability density function in this case can be represented by a set of delta functions such that at the point x_i, the probability density is given by $p_i \, \delta(x - x_i)$.

The probability density function corresponding to the discrete distribution function in Fig. 3.7-2a is shown in Fig. 3.7-2b. Note that the delta function fits in with the analogy of mass distribution, since a discrete mass located at a point results in an infinite mass density. In terms of delta functions, the probability density function and probability distribution function are given by

$$p(x) = \sum_i p_i \, \delta(x - x_i) \qquad (3.7\text{-}11)$$

and

$$\begin{aligned} P(x) &= \Pr\,(X \leqslant x) \\ &= \sum_i p_i \qquad \text{for } i \text{ such that } x_i \leqslant x \end{aligned} \qquad (3.7\text{-}12)$$

3.8. Joint Probability Distribution and Density Functions

The joint probability distribution function $P(x, y)$ of two random variables X and Y is defined for all x and y on the real line, $-\infty < x, y < +\infty$, by

$$P(x, y) = \Pr\,(X \leqslant x, \, Y \leqslant y) \qquad (3.8\text{-}1)$$

The joint probability density $p(x, y)$ of random variables X and Y is defined by

$$p(x, y) = \frac{\partial^2 P(x, y)}{\partial x \, \partial y} \qquad (3.8\text{-}2)$$

An example of a joint probability density function is shown in Fig. 3.8-1.

Equation (3.8-2) can be expressed in integral form as

$$P(x, y) = \int_{-\infty}^{x} \int_{-\infty}^{y} p(u, v) \, du \, dv \qquad (3.8\text{-}3)$$

From definition (3.8-1) and Eq. (3.8-3) the following useful relations can be demonstrated:

$$P(\infty, \infty) = \int_{-\infty}^{\infty} \int_{-\infty}^{\infty} p(x, y) \, dx \, dy = 1 \qquad (3.8\text{-}4)$$

†A. Papoulis, "The Fourier Integral and Its Applications," McGraw-Hill, New York, 1962.

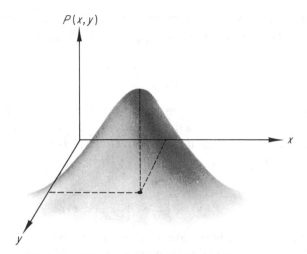

Fig. 3.8-1. Joint probability density function.

and

$$P(x, -\infty) = P(-\infty, y) = P(-\infty, -\infty) = 0 \qquad (3.8\text{-}5)$$

It follows from Eq. (3.6-1) that for two statistically independent random variables, X, Y, the joint probability distribution $P(x, y)$ is given by

$$P(x, y) = \Pr(X \leqslant x, Y \leqslant y) = \Pr(X \leqslant x) \cdot \Pr(Y \leqslant y) \qquad (3.8\text{-}6a)$$
$$= P_1(x) \cdot P_2(y) \qquad (3.8\text{-}6b)$$

where $P_1(x)$ and $P_2(y)$ are the probability distributions of X and Y, respectively. An analogous result is obtained for the joint probability density function $p(x, y)$ of statistically independent random variables by differentiating both sides of Eqs. (3.8-6a) and (3.8-6b) with respect to x and y:

$$p(x, y) = \frac{\partial^2 P(x, y)}{\partial x \, \partial y} = \frac{\partial P_1(x)}{\partial x} \cdot \frac{\partial P_2(y)}{\partial y} \qquad (3.8\text{-}7a)$$

$$= p_1(x) \cdot p_2(y) \qquad (3.8\text{-}7b)$$

where $p_1(x)$ and $p_2(y)$ are the derivatives with respect to x and y, respectively, of $P_1(x)$ and $P_2(y)$.

The relations above can easily be extended to more than two random variables. They apply also to discrete joint probability density functions, provided that an n-dimensional delta function is defined. We make this assumption in later chapters, so that our results will be applicable to discrete or continuous sample spaces (unless stated otherwise).

The probability distribution $P_1(x)$ of the random variable X can be evaluated from knowledge of the joint probability distribution of X and Y, $P(x, y)$, or computed from the joint probability density function $p(x, y)$.

Since $\Pr(X \leqslant x)$ is the same as $\Pr(X \leqslant x, Y \leqslant \infty)$, we may write

$$P_1(x) = \Pr(X \leqslant x) = \Pr(X \leqslant x, Y \leqslant \infty) = P(x, \infty) \qquad (3.8\text{-}8)$$

or

$$P_1(x) = \int_{-\infty}^{\infty} dy \int_{-\infty}^{x} p(u, y)\, du \qquad (3.8\text{-}9)$$

The function $P_1(x)$ computed in this manner is also known as a *marginal* probability distribution function. In analogous fashion it can be shown that the marginal probability distribution $P_2(y)$ is given either by

$$P_2(y) = P(\infty, y) \qquad (3.8\text{-}10)$$

or by

$$P_2(y) = \int_{-\infty}^{\infty} dx \int_{-\infty}^{y} p(x, v)\, dv \qquad (3.8\text{-}11)$$

By differentiating both sides of (3.8-9) with respect to x, and (3.8-11) with respect to y, we obtain the marginal probability density functions

$$p_1(x) = \int_{-\infty}^{\infty} p(x, y)\, dy \qquad (3.8\text{-}12)$$

and

$$p_2(y) = \int_{-\infty}^{\infty} p(x, y)\, dx \qquad (3.8\text{-}13)$$

To obtain (3.8-12) and (3.8-13) a well-known result† in calculus was used, which states that the derivative of an integral with respect to its upper limit is equal to the integrand evaluated at this limit.

These results can be extended to joint density functions of more than two random variables. For example, the marginal probability density function of the first k random variables in a set n may be found from the n-dimensional joint probability density function $p(x_1, x_2, \ldots, x_k, \ldots, x_n)$ as follows:

$$p(x_1, x_2, \ldots, x_k)$$
$$= \int_{-\infty}^{\infty} dx_{k+1} \int_{-\infty}^{\infty} dx_{k+2} \ldots \int_{-\infty}^{\infty} p(x_1, x_2, \ldots, x_k, \ldots, x_n)\, dx_n \qquad (3.8\text{-}14)$$

3.9. Sums of Random Variables

The sum of a number of real random variables is also a real random variable. The probability distribution and probability density function of the sum can be found if the joint probability density function of the individual random variables is known. Consider first the case of two random variables X and Y with joint probability density function $p(x, y)$. We wish to determine the probability density function $p_3(z)$ of the random variable $Z = X + Y$.

---

†P. Franklin, "Methods of Advanced Calculus," McGraw-Hill, New York, 1944.

From the definition of a distribution function

$$P_3(z) = \Pr(Z \leqslant z) = \Pr(X + Y \leqslant z) \tag{3.9-1}$$

Since the information available is in the form of the joint probability density function $p(x, y)$ of random variables X and Y, it is desirable to restate Eq. (3.9-1) as

$$\Pr(X + Y \leqslant z) = \Pr(Y \leqslant z - X, X < \infty) \tag{3.9-2}$$

From Eq. (3.9-1) and by definition of the quantity on the right in Eq. (3.9-2),

$$P_3(z) = \int_{-\infty}^{\infty} dx \int_{-\infty}^{z-x} p(x, y)\, dy \tag{3.9-3}$$

The joint sample space for real random variables X and Y is shown in Fig. 3.9-1. Upon differentiating Eq. (3.9-3), we get

$$p_3(z) = \frac{dP_3(z)}{dz} = \int_{-\infty}^{\infty} p(x, y = z - x)\, dx \tag{3.9-4}$$

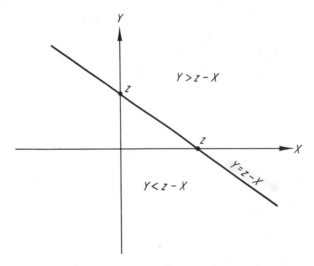

Fig. 3.9-1. Joint sample space for sum of two real random variables.

which is the desired result. For the special case in which X and Y are statistically independent random variables, their joint probability density function $p(x, y)$ can be expressed as the product $p_1(x) \cdot p_2(y)$. Then Eq. (3.9-4) becomes

$$p_3(z) = \int_{-\infty}^{\infty} p_1(x) \cdot p_2(z - x)\, dx \tag{3.9-5}$$

The right side of (3.9-5) is the convolution integral, which can be written as $p_1(x) * p_2(x)$. The probability density function of the sum of n independent random variables,

$$Z = \sum_{i=1}^{n} X_i$$

is simply an extension of the result for the sum of two random variables and is given by

$$p(z) = p_1(x) * p_2(x) * \ldots * p_n(x) \tag{3.9-6}$$

Equation (3.9-6) states that $p(z)$ is found by convolving $p_1(x)$ with $p_2(x)$ and the result with $p_3(x)$, and so on until all the random variables have been successively accounted for.

3.10. Transformations of Random Variables

A problem often encountered is the determination of the probability density function of a set of random variables that are related to an old set (with known density function) by a one-to-one mapping. Consider first the case of two random variables X and Y, where $Y = f(X)$ is a single-valued function of X with known probability density function $p_1(x)$. The probability that random variable Y lies in an interval $(y, y + dy)$ is equal to the probability that X lies in the interval $(x, x + dx)$. Thus,
Pr $(y < Y \leqslant y + dy) = p_2(y)\,dy$

$$= \text{Pr}\,(x < X \leqslant x + dx) = p_1(x)\,dx \tag{3.10-1}$$

When $f(x)$ is differentiable and its derivative vanishes only at a countable number of isolated points, Eq. (3.10-1) can be solved for $p_2(y)$:

$$p_2(y) = \frac{p_1(x)}{|dy/dx|} = \left[\frac{p_1(x)}{|df(x)/dx|}\right]_{x=g(y)} \tag{3.10-2}$$

where $X = g(Y)$ is the inverse relationship between X and Y. Since a probability density function is always positive, it can be demonstrated that absolute-magnitude signs are required in the denominator of (3.10-2) as shown. Result (3.10-2) can be modified when each value of Y corresponds to m values of X, that is, the inverse function $X = g(Y)$ is multivalued. The result is

$$p_2(y) = \left[\frac{mp_1(x)}{|df(x)/dx|}\right]_{x=g(y)} \tag{3.10-3}$$

provided $|df(x)/dx|$ is equal at each of the m values.

Consider next the case of two random variables, U and V, which are single-valued functions of two other variables, X and Y:

$$U = U(X, Y), \quad V = V(X, Y) \tag{3.10-4}$$

It is assumed that the joint probability density function $p_1(x, y)$ of variables

X and Y is known, and it is desired to find $p_2(u, v)$, the joint probability density function of the new variables U and V. Because of their one-to-one relationship the probability that X and Y lie in an area element $dx\,dy$ is equal to the probability that U and V lie in an area element $du\,dv$, so that

$$p_2(u, v)\,du\,dv = p_1(x, y)\,dx\,dy \tag{3.10-5}$$

It is well known† that elementary areas in each of two planes are related to each other by the magnitude of the *Jacobian J* of the transformation as follows:

$$du\,dv = |J|\,dx\,dy \tag{3.10-6}$$

where J is a determinant, given in this case by

$$J = \begin{vmatrix} \dfrac{\partial u}{\partial x} & \dfrac{\partial u}{\partial y} \\[2mm] \dfrac{\partial v}{\partial x} & \dfrac{\partial v}{\partial y} \end{vmatrix} = \dfrac{\partial u}{\partial x}\dfrac{\partial v}{\partial y} - \dfrac{\partial u}{\partial y}\dfrac{\partial v}{\partial x} \tag{3.10-7}$$

Combining (3.10-5) and (3.10-6) yields

$$p_2(u, v) = \left[\dfrac{p_1(x, y)}{|J|} \right]_{\substack{x=f(u,v) \\ y=g(u,v)}} \tag{3.10-8}$$

where $X = f(U, V)$, $Y = g(U, V)$ are the inverse functions of bi-uniform transformation (3.10-4).

Equation (3.10-8) can be formally extended to any number of random variables; for example, when $k = 3$,

$$p_2(u, v, w) = \left[\dfrac{p_1(x, y, z)}{|J|} \right]_{\substack{x=f_1(u,v,w) \\ y=f_2(u,v,w) \\ z=f_3(u,v,w)}} \tag{3.10.9}$$

where the Jacobian J is given by

$$J = \begin{vmatrix} \dfrac{\partial u}{\partial x} & \dfrac{\partial u}{\partial y} & \dfrac{\partial u}{\partial z} \\[2mm] \dfrac{\partial v}{\partial x} & \dfrac{\partial v}{\partial y} & \dfrac{\partial v}{\partial z} \\[2mm] \dfrac{\partial w}{\partial x} & \dfrac{\partial w}{\partial y} & \dfrac{\partial w}{\partial z} \end{vmatrix} \tag{3.10-10}$$

3.11. Statistical Averages

Suppose that the outcome of a particular experiment is denoted by the random variable X, which can take on values in the set $\{x_i\}$, $i = 1, 2, \ldots, m$.

---

†P. Franklin, "Methods of Advanced Calculus," McGraw-Hill, New York, 1944.

When this experiment is repeated in a sequence of n trials, assume that outcome $(X = x_i)$ occurs v_i times, $i = 1, 2, \ldots, m$. Then the *average* value of the random variable X in this sequence of trials is given by

[average value of X in n trials]

$$= (x_1)\left(\frac{v_1}{n}\right) + (x_2)\left(\frac{v_2}{n}\right) + \ldots + (x_m)\left(\frac{v_m}{n}\right) \qquad (3.11\text{-}1)$$

If n becomes indefinitely large in (3.11-1), the average value of X so computed is denoted the *mathematical expectation* of X, symbolically described by $E[X]$.

$$E[X] = \lim_{n\to\infty} \left\{(x_1)\left(\frac{v_1}{n}\right) + (x_2)\left(\frac{v_2}{n}\right) + \ldots + (x_m)\left(\frac{v_m}{n}\right)\right\} \qquad (3.11\text{-}2)$$

From the frequency definition of probability given in Eq. (3.2-1), we can replace $\lim_{n\to\infty} (v_i/n)$ in Eq. (3.11-2) by $p(x_i)$ so that the mathematical expectation of X can be more concisely expressed as

$$E[X] = x_1 p(x_1) + x_2 p(x_2) + \ldots + x_m p(x_m) \qquad (3.11\text{-}3\text{a})$$

$$E[X] = \sum_{i=1}^{m} x_i p(x_i) \qquad (3.11\text{-}3\text{b})$$

The quantity $E[X]$ is also known in the literature as the *statistical average*, *mean*, or *ensemble average* of the random variable X. When it is convenient, the mathematical expectation of X will be denoted by $\bar{X}$.

The definition of mathematical expectation can be extended to include continuous sample spaces in several ways. A simple heuristic procedure satisfactory for our purpose is to let m in Eq. (3.11-3b) approach infinity. To be consistent let $x_i \to x$, $p(x_i) \to p(x)\,dx$, and replace summations by integrals. When these modifications are introduced into (3.11-3b), the mathematical expectation of the random variable X for continuous sample spaces becomes

$$E[X] = \int_{-\infty}^{\infty} x\, p(x)\, dx \qquad (3.11\text{-}4)$$

If $p(x)$ is permitted to include delta functions, Eq. (3.11-4) is applicable to both discrete and continuous sample spaces. Hence Eq. (3.11-4) can be considered as the definition of $E[X]$.

Suppose $f(X)$ is a single-valued function of the random variable X. Clearly $f(X)$ is also a random variable, which occurs with the same relative frequency as X. For example, if in a particular sequence of experiments outcome $(X = x_i)$ occurs v_i times, then $f(X = x_i)$ also occurs v_i times. By an argument which parallels that described by Eqs. (3.11-1) through (3.11-4), the following expression for the mathematical expectation of $f(X)$ results:

$$E[f(X)] = \int_{-\infty}^{\infty} f(x)\, p(x)\, dx \qquad (3.11\text{-}5)$$

where $p(x)$ is the probability density function of the random variable X.

The mathematical expectation of the function $f(X) = X^n$ is of particular significance. It is conventionally referred to as the nth *moment* of random variable X and is denoted by m_n. Substituting into Eq. (3.11-5), we obtain for the nth moment:

$$m_n = E[X^n] = \int_{-\infty}^{\infty} x^n p(x)\, dx \qquad (3.11\text{-}6)$$

Of special concern in future discussions are the first and second moments of a random variable. The first moment $m_1 = E[X]$ is the mean value $\bar{X}$ of random variable X. The second moment $m_2 = E[X^2]$ is often called the *mean square value* of X and is denoted by $\overline{X^2}$.

The *central moments* of the random variable X are also very useful. The nth central moment μ_n is defined as the nth moment of the random variable $(X - \bar{X})$, or equivalently, the nth moment of the random variable X about its mean value $\bar{X}$. Mathematically μ_n is described by

$$\mu_n = E[(X - \bar{X})^n] = \int_{-\infty}^{\infty} (x - \bar{X})^n p(x)\, dx \qquad (3.11\text{-}7)$$

It is easily seen from Eq. (3.11-7) that the first central moment μ_1 is identically zero. The second central moment μ_2 is of major importance, since it is a measure of the *spread* or *dispersion* in values assumed by random variable X about its mean value $\bar{X}$. The second central moment has been given the special name, *variance*, and is often denoted by σ^2. The square root of the variance σ is frequently referred to as the *standard deviation* of the random variable or more simply as the *root-mean-square* (rms) value of X.

When calculating mathematical expectations of functions of random variables, it is very useful to recognize that $E[\ ]$ is a *linear operator*. The property of linearity implies that the expectation of a sum of random variables is the sum of the expectations of each random variable. For example,

$$E[X + Y] = E[X] + E[Y] = \bar{X} + \bar{Y} \qquad (3.11\text{-}8)$$

This relation can be extended to any number of random variables. A similar result can be obtained for the mathematical expectation of polynomials of random variables. For example,

$$E[X^2 + X - \bar{X}] = E[X^2] + E[X] - E[\bar{X}]$$
$$= \overline{X^2} + \bar{X} - \bar{X} = \overline{X^2} \qquad (3.11\text{-}9)$$

Let $f(X, Y)$ be a single-valued function of two random variables X and Y. The mathematical expectation of $f(X, Y)$ is defined by

$$\overline{f(X, Y)} = \int_{-\infty}^{\infty} dx \int_{-\infty}^{\infty} f(x, y) p(x, y)\, dy \qquad (3.11\text{-}10)$$

where $p(x, y)$ is the joint probability density function of random variables X and Y.

If $f(X, Y) = X^i Y^k$ and this quantity is substituted into Eq. (3.11-10), the defining relation for the (i, k)th (joint) moment of random variables X and Y is obtained, which is denoted by m_{ik}; that is,

$$m_{ik} = \overline{X^i Y^k} = \int_{-\infty}^{\infty} dx \int_{-\infty}^{\infty} x^i y^k p(x, y)\, dy \qquad (3.11\text{-}11)$$

The corresponding (joint) central moments, μ_{ik}, which are defined as the moments about the point $(\bar{X}, \bar{Y})$, are given by

$$\mu_{ik} = \overline{(X - \bar{X})^i (Y - \bar{Y})^k} \qquad (3.11\text{-}12)$$

Of particular interest are the joint central moments where i, k each take on the values 0, 1, 2. From (3.11-12) the following results are easily obtained:

$$\mu_{10} = \mu_{01} = 0 \qquad (3.11\text{-}13)$$

$$\mu_{20} = \overline{X^2} - \bar{X}^2 = \sigma_X^2 \qquad (3.11\text{-}14)$$

$$\mu_{02} = \overline{Y^2} - \bar{Y}^2 = \sigma_Y^2 \qquad (3.11\text{-}15)$$

$$\mu_{11} = \overline{XY} - \bar{X}\bar{Y} \qquad (3.11\text{-}16)$$

The quantities μ_{20} and μ_{02} are respectively equal to the individual variances σ_X^2 and σ_Y^2 of random variables X and Y as shown by Eqs. (3.11-14) and (3.11-15). The quantity μ_{11} is an important quantity, which is normally referred to as the *covariance* of X and Y.

If X and Y are statistically independent random variables, their joint probability density function $p(x, y)$ factors into the product $p_1(x)p_2(y)$. Then $\overline{XY}$, according to Eq. (3.11-10), can be written as

$$\overline{XY} = \int_{-\infty}^{\infty} x\, p_1(x)\, dx \int_{-\infty}^{\infty} y\, p_2(y)\, dy \qquad (3.11\text{-}17)$$

from which it follows that

$$\overline{XY} = \bar{X} \cdot \bar{Y} \qquad (3.11\text{-}18)$$

Result (3.11-18) can be directly extended to n statistically independent random variables $X_1, X_2, \ldots, X_n$ as follows:

$$\overline{X_1 X_2 X_3 \ldots X_n} = \bar{X}_1 \cdot \bar{X}_2 \cdot \bar{X}_3 \cdot \ldots \cdot \bar{X}_n \qquad (3.11\text{-}19)$$

If Eq. (3.11-18) is substituted into Eq. (3.11-16), we see that the covariance function of two statistically independent random variables is zero. On the other hand, if two random variables are statistically dependent, a measure of their interdependence is provided by the covariance function. A normalized measure of this interdependence is furnished by the *correlation coefficient* ρ. This quantity is also called the *normalized covariance* of X and Y and is given by

$$\rho = \frac{\mu_{11}}{\sqrt{\mu_{20}\mu_{02}}} = \frac{\mu_{11}}{\sigma_X \sigma_Y} \qquad (3.11\text{-}20)$$

where σ_X^2 and σ_Y^2 are respectively the variances of X and Y. It can be shown that the correlation coefficient is less than (or equal to) one in magnitude, $-1 \leqslant \rho \leqslant 1$. In particular, values of $|\rho|$ close to unity indicate that X and Y are highly *correlated*. Values of $|\rho|$ close to zero indicate low correlation.

Suppose that the performance of an experiment involving random variables X and Y results in some observed value of X. Nonzero correlation between X and Y implies Y may be estimated with some degree of accuracy from X; that is, there is a functional relation between X and Y. The simplest form of dependence between two random variables is *linear dependence*. Thus, the estimate Y_e of Y can be related to X by

$$Y_e = a + bX \tag{3.11-21}$$

The values of intercept a and slope b that minimize the mean-square difference between Y and Y_e define a line known as the *linear mean-square regression line*. (Higher-degree regression lines can be similarly defined.)

The values of a and b that determine the linear regression line are obtained by differentiating the mean-square difference between Y and Y_e

$$E[(Y - Y_e)^2] = E\{[Y - (a + bX)]^2\} \tag{3.11-22}$$

with respect to a and b and equating the results to zero. This results in

$$b = \frac{E[(X - \bar{X})(Y - \bar{Y})]}{\sigma_X^2} = \frac{\mu_{11}}{\sigma_X^2} \tag{3.11-23}$$

and

$$a = \bar{Y} - \frac{\mu_{11}}{\sigma_X^2} \bar{X} \tag{3.11-24}$$

Substituting (3.11-23) and (3.11-24) into Eq. (3.11-21) yields

$$Y_e = \bar{Y} + \frac{\mu_{11}}{\sigma_X^2}(X - \bar{X}) \tag{3.11-25}$$

It is often convenient to replace random variables X and Y by the standardized random variables X_s and Y_s, which are defined by

$$X_s = \frac{X - \bar{X}}{\sigma_X} \tag{3.11-26}$$

$$Y_s = \frac{Y - \bar{Y}}{\sigma_Y} \tag{3.11-27}$$

Definitions (3.11-26) and (3.11-27) lead to

$$\bar{X}_s = 0 \tag{3.11-28a}$$
$$\bar{Y}_s = 0 \tag{3.11-28b}$$

and

$$\sigma_{X_s}^2 = 1 \tag{3.11-29a}$$
$$\sigma_{Y_s}^2 = 1 \tag{3.11-29b}$$

Denoting the estimate of Y_s by Y_{s_e}, we can rewrite Eq. (3.11-25) as

$$Y_{s_e} = \rho X_s \qquad (3.11\text{-}30)$$

where ρ is the correlation coefficient previously defined by Eq. (3.11-20). Equation (3.11-30) shows that ρ is the slope of the linear regression line relating standardized variables Y_s and X_s. Hence, random variables with nonzero correlation coefficient are said to be *linearly dependent*. When ρ is identically zero—that is, X and Y are uncorrelated—they are said to be *linearly independent*. Note that if X and Y are statistically independent random variables, $\mu_{11} = 0$; from (3.11-20) it follows that they are uncorrelated or linearly independent. But the reverse may not be true, that is, a zero correlation coefficient does not insure that two random variables are statistically independent. However, for the case of Gaussian random variables, zero correlation is both necessary and sufficient for statistical independence.

3.12. Characteristic Functions

Another statistical average that is extensively used is the *characteristic function* of the random variable X, $C_X(\xi)$, which is defined as the mathematical expectation of $\exp(j\xi X)$—namely,

$$C_X(\xi) = E[e^{j\xi X}] \qquad (3.12\text{-}1a)$$

$$C_X(\xi) = \int_{-\infty}^{\infty} e^{j\xi x} p(x)\, dx \qquad (3.12\text{-}1b)$$

where $p(x)$ is the probability density function of X. The right side of Eq. (3.12-1b) is the Fourier transform of $p(x)$ with the transform variable ξ replaced by $-\xi$. It is easy to show that a probability density function always satisfies the requirements discussed in Chapter 2 for the (transform) characteristic function to exist. Since Fourier transform pairs are uniquely related, the characteristic function $C_X(\xi)$ uniquely characterizes the density function $p(x)$ and vice versa. Function $p(x)$ can be obtained from the inverse Fourier transform of $C_X(\xi)$ with ξ replaced by $(-\xi)$. This function, which is called the *anticharacteristic function* in statistics, is given by

$$p(x) = \frac{1}{2\pi} \int_{-\infty}^{\infty} C_X(\xi) e^{-jx\xi}\, d\xi \qquad (3.12\text{-}2)$$

The characteristic function and anticharacteristic function can be found for a wide variety of probability distributions from standard tables of Fourier transforms such as that compiled by Campbell and Foster [1], which we shall cite again in later chapters.

    The chief reason for the importance of the characteristic function is that in many practical problems it is much simpler to work with this function than with the probability density function. As a simple example, consider once again the problem of calculating the probability density function of a sum of n independent random variables,

$$Z = \sum_{i=1}^{n} X_i$$

It was shown that the result, Eq. (3.9-6), could be obtained by performing n successive convolutions of the probability density functions of random variables X_i. An alternate procedure is to find the characteristic function of Z. By Eq. (3.12-1b),

$$C_Z(\xi) = E[e^{j\xi Z}] = E[\exp(j\xi \sum_{i=1}^{n} X_i)]$$

$$= E\left[\prod_{i=1}^{n} e^{j\xi X_i}\right] = \prod_{i=1}^{n} E[e^{j\xi X_i}] \qquad (3.12\text{-}3)$$

$$= \prod_{i=1}^{n} C_{X_i}(\xi)$$

Equation (3.12-3) states that the characteristic function of the sum random variable Z is the product of the individual characteristic functions. The inversion formula (3.12-2) can then be employed to obtain $p(z)$. When the individual random variables are independent but have the same probability density function, Eq. (3.12-3) simplifies further to

$$C_Z(\xi) = [C_{X_i}(\xi)]^n \qquad (3.12\text{-}4)$$

The characteristic function is also very useful for finding the moments of a random variable. When $C_X(\xi)$ is expanded in a MacLaurin series,

$$C_X(\xi) = C_X(0) + \dot{C}_X(0)\xi + \ddot{C}_X(0)\frac{\xi^2}{2!} + \ldots + \overset{(n)}{C_X}(0)\frac{\xi^n}{n!} + \ldots \qquad (3.12\text{-}5)$$

it is a straightforward calculation to show that

$$\overset{(n)}{C_X}(0) = \int_{-\infty}^{\infty} \left(\frac{\partial^n}{\partial \xi^n} e^{j\xi x}\right)_{\xi=0} p(x)\, dx = j^n m_n \qquad (3.12\text{-}6)$$

where m_n are the moments of the random variable X. Equation (3.12-6) states that the coefficients of the MacLaurin expansion of the characteristic function are simply related to the moments of X (provided a valid Mac-Laurin expansion exists†).

The *joint characteristic function*, $C(\xi_1, \xi_2)$, is defined for two random variables X and Y with joint probability density function $p(x, y)$ by

$$C(\xi_1, \xi_2) = E[e^{j\xi_1 X + j\xi_2 Y}] = \int_{-\infty}^{\infty} \int_{-\infty}^{\infty} e^{j\xi_1 x + j\xi_2 y} p(x, y)\, dx\, dy \qquad (3.12\text{-}7)$$

The right side of (3.12-7) is the two-dimensional Fourier transform of the

---

†A valid expansion exists if the moments are finite and series (3.12-5) converges absolutely in the neighborhood of $\xi = 0$.

density function $p(x, y)$ (with variables ξ_1 and ξ_2 replaced by $-\xi_1$ and $-\xi_2$). The anticharacteristic function, as in the single variable case, is simply related to the inverse Fourier transform and is given by

$$p(x, y) = \frac{1}{(2\pi)^2} \int_{-\infty}^{\infty} \int_{-\infty}^{\infty} C(\xi_1, \xi_2) e^{-j(x\xi_1 + y\xi_2)} \, d\xi_1 \, d\xi_2 \qquad (3.12\text{-}8)$$

As before, the various joint moments of the random variables X and Y can be obtained from a power series expansion of $C(\xi_1, \xi_2)$ in terms of ξ_1 and ξ_2.

The joint characteristic function of a set of random variables $X_i, i = 1, 2, \ldots, n$, is by definition

$$C(\xi_1, \xi_2, \ldots, \xi_n) = E\left[\exp\left(j \sum_{i=1}^{n} \xi_i X_i \right) \right] = E\left[\prod_{i=1}^{n} e^{j\xi_i X_i} \right] \qquad (3.12\text{-}9)$$

If the $X_i, i = 1, 2, \ldots, n$, are statistically independent random variables, then the functions of the X_i, $\exp(j\xi_i X_i), i = 1, 2, \ldots, n$, are also statistically independent. By Eq. (3.11-19) the expectation of a product of independent random variables is equal to the product of the respective expectations. Hence Eq. (3.12-9) can be rewritten as

$$C(\xi_1, \xi_2, \ldots, \xi_n) = \prod_{i=1}^{n} E[e^{j\xi_i X_i}] = \prod_{i=1}^{n} C(\xi_i) \qquad (3.12\text{-}10)$$

It can be shown† that Eq. (3.12-10) is both a necessary and sufficient condition in order that n random variables be statistically independent—that is, the joint characteristic function must be expressible as the product of the individual characteristic functions.

3.13. The Gaussian (Normal) Distribution and the Central Limit Theorem

One of the most important distributions in noise theory is the *Gaussian* or *normal* probability density function, of which an example is shown in Fig. 3.7-1b. This density function is uniquely characterized by mean m and variance σ^2 of Gaussian random variable X and is given by

$$p(x) = \frac{1}{\sqrt{2\pi}\,\sigma} e^{-(x-m)^2/2\sigma^2} \qquad (3.13\text{-}1)$$

The corresponding characteristic function is

$$C_X(\xi) = e^{j\xi m - (\sigma^2 \xi^2)/2} \qquad (3.13\text{-}2)$$

The importance of this distribution is closely related to a famous theorem in statistics called the *central limit theorem* which states that under very general conditions the probability density function of the sum of n

---

†H. Cramer, "Mathematical Methods of Statistics," Princeton University Press, Princeton, N.J., 1946.

statistically independent random variables,

$$Z = \sum_{i=1}^{n} X_i$$

approaches the Gaussian density function as $n \to \infty$, no matter what the individual probability density functions may be. A plausibility argument is presented here for the case in which the random variables are independent with identical distributions.

The characteristic function of the sum, $C_Z(\xi)$, is related to the individual characteristic function $C_X(\xi)$ by Eq. (3.12-4):

$$C_Z(\xi) = [C_X(\xi)]^n \tag{3.13-3}$$

It is more convenient to treat the standardized sum variable Z_s,

$$Z_s = \frac{Z - \bar{Z}}{\sigma_Z} \tag{3.13-4}$$

which has mean zero and a variance of one that is independent of n (and the same type of density function as Z). The characteristic function $C_{Z_s}(\xi)$ of standardized variable Z_s is easily found to be

$$C_{Z_s}(\xi) = E[e^{j\xi[(Z-\bar{Z})/\sigma_z]}] = e^{-j(\bar{Z}\xi/\sigma_z)} C_Z\left(\frac{\xi}{\sigma_Z}\right) \tag{3.13-5}$$

Substituting (3.13-3) into (3.13-5) results in

$$C_{Z_s}(\xi) = e^{-j(\bar{Z}\xi/\sigma_z)} \left[C_X\left(\frac{\xi}{\sigma_Z}\right) \right]^n \tag{3.13-6}$$

Assuming that the characteristic function $C_X(\xi)$ can be expanded in a MacLaurin series, with coefficients expressed in terms of the moments of the random variable X as shown in Eqs. (3.12-5) and (3.12-6), the first four terms are

$$C_X(\xi) = 1 + j\bar{X}\xi - \overline{X^2}\frac{\xi^2}{2!} - j\overline{X^3}\frac{\xi^3}{3!} + \dots \tag{3.13-7}$$

For simplicity let the mean $\bar{X}$ of each variable be zero, let the variance be denoted by σ_X^2 and the higher central moments by μ_{k0}. In this case $C_X(\xi)$ simplifies to

$$C_X(\xi) = 1 - \frac{\sigma_X^2}{2}\xi^2 - j\frac{\mu_{30}}{3!}\xi^3 + \dots \tag{3.13-8}$$

Substituting (3.13-8) into (3.13-6) gives

$$C_{Z_s}(\xi) = \left(1 - \frac{\sigma_X^2}{2\sigma_Z^2}\xi^2 - j\frac{\mu_{30}}{3!\sigma_Z^3}\xi^3 + \dots \right)^n \tag{3.13-9}$$

Since $\sigma_Z = \sqrt{n}\,\sigma_X$, Eq. (3.13-9) can be rewritten as

$$C_{Z_s}(\xi) = \left[1 - \frac{1}{2n}\xi^2 - j\frac{\mu_{30}}{3!n}\left(\frac{\xi^3}{n^{1/2}}\right) + \dots \right]^n \tag{3.13-10}$$

For fixed ξ, the third and higher-order terms in the bracket in (3.13-10) tend to zero more rapidly than the second term as $n \rightarrow \infty$. Hence,

$$C_{Z_s}(\xi) = \lim_{n \to \infty} \left(1 - \frac{1}{2n}\xi^2\right)^n \qquad (3.13\text{-}11a)$$

$$= e^{-\xi^2/2} \qquad (3.13\text{-}11b)$$

Since the right side of (3.13-11b) is the characteristic function of a standardized Gaussian random variable [see Eq. (3.13-2) with $m = 0$ and $\sigma^2 = 1$], this demonstrates that the density function of the sum random variable Z approaches the Gaussian density function as $n \rightarrow \infty$.

The probability distribution function $P(x)$ of a Gaussian random variable X, shown in Fig. 3.7-1a, is also uniquely characterized by the mean and variance of the random variable. Thus, from (3.13-1),

$$P(x) = \int_{-\infty}^{x} \frac{1}{\sqrt{2\pi}\,\sigma} e^{-(u-m)^2/2\sigma^2} \, du \qquad (3.13\text{-}12a)$$

$$= \frac{1}{2}\left(1 + \text{erf}\,\frac{x - m}{\sqrt{2}\,\sigma}\right) \qquad (3.13\text{-}12b)$$

where

$$\text{erf}\,z = \frac{2}{\sqrt{\pi}} \int_{0}^{z} e^{-y^2} \, dy \qquad (3.13\text{-}13)$$

Calculation of the central moments of a Gaussian variable X is simplified if we assume zero mean. No loss in generality is suffered since, by definition, the central moments are moments about the mean. With this assumption,

$$\mu_n = \overline{X^n} = \frac{1}{\sqrt{2\pi}\,\sigma} \int_{-\infty}^{\infty} x^n e^{-x^2/2\sigma^2} \, dx \qquad (3.13\text{-}14)$$

When n is an odd integer, the integrand in (3.13-14) is also odd and the integral has a zero value. It follows that all odd central moments of a Gaussian random variable are identically zero:

$$\mu_n = 0 \qquad (n \text{ odd}) \qquad (3.13\text{-}15)$$

When n is even, Eq. (3.13-14) can be evaluated and yields

$$\mu_n = \sigma^n[1 \cdot 3 \cdot 5 \cdot \ldots \cdot (n-1)], \qquad n \geqslant 2 \quad (n \text{ even}) \qquad (3.13\text{-}16)$$

3.14. The Multidimensional Gaussian (Normal) Distribution

It is often convenient to use matrix notation when dealing with many random variables. Define **X** to be the column matrix of a set of random variables $X_1, X_2, \ldots, X_n$,

$$\mathbf{X} = \begin{bmatrix} X_1 \\ X_2 \\ X_3 \\ \cdot \\ \cdot \\ \cdot \\ X_n \end{bmatrix} \tag{3.14-1}$$

Let the matrix of the means of the random variables be described by $\bar{\mathbf{X}}$:

$$\bar{\mathbf{X}} = \begin{bmatrix} \bar{X}_1 \\ \bar{X}_2 \\ \bar{X}_3 \\ \cdot \\ \cdot \\ \cdot \\ \bar{X}_n \end{bmatrix} \tag{3.14-2}$$

and $\mathbf{X} - \bar{\mathbf{X}}$:

$$\mathbf{X} - \bar{\mathbf{X}} = \begin{bmatrix} X_1 - \bar{X}_1 \\ X_2 - \bar{X}_2 \\ \cdot \\ \cdot \\ \cdot \\ X_n - \bar{X}_n \end{bmatrix} \tag{3.14-3}$$

The covariance function of two random variables X_1 and X_2 was shown previously to be $\overline{(X_1 - \bar{X}_1)(X_2 - \bar{X}_2)}$. In order to concisely represent all possible covariance functions of a set of random variables $X_1, X_2, \ldots, X_n$, we define a square matrix Λ, known as the *covariance matrix*, by

$$\Lambda = \overline{(\mathbf{X} - \bar{\mathbf{X}})(\mathbf{X} - \bar{\mathbf{X}})^T} \tag{3.14-4}$$

where the matrix $(\mathbf{X} - \bar{\mathbf{X}})^T$ is the transpose of the matrix in Eq. (3.14-3). If we perform the indicated matrix multiplication and statistical averaging, the matrix becomes

$$\Lambda = \begin{bmatrix} \lambda_{11} & \lambda_{12} & \lambda_{13} & \ldots & \lambda_{1n} \\ \lambda_{21} & \lambda_{22} & \lambda_{23} & \ldots & \lambda_{2n} \\ \lambda_{31} & \lambda_{32} & \lambda_{33} & \ldots & \lambda_{3n} \\ \cdot \\ \cdot \\ \cdot \\ \lambda_{n1} & \lambda_{n2} & \lambda_{n3} & \ldots & \lambda_{nn} \end{bmatrix} \tag{3.14-5}$$

where

$$\lambda_{ik} = \lambda_{ki} = E[(X_i - \bar{X}_i)(X_k - \bar{X}_k)] \tag{3.14-6}$$

Thus λ_{ik} is the covariance of random variables X_i and X_k. Note that by Eq. (3.14-6) matrix elements on the diagonal of Λ, λ_{ii}, $i = 1, 2, \ldots, n$, are simply the variances of the random variables X_i.

In terms of the matrices defined in Eqs. (3.14-1) through (3.14-6), it can be shown that the joint probability density function of n Gaussian random variables $X_1, X_2, \ldots, X_n$ can be expressed by

$$p(x_1, x_2, \ldots, x_n)$$
$$= \frac{1}{(2\pi)^{n/2} |\Lambda|^{1/2}} \exp\left[-\frac{1}{2}(\mathbf{x} - \bar{\mathbf{X}})^T \Lambda^{-1}(\mathbf{x} - \bar{\mathbf{X}})\right] \qquad (3.14\text{-}7)$$

where $|\Lambda|$ indicates the determinant of the covariance matrix Λ, Λ^{-1} is the inverse matrix of Λ, and $(\mathbf{x} - \bar{\mathbf{X}})$ is a column matrix with elements $(x_i - \bar{X}_i)$. If the indicated matrix multiplications in Eq. (3.14-7) are carried out, the joint probability density function is also expressible as

$$p(x_1, x_2, \ldots, x_n)$$
$$= \frac{1}{(2\pi)^{n/2} |\Lambda|^{1/2}} \exp\left[-\frac{1}{2|\Lambda|} \sum_{i=1}^{n} \sum_{k=1}^{n} |\Lambda|_{ik}(x_i - \bar{X}_i)(x_k - \bar{X}_k)\right] \qquad (3.14\text{-}8)$$

where $|\Lambda|_{ik}$ is the cofactor of element λ_{ik} in the covariance matrix.

Suppose that the random variables are uncorrelated. This implies that $\lambda_{ik} = 0$ for all $i \neq k$, that is, only the diagonal terms of the covariance matrix for which $i = k$ are nonzero. It is also easy to show that when λ_{ik} is zero $(i \neq k)$, so is the cofactor $|\Lambda|_{ik}$. For this case Eq. (3.14-8) reduces to

$$p(x_1, x_2, \ldots, x_n) = \frac{1}{(2\pi)^{n/2} \prod_{i=1}^{n} \sigma_i} \exp\left[-\sum_{i=1}^{n} \frac{(x_i - \bar{X}_i)^2}{2\sigma_i^2}\right] \qquad (3.14\text{-}9a)$$

$$= \prod_{i=1}^{n} \frac{1}{\sqrt{2\pi}\,\sigma_i}\, e^{-(x_i - \bar{X}_i)^2/2\sigma_i^2} \qquad (3.14\text{-}9b)$$

The right side of (3.14-9b) is simply the product of the density functions of Gaussian random variables $X_1, X_2, \ldots,$ and X_n. Hence (3.14-9b) may be rewritten as

$$p(x_1, x_2, \ldots, x_n) = p(x_1) \cdot p(x_2) \cdot \ldots \cdot p(x_n) \qquad (3.14\text{-}10)$$

It follows from Eq. (3.14-10) that if n Gaussian random variables are uncorrelated, they are also statistically independent.

Another result of considerable interest, which we shall not prove, is that if n Gaussian random variables are transformed into a new set of n random variables by a linear transformation, the new variables are also Gaussian. Furthermore if the old set are statistically dependent, it is always possible to effect a transformation such that the new variables are statistically independent.

3.15. Summary

(a) For statistically independent events $A, B, C, \ldots$,

$$p(A, B, C, \ldots) = p(A) \cdot p(B) \cdot p(C) \ldots \tag{3.15-1}$$

(b) For dependent events A and B,

$$p(A, B) = p(A) \cdot p(B \mid A) \tag{3.15-2a}$$

or

$$p(A, B) = p(B) \cdot p(A \mid B) \tag{3.15-2b}$$

(c) Any function that assigns a real number to each point in a sample space of events is defined as a random variable.

(d) Random variables $X_1, X_2, \ldots, X_n$ are statistically independent if, and only if, for all sets of real numbers S_i,

$$\text{Pr}\,(X_i \text{ in } S_i, X_j \text{ in } S_j) = \text{Pr}\,(X_i \text{ in } S_i) \cdot \text{Pr}\,(X_j \text{ in } S_j)$$

$$i, j = 1, 2, \ldots, n, \quad i \neq j$$

. . . .

$$\text{Pr}\,(X_1 \text{ in } S_1, X_2 \text{ in } S_2, \ldots, X_n \text{ in } S_n)$$
$$= \text{Pr}\,(X_1 \text{ in } S_1) \cdot \text{Pr}\,(X_2 \text{ in } S_2) \cdot \ldots \cdot \text{Pr}\,(X_n \text{ in } S_n) \tag{3.15-3}$$

(e) A probability distribution function $P(x)$ of a random variable X is defined by

$$P(x) = \text{Pr}\,(X \leqslant x) \tag{3.15-4}$$

where x is any point on the real line between plus and minus infinity.

(f) When the probability distribution $P(x)$ of a random variable X is a differentiable function of x with a finite number of discontinuities at which the derivative of $P(x)$ is defined, the probability density function of random variable X is given by

$$p(x) = \frac{dP(x)}{dx} \tag{3.15-5}$$

(g) The joint probability distribution function $P(x, y)$ of random variables X and Y is defined for $-\infty < x, y < \infty$ by

$$P(x, y) = \text{Pr}\,(X \leqslant x, Y \leqslant y) \tag{3.15-6}$$

When X and Y are statistically independent,

$$P(x, y) = P_1(x) \cdot P_2(y) \tag{3.15-7}$$

(h) When $P(x, y)$ is a differentiable function with respect to x and y (appropriately defined at discontinuities), the joint probability density function $p(x, y)$ of random variables X and Y is given by

$$p(x, y) = \frac{\partial^2 P(x, y)}{\partial x \, \partial y} \tag{3.15-8}$$

(i) When X and Y are statistically independent random variables,

$$p(x, y) = p_1(x) \cdot p_2(y) \tag{3.15-9}$$

(j) When X and Y are not statistically independent, the marginal probability density functions $p_1(x)$ and $p_2(y)$ of X and Y are computed as follows:

$$p_1(x) = \int_{-\infty}^{\infty} p(x, y) \, dy \tag{3.15-10}$$

$$p_2(y) = \int_{-\infty}^{\infty} p(x, y) \, dx \tag{3.15-11}$$

For n random variables $X_1, X_2, \ldots, X_k, \ldots, X_n$, the joint marginal probability density function $p(x_1, x_2, \ldots, x_k)$ of random variables $X_1, X_2, \ldots, X_k$ is found from the joint probability density function $p(x_1, x_2, \ldots, x_n)$ by

$$p(x_1, x_2, \ldots, x_k)$$
$$= \int_{-\infty}^{\infty} dx_{k+1} \int_{-\infty}^{\infty} dx_{k+2} \ldots \int_{-\infty}^{\infty} p(x_1, x_2, \ldots, x_n) \, dx_n \tag{3.15-12}$$

(k) The probability density function $p_2(y_1, y_2, \ldots, y_n)$ of a set of n random variables $Y_1, Y_2, \ldots, Y_n$, which are related by a one-to-one mapping to another set of random variables $X_1, X_2, \ldots, X_n$, can be found by the relation

$$p_2(y_1, y_2, \ldots, y_n) = \left[\frac{p_1(x_1, x_2, \ldots, x_n)}{|J|} \right]_{\substack{x_i = g_i(y_1, y_2, \ldots, y_n) \\ i = 1, 2, \ldots, n}} \tag{3.15-13}$$

where $p_1(x_1, x_2, \ldots, x_n)$ is the joint probability density function of the old random variables, $g_i(Y_1, \ldots, Y_n)$ is the inverse function that relates the X_i to the Y_i, and J is the determinant of the Jacobian of the transformation relating the variables Y_i to the variables X_i,

$$J = \begin{vmatrix} \dfrac{\partial y_1}{\partial x_1} & \dfrac{\partial y_1}{\partial x_2} & \cdots & \dfrac{\partial y_1}{\partial x_n} \\[2ex] \dfrac{\partial y_2}{\partial x_1} & \dfrac{\partial y_2}{\partial x_2} & \cdots & \dfrac{\partial y_2}{\partial x_n} \\[2ex] \vdots & & & \\[2ex] \dfrac{\partial y_n}{\partial x_1} & \dfrac{\partial y_n}{\partial x_2} & \cdots & \dfrac{\partial y_n}{\partial x_n} \end{vmatrix} \tag{3.15-14}$$

(l) The statistical average or expectation (mean value) of random variable X with probability density function $p(x)$ is defined as

$$E[X] = \int_{-\infty}^{\infty} x \, p(x) \, dx = \bar{X} \tag{3.15-15}$$

(m) The expectation of a function $f(X)$ of a random variable X with probability density function $p(x)$ is given by

$$E[f(X)] = \int_{-\infty}^{\infty} f(x)p(x)\,dx = \overline{f(X)} \qquad (3.15\text{-}16)$$

(n) The mathematical expectation of $f(X) = X^n$ is called the nth moment of random variable X, designated by m_n. When $n = 1$, the first moment $\bar{X}$ is also known as the mean value of X. When $n = 2$, $\overline{X^2}$ is often referred to as the mean square value of X.

(o) The mathematical expectation of $f(X) = (X - \bar{X})^n$ is called the nth central moment of random variable X. When $n = 1$, the first central moment is identically zero. When $n = 2$, $\overline{(X - \bar{X})^2}$ is called the variance of random variable X and is denoted by σ^2.

(p) The mathematical expectation of a random variable is a linear operation. Hence,

$$E[X_1 + X_2 + \ldots] = E[X_1] + E[X_2] + \ldots \qquad (3.15\text{-}17)$$

(q) If $f(X, Y)$ is a single-valued function of random variables X and Y with joint probability density function $p(x, y)$, it is also a random variable whose mathematical expectation can be found by

$$\overline{f(X, Y)} = \int_{-\infty}^{\infty} \int_{-\infty}^{\infty} f(x, y)p(x, y)\,dx\,dy \qquad (3.15\text{-}18)$$

(r) The mathematical expectation of $f(X, Y) = X^i Y^k$ is called the (i, k)th joint moment of X and Y and is denoted by m_{ik}. If X and Y are statistically independent,

$$m_{ik} = \overline{X^i Y^k} = \overline{X^i} \cdot \overline{Y^k} \qquad (3.15\text{-}19)$$

(s) The mathematical expectation of $f(X, Y) = (X - \bar{X})^i(Y - \bar{Y})^k$ is called the (i, k)th joint central moment of X and Y and is denoted by μ_{ik}. When $i = k = 1$, the joint central moment μ_{11} is referred to as the covariance of random variables X and Y and is given by

$$\mu_{11} = \overline{(X - \bar{X})(Y - \bar{Y})} = \overline{XY} - \bar{X} \cdot \bar{Y} \qquad (3.15\text{-}20)$$

(t) The correlation coefficient ρ of X and Y (or the normalized covariance of X and Y) is given by

$$\rho = \frac{\mu_{11}}{\sqrt{\mu_{20}\mu_{02}}} \qquad (3.15\text{-}21)$$

(u) When X and Y are statistically independent, it follows from (3.15-20) and (3.15-21) that they are also uncorrelated since their covariance μ_{11} is identically zero. When two random variables are uncorrelated, however, they are not necessarily statistically independent. For Gaussian random variables, zero correlation is both a necessary and sufficient condition for statistical independence.

(v) The characteristic function $C_X(\xi)$ of a random variable X is defined to be the mathematical expectation of $\exp(j\xi X)$:

$$C_X(\xi) = E[e^{j\xi X}] = \int_{-\infty}^{\infty} e^{j\xi x} p(x)\, dx \qquad (3.15\text{-}22)$$

where $p(x)$ is the probability density function of X.

(w) The probability density function $p(z)$ of a random variable Z, which is the sum of n statistically independent random variables $X_1, X_2, \ldots, X_n$ with probability density functions $p_1(x), p_2(x), \ldots, p_n(x)$, respectively, can be found by successive convolutions as follows:

$$p(Z) = p_1(x) * p_2(x) * \ldots * p_n(x) \qquad (3.15\text{-}23)$$

The density function $p(z)$ can also be found by calculating first the characteristic function $C_Z(\xi)$ of the random variable Z and then the anticharacteristic function of $C_Z(\xi)$, which is equal to $p(z)$. $C_Z(\xi)$ may be calculated in terms of the characteristic functions $C_{X_i}(\xi)$, $i = 1, 2, \ldots, n$, of random variables $X_1, X_2, \ldots, X_n$ as follows:

$$C_Z(\xi) = \prod_{i=1}^{n} C_{X_i}(\xi) \qquad (3.15\text{-}24)$$

The anticharacteristic function, equal to $p(z)$, is then found by

$$p(z) = \frac{1}{2\pi} \int_{-\infty}^{\infty} C_Z(\xi) e^{-j\xi z}\, d\xi \qquad (3.15\text{-}25)$$

(x) *Central Limit Theorem:* Under very general conditions, the probability density function of the sum of n statistically independent random variables approaches the Gaussian density function as $n \to \infty$, no matter what the individual probability density functions may be.

(y) If n Gaussian random variables are uncorrelated, they are also statistically independent.

(z) If n Gaussian random variables are transformed into a new set of n random variables by a linear transformation, the new variables are also Gaussian. It is always possible to define a linear transformation that transforms n statistically dependent Gaussian random variables into n statistically independent Gaussian random variables.

PROBLEMS

3.1. An urn contains three red balls, two white balls, and four black balls. What is the probability that two balls drawn without replacement from the urn will both be white?

3.2. Find the probability of obtaining an ordered sequence of two heads and one tail on three independent tosses of an unbiased coin. What is the probability of two heads and a tail in any order?

3.3. What is the probability that four cards drawn in succession from an ordinary bridge deck of cards will all be of the same suit?

3.4. Calculate the probability of obtaining the number n on a single toss of two unbiased dice for $n = 2, 3, \ldots, 12$. Calculate the probability that the result of a single toss will be equal to or less than n.

3.5. Calculate the probability that three tails will occur in succession in five tosses of an unbiased coin.

3.6. Suppose that in a collection of pencils, $\frac{1}{4}$ of the total are red, $\frac{2}{3}$ of the total have erasers, and $\frac{4}{5}$ of the red pencils contain erasers. If a pencil drawn at random has an eraser, what is the probability it is also red?

3.7. Three cards are drawn from a deck. Given that all three have different suits, what is the probability that at least two cards will be kings?

3.8. Let X_i, $i = 1, 2, \ldots, n$, be statistically independent real random variables such that $\overline{X_i} = m$ and $\overline{X_i^2} = \sigma^2$ for all i. Define

$$Y = \frac{1}{n} \sum_{i=1}^{n} X_i$$

Calculate the mean value and the variance of Y.

3.9. If $E[X] = 1$ and $E[X^2] = 2$, find the mean and variance of $Y = 3X - 1$.

3.10. If X and Y are statistically independent real random variables with arbitrary means, and variances σ_X^2 and σ_Y^2, respectively, find the variance of $(X - Y)$.

3.11. If Y is a real random variable with probability density function

$$p(y) = \frac{y}{a} \exp\left(-\frac{y^2}{2a}\right), \qquad y \geqslant 0$$

find its mean and variance.

3.12. For the uniform probability density function shown in Fig. 3P-12, find the mean and variance of X.

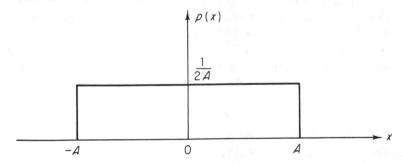

Fig. 3P-12. Uniform probability density function.

3.13. If $Z = X_1 + X_2 + X_3$, where the X_i are real, independent random variables with probability density functions $p(x_i) = 1/T$ for $|x_i| < T/2$, calculate the probability density function of the sum by the method of convolution.

3.14. Show that the joint density function of the random variables $X_1, X_2, \ldots,$ X_n can be expressed in terms of the conditional probabilities by the relation

$$p(X_1, X_2, \ldots, X_n)$$
$$= p(X_n \mid X_1, \ldots, X_{n-1}) \cdot \ldots \cdot p(X_3 \mid X_1, X_2) \cdot p(X_2 \mid X_1) \cdot p(X_1)$$

3.15. If X is a real Gaussian random variable with probability density function

$$p(x) = \frac{1}{\sqrt{2\pi}\,\sigma} \exp\left[-\frac{(x-m)^2}{2\sigma^2}\right]$$

verify by integration that $\bar{X} = m$ and $\overline{X^2} - \bar{X}^2 = \sigma^2$.

3.16. If random variable X has a Cauchy density function,

$$p(x) = \frac{1}{\pi} \frac{a}{a^2 + x^2}$$

show that the characteristic function is given by

$$C_X(\xi) = e^{-a|\xi|}$$

Plot the density function and distribution function of X.

3.17. For independent random variables $X_1, X_2, \ldots, X_n$, each with identical Cauchy densities,

$$p(x) = \frac{1}{\pi(1 + x^2)}$$

show that the density of the sample mean

$$\bar{X} = \frac{1}{n} \sum_{i=1}^{n} X_i$$

is also Cauchy. Explain why the central limit theorem does not hold in this case.

3.18. Chebyshev's inequality states that if X is a random variable with mean $\bar{X}$ and variance σ^2, then for any $t > 0$, the probability that $|X - \bar{X}| \geq t$ is equal to or less than σ^2/t^2. Calculate the probability that $|X - \bar{X}|$ is greater than 3σ (using tables) if X is a Gaussian random variable and compare it with the bound given by Chebyshev's inequality.

3.19. If X and Y are statistically independent real Gaussian random variables each with zero mean and variance σ^2, show that $Z = (X^2 + Y^2)^{1/2}$ is a random variable with a Rayleigh probability density function

$$p(z) = \frac{z}{\sigma^2} \exp\left(-\frac{z^2}{2\sigma^2}\right)$$

Calculate the probability density function of $W = X^2 + Y^2 = Z^2$.

3.20. For the random variable W defined in problem 3.19, find $\bar{W}$ and $\overline{W^2}$.

3.21. A positive random variable X has a probability density function that is proportional to $\exp(-ax)$ for $x \geq 0$. Calculate the constant of proportionality. Find the first and second moments and the variance of the random variable.

3.22. Find the probability density $p(x)$ corresponding to the characteristic function

$$C_X(\xi) = \frac{1}{1 + \xi^2}$$

3.23. If $Z = X + Y$, where X and Y are statistically independent real Gaussian random variables with means $\bar{X}$, $\bar{Y}$ and variances σ_X^2, σ_Y^2, respectively, find the characteristic function of Z.

3.24. Let X_1 and X_2 be real random variables with zero means and variances σ_1^2, σ_2^2, respectively, and with a joint Gaussian probability density function. Show that if the random variables Y_1 and Y_2 are defined by a rotational transformation,

$$Y_1 = X_1 \cos \theta + X_2 \sin \theta$$
$$Y_2 = -X_1 \sin \theta + X_2 \cos \theta$$

then Y_1, Y_2 are statistically independent when θ is chosen such that $\tan 2\theta = 2\overline{X_1 X_2}/(\sigma_1^2 - \sigma_2^2)$.

3.25. Superimpose a plot of the probability density function of problem 3.13 on a plot of a Gaussian density function with the same variance. In what region does the sum variable rapidly approach a Gaussian distribution? This is a demonstration of the central limit theorem.

3.26. Let $\mathbf{C}$ be an $n \times n$ orthogonal matrix satisfying the condition $\mathbf{C}^T\mathbf{C} = \mathbf{I}$, where $\mathbf{I}$ is the *identity* matrix. The orthogonality condition implies

$$\sum_{j=1}^{n} c_{ji}c_{jk} = 1$$

for $i = k$, and 0 otherwise. Using this result, show that a new set of random variables Y_i, $i = 1, 2, \ldots, n$, are independent and Gaussian when they are related to an old set of independent Gaussian random variables $X_1, X_2, \ldots, X_n$ by $\mathbf{Y} = \mathbf{CX}$, where the X_i have different means and the same variance σ^2.

REFERENCE

1. Campbell, G. A., and Foster, R. M.: "Fourier Integrals," Van Nostrand, Princeton, N.J., 1948.

RECOMMENDED BIBLIOGRAPHY

1. Cramer, H.: "Mathematical Methods of Statistics," Princeton University Press, Princeton, N.J., 1945.

2. Davenport, W. B., and Root, W. L.: "An Introduction to the Theory of Random Signals and Noise," McGraw-Hill, New York, 1958.

3. Feller, W.: "Probability Theory and Its Applications," Wiley, New York, 1950.

4. Laning, J. H., and Battin, R. H.: "Random Processes in Automatic Control," McGraw-Hill, New York, 1956.

5. Papoulis, A.: "Probability, Random Variables and Stochastic Processes," McGraw-Hill, New York, 1965.

4

RANDOM PROCESSES

4.1. Introduction

In radar and communication, signal reception is made difficult by randomly fluctuating waveforms that modulate or add to the desired signal. These waveforms, which change unpredictably as a function of time, are conveniently described statistically. They are observables of physical processes that are apparently controlled by a random mechanism; such functions of time are called *random processes*. There are many random processes in nature, such as the voltage generated across a resistor by thermally excited electrons, the spatial position of a particle undergoing Brownian motion, atmospheric pressure fluctuations, and so on.

The mathematical representation of random processes introduced in this chapter is one of the tools required in the remainder of this book. Those wishing a broader and deeper understanding of this subject may consult the bibliography at the end of the chapter.

4.2. Characterization of a Random Process

The fluctuation in the rate of flow of electrons through a vacuum tube is a random process. This random process is often referred to as *shot noise*. Figure 4.2-1 illustrates typical current fluctuations in vacuum tubes of the same type as a function of time. A similar ensemble of waveforms will be generated by the current fluctuations in a single vacuum tube at different times. In either case, every waveform realization in the ensemble can be interpreted as an event in an infinite waveform sample space. The sample space characterizes all of the possible outcomes of a shot-noise experiment.

Experiments of the type just described lead to the following mathematical characterization of a random process. *A random process is described*

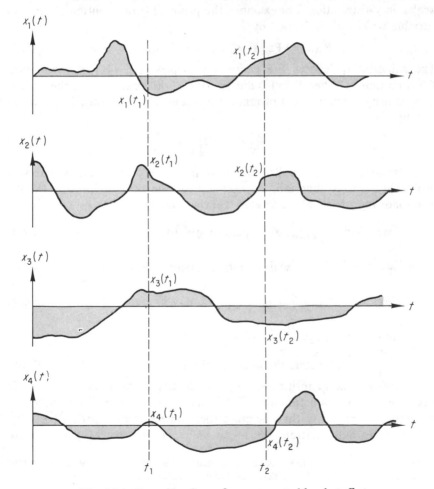

Fig. 4.2-1. Ensemble of waveforms generated by shot effect.

by a set or ensemble of functions $\{x(t)\}$ *that contains every possible realization of the random process.* It is useful to picture the ensemble of such waveforms as being displayed simultaneously, one under the other with coincident time origins, as in Fig. 4.2-1. At a selected instant of time t_1, there is a set of values $\{x(t_1)\}$ whose frequency of occurrence in the ensemble is governed by a probability density function. Hence $x(t_1) = X_{t_1}$ may be identified as a random variable that conforms in all ways to the previous description of a random variable. As the result of a particular experiment, X_{t_1} takes on a value in the set $\{x(t_1)\}$. Some members of this set are illustrated in Fig. 4.2-1. Clearly, $x(t_2) = X_{t_2}$ is also a random variable that takes on values in another set $\{x(t_2)\}$. Examples of members of this set are also illustrated in Fig. 4.2-1.

Each random variable $x(t_i)$ is characterized, in general, by a different probability distribution. For example, the probability distribution of random variable $x(t_1) = X_{t_1}$ is defined by

$$P_{t_1}(x) = \text{Pr}\,[X_{t_1} \leqslant x] = \text{Pr}\,[x(t_1) \leqslant x] \qquad (4.2\text{-}1)$$

The subscript t_1 denotes the parametric dependence of the distribution $P_{t_1}(x)$ on time t_1. When $P_{t_1}(x)$ is differentiable with respect to x, the probability density function $p_{t_1}(x)$ of random variable $x(t_1)$ is defined in the usual way by

$$p_{t_1}(x) = \frac{\partial P_{t_1}(x)}{\partial x} \qquad (4.2\text{-}2)$$

$P_{t_1}(x)$ and $p_{t_1}(x)$ are called the first probability distribution and first probability density function, respectively, of the random process $\{x(t)\}$.†
The following statistical parameters follow directly:

$$\overline{x(t_i)} = \int_{-\infty}^{\infty} x\,p_{t_i}(x)\,dx = \text{mean value of } x(t_i) \qquad (4.2\text{-}3)$$

$$\overline{x^2(t_i)} = \int_{-\infty}^{\infty} x^2\,p_{t_i}(x)\,dx = \text{mean-square value of } x(t_i) \qquad (4.2\text{-}4)$$

$$\overline{x^n(t_i)} = \int_{-\infty}^{\infty} x^n\,p_{t_i}(x)\,dx = n\text{th moment of } x(t_i) \qquad (4.2\text{-}5)$$

$$\overline{e^{j\xi x(t_i)}} = C_{x(t_i)}(\xi) = \int_{-\infty}^{\infty} e^{j\xi x}\,p_{t_i}(x)\,dx$$

$$= \text{characteristic function of } x(t_i) \qquad (4.2\text{-}6)$$

In defining the quantities above, the identity of each random variable has been retained as a function of its time of occurrence. Frequently, the subscript i is omitted so that the quantities above appear as $\overline{x(t)}$, $\overline{x^2(t)}$, $\overline{x^n(t)}$, and $\overline{e^{j\xi x(t)}}$. In this notation, the mean value, mean square value, the nth

---

†In recent years, usage has practically eliminated the distinction between a random process and the ensemble $\{x(t)\}$ that mathematically characterizes it.

moment, and the characteristic function are, in general, functions of time.

Random variables chosen from the ensemble $\{x(t)\}$ can be statistically related. For example, the joint probability distribution of two random variables $x(t_1)$ and $x(t_2)$ is defined by

$$P_{t_1, t_2}(x_1, x_2) = \Pr\left[X_{t_1} \leqslant x_1, X_{t_2} \leqslant x_2\right] \tag{4.2-7a}$$

$$= \Pr\left[x(t_1) \leqslant x_1, x(t_2) \leqslant x_2\right] \tag{4.2-7b}$$

The corresponding probability density function (when it exists) is given by

$$p_{t_1, t_2}(x_1, x_2) = \frac{\partial P_{t_1, t_2}(x_1, x_2)}{\partial x_1 \partial x_2} \tag{4.2-8}$$

$P_{t_1, t_2}(x_1, x_2)$ and $p_{t_1, t_2}(x_1, x_2)$ are called the second probability distribution and density functions, respectively, of the random process $\{x(t)\}$.

The (i, k)th joint moment m_{ik} of random variables $x(t_1)$ and $x(t_2)$ is defined in the usual way by

$$m_{ik} = \overline{x^i(t_1)x^k(t_2)} \tag{4.2-9}$$

Moment m_{11}, which is of particular importance, is known as the *autocorrelation* function and is denoted by $\phi_{xx}(t_1, t_2)$. From Eq. (4.2-9)

$$\phi_{xx}(t_1, t_2) = \overline{x(t_1)x(t_2)} \tag{4.2-10}$$

The subscript xx indicates that random variables $x(t_1)$ and $x(t_2)$ are samples of the same random process $\{x(t)\}$. The explicit appearance of t_1 and t_2 in $\phi_{xx}(t_1, t_2)$ indicates the dependence of the autocorrelation function on these two parameters. It follows from Eq. (4.2-10) that $\phi_{xx}(t_1, t_2)$ is symmetric in t_1 and t_2—that is, $\phi_{xx}(t_1, t_2) = \phi_{xx}(t_2, t_1)$. For the special case $t_1 = t_2 = t$,

$$\phi_{xx}(t, t) = \overline{x^2(t)} \tag{4.2-11}$$

which is the mean-square value of the random process $\{x(t)\}$.

In addition to the first and second probability distributions, the higher-order probability distributions further describe the random process $\{x(t)\}$. For example, corresponding to $t_1, t_2, \ldots, t_n$ there are random variables $x(t_1) = X_{t_1}$, $x(t_2) = X_{t_2}, \ldots, x(t_n) = X_{t_n}$ whose joint probability distribution $P_{t_1, t_2, \ldots, t_n}(x_1, x_2, \ldots, x_n)$ is defined by

$$P_{t_1, t_2, \ldots, t_n}(x_1, x_2, \ldots, x_n) = \Pr[X_{t_1} \leqslant x_1, X_{t_2} \leqslant x_2, \ldots, X_{t_n} \leqslant x_n] \tag{4.2-12}$$

A complete statistical description of a random process requires knowledge of the joint probability distributions of all orders. In many physical problems, however, only the first and second probability distributions are known. In some cases even these distributions are unavailable and all that is known are such statistical averages as the mean and the autocorrelation function of the process. Surprisingly, despite the meagerness of this information, it is still possible to obtain useful solutions to many problems.

4.3. Statistically Independent Random Processes

Suppose $\{x(t)\}$ and $\{y(t)\}$ characterize two random processes. Their statistical interrelatedness may be described by extending concepts already introduced. For example, the joint probability distribution of order m, n of random processes $\{x(t)\}$ and $\{y(t)\}$ is defined by

$$P_{t_1, t_2, \ldots, t_m, t_1', t_2', \ldots, t_n'}(x_1, x_2, \ldots, x_m, y_1, y_2, \ldots, y_n)$$
$$= \Pr[X_{t_1} \leqslant x_1, X_{t_2} \leqslant x_2, \ldots, X_{t_m} \leqslant x_m, Y_{t_1'} \leqslant y_1, Y_{t_2'} \leqslant y_2, \ldots, Y_{t_n'} \leqslant y_n]$$
$$(4.3\text{-}1)$$

where $t_1, t_2, \ldots, t_m, t_1', t_2', \ldots, t_n'$ are arbitrary values of t. The joint probability density of order m, n is related to the joint probability distribution in the usual way by

$$p_{t_1, t_2, \ldots, t_m, t_1', t_2' \ldots, t_n'}(x_1, x_2, \ldots, x_m, y_1, y_2, \ldots, y_n)$$
$$= \frac{\partial^{m+n} P_{t_1, t_2, \ldots, t_m, t_1', t_2', \ldots, t_n'}(x_1, x_2, \ldots, x_m, y_1, y_2, \ldots, y_n)}{\partial x_1 \, \partial x_2 \ldots \partial x_m \, \partial y_1 \, \partial y_2 \ldots \partial y_n} \qquad (4.3\text{-}2)$$

Definitions (4.3-1) and (4.3-2) can be extended to three or more random processes.

In practical applications it is difficult to obtain the joint probability density function relating two random processes for any order greater than $m = n = 1$. Of frequent concern is the *cross correlation* $\phi_{xy}(t_1, t_2)$ between two random processes $\{x(t)\}$ and $\{y(t)\}$, which is defined by

$$\phi_{xy}(t_1, t_2) = \overline{x(t_1) y(t_2)} \qquad (4.3\text{-}3a)$$

$$= \int_{-\infty}^{\infty} \int_{-\infty}^{\infty} xy p_{t_1, t_2}(x, y) \, dx \, dy \qquad (4.3\text{-}3b)$$

It follows from (4.3-3b) that

$$\phi_{xy}(t_1, t_2) = \phi_{yx}(t_2, t_1) \qquad (4.3\text{-}4)$$

so that $\phi_{xy}(t_1, t_2)$, in general, is not symmetric in t_1, t_2.

To illustrate the use of the cross-correlation function, consider the autocorrelation function $\phi_{zz}(t_1, t_2)$ of a random process $\{z(t)\}$ that is defined as the sum of processes $\{x(t)\}$ and $\{y(t)\}$; it is given by

$$\phi_{zz}(t_1, t_2) = \phi_{xx}(t_1, t_2) + \phi_{xy}(t_1, t_2) + \phi_{yx}(t_1, t_2) + \phi_{yy}(t_1, t_2) \qquad (4.3\text{-}5)$$

Equation (4.3-5) states that the autocorrelation of $\{z(t)\}$ is the sum of the autocorrelations of $\{x(t)\}$ and $\{y(t)\}$ and their cross correlations. If $\{x(t)\}$ and $\{y(t)\}$ are statistically independent, and one or both random processes have a zero mean at t_1 and t_2, then it follows from Eq. (4.3-3a) that cross correlations ϕ_{xy} and ϕ_{yx} are zero, in which case $\phi_{zz}(t_1, t_2)$ reduces to

$$\phi_{zz}(t_1, t_2) = \phi_{xx}(t_1, t_2) + \phi_{yy}(t_1, t_2) \qquad (4.3\text{-}6)$$

4.4. Stationarity and Ergodicity

Random processes are conveniently classified into three categories denoting increasing degrees of statistical regularity. They are conventionally referred to as: (1) nonstationary processes, (2) stationary processes (strict and wide sense), (3) ergodic processes.

A *strict-sense stationary random process* is characterized by a set of nth-order probability distributions, $n = 1, 2, \ldots$, that are invariant with time shift. Thus

$$P_{t_1}(x_1) = P_{t_1 + t}(x_1) \tag{4.4-1}$$

$$P_{t_1, t_2}(x_1, x_2) = P_{t_1 + t, t_2 + t}(x_1, x_2) \tag{4.4-2}$$

$$P_{t_1, t_2, t_3}(x_1, x_2, x_3) = P_{t_1 + t, t_2 + t, t_3 + t}(x_1, x_2, x_3) \tag{4.4-3}$$

and so on. An example of a strict-sense stationary random process is the noise voltage across a resistor at constant temperature.

A direct consequence of strict-sense stationarity is that

$$\overline{x(t_1)} = \overline{x(t_1 + t)} \qquad \text{for all } t \text{ and } t_1 \tag{4.4-4}$$

For $t_1 = 0$, it follows that

$$\overline{x(t)} = \overline{x(0)} = \text{constant} \qquad \text{for all } t \tag{4.4-5}$$

which proves that the mean value of the process is invariant with time. Another consequence of strict-sense stationarity is that

$$\phi_{xx}(t_1, t_2) = \phi_{xx}(t_1 + t, t_2 + t) \qquad \text{for all } t \tag{4.4-6}$$

For $t = -t_1$,

$$\phi_{xx}(t_1, t_2) = \phi_{xx}(0, t_2 - t_1) \tag{4.4-7}$$

Equation (4.4-7) indicates that the autocorrelation function is a function only of the time difference $\tau = t_2 - t_1$, so that it is often written in the following notation:

$$\phi_{xx}(t_1, t_2) = \phi_{xx}(t_2 - t_1) = \phi_{xx}(\tau) \tag{4.4-8}$$

Since $\phi_{xx}(t_1, t_2)$ is symmetric in t_1, t_2, it is easy to show that

$$\phi_{xx}(\tau) = \phi_{xx}(-\tau) \tag{4.4-9}$$

from which it follows that $\phi_{xx}(\tau)$ is an even function of τ.

There are many random processes that are not stationary in the strict sense, yet their means and autocorrelation functions satisfy Eqs. (4.4-5) and (4.4-8). Such processes are said to be *stationary in the wide sense*. It follows that a strict-sense stationary random process is also stationary in the wide sense.

Random processes that are not stationary in the strict sense or in the wide sense are said to be *nonstationary*. The noise waveform generated by a resistor will be nonstationary if its resistance, or its temperature, changes with time.

The cross correlation $\phi_{xy}(t_1, t_2)$ between two separately and jointly stationary random processes $\{x(t)\}$ and $\{y(t)\}$ may be written as

$$\phi_{xy}(t_1, t_2) = \phi_{xy}(t_1 + t, t_2 + t) \tag{4.4-10}$$

For $t = -t_1$, Eq. (4.4-10) can be rewritten as

$$\phi_{xy}(t_1, t_2) = \phi_{xy}(t_2 - t_1) = \phi_{xy}(\tau) \tag{4.4-11}$$

from which it follows that $\phi_{xy}(\tau)$ can also be expressed as

$$\phi_{xy}(\tau) = \overline{x(t)y(t + \tau)} \tag{4.4-12}$$

Suppose a representative waveform $x(t)$ is selected from the set $\{x(t)\}$. Its *time average* $\langle x(t) \rangle$ is defined by

$$\langle x(t) \rangle = \lim_{T \to \infty} \frac{1}{2T} \int_{-T}^{T} x(t)\, dt \tag{4.4-13}$$

Although this time average may not exist in general, it does exist for every sample function of a strict-sense stationary random process except for a set of probability zero.

If a stationary random process is *ergodic*, then the ensemble average $\overline{x(t)}$ is equal with probability one to the time average $\langle x(t) \rangle$. More generally, if $g(X_t)$ denotes a random variable that is a function of random variable $x(t) = X_t$, then random process $\{x(t)\}$ is said to be ergodic if (a) it is strict-sense stationary, and (b) for every sample function $x(t)$, except for a set of probability zero, the following condition is satisfied:

$$\overline{g(X_t)} = \lim_{T \to \infty} \frac{1}{2T} \int_{-T}^{T} g[x(t + \tau)]\, dt \tag{4.4-14}$$

over all translations in time τ of each sample function $x(t)$. Equation (4.4-14) usually simplifies in practice to

$$\overline{g(X_t)} = \lim_{T \to \infty} \frac{1}{2T} \int_{-T}^{T} g[x(t)]\, dt \tag{4.4-15}$$

A number of relations follow from (4.4-15) for an ergodic random process $\{x(t)\}$:

$$\overline{x(t)} = \langle x(t) \rangle = \lim_{T \to \infty} \frac{1}{2T} \int_{-T}^{T} x(t)\, dt \tag{4.4-16}$$

$$\overline{x^2(t)} = \langle x^2(t) \rangle = \lim_{T \to \infty} \frac{1}{2T} \int_{-T}^{T} x^2(t)\, dt \tag{4.4-17}$$

and

$$\phi_{xx}(\tau) = \overline{x(t)x(t + \tau)} \tag{4.4-18a}$$

$$= \langle x(t)x(t + \tau) \rangle = \lim_{T \to \infty} \frac{1}{2T} \int_{-T}^{T} x(t)x(t + \tau)\, dt \tag{4.4-18b}$$

The function $\langle x(t)x(t + \tau)\rangle$ defined by (4.4-18b) is conventionally referred to as the time autocorrelation function of $x(t)$, for $x(t)$ of finite average power. It is not limited to random waveforms but applies equally well to deterministic waveforms. When $x(t)$ is a member of a random process, the time autocorrelation function is denoted by $R_{xx}(\tau)$:

$$R_{xx}(\tau) = \langle x(t)x(t + \tau)\rangle \tag{4.4-19}$$

Thus for a (stationary) ergodic random process $\{x(t)\}$, Eqs. (4.4-18) may be rewritten more compactly as

$$\phi_{xx}(\tau) = R_{xx}(\tau) \tag{4.4-20}$$

which states that the time and ensemble autocorrelation functions are equal with probability one.

The following important result can be established for stationary and ergodic random processes. If the input to a time-invariant filter (not necessarily linear) consists of a stationary and/or ergodic process, then the output possesses the same property. This result, although not explicitly stated, is assumed quite often.

4.5. Power Spectral Density

The instantaneous power associated with an arbitrary real function of time $x(t)$ is proportional to $x^2(t)$. It is customary in many physical situations to choose a unity constant of proportionality for simplicity. The energy in $x(t)$ is then given by

$$[\text{energy in } x(t)] = \int_{-\infty}^{\infty} x^2(t)\, dt \tag{4.5-1}$$

When the energy in $x(t)$ is finite, the Fourier transform $X(\omega)$ of $x(t)$ exists† and is given by

$$X(\omega) = \int_{-\infty}^{\infty} x(t)e^{-j\omega t}\, dt \tag{4.5-2}$$

From Eq. (4.5-1) and Parseval's relation [Eq. (2.2-21) in Chapter 2] the energy in $x(t)$ may also be expressed in terms of $X(\omega)$ by

$$[\text{energy in } x(t)] = \frac{1}{2\pi} \int_{-\infty}^{\infty} |X(\omega)|^2\, d\omega \tag{4.5-3}$$

The form of Eq. (4.5-3) implies that $|X(\omega)|^2$ can be interpreted as the distribution of energy in $x(t)$ as a function of (radian) frequency. For this reason, it is referred to as the *energy spectral density* of function $x(t)$.

---

†At least as a limit in the mean [see Section 2.2 (page 45)].

Consider an arbitrary bounded real function $x(t)$ of infinite duration with finite average power, which is defined, when the indicated limit exists, by

$$[\text{average power in } x(t)] = \lim_{T \to \infty} \frac{1}{2T} \int_{-T}^{T} x^2(t)\, dt \qquad (4.5\text{-}4)$$

Since the energy contained in $x(t)$ is infinite, its Fourier transform does not, in general, exist. Consider the function $x_T(t)$, which is related to $x(t)$ by

$$x_T(t) = \begin{cases} x(t), & -T \leqslant t \leqslant T \\ 0, & \text{otherwise} \end{cases} \qquad (4.5\text{-}5)$$

Clearly $x_T(t)$ has finite energy for finite T, and its Fourier transform $X_T(\omega)$ is given by

$$X_T(\omega) = \int_{-\infty}^{\infty} x_T(t)e^{-j\omega t}\, dt \qquad (4.5\text{-}6)$$

Since $x_T(t) = x(t)$ in the interval $(-T, T)$ and is zero elsewhere, Eq. (4.5-4) can be rewritten using Parseval's relation as

$$[\text{average power in } x(t)] = \lim_{T \to \infty} \frac{1}{2T} \left[\int_{-\infty}^{\infty} x_T^2(t)\, dt \right] \qquad (4.5\text{-}7a)$$

$$= \lim_{T \to \infty} \frac{1}{2T} \left[\frac{1}{2\pi} \int_{-\infty}^{\infty} |X_T(\omega)|^2\, d\omega \right] \qquad (4.5\text{-}7b)$$

$$= \lim_{T \to \infty} \frac{1}{2\pi} \int_{-\infty}^{\infty} \frac{|X_T(\omega)|^2}{2T}\, d\omega \qquad (4.5\text{-}7c)$$

$$= \lim_{T \to \infty} \frac{1}{2\pi} \int_{-\infty}^{\infty} G_T^{(x)}(\omega)\, d\omega \qquad (4.5\text{-}7d)$$

where, by definition,

$$G_T^{(x)}(\omega) = \frac{|X_T(\omega)|^2}{2T} \qquad (4.5\text{-}8)$$

A comparison of Eq. (4.5-7d) with Eq. (4.5-3) leads to the definition of $G^{(x)}(\omega)$,

$$G^{(x)}(\omega) = \lim_{T \to \infty} G_T^{(x)}(\omega) \qquad (4.5\text{-}9)$$

as the power spectral density of a function $x(t)$ with finite average power, when the indicated limit exists.

The notion of power spectral density can be extended to random processes. If $x_T(t)$ is a $2T$-second sample of a member of ensemble $\{x(t)\}$, it may be considered as one possible outcome of an experiment. Hence $x_T(t)$ is a random variable and its Fourier transform $X_T(\omega)$ is a related random variable. From Eq. (4.5-8) $G_T^{(x)}(\omega)$ is also a random variable. The power spectral density of a random process $\{x(t)\}$, denoted by $G_{xx}(\omega)$, is defined as the limit, as $T \to \infty$, of the expected value of $G_T^{(x)}(\omega)$,

$$G_{xx}(\omega) = \lim_{T \to \infty} \overline{G_T^{(x)}(\omega)} \qquad (4.5\text{-}10)$$

4.6. Wiener-Khintchine Theorem

The Wiener-Khintchine theorem states that the autocorrelation function $\phi_{xx}(\tau)$ and the power spectral density $G_{xx}(\omega)$ of a wide-sense stationary random process $\{x(t)\}$ are Fourier transforms of each other. Thus

$$G_{xx}(\omega) = \int_{-\infty}^{\infty} \phi_{xx}(\tau)e^{-j\omega\tau}\,d\tau \qquad (4.6\text{-}1)$$

and

$$\phi_{xx}(\tau) = \frac{1}{2\pi}\int_{-\infty}^{\infty} G_{xx}(\omega)e^{j\omega\tau}\,d\omega \qquad (4.6\text{-}2)$$

To demonstrate the Wiener-Khintchine theorem, define a function $R(\tau)$ by

$$R(\tau) = \lim_{T\to\infty}\frac{1}{2T}\int_{-T}^{T} x(t)x(t+\tau)\,dt \qquad (4.6\text{-}3)$$

where $x(t)$ is an arbitrary real function with finite average power. Define another function $R_T(\tau)$ by

$$R_T(\tau) = \frac{1}{2T}\int_{-\infty}^{\infty} x_T(t)x_T(t+\tau)\,dt \qquad (4.6\text{-}4)$$

where $x_T(t)$ is related to $x(t)$ by Eq. (4.5-5), and $X_T(\omega)$, the Fourier transform of $x_T(t)$, is given by (4.5-6). It is assumed that $R(\tau)$, $R_T(\tau)$, and $X_T(\omega)$ all exist. From (4.5-5), Eq. (4.6-4) for $R_T(\tau)$ can be rewritten for $0 \leqslant \tau \leqslant T$ as

$$R_T(\tau) = \frac{1}{2T}\int_{-T}^{T-\tau} x(t)x(t+\tau)\,dt \qquad (4.6\text{-}5a)$$

$$= \frac{1}{2T}\int_{-T}^{T} x(t)x(t+\tau)\,dt - \frac{1}{2T}\int_{T-\tau}^{T} x(t)x(t+\tau)\,dt \qquad (4.6\text{-}5b)$$

A sufficient condition to assure that the value of the second integral in (4.6-5b) go to zero as $T\to\infty$ is that $x(t)$ be bounded† for all t by some value M. In this case,

$$\lim_{T\to\infty}\frac{1}{2T}\int_{T-\tau}^{T} x(t)x(t+\tau)\,dt \leqslant \lim_{T\to\infty}\frac{1}{2T}\int_{T-\tau}^{T} x^2(t)\,dt \qquad (4.6\text{-}6a)$$

$$\leqslant \lim_{T\to\infty}\frac{1}{2T}[M^2\tau] \to 0 \qquad (4.6\text{-}6b)$$

A similar argument can be made for $0 \geqslant \tau \geqslant -T$. It follows from Eqs. (4.6-3) through (4.6-6) that

$$\lim_{T\to\infty} R_T(\tau) = R(\tau) \qquad (4.6\text{-}7)$$

The Fourier transform of $R_T(\tau)$ (assuming all mathematical operations to be valid) is given by

†Except for a set of measure zero.

$$\int_{-\infty}^{\infty} R_T(\tau)e^{-j\omega\tau}\,d\tau = \frac{1}{2T}\int_{-\infty}^{\infty} e^{-j\omega\tau}\,d\tau \int_{-\infty}^{\infty} x_T(t)\,x_T(t+\tau)\,dt \tag{4.6-8a}$$

$$= \frac{1}{2T}\int_{-\infty}^{\infty} dt \int_{-\infty}^{\infty} [x_T(t)e^{j\omega t}][x_T(t+\tau)e^{-j\omega(t+\tau)}]\,d\tau$$

$$= \frac{1}{2T}\int_{-\infty}^{\infty} x_T(t)e^{j\omega t}\,dt \int_{-\infty}^{\infty} x_T(t+\tau)e^{-j\omega(t+\tau)}\,d\tau$$

$$= \frac{1}{2T}\int_{-\infty}^{\infty} x_T(t)e^{j\omega t}\,dt \int_{-\infty}^{\infty} x_T(v)e^{-j\omega v}\,dv$$

$$= \frac{1}{2T}[X_T^*(\omega)][X_T(\omega)]$$

$$= \frac{|X_T(\omega)|^2}{2T} = G_T^{(x)}(\omega) \tag{4.6-8b}$$

The last step in (4.6-8b) follows from the definition of $G_T^{(x)}(\omega)$ in (4.5-8). Combining (4.6-7) and (4.6-8b), the Fourier transform of $R(\tau)$ is given by

$$\int_{-\infty}^{\infty} R(\tau)\,e^{-j\omega\tau}\,d\tau = \lim_{T\to\infty} G_T^{(x)}(\omega) \tag{4.6-9}$$

If $x(t)$ belongs to ensemble $\{x(t)\}$, the power spectral density $G_{xx}(\omega)$ of the process is found by Eqs. (4.5-10) and (4.6-9) to be

$$G_{xx}(\omega) = \lim_{T\to\infty} \overline{G_T^{(x)}(\omega)} = \int_{-\infty}^{\infty} \overline{R(\tau)}\,e^{-j\omega\tau}\,d\tau \tag{4.6-10a}$$

$$= \lim_{T\to\infty} \frac{1}{2T}\int_{-\infty}^{\infty} e^{-j\omega\tau}\,d\tau \int_{-T}^{T} \overline{x(t)x(t+\tau)}\,dt$$

$$= \lim_{T\to\infty} \frac{1}{2T}\int_{-\infty}^{\infty} e^{-j\omega\tau}\,d\tau \int_{-T}^{T} \phi_{xx}(t,t+\tau)\,dt \tag{4.6-10b}$$

where $\phi_{xx}(t, t+\tau)$ is the ensemble autocorrelation function of process $\{x(t)\}$. For a wide-sense stationary process

$$\phi_{xx}(t, t+\tau) = \phi_{xx}(\tau) \tag{4.6-11}$$

Substituting (4.6-11) into (4.6-10b) results in

$$G_{xx}(\omega) = \int_{-\infty}^{\infty} \phi_{xx}(\tau)e^{-j\omega\tau}\,d\tau \left[\lim_{T\to\infty} \frac{1}{2T}\int_{-T}^{T} dt\right] \tag{4.6-12}$$

Since the bracketed quantity in (4.6-12) is unity, it follows that

$$G_{xx}(\omega) = \int_{-\infty}^{\infty} \phi_{xx}(\tau)e^{-j\omega\tau}\,d\tau \tag{4.6-13}$$

which is the first of the Wiener-Khintchine relations given by Eq. (4.6-1). The inverse relation follows from Fourier transform theory:

$$\phi_{xx}(\tau) = \frac{1}{2\pi}\int_{-\infty}^{\infty} G_{xx}(\omega)e^{j\omega\tau}\,d\omega \tag{4.6-14}$$

In some texts the Wiener-Khintchine relations differ by a factor of two from Eqs. (4.6-13) and (4.6-14). This difference arises when power spectral density is defined as nonzero only over the positive frequency spectrum.

To illustrate the Wiener-Khintchine relations, consider an example introduced earlier in which a random process $\{z(t)\}$ was defined to be the sum of two random processes $\{x(t)\}$ and $\{y(t)\}$. If $\{x(t)\}$ and $\{y(t)\}$ are both wide-sense stationary, then by Eq. (4.3-5), $\phi_{zz}(t, t + \tau)$ is given by

$$\phi_{zz}(t, t + \tau) = \phi_{xx}(\tau) + \phi_{yy}(\tau) + \phi_{xy}(t, t + \tau) + \phi_{yx}(t, t + \tau) \quad (4.6\text{-}15)$$

If, in addition, random processes $\{x(t)\}$ and $\{y(t)\}$ are separately and jointly wide-sense stationary, then the sum random process $\{z(t)\}$ is also wide-sense stationary. By the Wiener-Khintchine theorem, the power spectral density $G_{zz}(\omega)$ of random process $\{z(t)\}$ is found from the Fourier transform of Eq. (4.6-15) as follows:

$$G_{zz}(\omega) = G_{xx}(\omega) + G_{yy}(\omega) + \int_{-\infty}^{\infty} \phi_{xy}(\tau)e^{-j\omega\tau}\,d\tau$$

$$+ \int_{-\infty}^{\infty} \phi_{yx}(\tau)e^{-j\omega\tau}\,d\tau \quad (4.6\text{-}16)$$

where $G_{xx}(\omega)$ and $G_{yy}(\omega)$ are the power spectral densities of stationary processes $\{x(t)\}$ and $\{y(t)\}$, respectively. The form of Eq. (4.6-16) suggests that the third and fourth terms on the right be defined as the cross-spectral (power) densities, $G_{xy}(\omega)$ and $G_{yx}(\omega)$, respectively, of random processes $\{x(t)\}$ and $\{y(t)\}$. By a development analogous to that which led to the definition of the power spectral density of a stationary random process, it can be shown that

$$G_{xy}(\omega) = \lim_{T \to \infty} \overline{G_T^{(xy)}(\omega)} = \lim_{T \to \infty} \overline{\left[\frac{X_T^*(\omega) Y_T(\omega)}{2T} \right]} \quad (4.6\text{-}17)$$

and

$$G_{yx}(\omega) = \lim_{T \to \infty} \overline{G_T^{(yx)}(\omega)} = \lim_{T \to \infty} \overline{\left[\frac{Y_T^*(\omega) X_T(\omega)}{2T} \right]} \quad (4.6\text{-}18)$$

In the above relations $X_T(\omega)$ and $Y_T(\omega)$ are the Fourier transforms of $x_T(t)$ and $y_T(t)$, where

$$x_T(t) = \begin{cases} x(t), & -T \leqslant t \leqslant T \\ 0, & \text{otherwise} \end{cases} \quad (4.6\text{-}19)$$

$$y_T(t) = \begin{cases} y(t), & -T \leqslant t \leqslant T \\ 0, & \text{otherwise} \end{cases} \quad (4.6\text{-}20)$$

The Wiener-Khintchine theorem can be extended to show that $\phi_{xy}(\tau)$ and $\phi_{yx}(\tau)$ can be obtained from inverse Fourier transforms of $G_{xy}(\omega)$ and $G_{yx}(\omega)$, respectively. If random processes $\{x(t)\}$ and $\{y(t)\}$ are uncorrelated and have zero means, their cross-spectral densities are zero, in which case Eq. (4.6-16) simplifies to

$$G_{zz}(\omega) = G_{xx}(\omega) + G_{yy}(\omega) \qquad (4.6\text{-}21)$$

If random processes $\{x(t)\}$ and $\{y(t)\}$ are each ergodic, and jointly ergodic, then with probability one: $\phi_{xx}(\tau)$, $\phi_{yy}(\tau)$, $\phi_{xy}(\tau)$, and $\phi_{yx}(\tau)$ can be replaced by $R_{xx}(\tau)$, $R_{yy}(\tau)$, $R_{xy}(\tau)$, and $R_{yx}(\tau)$, respectively, where $R_{xx}(\tau)$ and $R_{yy}(\tau)$ are defined as in Eq. (4.4-19) and $R_{xy}(\tau)$ and $R_{yx}(\tau)$ are defined by

$$R_{xy}(\tau) = \langle x(t)y(t + \tau) \rangle \qquad (4.6\text{-}22)$$

and

$$R_{yx}(\tau) = \langle y(t)x(t + \tau) \rangle \qquad (4.6\text{-}23)$$

If $\{x(t)\}$ and $\{y(t)\}$ are also uncorrelated and have zero means, then $R_{xy}(\tau)$ and $R_{yx}(\tau)$ are both zero.

The instantaneous power of a random process $\{x(t)\}$ is defined as

$$\left(\begin{matrix} \text{instantaneous power} \\ \text{of } \{x(t)\} \end{matrix} \right) = \overline{x^2(t)} \qquad (4.6\text{-}24)$$

The instantaneous fluctuation power $\sigma^2(t)$ about the mean of the process is defined by

$$\left(\begin{matrix} \text{instantaneous fluctuation} \\ \text{power of } \{x(t)\} \end{matrix} \right) \equiv \sigma^2(t) = \overline{[x(t) - \bar{x}(t)]^2} \qquad (4.6\text{-}25\text{a})$$

$$= \overline{x^2(t)} - [\bar{x}(t)]^2 \qquad (4.6\text{-}25\text{b})$$

For a wide-sense stationary process, $\overline{x^2(t)}$ and $\sigma^2(t)$ are both independent of time. It follows from the definition of the autocorrelation function $\phi_{xx}(\tau)$ of a stationary process that

$$\overline{x^2} = \phi_{xx}(0) \qquad (4.6\text{-}26)$$

Combining Eqs. (4.6-25) and (4.6-26) yields

$$\sigma^2 = \phi_{xx}(0) - \bar{x}^2 \qquad (4.6\text{-}27)$$

With Wiener-Khintchine relation (4.6-14), Eqs. (4.6-26) and (4.6-27) can be rewritten as

$$\overline{x^2} = \frac{1}{2\pi} \int_{-\infty}^{\infty} G_{xx}(\omega) \, d\omega \qquad (4.6\text{-}28)$$

and

$$\sigma^2 = \frac{1}{2\pi} \int_{-\infty}^{\infty} G_{xx}(\omega) \, d\omega - \bar{x}^2 \qquad (4.6\text{-}29)$$

From the definition of the autocorrelation function $\phi_{xx}(\tau)$ of a complex wide-sense stationary random process $\{x(t)\}$ and the Wiener-Khintchine relations, a number of correspondences can be established between $\{x(t)\}$, $\phi_{xx}(\tau)$, and power spectral density $G_{xx}(\omega)$, as shown in Table 4.6-1. Note that the first entry in the table extends the definition of the autocorrelation function to complex random processes.

TABLE 4.6-1

$\{x(t)\}$	$\phi_{xx}(\tau)$	$G_{xx}(\omega)$
$\{x(t)\}$	$E[x^*(t)x(t + \tau)]$	$\int_{-\infty}^{\infty} \phi_{xx}(\tau)e^{-j\omega\tau}\,d\tau$
$\{ax(t)\}$	$|a|^2\phi_{xx}(\tau)$	$|a|^2 G_{xx}(\omega)$
$\{x(t)e^{-j\omega_0 t}\}$	$\phi_{xx}(\tau)e^{-j\omega_0\tau}$	$G_{xx}(\omega + \omega_0)$
$\left\{\dfrac{d}{dt}x(t)\right\}$	$-\dfrac{d^2}{d\tau^2}\phi_{xx}(\tau)$	$\omega^2 G_{xx}(\omega)$
$\left\{\dfrac{d^n}{dt^n}x(t)\right\}$	$(-1)^n\dfrac{d^{2n}}{d\tau^{2n}}\phi_{xx}(\tau)$	$\omega^{2n}G_{xx}(\omega)$

4.7. Shape of Power Spectral Density

A stationary random process is often further classified by the shape of its power spectral density. A *white* process denotes one whose power spectral density is flat over the entire frequency range from $-\infty$ to $+\infty$ (see Fig. 4.7-1a). In practice, a process is considered white if its spectrum is flat over a spectral band that is large compared to the actual band being considered. The term "white" originates from the designation of electromagnetic radiation with approximately constant spectral density over the visible portion of the spectrum as "white light." Examples of (almost) white random processes are shot noise current flow through thermionic vacuum tubes and the thermal noise waveforms observed across resistors. The spectra of these processes are flat from dc to frequencies well above the microwave band.

Random processes whose spectra are not white are referred to as *colored.* Transistor noise, for example, at low frequencies is colored. Because the solutions of problems with colored noise are more difficult than with white noise, colored random processes are often approximated by a white process, if it is at all reasonable to do so. For example, filtered shot noise can be considered white with respect to another filter whose spectral transfer function lies in the flat portion of the transmission characteristic of the first filter.

Figure 4.7-1 shows the power spectral density corresponding to several theoretically important, as well as practically significant, autocorrelation functions. The correspondences can be established by Fourier transform theory.

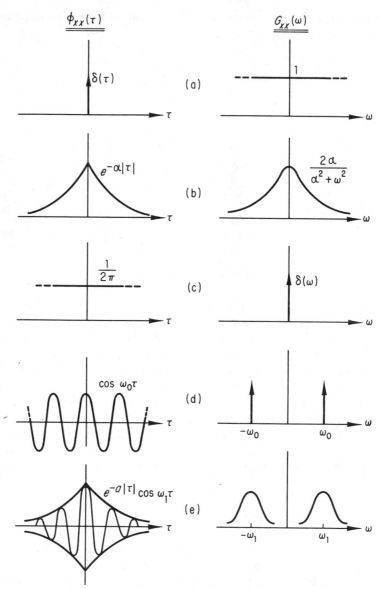

Fig. 4.7-1. Correspondences between $\phi_{xx}(\tau)$ and $G_{xx}(\omega)$.

4.8. The Gaussian Random Process

The Gaussian random process is of particular importance because of its frequent occurrence in physical processes. It appears often by reason of its close relationship to the central limit theorem, which states that the probability density function of a sum of independent random variables tends to become Gaussian as the number of variables becomes indefinitely large. Thus, physical processes whose instantaneous output is the result of the superposition of a large number of independent random events tend to have Gaussian statistics.

By definition, a random process is Gaussian if, for every set of sample instants $t_1, t_2, \ldots, t_n$, the random variables $x(t_1) = X_{t_1}, x(t_2) = X_{t_2}, \ldots, x(t_n) = X_{t_n}$ have a multivariate Gaussian probability density function. This probability density function was previously described by Eq. (3.14-7) in Chapter 3 and is repeated here in the present notation:

$$p(x_{t_1}, x_{t_2}, \ldots, x_{t_n}) = \frac{1}{(2\pi)^{n/2} |\Lambda|^{1/2}} \exp\left\{-\frac{1}{2}(\mathbf{x} - \bar{\mathbf{X}})^T \Lambda^{-1}(\mathbf{x} - \bar{\mathbf{X}})\right\}$$

$$(4.8\text{-}1)$$

where $\mathbf{x}$ and $\bar{\mathbf{X}}$ are column vectors given by

$$\mathbf{x} = \begin{bmatrix} x_{t_1} \\ x_{t_2} \\ \cdot \\ \cdot \\ \cdot \\ x_{t_n} \end{bmatrix} \tag{4.8-2}$$

$$\bar{\mathbf{X}} = \begin{bmatrix} \bar{X}_{t_1} \\ \bar{X}_{t_2} \\ \cdot \\ \cdot \\ \cdot \\ \bar{X}_{t_n} \end{bmatrix} \tag{4.8-3}$$

and Λ is the covariance matrix, defined by

$$\Lambda = \overline{(\mathbf{X} - \bar{\mathbf{X}})(\mathbf{X} - \bar{\mathbf{X}})^T} = \begin{bmatrix} \lambda_{11} & \lambda_{12} & \ldots & \lambda_{1n} \\ \lambda_{21} & \lambda_{22} & \ldots & \lambda_{2n} \\ \cdot & & & \\ \cdot & & & \\ \cdot & & & \\ \lambda_{n1} & \lambda_{n2} & \ldots & \lambda_{nn} \end{bmatrix} \tag{4.8-4}$$

with elements λ_{ij} given by

$$\lambda_{ij} = \overline{(X_{t_i} - \bar{X}_{t_i})(X_{t_j} - \bar{X}_{t_j})} \tag{4.8-5a}$$

$$= \overline{X_{t_i}X_{t_j}} - \bar{X}_{t_i}\bar{X}_{t_j} \tag{4.8-5b}$$

$$= \phi_{xx}(t_i, t_j) - \bar{X}_{t_i}\bar{X}_{t_j} \tag{4.8-5c}$$

In passing from (4.8-5b) to (4.8-5c) we used the definition of the autocorrelation function $\phi_{xx}(t_1, t_2)$ of a random process. When the random process is strict- or wide-sense stationary, $\phi_{xx}(t_1, t_2)$ may be written as $\phi_{xx}(t_2 - t_1)$ and is related to the power spectral density of the random process by the Wiener-Khintchine theorem.

4.9. Fourier Series Representation of a Gaussian Random Process

In many applications it is convenient to represent a stationary Gaussian random process by means of a Fourier series. If $x(t)$ is a typical member of the ensemble that characterizes the process, it can be represented over an interval $-T/2 \leqslant t \leqslant T/2$, as discussed in Section 3 of Chapter 2, by

$$x(t) = \frac{a_0}{2} + \sum_{k=1}^{\infty} a_k \cos \omega_k t + b_k \sin \omega_k t, \qquad -\frac{T}{2} \leqslant t \leqslant \frac{T}{2} \qquad (4.9\text{-}1)$$

where $\omega_k = 2\pi k/T$ and

$$a_k = \frac{2}{T} \int_{-T/2}^{T/2} x(t) \cos \omega_k t \, dt \qquad (4.9\text{-}2)$$

$$b_k = \frac{2}{T} \int_{-T/2}^{T/2} x(t) \sin \omega_k t \, dt \qquad (4.9\text{-}3)$$

At every instant of time t, $-T/2 \leqslant t \leqslant T/2$, $x(t)$ is a Gaussian random variable. It follows from Eqs. (4.9-2) and (4.9-3) that, if an integral is interpreted as the limit of a sum, a_k and b_k are each the sum of weighted Gaussian random variables and hence also Gaussian random variables. The principal results of this section are that a_i, b_k become uncorrelated for all i, k as T approaches infinity, and also that a_i, a_k and b_i, b_k become uncorrelated for all $i \neq k$ as T approaches infinity. Since each variable is Gaussian, zero correlation is sufficient to insure that these variables will also be statistically independent as $T \to \infty$.

Because the values of the Fourier coefficients given by Eqs. (4.9-2) and (4.9-3) are each dependent on T, to demonstrate the desired results we shall show that the normalized covariance of each pair of coefficients approaches zero as T approaches infinity. It is convenient, in the present instance, to multiply the numerator and denominator of each normalized covariance function by T. Thus, it will be shown that for $T \to \infty$,

$$\frac{T \overline{a_i a_k}}{(T \overline{a_i^2})^{1/2} (T \overline{a_k^2})^{1/2}} \longrightarrow 0, \qquad i \neq k \qquad (4.9\text{-}4)$$

$$\frac{T \overline{b_i b_k}}{(T \overline{b_i^2})^{1/2} (T \overline{b_k^2})^{1/2}} \longrightarrow 0, \qquad i \neq k \qquad (4.9\text{-}5)$$

$$\frac{T \overline{a_i b_k}}{(T \overline{a_i^2})^{1/2} (T \overline{b_k^2})^{1/2}} \longrightarrow 0, \qquad \text{all } i, k \qquad (4.9\text{-}6)$$

If a rectangular gating pulse $g_T(t)$ is defined by

$$g_T(t) = \begin{cases} 1, & |t| \leqslant T/2 \\ 0, & |t| > T/2 \end{cases} \tag{4.9-7}$$

then Eqs. (4.9-2) and (4.9-3) can be replaced by

$$a_k = \frac{2}{T} \int_{-\infty}^{\infty} x(t) g_T(t) \cos \omega_k t \, dt \tag{4.9-8}$$

$$b_k = \frac{2}{T} \int_{-\infty}^{\infty} x(t) g_T(t) \sin \omega_k t \, dt \tag{4.9-9}$$

The gating function $g_T(t)$ is used to simplify calculations.

Evaluate first the quantity $\overline{Ta_i a_k}$. By direct substitution of Eqs. (4.9-8) and (4.9-9) for a_i and a_k,

$$\overline{Ta_i a_k} = T \cdot E\left[\frac{4}{T^2} \int_{-\infty}^{\infty} dt_2 \int_{-\infty}^{\infty} x(t_1)x(t_2) g_T(t_1) g_T(t_2) \cos \omega_i t_1 \cos \omega_k t_2 \, dt_1 \right] \tag{4.9-10a}$$

$$= \frac{4}{T} \int_{-\infty}^{\infty} dt_2 \int_{-\infty}^{\infty} \overline{x(t_1)x(t_2)} \, g_T(t_1) g_T(t_2) \cos \omega_i t_1 \cos \omega_k t_2 \, dt_1 \tag{4.9-10b}$$

Since $\{x(t)\}$ characterizes a stationary process with an assumed mean value of zero, it follows that

$$\overline{x(t_1)x(t_2)} = \phi_{xx}(t_1, t_2) = \phi_{xx}(t_2 - t_1) \tag{4.9-11}$$

where $\phi_{xx}(t_2 - t_1)$ is the autocorrelation function of random process $\{x(t)\}$. Substituting Eq. (4.9-11) into (4.9-10b),

$$\overline{Ta_i a_k} = \frac{4}{T} \int_{-\infty}^{\infty} dt_2 \int_{-\infty}^{\infty} \phi_{xx}(t_2 - t_1) g_T(t_1) g_T(t_2) \cos \omega_i t_1 \cos \omega_k t_2 \, dt_1 \tag{4.9-12}$$

If $t_2 - t_1$ is replaced by τ and the order of integration is interchanged, Eq. (4.9-12) becomes

$$\overline{Ta_i a_k} = \frac{4}{T} \int_{-\infty}^{\infty} \phi_{xx}(\tau) d\tau \int_{-\infty}^{\infty} g_T(t_2 - \tau) g_T(t_2) \cos \omega_i(t_2 - \tau) \cos \omega_k t_2 \, dt_2 \tag{4.9-13}$$

For $0 \leqslant \tau \leqslant T$, we can eliminate the two gating functions in Eq. (4.9-13) by modifying the limits on the inner integral as follows:

$$\overline{Ta_i a_k} = \frac{4}{T} \int_{-\infty}^{\infty} \phi_{xx}(\tau) \, d\tau \int_{\tau - T/2}^{T/2} \cos \omega_i(t_2 - \tau) \cos \omega_k t_2 \, dt_2 \tag{4.9-14}$$

The inner integral of Eq. (4.9-14) can also be rewritten as

$$\int_{\tau - T/2}^{T/2} \cos \omega_i(t_2 - \tau) \cos \omega_k t_2 \, dt_2$$

$$= \int_{-T/2}^{T/2} \cos \omega_i(t_2 - \tau) \cos \omega_k t_2 \, dt_2 - \int_{-T/2}^{\tau - T/2} \cos \omega_i(t_2 - \tau) \cos \omega_k t_2 \, dt_2 \tag{4.9-15}$$

The first integral on the right side of Eq. (4.9-15) can be evaluated, and yields

$$\int_{-T/2}^{T/2} \cos \omega_i(t_2 - \tau) \cos \omega_k t_2 \, dt_2 = \begin{cases} (T/2) \cos \omega_k \tau, & i = k \\ 0, & i \neq k \end{cases} \tag{4.9-16}$$

A bound on the magnitude of the second integral on the right side of Eq. (4.9-15) can be obtained as follows:

$$\left| \int_{-T/2}^{\tau - T/2} \cos \omega_i(t_2 - \tau) \cos \omega_k t_2 \, dt_2 \right| \leqslant \int_{-T/2}^{\tau - T/2} |\cos \omega_i(t_2 - \tau) \cos \omega_k t_2| \, dt_2 \leqslant \tau \tag{4.9-17}$$

For $\tau < 0$, similar results are obtained. From Eqs. (4.9-14) through (4.9-17), a bound is obtained on $|\overline{Ta_i a_k}|$ for $i \neq k$:

$$|\overline{Ta_i a_k}| \leqslant \frac{4}{T} \int_{-\infty}^{\infty} |\tau| \, \phi_{xx}(\tau) \, d\tau \tag{4.9-18}$$

Provided the integral in (4.9-18) is finite, it follows that for $i \neq k$,

$$|\overline{Ta_i a_k}| = O\left(\frac{1}{T}\right) \tag{4.9-19}$$

To show that the integral in (4.9-18) is finite, consider a band-limited stationary random process with a rectangular power spectral density. By the Wiener-Khintchine theorem

$$\phi_{xx}(\tau) = a \frac{\sin b |\tau|}{|\tau|} \tag{4.9-20}$$

Substituting (4.9-20) into the integral in (4.9-18) results in

$$\left| \int_{-\infty}^{\infty} a \sin b |\tau| \, d\tau \right| \leqslant \frac{4a}{b} \tag{4.9-21}$$

The spectral densities of most physical random processes have tails that fall off more slowly than the band-limited case cited above. From Fourier transform theory it can be shown that for such spectra the autocorrelation function $\phi_{xx}(\tau)$ decays more rapidly for large τ than the autocorrelation function given in Eq. (4.9-20); this demonstrates the finiteness of the integral in (4.9-18).

An argument similar to that which led to (4.9-19) gives, for $i = k$,

$$\overline{Ta_k^2} = 2 \int_{-\infty}^{\infty} \phi_{xx}(\tau) \cos \omega_k \tau \, d\tau + O\left(\frac{1}{T}\right) \tag{4.9-22}$$

Substituting Eqs. (4.9-19) and (4.9-22) into Eq. (4.9-4), we get

$$\frac{\overline{Ta_i a_k}}{(\overline{Ta_i^2})^{1/2} (\overline{Ta_k^2})^{1/2}} = O\left(\frac{1}{T}\right), \qquad i \neq k \tag{4.9-23}$$

which tends to zero as $T \to \infty$. By a similar argument it can be shown that the normalized covariance of b_i and b_k and of a_i and b_k both tend to zero as $T \to \infty$. Since these coefficients are Gaussian and uncorrelated for

$T \rightarrow \infty$, they are also statistically independent. This completes the proof.

With the above results, the variances $\overline{a_k^2}$ and $\overline{b_k^2}$ can be related to the power spectral density $G_{xx}(\omega)$ of random process $\{x(t)\}$. Equation (4.9-22) can be rewritten by means of the Wiener-Khintchine relation:

$$\overline{a_k^2} = \frac{2}{T} G_{xx}(\omega_k) + O\left(\frac{1}{T^2}\right) \qquad (4.9\text{-}24)$$

A similar result can be obtained for $\overline{b_k^2}$. Since

$$\omega_k = \frac{2\pi k}{T} \qquad (4.9\text{-}25)$$

$\Delta\omega$ can be defined by

$$\Delta\omega = \omega_{k+1} - \omega_k \qquad (4.9\text{-}26\text{a})$$

$$= \frac{2\pi}{T} \qquad (4.9\text{-}26\text{b})$$

Then, for large T, Eq. (4.9-24) simplifies to

$$\overline{a_k^2} = \overline{b_k^2} \approx 2 G_{xx}(\omega_k) \frac{\Delta\omega}{2\pi} = \frac{1}{\pi} G_{xx}(\omega_k)\Delta\omega \qquad (4.9\text{-}27)$$

Equation (4.9-27) can be interpreted in the following way. The power associated with frequency ω_k in the Fourier series of an arbitrary function $x(t)$ is given by $\frac{1}{2}(a_k^2 + b_k^2)$. The average power at frequency ω_k, when $x(t)$ is a member of a random process $\{x(t)\}$, is simply $\frac{1}{2}(\overline{a_k^2} + \overline{b_k^2})$, which, by Eq. (4.9-27), is equal to twice the average power of the random process in the interval $\Delta\omega$ about frequency ω_k.

4.10. (Karhunen-Loéve) Orthonormal Expansion of a Gaussian Random Process

In Section 4.9 it was demonstrated that the coefficients of a Fourier series expansion of a T-second waveform sample characterizing a stationary Gaussian random process are random variables that become uncorrelated only as $T \rightarrow \infty$. However, it is possible to describe a finite waveform sample by an orthonormal expansion whose coefficients are uncorrelated Gaussian random variables. For this purpose a waveform representation in terms of orthogonal functions is employed. As stated in Chapter 2, Section 2.7, a sufficient condition for the solutions to integral equation (2.7-5) to form a complete orthogonal set (which can be used to describe any real function over the interval T) is that the kernel of the integral equation be symmetric and related to a power spectral density function by a Fourier transformation.

The autocorrelation function $\phi_{xx}(s, t)$ of a real stationary random process $\{x(t)\}$ satisfies both conditions, since by stationarity, $\phi_{xx}(s, t) = \phi_{xx}(t - s)$

$= \phi_{xx}(\tau)$, and by the Wiener-Khintchine theorem it is the inverse Fourier transform of the power spectral density of random process $\{x(t)\}$. Thus (normalized) solutions $\psi_k(t)$ of the integral equation

$$\int_{-T/2}^{T/2} \phi_{xx}(t - s)\psi_k(s)\,ds = \lambda_k\psi_k(t), \qquad -\frac{T}{2} \leqslant t,\, s \leqslant \frac{T}{2} \qquad (4.10\text{-}1)$$

form a complete orthonormal set over the interval $(-T/2, T/2)$. Hence every member function $x(t)$ in the ensemble $\{x(t)\}$ is expressible as

$$x(t) = \sum_{k=1}^{\infty} x_k\psi_k(t), \qquad -\frac{T}{2} \leqslant t \leqslant \frac{T}{2} \qquad (4.10\text{-}2)$$

where

$$x_k = \int_{-T/2}^{T/2} x(t)\psi_k(t)\,dt \qquad (4.10\text{-}3)$$

It follows from Eq. (4.10-3) that the coefficients x_k are Gaussian random variables, since $x(t)$ is a member function of an ensemble characterizing a Gaussian random process. To show that these random variables are also uncorrelated and hence statistically independent, the covariance $\overline{x_i x_k}$ is evaluated as follows:

$$\overline{x_i x_k} = E\left[\int_{-T/2}^{T/2} x(t_1)\psi_i(t_1)\,dt_1 \int_{-T/2}^{T/2} x(t_2)\psi_k(t_2)\,dt_2 \right] \qquad (4.10\text{-}4\text{a})$$

$$= \int_{-T/2}^{T/2} \int_{-T/2}^{T/2} \overline{x(t_1)x(t_2)}\psi_i(t_1)\psi_k(t_2)\,dt_1\,dt_2 \qquad (4.10\text{-}4\text{b})$$

$$= \int_{-T/2}^{T/2} \psi_k(t_2)\,dt_2 \int_{-T/2}^{T/2} \phi_{xx}(t_2 - t_1)\psi_i(t_1)\,dt_1 \qquad (4.10\text{-}4\text{c})$$

From Eq. (4.10-1) the inner integral in Eq. (4.10-4c) may be replaced by $\lambda_i\psi_i(t_2)$, so that

$$\overline{x_i x_k} = \lambda_i \int_{-T/2}^{T/2} \psi_i(t_2)\psi_k(t_2)\,dt_2 \qquad (4.10\text{-}5)$$

The orthonormality condition

$$\int_{-T/2}^{T/2} \psi_i(t)\psi_k(t) = \begin{cases} 1, & i = k \\ 0, & i \neq k \end{cases} \qquad (4.10\text{-}6)$$

results in (4.10-5) becoming

$$\overline{x_i x_k} = \begin{cases} \lambda_k, & i = k \\ 0, & i \neq k \end{cases} \qquad (4.10\text{-}7)$$

Equation (4.10-7) demonstrates that the coefficients x_k are uncorrelated and have variance equal to λ_k. As noted in Chapter 2, the $\psi_k(t)$ are conventionally known as eigenfunctions and the λ_k as eigenvalues of integral equation (4.10-1).

The usefulness of orthogonal expansion (4.10-2) is limited by the fact that solutions to integral equation (4.10-1), when T is finite, are not known except in special cases. In addition, even when a decomposition is possible, it is difficult to perform calculations in specific applications, since ortho-normal expansion (4.10-2) generally contains an infinite number of terms. However, this representation is very useful in formulating and solving general theoretical problems.

4.11. The Narrowband Gaussian Random Process

This concluding section presents some important results concerning narrow-band Gaussian random processes. This topic is important because a narrowband Gaussian process describes the noise in the radio-frequency and intermediate-frequency stages of a radar receiver. Results are obtained concerning the first- and second-order statistics of narrowband Gaussian noise. These results are extended in later chapters to include a signal in additive Gaussian noise.

In Chapter 2, Section 2.5, several waveform representations were de-veloped for a deterministic narrowband waveform. These representations can also describe a sample function in ensemble $\{z(t)\}$ of a narrowband Gaussian random process. Thus, narrowband sample waveform $z(t)$ can be represented in terms of an envelope function $V(t)$ and phase function $\phi(t)$ by

$$z(t) = V(t) \cos\left[\omega_0 t + \phi(t)\right] \tag{4.11-1}$$

By definition ω_0 is the center frequency of the positive portion of the narrow-band spectral density function (see Fig. 4.11-1). Sample waveform $z(t)$ can also be expressed in terms of in-phase and quadrature sample waveforms, $z_I(t)$ and $z_Q(t)$, respectively, by

$$z(t) = z_I(t) \cos \omega_0 t - z_Q(t) \sin \omega_0 t \tag{4.11-2}$$

From Eqs. (4.11-1) and (4.11-2) it is easy to show that

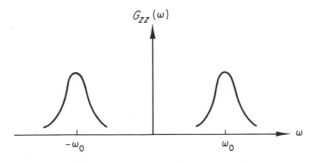

Fig. 4.11-1. Narrowband noise process spectral density.

$$z_I(t) = V(t) \cos \phi(t) \qquad \text{(4.11-3a)}$$
$$z_Q(t) = V(t) \sin \phi(t) \qquad \text{(4.11-3b)}$$

In this section some of the statistical relations among random variables $z_I(t)$, $z_Q(t)$, $V(t)$, and $\phi(t)$ will be given. For this purpose consider the Fourier series expansion of a narrowband waveform defined over an interval T (see Chapter 2):

$$z(t) = \sum_{k=1}^{\infty} \alpha_k \cos \omega_k t + \beta_k \sin \omega_k t \qquad \text{(4.11-4)}$$

where $\omega_k = 2\pi k/T$ and α_k and β_k are given by

$$\alpha_k = \frac{2}{T} \int_0^T z(t) \cos \omega_k t \, dt \qquad \text{(4.11-5a)}$$

$$\beta_k = \frac{2}{T} \int_0^T z(t) \sin \omega_k t \, dt \qquad \text{(4.11-5b)}$$

If $\{z(t)\}$ characterizes a Gaussian random process with zero mean, it follows from Eqs. (4.11-5) that α_k and β_k are Gaussian random variables with zero mean. Further, in Section 4.9 it was shown that for $T \to \infty$, α_i and β_k are statistically independent for all i and k, and also that α_i, α_k and β_i, β_k are statistically independent for all $i \neq k$. In addition, it was shown that the variance of α_k is equal to the variance of β_k. These results are useful in the following paragraphs.

The sample waveform quadrature components $z_I(t)$ and $z_Q(t)$ can be described in terms of the Fourier coefficients defined in (4.11-4) (see Chapter 2, Section 2.5) by

$$z_I(t) = \sum_{k=1}^{\infty} \alpha_k \cos(\omega_k - \omega_0)t + \beta_k \sin(\omega_k - \omega_0)t, \qquad 0 \leqslant t \leqslant T \quad \text{(4.11-6)}$$

$$z_Q(t) = \sum_{k=1}^{\infty} \alpha_k \sin(\omega_k - \omega_0)t - \beta_k \cos(\omega_k - \omega_0)t, \qquad 0 \leqslant t \leqslant T \quad \text{(4.11-7)}$$

It follows from Eqs. (4.11-6) and (4.11-7) that, for any $t = t_1$, $z_I(t)$ and $z_Q(t)$ are sums of Gaussian random variables and therefore are also Gaussian random variables. Their mean values are zero, since by Eqs. (4.11-6) and (4.11-7)

$$\overline{z_I(t_1)} = \sum_{k=1}^{\infty} \overline{\alpha_k} \cos(\omega_k - \omega_0)t_1 + \overline{\beta_k} \sin(\omega_k - \omega_0)t_1 = 0 \quad \text{(4.11-8a)}$$

$$\overline{z_Q(t_1)} = \sum_{k=1}^{\infty} \overline{\alpha_k} \sin(\omega_k - \omega_0)t_1 - \overline{\beta_k} \cos(\omega_k - \omega_0)t_1 = 0 \quad \text{(4.11-8b)}$$

With $(\omega_k - \omega_0) = \omega_k'$, the variances of $z_I(t_1)$ and $z_Q(t_1)$ can be computed as follows:

$$\overline{z_I^2(t_1)} = E\left\{ \sum_{i=1}^{\infty} \sum_{k=1}^{\infty} (\alpha_i \cos \omega_i' t_1 + \beta_i \sin \omega_i' t_1)(\alpha_k \cos \omega_k' t_1 + \beta_k \sin \omega_k' t_1) \right\}$$

$$\text{(4.11-9a)}$$

$$= \sum_{i=1}^{\infty} \sum_{k=1}^{\infty} \overline{\alpha_i \alpha_k} \cos \omega_i' t_1 \cos \omega_k' t_1 + \overline{\beta_i \beta_k} \sin \omega_i' t_1 \sin \omega_k' t_1$$

$$+ \overline{\alpha_i \beta_k} \cos \omega_i' t_1 \sin \omega_k' t_1 + \overline{\alpha_k \beta_i} \cos \omega_k' t_1 \sin \omega_i' t_1 \qquad (4.11\text{-}9b)$$

For the limiting case $T \to \infty$, Eq. (4.11-9b) becomes

$$\lim_{T \to \infty} \overline{z_I^2(t_1)} = \sum_{k=1}^{\infty} \overline{\alpha_k^2} \cos^2 \omega_k' t_1 + \overline{\beta_k^2} \sin^2 \omega_k' t_1 \qquad (4.11\text{-}10a)$$

$$= \sum_{k=1}^{\infty} \overline{\alpha_k^2} (\cos^2 \omega_k' t_1 + \sin^2 \omega_k' t_1)$$

$$= \sum_{k=1}^{\infty} \overline{\alpha_k^2} \qquad (4.11\text{-}10b)$$

A similar result is obtained for $\overline{z_Q^2(t_1)}$:

$$\lim_{T \to \infty} \overline{z_Q^2(t_1)} = \sum_{k=1}^{\infty} \overline{\alpha_k^2} \qquad (4.11\text{-}11)$$

Inspection of Eqs. (4.11-4), (4.11-6), and (4.11-7) shows that the Fourier expansion of sample waveform $z(t)$ is identical to that of $z_I(t)$ and $z_Q(t)$ except that the argument of the cosine function differs by the carrier frequency ω_0. Since this difference does not affect the development that led to Eqs. (4.11-10b) and (4.11-11), it follows that for any $t = t_1$

$$\lim_{T \to \infty} \overline{z^2(t_1)} = \sum_{k=1}^{\infty} \overline{\alpha_k^2} \qquad (4.11\text{-}12)$$

From (4.11-10b) through (4.11-12) (and for all t),

$$\lim_{T \to \infty} \overline{z^2(t)} = \lim_{T \to \infty} \overline{z_I^2(t)} = \lim_{T \to \infty} \overline{z_Q^2(t)} = \sigma^2 \qquad (4.11\text{-}13)$$

Hence, random variables $z(t)$, $z_I(t)$, and $z_Q(t)$ all have mean zero and equal variances. Further, since Eqs. (4.9-1) and (4.11-4) are similar, it follows from Eq. (4.9-27) that the variance σ^2 can be related to the power spectral density $G_{zz}(\omega)$ of the random process $z(t)$ by

$$\sigma^2 = \lim_{T \to \infty} 2 \sum_{k=1}^{\infty} G_{zz}(\omega_k) \frac{\Delta \omega}{2\pi} \qquad (4.11\text{-}14)$$

Note that $(\omega_k - \omega_{k-1}) = \Delta \omega = 2\pi/T$; for $T \to \infty$, let

$$\omega_k \longrightarrow \omega$$
$$\Delta \omega \longrightarrow d\omega \qquad (4.11\text{-}15)$$

Thus, as T becomes very large, Eq. (4.11-14) may be rewritten as

$$\sigma^2 = \frac{1}{\pi} \int_0^{\infty} G_{zz}(\omega) \, d\omega \qquad (4.11\text{-}16)$$

The covariance of quadrature components $z_I(t_1)$ and $z_Q(t_1)$, at the same instant of time t_1, are found in similar fashion:

$$\overline{z_I(t_1) z_Q(t_1)} = E \left\{ \sum_{i=1}^{\infty} \sum_{k=1}^{\infty} (\alpha_i \cos \omega_i' t_1 + \beta_i \sin \omega_i' t_1)(\alpha_k \sin \omega_k' t_1 - \beta_k \cos \omega_k' t_1) \right\}$$

$$= \sum_{i=1}^{\infty} \sum_{k=1}^{\infty} \overline{\alpha_i \alpha_k} \cos \omega_i' t_1 \sin \omega_k' t_1 - \overline{\beta_i \beta_k} \sin \omega_i' t_1 \cos \omega_k' t_1$$

$$- \overline{\alpha_i \beta_k} \cos \omega_i' t_1 \cos \omega_k' t_1 + \overline{\alpha_k \beta_i} \sin \omega_k' t_1 \sin \omega_i' t_1 \qquad (4.11\text{-}17)$$

where $(\omega_k - \omega_0)$ is replaced by ω_k' as before. For $T \to \infty$, Eq. (4.11-17) simplifies to

$$\overline{z_I(t_1)z_Q(t_1)} = \sum_{k=1}^{\infty} (\overline{\alpha_k^2} - \overline{\beta_k^2}) \cos \omega_k't_1 \sin \omega_k't_1 = 0 \qquad (4.11\text{-}18)$$

Since random variables $z_I(t_1)$ and $z_Q(t_1)$ are Gaussian and uncorrelated for $T \to \infty$, they are also statistically independent. Thus, sample values of the in-phase and quadrature waveforms at the same instant of time are independent.

The statistical properties of $V(t)$ and $\phi(t)$ can be deduced from the results just obtained. For $T \to \infty$, the joint probability density function of Gaussian random variables $z_I(t_1) = z_{I\text{-}t_1}$ and $z_Q(t_1) = z_{Q\text{-}t_1}$, each with mean zero and variance σ^2, is given by

$$p(z_{I\text{-}t_1}, z_{Q\text{-}t_1}) = p(z_{I\text{-}t_1}) \cdot p(z_{Q\text{-}t_1}) = \frac{1}{2\pi\sigma^2} \exp\left(-\frac{z_{I\text{-}t_1}^2 + z_{Q\text{-}t_1}^2}{2\sigma^2}\right) \qquad (4.11\text{-}19)$$

We express the new variables V_{t_1} and ϕ_{t_1} in terms of variables $z_{I\text{-}t_1}$ and $z_{Q\text{-}t_1}$ by inverting Eqs. (4.11-3):

$$V_{t_1} = (z_{I\text{-}t_1}^2 + z_{Q\text{-}t_1}^2)^{1/2} \qquad (4.11\text{-}20\text{a})$$

$$\phi_{t_1} = \tan^{-1} \frac{z_{Q\text{-}t_1}}{z_{I\text{-}t_1}} \qquad (4.11\text{-}20\text{b})$$

From the definition of the Jacobian,

$$J = \begin{vmatrix} \dfrac{\partial V_{t_1}}{\partial z_{I\text{-}t_1}} & \dfrac{\partial V_{t_1}}{\partial z_{Q\text{-}t_1}} \\[2mm] \dfrac{\partial \phi_{t_1}}{\partial z_{I\text{-}t_1}} & \dfrac{\partial \phi_{t_1}}{\partial z_{Q\text{-}t_1}} \end{vmatrix} = \frac{1}{(z_{I\text{-}t_1}^2 + z_{Q\text{-}t_1}^2)^{1/2}} = \frac{1}{V_{t_1}} \qquad (4.11\text{-}21)$$

Substituting (4.11-21) into

$$p(V_{t_1}, \phi_{t_1}) = \left[\frac{p(z_{I\text{-}t_1}, z_{Q\text{-}t_1})}{|J|}\right]_{\substack{z_{I\text{-}t_1} = V_{t_1} \cos \phi_{t_1} \\ z_{Q\text{-}t_1} = V_{t_1} \sin \phi_{t_1}}} \qquad (4.11\text{-}22\text{a})$$

gives (for $T \to \infty$),

$$p(V_{t_1}, \phi_{t_1}) = \begin{cases} \dfrac{V_{t_1}}{2\pi\sigma^2} \exp\left(\dfrac{V_{t_1}^2}{2\sigma^2}\right), & \text{for } \begin{cases} V_{t_1} \geqslant 0 \\ 0 \leqslant \phi_{t_1} \leqslant 2\pi \end{cases} \\ 0, & \text{otherwise} \end{cases} \qquad (4.11\text{-}22\text{b})$$

The marginal probability density functions $p(V_{t_1})$ may be obtained from Eq. (4.11-22b) by integrating the joint probability density function with respect to ϕ_{t_1} over its range of values. Thus, for $T \to \infty$,

$$p(V_{t_1}) = \int_0^{2\pi} p(V_{t_1}, \phi_{t_1}) \, d\phi_{t_1} = \begin{cases} \dfrac{V_{t_1}}{\sigma^2} \exp\left(\dfrac{V_{t_1}^2}{2\sigma^2}\right), & V_{t_1} \geqslant 0 \\ 0, & \text{otherwise} \end{cases} \qquad (4.11\text{-}23)$$

In a similar manner, the marginal probability density $p(\phi_{t_1})$ is found to be

$$p(\phi_{t_1}) = \int_0^\infty p(V_{t_1}, \phi_{t_1})\, dV_{t_1} = \begin{cases} \dfrac{1}{2\pi}, & 0 \leqslant \phi_{t_1} \leqslant 2\pi \\ 0, & \text{otherwise} \end{cases} \qquad (4.11\text{-}24)$$

The function $p(V_{t_1})$ is a Rayleigh probability density function (see Fig. 4.11-2) while $p(\phi_{t_1})$ is a uniform probability density function (see Fig. 4.11-3). Further, it is easy to show that

$$p(V_{t_1}, \phi_{t_1}) = p(V_{t_1}) \cdot p(\phi_{t_1}) \qquad (4.11\text{-}25)$$

so that, for $T \to \infty$, the envelope and phase of a Gaussian narrowband process are statistically independent at the same instant of time.

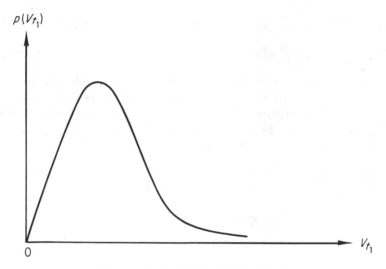

Fig. 4.11-2. Rayleigh probability density.

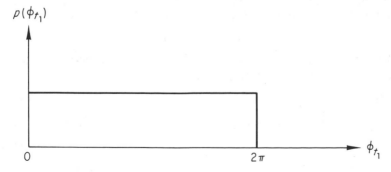

Fig. 4.11-3. Uniform probability density.

The previous results can be extended to obtain the second-order statistics of quadrature random processes $\{z_I(t)\}$ and $\{z_Q(t)\}$. From the definition of the autocorrelation function,

$$\phi_{z_I z_I}(t_1, t_2) = \overline{z_I(t_1) z_I(t_2)} \qquad (4.11\text{-}26a)$$

$$= E\left[\sum_{i=1}^{\infty} \sum_{k=1}^{\infty} (\alpha_i \cos \omega_i' t_1 + \beta_i \sin \omega_i' t_1)(\alpha_k \cos \omega_k' t_2 + \beta_k \sin \omega_k' t_2)\right]$$
$$(4.11\text{-}26b)$$

$$= \sum_{i=1}^{\infty} \sum_{k=1}^{\infty} \overline{\alpha_i \alpha_k} \cos \omega_i' t_1 \cos \omega_k' t_2 + \overline{\beta_i \beta_k} \sin \omega_i' t_1 \sin \omega_k' t_2$$
$$+ \overline{\alpha_i \beta_k} \cos \omega_i' t_1 \sin \omega_k' t_2 + \overline{\alpha_k \beta_i} \cos \omega_k' t_2 \sin \omega_i' t_1 \qquad (4.11\text{-}26c)$$

where $(\omega_k - \omega_0)$ is again replaced by ω_k'. For $T \to \infty$, Eq. (4.11-26c) simplifies to

$$\lim_{T \to \infty} \phi_{z_I z_I}(t_1, t_2) = \lim_{T \to \infty} \sum_{k=1}^{\infty} \overline{\alpha_k^2}(\cos \omega_k' t_1 \cos \omega_k' t_2 + \sin \omega_k' t_1 \sin \omega_k' t_2) \qquad (4.11\text{-}27a)$$

$$= \lim_{T \to \infty} \sum_{k=1}^{\infty} \overline{\alpha_k^2} \cos \omega_k'(t_2 - t_1) \qquad (4.11\text{-}27b)$$

Note that the right side of Eq. (4.11-27b) is a function of the difference $(t_2 - t_1)$, from which it follows that $\{z_I(t)\}$ is a stationary random process. Let $t_2 - t_1 = \tau$ and replace ω_k' by $(\omega_k - \omega_0)$ and $\overline{\alpha_k^2}$ by $(1/\pi)G_{zz}(\omega_k)\,\Delta\omega$; then Eq. (4.11-27b) can be rewritten as [see development of Eqs. (4.11-12) through (4.11-16)]

$$\lim_{T \to \infty} \phi_{z_I z_I}(t_1, t_2) = \phi_{z_I z_I}(\tau) = \lim_{T \to \infty} \frac{1}{\pi} \sum_{k=1}^{\infty} G_{zz}(\omega_k) \cos[(\omega_k - \omega_0)\tau]\,\Delta\omega$$
$$(4.11\text{-}28a)$$

$$= \frac{1}{\pi} \int_0^{\infty} G_{zz}(\omega) \cos[(\omega - \omega_0)\tau]\,d\omega \qquad (4.11\text{-}28b)$$

If $\tau = 0$,

$$\phi_{z_I z_I}(0) = \sigma^2 = \frac{1}{\pi} \int_0^{\infty} G_{zz}(\omega)\,d\omega \qquad (4.11\text{-}29)$$

which confirms the previous result for the variance of $z_I(t)$ given in Eq. (4.11-16). An identical result is obtained for random process $\{z_Q(t)\}$, so that

$$\phi_{z_I z_I}(\tau) = \phi_{z_Q z_Q}(\tau) \qquad (4.11\text{-}30)$$

The cross correlation $\phi_{z_I z_Q}(t_1, t_2)$ of quadrature processes $\{z_I(t)\}$ and $\{z_Q(t)\}$ is found in the same way. Thus,

$$\phi_{z_I z_Q}(t_1, t_2) = \overline{z_I(t_1) z_Q(t_2)} \qquad (4.11\text{-}31a)$$

$$= E\left[\sum_{i=1}^{\infty} \sum_{k=1}^{\infty} (\alpha_i \cos \omega_i' t_1 + \beta_i \sin \omega_i' t_1)(\alpha_k \sin \omega_k' t_2 - \beta_k \cos \omega_k' t_2)\right]$$
$$(4.11\text{-}31b)$$

$$= \sum_{i=1}^{\infty} \sum_{k=1}^{\infty} \overline{\alpha_i \alpha_k} \cos \omega_i' t_1 \sin \omega_k' t_2 - \overline{\beta_i \beta_k} \sin \omega_i' t_1 \cos \omega_k' t_2$$
$$- \overline{\alpha_i \beta_k} \cos \omega_i' t_1 \cos \omega_k' t_2 + \overline{\alpha_k \beta_i} \sin \omega_i' t_1 \sin \omega_k' t_2 \qquad (4.11\text{-}31c)$$

For $T \to \infty$, Eq. (4.11-31c) simplifies to

$$\lim_{T \to \infty} \phi_{z_I z_Q}(t_1, t_2) = \lim_{T \to \infty} \sum_{k=1}^{\infty} \overline{\alpha_k^2} (\cos \omega_k' t_1 \sin \omega_k' t_2 - \sin \omega_k' t_1 \cos \omega_k' t_2) \qquad (4.11\text{-}32\text{a})$$

$$= \lim_{T \to \infty} \sum_{k=1}^{\infty} \overline{\alpha_k^2} \sin \omega_k'(t_2 - t_1) \qquad (4.11\text{-}32\text{b})$$

Equation (4.11-32b) indicates that random processes $\{z_I(t)\}$ and $\{z_Q(t)\}$ are jointly stationary, as well as separately stationary. Similar to the development of Eq. (4.11-28b), it can be shown that

$$\lim_{T \to \infty} \phi_{z_I z_Q}(t_1, t_2) = \phi_{z_I z_Q}(\tau) \qquad (4.11\text{-}33\text{a})$$

$$= \frac{1}{\pi} \int_0^{\infty} G_{zz}(\omega) \sin [(\omega - \omega_0)\tau] \, d\omega \qquad (4.11\text{-}33\text{b})$$

Note that if spectral density $G_{zz}(\omega)$ is narrowband and symmetrical about ω_0, then the integrand in Eq. (4.11-33b) is an odd function about ω_0 and $\phi_{z_I z_Q}(\tau)$ is essentially zero for all τ. Only for this case are random processes $\{z_I(t)\}$ and $\{z_Q(t)\}$ statistically independent. In general, if $G_{zz}(\omega)$ is asymmetrical, random variables $z_I(t_1)$ and $z_Q(t_2)$ are statistically dependent except, as shown previously, at the same instant of time $t_1 = t_2$ (for the limiting case $T \to \infty$).

Next, the joint statistics of random variables $V(t_1)$, $V(t_2)$, $\phi(t_1)$, and $\phi(t_2)$ are obtained by first forming the joint probability of random variables $z_I(t_1)$, $z_Q(t_1)$, $z_I(t_2)$, $z_Q(t_2)$ and then performing a transformation of variables. The steps are briefly outlined here. The covariance matrix Λ of random variables $z_I(t_1)$, $z_Q(t_1)$, $z_I(t_2)$, $z_Q(t_2)$ for $T \to \infty$ can be immediately written using previous results and the relations $\phi_{z_I z_I}(\tau) = \phi_{z_Q z_Q}(\tau)$ and $\phi_{z_I z_Q}(\tau) = -\phi_{z_Q z_I}(\tau)$ (see problem 4.13). Thus,

$$\Lambda = \begin{bmatrix} \sigma^2 & 0 & \phi_{z_I z_I}(\tau) & \phi_{z_I z_Q}(\tau) \\ 0 & \sigma^2 & -\phi_{z_I z_Q}(\tau) & \phi_{z_I z_I}(\tau) \\ \phi_{z_I z_I}(\tau) & -\phi_{z_I z_Q}(\tau) & \sigma^2 & 0 \\ \phi_{z_I z_Q}(\tau) & \phi_{z_I z_I}(\tau) & 0 & \sigma^2 \end{bmatrix} \qquad (4.11\text{-}34)$$

where $\tau = t_2 - t_1$. By means of (4.11-34) the determinant of the covariance matrix and its inverse matrix can be computed and substituted into Eq. (4.8-1) to obtain the joint probability density function,

$$p(z_{I\text{-}t_1}, z_{Q\text{-}t_1}, z_{I\text{-}t_2}, z_{Q\text{-}t_2})$$

$$= \frac{1}{4\pi \, |\Lambda|^{1/2}} \exp \left\{ -\frac{1}{2 \, |\Lambda|^{1/2}} \left[\sigma^2 (z_{I\text{-}t_1}^2 + z_{Q\text{-}t_1}^2 + z_{I\text{-}t_2}^2 + z_{Q\text{-}t_2}^2) \right. \right.$$

$$\left. \left. - 2\phi_{z_I z_I}(\tau)(z_{I\text{-}t_1} z_{I\text{-}t_2} + z_{Q\text{-}t_1} z_{Q\text{-}t_2}) - 2\phi_{z_I z_Q}(\tau)(z_{I\text{-}t_1} z_{Q\text{-}t_2} - z_{Q\text{-}t_1} z_{I\text{-}t_2}) \right] \right\}$$

$$(4.11\text{-}35)$$

where

$$|\Lambda|^{1/2} = \sigma^4 - \phi_{z_I z_I}^2(\tau) - \phi_{z_I z_Q}^2(\tau) \qquad (4.11\text{-}36)$$

After a transformation of variables, Eq. (4.11-35) becomes

$$
\begin{aligned}
p(V_{t_1}, V_{t_2}, \phi_{t_1}, \phi_{t_2}) = {} & \frac{V_{t_1} V_{t_2}}{4\pi^2 |\Lambda|^{1/2}} \exp\left\{ -\frac{1}{2|\Lambda|^{1/2}} \left[\sigma^2 (V_{t_1}^2 + V_{t_2}^2) \right.\right. \\
& - 2\phi_{z_I z_I}(\tau) V_{t_1} V_{t_2} \cos(\phi_{t_2} - \phi_{t_1}) \\
& \left.\left. - 2\phi_{z_I z_Q}(\tau) V_{t_1} V_{t_2} \sin(\phi_{t_2} - \phi_{t_1}) \right] \right\} \\
& \text{for} \quad \begin{cases} V_{t_1}, V_{t_2} \geqslant 0 \\ 0 \leqslant \phi_{t_1}, \phi_{t_2} \leqslant 2\pi \end{cases}
\end{aligned} \qquad (4.11\text{-}37)
$$

and zero otherwise. Equation (4.11-37) in general is not separable into a product of $p(V_{t_1}, V_{t_2})$ and $p(\phi_{t_1}, \phi_{t_2})$, so that the envelope and phase random processes are not statistically independent.

This concludes the discussion of the narrowband Gaussian random process. More extensive results may be found in a paper by Rice and in other sources in the bibliography that follows.

PROBLEMS

4.1. Suppose a random process has ensemble members given by the set of functions

$$x(t) = A \cos(\omega_0 t + \phi)$$

where A and ω_0 are constants and ϕ is a random variable with probability density function $p(\phi)$,

$$p(\phi) = \frac{1}{2\pi}, \qquad 0 \leqslant \phi \leqslant 2\pi$$

Calculate the (ensemble) autocorrelation function of the random process.

4.2. Repeat problem 4.1 assuming that A and ϕ are independent random variables with ϕ distributed as before and A a Gaussian random variable with mean zero and variance σ^2.

4.3. Suppose the random process $\{x(t)\}$ has autocorrelation function $\phi_{xx}(t_1, t_2)$. Calculate the autocorrelation function of the derivative of the process in terms of derivatives of $\phi_{xx}(t_1, t_2)$. Also calculate the cross correlation of $\{x(t)\}$ and its derivative in terms of derivatives of $\phi_{xx}(t_1, t_2)$. (See Table 4.6-1.)

4.4. If random process $\{z(t)\}$ is the sum of two statistically independent random processes $\{x(t)\}$ and $\{y(t)\}$ with zero means, show that the variance of $z(t)$ is the sum of the variances of $x(t)$ and $y(t)$ for any t.

4.5. Let $\{x(t)\}$ be a wide-sense stationary random process with autocorrelation function $\phi_{xx}(\tau)$. Show that

$$|\phi_{xx}(\tau)| \leqslant \phi_{xx}(0)$$

using the fact that

$$[x(t) \pm x(t + \tau)]^2 \geqslant 0$$

4.6. Show that the Fourier transform of an autocorrelation function is non-negative starting with the fact that for any $T > 0$

$$\left| \int_0^T x(t)e^{-j\omega t} \, dt \right|^2 \geqslant 0$$

4.7. Suppose a random process $\{y(t)\}$ has sample functions

$$y(t) = x(t) \cos (\omega_0 t + \phi)$$

where $x(t)$ is a sample function of a wide-sense stationary random process, ϕ is a random variable independent of $x(t)$ which is uniformly distributed over the interval $0 \leqslant \phi \leqslant 2\pi$, and ω_0 is a constant. Find the autocorrelation function of random process $\{y(t)\}$ in terms of the autocorrelation function of $\{x(t)\}$. Is it stationary? Find the spectral density of $\{y(t)\}$ in terms of the spectral density of $\{x(t)\}$.

4.8. Calculate the spectral density of the following autocorrelation function

$$\phi_{xx}(\tau) = \phi_{xx}(0)e^{-a|\tau|} \cos b\tau$$

4.9. If stationary white noise of power spectral density $N_0/2$ is passed through an ideal low-pass filter whose bandwidth is $(-\omega_c, \omega_c)$, calculate the autocorrelation function of the noise at the filter output.

4.10. Let $\{y(t)\}$ be a random process whose sample functions are related to the sample functions of a real stationary Gaussian process $\{x(t)\}$ by

$$y(t) = x^2(t)$$

Find the autocorrelation function of $\{y(t)\}$ in terms of the autocorrelation function $\phi_{xx}(\tau)$ of $\{x(t)\}$.

4.11. Suppose that $\{x(t)\}$ is a stationary Gaussian random process with autocorrelation function

$$\phi_{xx}(\tau) = e^{-a|\tau|}$$

Let the random variables X_1, X_2, and X_3 represent samples of the process $\{x(t)\}$ at times t_1, t_2, and t_3, respectively. Show that

$$p(x_3 \,|\, x_2, x_1) = p(x_3 \,|\, x_2)$$

That is, the probability density function of X_3 is conditional only on the previous sample X_2 and is independent of X_1. Find spectral density $G_{xx}(\omega)$ and plot.

4.12. Let $\{z(t)\}$ describe a narrowband real Gaussian random process with sample functions

$$z(t) = x(t) \cos \omega_0 t - y(t) \sin \omega_0 t$$

where $\{x(t)\}$ and $\{y(t)\}$ are separately and jointly stationary random processes. Show that $\phi_{zz}(\tau)$, the autocorrelation function of $\{z(t)\}$, is given by

$$\phi_{zz}(\tau) = \phi_{xx}(\tau) \cos \omega_0 \tau - \phi_{xy}(\tau) \sin \omega_0 \tau$$

where $\phi_{xx}(\tau)$ is the autocorrelation of $\{x(t)\}$ and $\phi_{xy}(\tau)$ is the cross-correlation of random processes $\{x(t)\}$ and $\{y(t)\}$.

4.13. If $\{x(t)\}$ and $\{y(t)\}$ are separately and jointly stationary, show that

$$\phi_{xy}(\tau) = -\phi_{yx}(\tau)$$

4.14. (a) Verify the covariance matrix in Eq. (4.11-34). Find the determinant and inverse of this matrix

(b) Verify Eqs. (4.11-35) and (4.11-37).

(c) Find $p(V_{t_1}, V_{t_2})$ from (4.11-37) (by integrating with respect to ϕ_{t_1} and ϕ_{t_2}). Are V_{t_1} and V_{t_2} statistically independent?

4.15. Let $n(t)$ be a sample function of a stationary Gaussian random process and let

$$v(t) = A \cos(\omega_0 t + \phi) + n(t)$$

where A and ω_0 are known constants and ϕ is an independent random variable that is uniformly distributed over the interval $(0, 2\pi)$. Using Eq. (4.11-2) to describe $n(t)$, express $v(t)$ in terms of its in-phase and quadrature components $v_I(t)$ and $v_Q(t)$.

(a) Find the joint density $p(v_{I\text{-}t_1}, v_{Q\text{-}t_1} | \phi)$ at time t_1.

(b) Find $p(v_{I\text{-}t_1}, v_{Q\text{-}t_1}, \phi)$ at time t_1.

(c) Express $v(t)$ in terms of an envelope and phase function and find the joint density of the envelope, phase, and ϕ at time t_1, assuming a narrow-band process. (See page 166 in ref. [2].)

RECOMMENDED BIBLIOGRAPHY

1. W. R. Bennett, Methods of Solving Noise Problems, *Proc. IRE*, **44**: 609–638 (May, 1956).

2. W. B. Davenport and W. L. Root, "An Introduction to the Theory of Random Signals and Noise," McGraw-Hill, New York, 1958.

3. J. H. Laning and R. H. Battin, "Random Processes in Automatic Control," McGraw-Hill, New York, 1956.

4. S. O. Rice, Mathematical Analysis of Random Noise, *Bell System Tech. J.*, **23**: 282–332 (1944); **24**: 46–156 (1945).

5. J. S. Bendat, "Principles and Applications of Random Noise Theory," Wiley, New York, 1958.

6. A. Papoulis, "Probability, Random Variables and Stochastic Processes," McGraw-Hill, New York, 1965.

7. D. Middleton, "Statistical Communication Theory," McGraw-Hill, New York, 1960.

PART III

OPTIMUM RECEPTION OF SIGNALS IN NOISE

In Part III we consider the principal approaches for optimizing the reception of signals immersed in noise that have evolved over the last few decades. They are of more than just historical interest, for each provides a different and complementary insight into the problem of optimum reception.

In Chapters 5 and 6 the criterion for optimum reception is assumed to be the maximization of output signal-to-noise ratio. This was one of the earliest criteria investigated. In this approach noise is considered to be an undesirable interfering signal, In the early days of the receiver art, band-pass filtering techniques were developed to effect discrimination between a desired signal and an interfering signal with adjacent but nonoverlapping spectra. A filter that maximizes signal-to-noise ratio may be regarded as an extension of this concept in order to discriminate between signals with overlapping but dissimilar spectra. Matched filter theory is one of the important results of this criterion [1].

During the Second World War another optimization criterion for radar detection was investigated at the MIT Radiation Laboratory. Researchers there, recognizing that receiver noise can be characterized mathematically as a random process, developed the theory of the "ideal observer." Since the random fluctuations of noise mask the presence of a signal in an unpredictable way, it is necessary for the radar observer to guess the presence or absence of a signal. Only when the signal is much greater than the noise does the probability of detecting a target echo approach certainty. The observer may commit two types of errors: he may err in deciding that a noise waveform contains a signal, or he may err in not identifying a signal in noise. The researchers defined an "ideal observer" as one who minimizes the probability of both types of errors. These early ideas are described by J. L. Lawson and G. E. Uhlenbeck [2] and are discussed in Chapter 8 within the broader framework of statistical decision theory.

Information theory, introduced by Hartley [3] in 1928, was advanced significantly after the war by C. E. Shannon in a series of classic papers, the first of which appeared in 1948 [4]. Publication of these papers stimulated many scientists in the fields of communications and radar to evaluate receiver design philosophy with respect to new criteria based on Shannon's work. In particular P. M. Woodward and I. L. Davies [5] poineered the application of information theory to radar. The basic notions of information measure, inverse probability, and the concept of the *ideal receiver* as proposed by Woodward and Davies are discussed in Chapter 7.

Because of its power and generality, statistical decision theory has been applied by researchers in communications and radar to a wide variety of problems. Chapter 8 introduces the application of statistical decision theory to radar reception. Decision theory provides a unifying theoretical framework for the study of both optimum detection and parameter estimation of radar signals. The formulation and treatment of the radar problem in this text is patterned after the exposition of the general communication problem by Middleton and Van Meter [6, 7], which is based on Wald's original work in decision theory [8,9].

REFERENCES

1. North, D. O.: Analysis of Factors which Determine Signal-Noise Discrimination in Pulsed-Carrier Systems, *RCA Tech. Report*, PTR-6C (June, 1943); reprinted in *Proc. IEEE*, **51**: (7), 1015–1027 (July, 1963).
2. Lawson, J. L., and Uhlenbeck, G. E.: "Threshold Signals," MIT Radiation Laboratory Series, Vol. 24, McGraw-Hill, New York, 1950.
3. Hartley, R. V. L.: The Transmission of Information, *Bell System Tech. J.*, **7**: 535–563 (1928).
4. Shannon, C. E.: A Mathematical Theory of Communication, *Bell System Tech. J.*, **27**: 379–423, 623–656 (1948).

5. Woodward, P. M., and Davies, I. L.: Information Theory and Inverse Probability in Telecommunications, *J. IEE (London)*, **99** (*Pt. III*): 37–44 (1952).

6. Van Meter, D., and Middleton, D.: Modern Statistical Approaches to Reception in Communication Theory, *Trans. IRE Prof. Group on Information Theory*, **PGIT-4:** 119–145 (September, 1954).

7. Middleton, D., and Van Meter, D.: Detection and Extraction of Signals in Noise from the Point of View of Statistical Decision Theory, *J. Soc. Ind. Appl. Math.*, **3:** 192 (1955); **4:** 86 (1956).

8. Wald, A.: Contributions to the Theory of Statistical Estimation and Testing of Hypotheses, *Ann. Math. Stat.*, **10:** 299 (1939).

9. Wald, A.: "Statistical Decision Functions," Wiley, New York, 1950.

5

MAXIMIZATION OF
SIGNAL-TO-NOISE RATIO
AND THE MATCHED FILTER

5.1. Introduction

In the infancy of radar, signal-to-noise ratio was a popular measure of the effectiveness of a radar receiving system for combating noise. Increased signal power, with respect to average noise power, made it easier to distinguish a signal from background noise. Further, the parameters of a target echo, such as amplitude, (range) delay, and doppler frequency shift, could be estimated more accurately with increased signal-to-noise ratio. It was apparent that both parameter estimation and detection were dependent on signal-to-noise ratio—at least in a qualitative sense.

Signal-to-noise ratio, however, is only indirectly related to such radar performance measures as the probability of detecting a target, false alarm rate, false-dismissal probability, range and doppler estimation accuracy, and so on. These performance measures are treated in later chapters within the broader framework of statistical decision theory. In many cases decision-theory leads to solutions identical to those obtained by maximizing signal-to-noise ratio.

5.2. Maximizing Signal-to-Noise Ratio†

Consider the problem of specifying the impulse response of a linear time-invariant filter, $h_{\text{opt}}(t)$, which maximizes the output signal-to-noise ratio at time $t = t_m$. Let the input waveform $v(t)$ be given, as shown in Fig. 5.2-1, by

$$v(t) = s_i(t) + n_i(t) \tag{5.2-1}$$

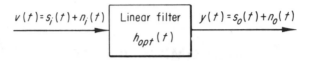

Fig. 5.2-1. Filter representation.

where $s_i(t)$ represents the signal component and $n_i(t)$ the noise component. If the waveform out of the optimum filter is denoted by

$$y(t) = s_o(t) + n_o(t) \tag{5.2-2}$$

the maximum signal-to-noise ratio is defined by

$$\chi_m(t_m) = \frac{s_o^2(t_m)}{n_o^2(t_m)} \tag{5.2-3}$$

It is assumed that $s_i(t)$ is exactly known and that $n_i(t)$ is a member function of a stationary random process. If the optimum filter has a finite memory T, it follows from linear filter theory that

$$s_o(t_m) = \int_0^T h_{\text{opt}}(\tau) s_i(t_m - \tau)\, d\tau \tag{5.2-4}$$

and

$$n_o(t_m) = \int_0^T h_{\text{opt}}(\tau) n_i(t_m - \tau)\, d\tau \tag{5.2-5}$$

To find $h_{\text{opt}}(t)$, a variational technique is employed. For this purpose it is convenient to define a real function $g(t)$ with the property

$$\int_0^T g(\tau) s_i(t_m - \tau)\, d\tau = 0 \tag{5.2-6}$$

Consider the response, at time t_m, of a nonoptimum‡ filter with impulse response $[h_{\text{opt}}(t) + \epsilon g(t)]$ to input waveform $v(t)$, where ϵ is an arbitrary real number. The output signal component $s_o'(t_m)$ and output noise component $n_o'(t_m)$ are then given by

---

†An excellent review may be found in G. L. Turin, An Introduction to Matched Filters, *IRE Trans. on Information Theory*, **IT-6**: (3), 311–329 (June, 1960).

‡A filter whose maximum output signal-to-noise ratio is less than that of the optimum filter.

$$s_o'(t_m) = \int_0^T [h_{\text{opt}}(\tau) + \epsilon g(\tau)] s_i(t_m - \tau)\, d\tau \qquad (5.2\text{-}7a)$$

$$= \int_0^T h_{\text{opt}}(\tau) s_i(t_m - \tau)\, d\tau + \epsilon \int_0^T g(\tau) s_i(t_m - \tau)\, d\tau \qquad (5.2\text{-}7b)$$

$$= s_o(t_m) \qquad (5.2\text{-}7c)$$

and

$$n_o'(t_m) = \int_0^T [h_{\text{opt}}(\tau) + \epsilon g(\tau)] n_i(t_m - \tau)\, d\tau \qquad (5.2\text{-}8a)$$

$$= \int_0^T h_{\text{opt}}(\tau) n_i(t_m - \tau)\, d\tau + \epsilon \int_0^T g(\tau) n_i(t_m - \tau)\, d\tau \qquad (5.2\text{-}8b)$$

$$= n_o(t_m) + \epsilon \int_0^T g(\tau) n_i(t_m - \tau)\, d\tau \qquad (5.2\text{-}8c)$$

The mean-square value of $n_o'(t_m)$ can be computed as follows,

$$\overline{[n_o'(t_m)]^2} = \overline{[n_o(t_m)]^2} + \overline{2\epsilon n_o(t_m) \int_0^T g(\tau) n_i(t_m - \tau)\, d\tau}$$
$$+ \overline{\epsilon^2 \left[\int_0^T g(\tau) n_i(t_m - \tau)\, d\tau \right]^2} \qquad (5.2\text{-}9)$$

Since $s_o'(t_m)$ is equal to $s_o(t_m)$ by (5.2-7c), it follows from the previous definitions of optimum and nonoptimum that

$$\overline{[n_o'(t_m)]^2} \geqslant \overline{[n_o(t_m)]^2} \qquad (5.2\text{-}10)$$

Combining Eqs. (5.2-9) and (5.2-10) yields the inequality

$$\overline{2\epsilon n_o(t_m) \int_0^T g(\tau) n_i(t_m - \tau)\, d\tau} + \overline{\epsilon^2 \left[\int_0^T g(\tau) n_i(t_m - \tau)\, d\tau \right]^2} \geqslant 0 \qquad (5.2\text{-}10a)$$

Since ϵ is an arbitrary constant and may be positive or negative, inequality (5.2-10a) can be satisfied for all ϵ only if the first term in (5.2-10a) is zero. Replacing $n_o(t_m)$ by Eq. (5.2-5), we may rewrite this condition as

$$\int_0^T \int_0^T g(\tau) h_{\text{opt}}(\sigma) \overline{n_i(t_m - \tau) n_i(t_m - \sigma)}\, d\tau\, d\sigma = 0 \qquad (5.2\text{-}11)$$

By definition, the autocorrelation function $\phi_{nn}(\tau)$ of the input noise process is given by

$$\overline{n_i(t_m - \tau) n_i(t_m - \sigma)} = \phi_{nn}(\tau - \sigma) \qquad (5.2\text{-}12)$$

so that (5.2-11) can be rewritten as

$$\int_0^T g(\tau)\, d\tau \int_0^T h_{\text{opt}}(\sigma) \phi_{nn}(\tau - \sigma)\, d\sigma = 0 \qquad (5.2\text{-}13)$$

Comparing Eq. (5.2-13) with Eq. (5.2-6), we have for all $g(\tau)$ satisfying (5.2-6),

$$\int_0^T h_{\text{opt}}(\sigma) \phi_{nn}(\tau - \sigma)\, d\sigma = k s_i(t_m - \tau), \qquad 0 \leqslant \tau \leqslant T \qquad (5.2\text{-}14)$$

where k is a constant. Equation (5.2-14) is the classic solution obtained by Zadeh and Ragazzini [1] for the impulse response $h_{\text{opt}}(t)$ of the optimum filter. The development leading to Eq. (5.2-14) shows that (5.2-14) is a necessary condition. It can also be shown that (5.2-14) is sufficient.[†]

Equation (5.2-14) can be interpreted as follows. If a signal identical to the noise autocorrelation function $\phi_{nn}(t)$ is applied to the input of filter $h_{\text{opt}}(t)$, its output is equal to $s_i(t)$ stretching backward in time from the instant t_m, within the interval defined by $0 \leqslant t \leqslant T$. This interpretation will be discussed later in greater detail. Mathematically, Eq. (5.2-14) is classified as an inhomogeneous Fredholm integral equation of the first kind; its solution is discussed in a number of texts.[‡]

The constant k may be evaluated by substituting Eq. (5.2-14) into (5.2-7a) with $\epsilon = 0$:

$$s_o(t_m) = \frac{1}{k} \int_0^T \int_0^T h_{\text{opt}}(\tau) h_{\text{opt}}(\sigma) \phi_{nn}(\tau - \sigma) \, d\tau \, d\sigma \qquad (5.2\text{-}15)$$

With Eq. (5.2-5) it is easy to show that the double integral in (5.2-15) is equal to $\overline{n_o^2(t_m)}$. Solving (5.2-15) for k,

$$k = \frac{\overline{n_o^2(t_m)}}{s_o(t_m)} \qquad (5.2\text{-}16)$$

When the region of integration in (5.2-14) can be extended to allow $T \longrightarrow \infty$, the Fourier transformation of Eq. (5.2-14) yields

$$H_{\text{opt}}(\omega) = \frac{k S_i^*(\omega) e^{-j\omega t_m}}{G_{nn}(\omega)} \qquad (5.2\text{-}17)$$

where $H_{\text{opt}}(\omega)$ is the Fourier transform of $h_{\text{opt}}(t)$, $S_i^*(\omega)$ is the complex conjugate of the input signal transform and $G_{nn}(\omega)$ is the spectral density of the stationary noise process, which by the Wiener-Khintchine theorem is the Fourier transform of $\phi_{nn}(\tau)$. Equation (5.2-17) demonstrates that the optimum filter weights most heavily those spectral components of the input signal which greatly exceed the noise spectral density and de-emphasizes the spectral regions in which the noise spectral density is much larger than the signal spectrum.

5.3. The Matched Filter

When the two-sided noise spectral density $G_{nn}(\omega)$ is white and equal to $N_0/2$ (watts/Hz), the noise autocorrelation function is given by $(N_0/2) \, \delta(\tau)$

†See W. Davenport and W. Root, "Random Signals and Noise," McGraw-Hill, New York, 1958.

‡See R. Courant and D. Hilbert, "Methods of Mathematical Physics," Wiley-Interscience, New York, 1953.

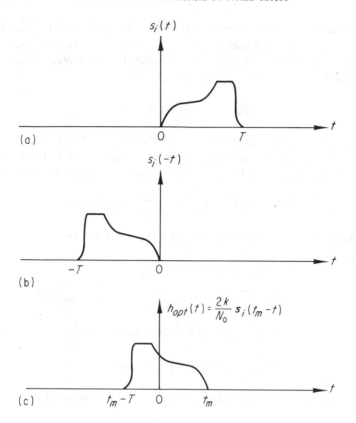

Fig. 5.3-1. (a) Arbitrary signal; (b) signal reversed in time; (c) matched filter impulsive response.

(see Fig. 4.7-1 in Chapter 4) where $\delta(\tau)$ is the Dirac delta function.† For this case Eq. (5.2-14) becomes

$$\frac{N_0}{2} \int_0^T h_{\text{opt}}(\tau)\, \delta(t - \tau)\, d\tau = \frac{N_0}{2} h_{\text{opt}}(t) = k s_i(t_m - t) \qquad (5.3\text{-}1)$$

or

$$h_{\text{opt}}(t) = \frac{2k}{N_0}\, s_i(t_m - t) \qquad \text{for} \quad 0 \leqslant t \leqslant T \qquad (5.3\text{-}2)$$

Equation (5.3-2) may be interpreted with reference to Fig. 5.3-1 as follows. For a known signal $s_i(t)$ of finite duration T, the impulsive response of the optimum linear filter is given by the signal run backwards in time from the

---

†For a discussion of the Dirac delta function as a limit of a distribution that has significance only under a transformation by an integral operator, see A. Papoulis, "The Fourier Integral," McGraw-Hill, New York, 1962.

instant t_m at which the maximum signal-to-noise ratio occurs. This filter is normally referred to as a *matched filter* and its impulse response designated by $h_{MF}(t)$. The matched filter spectral transfer function $H_{MF}(\omega)$ is obtained from the Fourier transform of Eq. (5.3-2),

$$H_{MF}(\omega) = \frac{2k}{N_0} S_i^*(\omega) e^{-j\omega t_m} \tag{5.3-3}$$

Hence, a matched filter is sometimes referred to as a *conjugate* filter.

The signal and noise out of a filter matched to the input signal (see Fig. 5.2-1) are found as follows. Consider first the signal term. Since

$$S_o(\omega) = H_{MF}(\omega) S_i(\omega) = \frac{2k}{N_0} |S_i(\omega)|^2 e^{-j\omega t_m} \tag{5.3-4}$$

where $S_o(\omega)$ is the Fourier transform of $s_o(t)$, it follows that

$$s_o(t) = \frac{2k}{N_0} \int_{-\infty}^{\infty} |S_i(\omega)|^2 e^{j\omega(t-t_m)} \frac{d\omega}{2\pi} \tag{5.3-5}$$

$$= \frac{2k}{N_0} \int_{-\infty}^{\infty} S_i^*(\omega) e^{j\omega(t-t_m)} \left[\int_{-\infty}^{\infty} s_i(v) e^{-j\omega v} \, dv \right] \frac{d\omega}{2\pi} \tag{5.3-6}$$

$$= \frac{2k}{N_0} \int_{-\infty}^{\infty} s_i(v) \, dv \int_{-\infty}^{\infty} S_i^*(\omega) e^{j\omega(t-t_m-v)} \frac{d\omega}{2\pi} \tag{5.3-7}$$

$$= \frac{2k}{N_0} \int_{-\infty}^{\infty} s_i(v) s_i(v - t + t_m) \, dv \tag{5.3-8}$$

Let the time autocorrelation function $r_{ss}(\tau)$ of $s_i(t)$ be defined by[†]

$$r_{ss}(\tau) = \int_{-\infty}^{\infty} s_i(v) s_i(v - \tau) \, dv \tag{5.3-9}$$

Then Eq. (5.3-8) can be rewritten as

$$s_o(t) = \frac{2k}{N_0} r_{ss}(t - t_m) \tag{5.3-10}$$

Thus the signal out of a matched filter is proportional to the autocorrelation function of the input signal, or—as shown in Eq. (5.3-5)—the inverse Fourier transform of the input signal energy spectral density.

The average noise power out of the matched filter with white noise input is given by[‡]

[†]The time autocorrelation function of arbitrary signals with finite energy, defined by (5.3-9) above, is analogous to the definition of the time autocorrelation function $R_{xx}(\tau)$ for random processes with finite average power [see Eq. (4.4-19) of Chapter 4].

[‡]The energy (or power) spectral density out of a linear time-invariant filter with transfer function $H(\omega)$ is simply $|H(\omega)|^2$ times the input energy (or power) spectral density. Equation (5.3-11a) follows from Parseval's formula and the assumed ergodicity of the noise process.

$$\overline{n_o^2(t)} = \frac{1}{2\pi} \int_{-\infty}^{\infty} |H_{\mathrm{MF}}(\omega)|^2 G_{nn}(\omega)\, d\omega \qquad (5.3\text{-}11\mathrm{a})$$

$$= \frac{1}{2\pi} \int_{-\infty}^{\infty} |H_{\mathrm{MF}}(\omega)|^2 \frac{N_0}{2}\, d\omega \qquad (5.3\text{-}11\mathrm{b})$$

Substituting (5.3-8) and (5.3-11b) into (5.2-3) yields the peak signal-to-noise ratio out of a matched filter at $t = t_m$:

$$\chi_{\mathrm{MF}}(t_m) = \frac{\left\{ \dfrac{2k}{N_0} \displaystyle\int_{-\infty}^{\infty} [s_i(v)]^2\, dv \right\}^2}{\displaystyle\int_{-\infty}^{\infty} \left(\dfrac{2k}{N_0} \right)^2 |S_i(\omega)|^2 \dfrac{N_0}{2} \dfrac{d\omega}{2\pi}} \qquad (5.3\text{-}12)$$

Since

$$\int_{-\infty}^{\infty} [s_i(v)]^2\, dv = \int_{-\infty}^{\infty} |S_i(\omega)|^2 \frac{d\omega}{2\pi} = E \qquad (5.3\text{-}13)$$

by Parseval's relation, where E is the input signal energy, Eq. (5.3-12) becomes

$$\chi_{\mathrm{MF}}(t_m) = \frac{2E}{N_0} \qquad (5.3\text{-}14)$$

Ratio $2E/N_0$ is a most important quantity, which appears many times in radar theory. The signal-to-noise ratio is less than $2E/N_0$ for all other values of t since, by Eq. (5.3-5), $s_o(t) < s_o(t_m)$ for all $t \neq t_m$.

When the input signal amplitude is a random variable, signal-to-noise ratio and signal energy are defined as averages with respect to the probability density function of the signal amplitude. It is conventional, however, to retain the same notation for χ and E in this situation.

5.4. Example I: Matched Filter Approximation for a Rectangular rf Pulse

In practice it is often difficult to implement a filter that is exactly matched to the transmitted radar waveform. This and succeeding sections treat a number of examples of frequently used radar waveforms that are processed by filters that approximate the matched filter characteristic.

Consider an input pulse $s_i(t)$ as shown in Fig. 5.4-1. The Fourier transform of this pulse, $S_i(\omega)$, is given by

$$S_i(\omega) = \int_{-\Delta_\tau/2}^{\Delta_\tau/2} A \cos \omega_0 t \, e^{-j\omega t}\, dt \qquad (5.4\text{-}1)$$

$$S_i(\omega) = \frac{A\,\Delta_\tau}{2} \left[\frac{\sin (\omega - \omega_0)\,\Delta_\tau/2}{(\omega - \omega_0)\,\Delta_\tau/2} + \frac{\sin (\omega + \omega_0)\,\Delta_\tau/2}{(\omega + \omega_0)\,\Delta_\tau/2} \right] \qquad (5.4\text{-}2)$$

$S_i(\omega)$ is illustrated in Fig. 5.4-2. The matched filter transfer function for this signal, which is obtained by substituting (5.4-2) into (5.3-3), is proportional

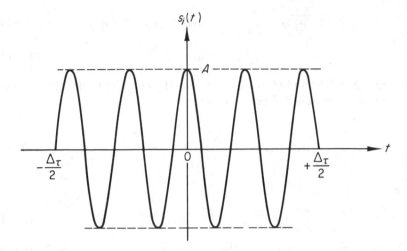

Fig. 5.4-1. Radio-frequency pulse with rectangular envelope.

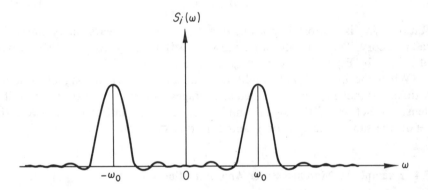

Fig. 5.4-2. Spectrum of rf rectangular pulse.

to the complex conjugate of $S_i(\omega)$. For simplicity t_m in Eq. (5.3-8) may be set equal to zero; in this case the signal out of the matched filter is given by

$$s_o(t) = \frac{2k}{N_0} \int_{-\infty}^{\infty} s(v)s(v - t)\, dv \tag{5.4-3}$$

Substituting $A \cos \omega_0 v$ for $s(v)$ in (5.4-3) yields for $0 < t < \Delta_\tau/2$

$$s_o(t) = \frac{2kA^2}{N_0} \int_{t-\Delta_\tau/2}^{\Delta_\tau/2} \cos \omega_0 v \cos \omega_0 (v - t)\, dv \tag{5.4-4}$$

or, using a trigonometric identity,

$$s_o(t) = \frac{kA^2}{N_0} \int_{t-\Delta_\tau/2}^{\Delta_\tau/2} [\cos \omega_0 t + \cos (2\omega_0 v - \omega_0 t)]\, dv \tag{5.4-5}$$

For very large ω_0 the second term approaches zero† and may be neglected with respect to the first term. Then

$$s_0(t) \cong \frac{kA^2}{N_0} [\Delta_\tau - t] \cos \omega_0 t \qquad \text{for} \quad 0 < t < \Delta_\tau \qquad (5.4\text{-}6)$$

Since the pulse is symmetrical, $s_0(t)$ can be written as

$$s_0(t) \cong \frac{kA^2}{N_0} [\Delta_\tau - |t|] \cos \omega_0 t \qquad (5.4\text{-}7)$$

for $|t| < \Delta_\tau$ and is zero elsewhere. Result (5.4-7) is sketched in Fig. 5.4-3. The pulse envelope, as indicated by Eq. (5.4-7), is triangular in shape.

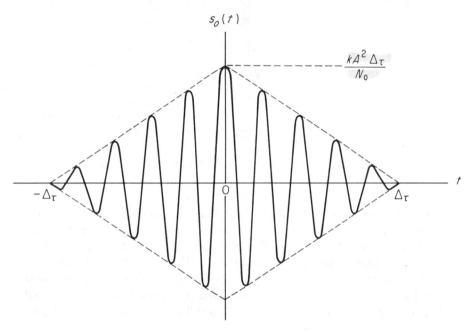

Fig. 5.4-3. Signal out of matched filter for rectangular rf input pulse.

The peak signal-to-noise ratio out of the matched filter is obtained by substituting $A \cos \omega_0 v$ into (5.3-13) and the result into (5.3-14):

$$\chi_{\text{MF}}(t_m) = \frac{2}{N_0} \left[\frac{A^2 \Delta_\tau}{2} \right] = \frac{A^2 \Delta_\tau}{N_0} \qquad (5.4\text{-}8)$$

Comparing (5.4-7) and (5.4-8) reveals that signal $s_0(t)$ out of the matched filter has a maximum value at $t = t_m = 0$ that is proportional to peak signal-

---

†This term after integration is of the form $(\sin \omega_0 x)/\omega_0$ evaluated at the appropriate limits.

to-noise ratio $\chi_{MF}(0)$. The scale factor k in (5.4-7) is not significant since, as Eq. (5.3-12) indicates, $\chi_{MF}(t_m)$ is unaffected by its value.†

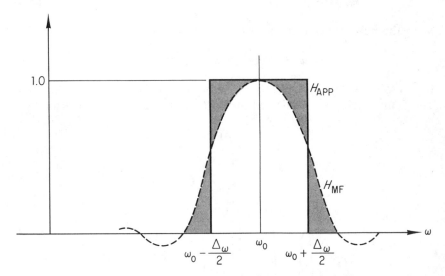

Fig. 5.4-4. Rectangular approximation to matched filter transfer function.

It is difficult to construct a matched filter with transfer function H_{MF} as shown in Fig. 5.4-4; however, considerable information exists on the design of approximations to a rectangular bandpass filter function. Consider the rectangular approximation to H_{MF} shown in Fig. 5.4-4 designated by H_{APP}. For large ω_0 the input signal spectrum in the right half plane is given principally by the first term in Eq. (5.4-2) and the signal spectrum in the left half plane principally by the second term. Hence the signal out of filter $H_{APP}(\omega)$ is approximately given by

$$s_o(t) = \frac{1}{2\pi} \int_{\omega_0-\Delta_\omega/2}^{\omega_0+\Delta_\omega/2} \frac{A\Delta_\tau}{2} \cdot \frac{\sin (\omega - \omega_0)\Delta_\tau/2}{(\omega - \omega_0)\Delta_\tau/2} e^{j\omega t} \, d\omega$$
$$+ \frac{1}{2\pi} \int_{-\omega_0-\Delta_\omega/2}^{-\omega_0+\Delta_\omega/2} \frac{A\Delta_\tau}{2} \cdot \frac{\sin (\omega + \omega_0)\Delta_\tau/2}{(\omega + \omega_0)\Delta_\tau/2} e^{j\omega t} \, d\omega$$

(5.4-9)

or, combining terms,

$$s_o(t) = \frac{A\Delta_\tau}{2\pi} \int_{\omega_0-\Delta_\omega/2}^{\omega_0+\Delta_\omega/2} \frac{\sin (\omega - \omega_0)\Delta_\tau/2}{(\omega - \omega_0)\Delta_\tau/2} \cos \omega t \, d\omega$$

(5.4-10)

For $\Delta_\omega < 4\pi/\Delta_\tau$ the $(\sin x)/x$ term in Eq. (5.4-10) is positive; hence, an upper bound on the value of $s_o(t)$ is given by

†Both signal and noise are modified by the same scale factor.

$$s_o(t) \leqslant \frac{A\Delta_\tau}{2\pi} \int_{\omega_0-\Delta\omega/2}^{\omega_0+\Delta\omega/2} |\cos \omega t| \frac{\sin (\omega - \omega_0)\Delta_\tau/2}{(\omega - \omega_0)\Delta_\tau/2} d\omega \qquad (5.4\text{-}11)$$

The maximum value of $s_o(t)$ occurs (with no approximation) at $t = 0$:

$$s_o(0) = \frac{A\Delta_\tau}{\pi} \int_0^{\omega_0+\Delta\omega/2} \frac{\sin (\omega - \omega_0)\Delta_\tau/2}{(\omega - \omega_0)\Delta_\tau/2} d\omega \qquad (5.4\text{-}12)$$

If $(\omega - \omega_0)\Delta_\tau/2$ is replaced by x, Eq. (5.4-12) can be rewritten as

$$s_o(0) = \frac{2A}{\pi} \int_0^{\Delta\omega \Delta\tau/4} \frac{\sin x}{x} dx = \frac{2A}{\pi} \text{Si} \left(\frac{\Delta_\omega \Delta_\tau}{4} \right) \qquad (5.4\text{-}13)$$

where $\text{Si}\,(\Delta_\omega\Delta_\tau/4)$ denotes the sine integral with upper limit $\Delta_\omega\Delta_\tau/4$.

The average noise power out of the rectangular bandpass filter approximation is obtained by replacing $|H(\omega)|^2$ in Eq. (5.3-11b) by unity for $\omega_0 - \Delta_\omega/2 \leqslant |\omega| \leqslant \omega_0 + \Delta_\omega/2$ and zero elsewhere. This results in

$$\overline{n_o^2(t)} = 2 \left[\frac{1}{2\pi} \int_{\omega_0-\Delta_\omega/2}^{\omega_0+\Delta_\omega/2} \frac{N_0}{2} d\omega \right] = \frac{N_0\Delta_\omega}{2\pi} \qquad (5.4\text{-}14)$$

The signal-to-noise ratio χ_{APP} out of the approximating filter at $t = 0$ is given by

$$\chi_{\text{APP}} = \frac{[s_o(0)]^2}{\overline{n_o^2(t)}} = \frac{8A^2}{\pi N_0\Delta_\omega} \text{Si}^2 \left(\frac{\Delta_\omega\Delta_\tau}{4} \right) \qquad (5.4\text{-}15)$$

The ratio of χ_{APP} at $t = 0$ to the peak value of χ_{MF} is found from Eqs. (5.4-8) and (5.4-15) to be

$$\frac{\chi_{\text{APP}}}{\chi_{\text{MF}}} = \frac{8}{\pi \Delta_\omega\Delta_\tau} \text{Si}^2 \left(\frac{\Delta_\omega\Delta_\tau}{4} \right) \qquad (5.4\text{-}16)$$

This equation is plotted in Fig. 5.4-5 as a function of the (radian) bandwidth-

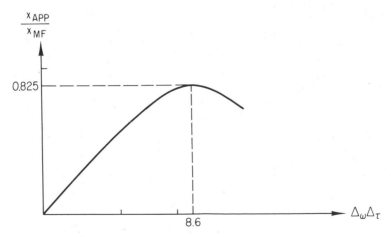

Fig. 5.4-5. Comparison of matched filter peak output signal-to-noise ratio with that of a rectangular passband filter.

duration product $\Delta_\omega \Delta_\tau$. The peak value on this curve occurs for $\Delta_\omega \Delta_\tau \cong 8.6$ or $\Delta_f \Delta_\tau \cong 1.37$, where $2\pi \Delta_f = \Delta_\omega$. At this point χ_{APP} is about 0.84 decibel below χ_{MF}. Thus when a pulsed signal of the type shown in Fig. 5.4-1, for which $\Delta_f \Delta_\tau \cong 1$, is processed by a rectangular bandpass filter approximation (of the true matched filter) with $\Delta_f \Delta_\tau \cong 1.4$, a loss of about 0.8 db is suffered.

If the reciprocal of Eq. (5.4-16) is plotted with logarithmic scales, a curve similar to the top curve shown in Fig. 5.4-6 will result. The curves in Fig.

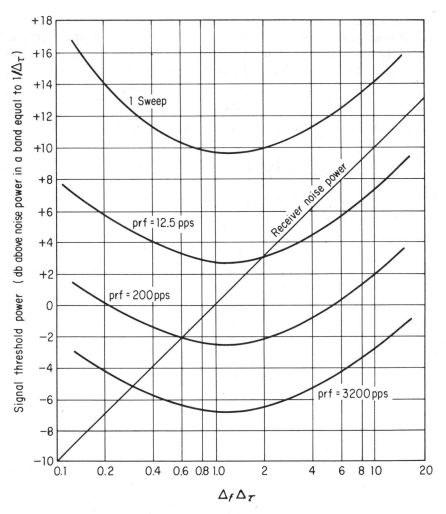

Fig. 5.4-6. Signal threshold power vs. i-f bandwidth times pulse length (*after Lawson and Uhlenbeck* [9], *courtesy of McGraw-Hill Book Company*).

5.4-6 were obtained from measurements by Sydoriak, Ashby, and Lawson.†
In these experiments the threshold signal power for reliable detection was
determined experimentally for a fixed pulse length as a function of receiver
bandwidth. A broad null is evident in the vicinity of $\Delta_f \Delta_\tau = 1.2$. Although
the experimental receiver bandpass shape was not rectangular, the experimental results indicate an optimum $\Delta_f \Delta_\tau$ product close to that shown in
Fig. 5.4-5.

In the early days of radio and radar, matched filtering was an unknown
concept. The fact that good performance could be obtained by designing receiver bandwidth Δ_f to be slightly greater than the reciprocal of pulse length
Δ_τ was of historical importance. In the next example significantly different
results are obtained when the signal is phase-coded.

5.5 Example II: Matched Filter Approximation for a Linear fm Pulse with Rectangular Envelope

In this example the input signal $s_i(t)$ is assumed to be a radio-frequency
pulse of amplitude A, whose angular frequency increases linearly during
the pulse [2, 3]:

$$s_i(t) = A \cos\left(\omega_0 t + \frac{1}{2} bt^2\right) \qquad \text{for} \quad |t| < \frac{\Delta_\tau}{2} \qquad (5.5\text{-}1)$$

The matched filter output $s_o(t)$ is obtained by substituting (5.5-1) into Eq.
(5.4-3). For $0 < t < \Delta_\tau$ Eq. (5.4-3) becomes

$$s_o(t) = \frac{2kA^2}{N_0} \int_{t-\Delta_\tau/2}^{\Delta_\tau/2} \cos\left(\omega_0 v + \frac{1}{2} bv^2\right) \cos\left[\omega_0(v-t) + \frac{1}{2} b(v-t)^2\right] dv$$

$$\text{for} \quad 0 < t < \Delta_\tau \qquad (5.5\text{-}2)$$

Expanding the integrand in (5.5-2) trigonometrically and discarding the
double frequency term, whose contribution to the value of the integral is small
for large ω_0, we obtain

$$s_o(t) = \frac{kA^2}{N_0} \int_{t-\Delta_\tau/2}^{\Delta_\tau/2} \cos\left[\omega_0 t + bvt - \frac{1}{2} bt^2\right] dv \qquad (5.5\text{-}3)$$

The integral in (5.5-3) is evaluated as follows:

$$s_o(t) = \frac{kA^2}{N_0} \cdot \frac{\sin\left(\omega_0 t - \frac{1}{2}bt^2 + bvt\right)}{bt} \Bigg|_{t-\Delta_\tau/2}^{\Delta_\tau/2} \qquad (5.5\text{-}4)$$

$$s_o(t) = \frac{kA^2}{N_0 bt} \left[\sin\left(\omega_0 t + \frac{bt\Delta_\tau}{2} - \frac{1}{2} bt^2\right) - \sin\left(\omega_0 t + \frac{bt^2}{2} - \frac{bt\Delta_\tau}{2}\right)\right] \qquad (5.5\text{-}5)$$

†See J. Lawson and G. Uhlenbeck, "Threshold Signals," p. 201, McGraw-Hill, New
York, 1950. These curves are the results of a small sample of observations but are representative of expected results.

For $0 < t < \Delta_\tau$,

$$s_o(t) = \frac{2kA^2}{N_0 bt} \sin\left[\frac{bt}{2}(\Delta_\tau - t)\right]\cos \omega_0 t \qquad (5.5\text{-}6)$$

Steps (5.5-2) through (5.5-6) can be repeated for $-\Delta_\tau < t \leqslant 0$. The final result is

$$s_o(t) = \frac{kA^2}{N_0} \cdot \frac{\sin\left[bt(\Delta_\tau - |t|)\right]/2}{bt/2}\cos \omega_0 t \qquad \text{for} \quad |t| < \Delta_\tau \qquad (5.5\text{-}7)$$

The output signal in the vicinity of $t = 0$ is given by

$$s_o(t) = \frac{kA^2 \Delta_\tau}{N_0}\left[\frac{\sin bt\, \Delta_\tau/2}{bt\, \Delta_\tau/2}\right]\cos \omega_0 t \qquad (5.5\text{-}8)$$

Equation (5.5-7) is plotted in Fig. 5.5-1 for the condition $b\,\Delta_\tau \gg 1$.

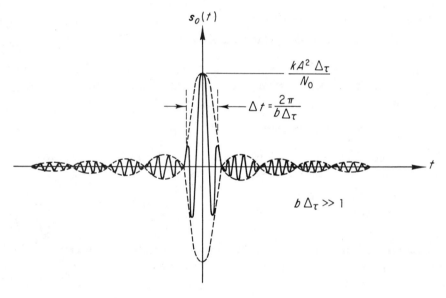

Fig. 5.5-1. Signal out of matched filter for linear fm rectangular input pulse.

The peak value of $s_o(t)$ occurs at $t = 0$, where (5.5-7) and (5.5-8) become (applying L'Hospital's rule)

$$s_o(0) = \frac{kA^2 \Delta_\tau}{N_0} \qquad (5.5\text{-}9)$$

The peak output signal-to-noise ratio χ_{MF} at $t = 0$ is obtained by substituting (5.5-1) into (5.3-13) and the result into (5.3-14):

$$\chi_{\text{MF}}(0) = \frac{2E}{N_0} = \frac{2}{N_0}\left[\frac{A^2 \Delta_\tau}{2}\right] = \frac{A^2 \Delta_\tau}{N_0} \qquad (5.5\text{-}10)$$

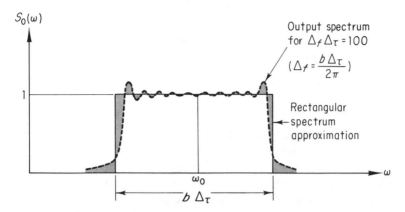

Fig. 5.5-2. Approximate signal spectrum out of matched filter for linear fm rectangular input pulse.

Observe that the peak output signal value in (5.5-9) is once again proportional to peak signal-to-noise ratio $\chi_{\text{MF}}(0)$ in Eq. (5.5-10).

The spectrum of $s_o(t)$ is approximately rectangular as shown in Fig. 5.5-2 for $b\,\Delta_\tau \gg 1$. The actual spectrum $S_o(\omega)$ of the output signal is sketched in dotted lines in Fig. 5.5-2 for $\Delta_\tau\,\Delta_f = 100$, where $\Delta_f = b\Delta_\tau/2\pi$. The output spectrum can be computed exactly by taking the Fourier transform of Eq. (5.5-7); this is a laborious task.† As the time-bandwidth product $\Delta_\tau\,\Delta_f$ of the input waveform increases, the output spectrum becomes progressively more rectangular. Since, from Eq. (5.3-4), the signal spectrum out of the matched filter is proportional to the squared magnitude of the input signal spectrum, the spectrum of a linear fm pulse is also very closely rectangular.

Instead of a matched filter, consider the output of a filter with a rectangular transfer function, as shown in Fig. 5.5-2, and with a linear phase spectrum over the bandpass region. Since the filter is essentially distortionless, the input and output signals are practically identical. Hence, for an input signal given by (5.5-1), the peak output signal power is equal to A^2. The average noise power out of the filter is calculated from Eq. (5.3-11b) as

$$\overline{n_o(t)^2} = \frac{1}{2\pi}\left(\frac{N_0}{2}\right)2\int_{\omega_0-b\Delta_\tau/2}^{\omega_0+b\Delta_\tau/2}d\omega = \frac{N_0 b\,\Delta_\tau}{2\pi} \tag{5.5-11}$$

With these results the peak signal-to-noise ratio out of the distortionless filter is given by

$$\chi = \frac{2\pi A^2}{N_0 b\,\Delta_\tau} \tag{5.5-12}$$

The ratio of χ to the peak signal-to-noise ratio out of a matched filter, given by Eq. (5.5-10), is

†For excellent pictures of actual spectra see Cook [2]; also Cook and Bernfeld [13].

$$\frac{\chi}{\chi_{MF}} = \frac{2\pi}{b\,\Delta_\tau^2} \tag{5.5-13}$$

Since the bandwidth Δ_f of the ideal filter is equal to $b\,\Delta_\tau/2\pi$, Eq. (5.5-13) may be rewritten as

$$\frac{\chi}{\chi_{MF}} = \frac{1}{\Delta_\tau\,\Delta_f} \tag{5.5-14}$$

In this case a severe penalty is exacted if we do not use a matched filter. To illustrate, for a signal with a time-bandwidth product of 100, the peak signal-to-noise ratio out of the distortionless filter is 20 db below χ_{MF}.

The reason for the disparity in performance between the two filters may be qualitatively understood as follows. A linear fm pulse of length Δ_τ is *squashed* or compressed at the matched filter output into a narrow pulse whose duration is of the order of $1/\Delta_f$, with a corresponding increase in pulse amplitude. The phase coding on the input pulse [see Eq. (5.5-1)] is eliminated in the output signal [see Eq. (5.5-7)]. In a sense, the matched filter decodes the input phase modulation to construct a narrow pulse output signal with high peak amplitude.

The time-bandwidth product $\Delta_\tau\,\Delta_f$ is an important parameter of a matched filter. The sampling theorem (see Chapter 2) states that a signal with bandwidth Δ_f and duration Δ_τ can be characterized by $2\Delta_\tau\,\Delta_f$ numbers. It follows that a filter matched to this signal can be specified by $2\Delta_\tau\,\Delta_f$ numbers. Thus, the time-bandwidth product is a measure of filter complexity and can be related to the number of elements or parameters of the network that must be specified.

A good approximation to a linear fm matched filter is provided by a filter with transfer function

$$H(\omega) = A(\omega)e^{j\beta(\omega)} \tag{5.5-15a}$$

where $A(\omega)$ is rectangular, as in the preceding case, and

$$\beta(\omega) = -\omega T_d + \frac{(\omega_i - \omega)^2}{2b} \tag{5.5-15b}$$

The second term in (5.5-15b) is a quadratic phase term; this differentiates this case from the distortionless filter previously considered. Phase characteristic (5.5-15b) is similar to the phase characteristic obtained from the Fourier transform of Eq. (5.5-7). The new filter transfer function is shown in Fig. 5.5-3.

To obtain the output signal, we first compute the Fourier transform of $s_i(t)$ in Eq. (5.5-1) as follows:

$$S_i(\omega) = \frac{A}{2}\int_{-\Delta_\tau/2}^{\Delta_\tau/2}\left\{\exp\left[j\left((\omega_0 - \omega)\tau + \frac{b\tau^2}{2}\right)\right] + \exp\left[-j\left((\omega_0 + \omega)\tau + \frac{b\tau^2}{2}\right)\right]\right\}d\tau$$

$$\tag{5.5-16}$$

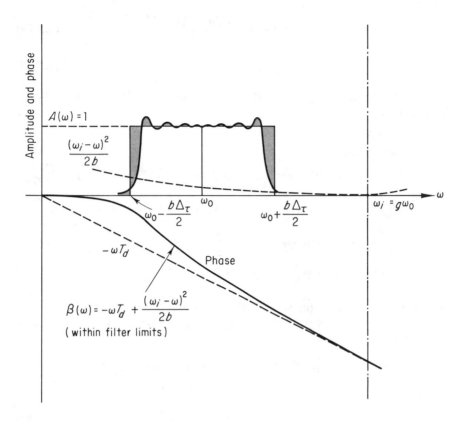

Fig. 5.5-3. Approximate filter transfer characteristic (*after DiFranco* [4], *courtesy of IEEE*).

The first term in the integral is the principal contribution to $S_i(\omega)$ in the vicinity of $\omega = \omega_0$ while the second term describes $S_i(\omega)$ in the region near $\omega = -\omega_0$. The signal out of the filter can be written as

$$s_o(t) = 2\,\mathrm{Re} \int_0^\infty S_i(\omega)H(\omega)e^{j\omega t}\frac{d\omega}{2\pi} \qquad (5.5.17)$$

Substituting (5.5-15) for $H(\omega)$ into (5.5-17) and approximating $S_i(\omega)$ by the first term in (5.5-16) results in

$$s_o(t) = A\,\mathrm{Re}\left\{\exp j\left[\omega_i(t - T_d) - \frac{b(t - T_d)^2}{2}\right]\right.$$
$$\left. \cdot \int_{-\Delta_\tau/2}^{\Delta_\tau/2}\left[\exp\,(jb\delta\tau)\int_{\omega_0 - b\Delta_\tau/2}^{\omega_0 + b\Delta_\tau/2}\exp\left\{j\frac{[\omega - \omega_0 + b(\delta - \tau)]^2}{2b}\right\}\frac{d\omega}{2\pi}\right]d\tau\right\}$$

$$(5.5\text{-}18)$$

where

$$\delta = t - T_d - \frac{(g-1)\omega_0}{b} \qquad (5.5\text{-}19)$$

$$g = \frac{\omega_i}{\omega_0} \qquad (5.5\text{-}20)$$

The double integral in (5.5-18) can be evaluated by integrating by parts. The solution is

$$s_o(\delta) = \frac{A\sqrt{\Delta_\tau \Delta_f}}{2\pi\sqrt{2\,\Delta_f \delta}}$$

$$\cdot \left\{ \sqrt{[S^{(1)}]^2 + [C^{(1)}]^2} \cos\left[(\omega_0 + \pi\,\Delta_f)\delta - \frac{\pi\Delta_f}{\Delta_\tau}\delta^2 + \phi - \tan^{-1}\frac{C^{(1)}}{S^{(1)}}\right] \right.$$

$$- \sqrt{[S^{(2)}]^2 + [C^{(2)}]^2} \cos\left[(\omega_0 - \pi\,\Delta_f)\delta - \frac{\pi\Delta_f}{\Delta_\tau}\delta^2 + \phi - \tan^{-1}\frac{C^{(2)}}{S^{(2)}}\right]$$

$$+ \sqrt{[S^{(3)}]^2 + [C^{(3)}]^2} \cos\left[\omega_0\delta + \frac{b\,\Delta_\tau}{2}\delta + \phi - \tan^{-1}\frac{C^{(3)}}{S^{(3)}}\right]$$

$$\left. - \sqrt{[S^{(3)}]^2 + [C^{(3)}]^2} \cos\left[\omega_0\delta - \frac{b\,\Delta_\tau}{2}\delta + \phi - \tan^{-1}\frac{C^{(3)}}{S^{(3)}}\right] \right\} \qquad (5.5\text{-}21)$$

where ϕ is a constant phase shift that appears in each term and can be neglected; $S^{(1)}, S^{(2)}, S^{(3)}, C^{(1)}, C^{(2)}, C^{(3)}$ are, respectively, the sine and cosine Fresnel integrals:

$$S^{(1)} = \int_{\alpha_1}^{\alpha_2} \sin\left(\frac{\pi}{2}u^2\right) du, \qquad C^{(1)} = \int_{\alpha_1}^{\alpha_2} \cos\left(\frac{\pi}{2}u^2\right) du \qquad (5.5\text{-}22)$$

in which

$$\alpha_1 = \sqrt{2\Delta_\tau \Delta_f}\left(\frac{\delta}{\Delta_\tau} - 1\right) \qquad (5.5\text{-}23)$$

$$\alpha_2 = \sqrt{2\Delta_\tau \Delta_f}\left(\frac{\delta}{\Delta_\tau}\right) \qquad (5.5\text{-}24)$$

$S^{(2)}$ and $C^{(2)}$ are the sine and cosine Fresnel integrals with upper and lower limits β_2 and β_1, where β_1 and β_2 are given, respectively, by adding $\sqrt{\Delta_\tau \Delta_f}$ to α_1 and α_2; and

$$S^{(3)} = \int_0^{v_1} \sin\left(\frac{\pi}{2}u^2\right) du, \qquad C^{(3)} = \int_0^{v_1} \cos\left(\frac{\pi}{2}u^2\right) du \qquad (5.5\text{-}25)$$

where

$$v_1 = \sqrt{2\Delta_\tau \Delta_f} \qquad (5.5\text{-}26)$$

Eq. (5.5-21) can be used to compute the output of the matched filter approximation for any time-bandwidth product $\Delta_\tau \Delta_f$. The effects of doppler shifts can also be easily accounted for.† The output waveform envelope for a linear

†See DiFranco [4].

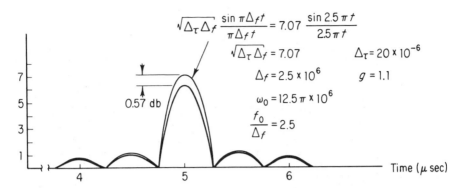

Fig. 5.5-4. Signals out of exact and approximate linear fm matched filters (*after DiFranco* [4], *courtesy of IEEE*).

fm input pulse with a time-bandwidth product of fifty is plotted in Fig. 5.5-4 and compared with the $(\sin x)/x$ response of a perfect matched filter. The output signal amplitude is approximately 0.6 db below the peak matched filter output. Thus if we add a quadratic phase term to a simple rectangular filter with a linear phase characteristic, the matched filter response is approached very closely.

The quadratic phase term corresponds to a group delay that is linearly dependent on frequency; by definition,[†]

$$\tau_{\text{gr}} = -\frac{d\beta}{d\omega} = T_d + \frac{\omega - \omega_i}{b} \tag{5.5-27}$$

Since group delay increases with frequency, the signal frequency components in the input signal that arrive last are delayed the least and those that arrive first are delayed the most. This results in squeezing the input signal spectral components into a shortened time interval at the filter output.

5.6. Example III: Matched Filter Approximation for Finite Pulse Train (Comb Filter)

In this section we consider a matched filter approximation for the finite rf pulse train shown in Fig. 5.6-1. For simplicity the interpulse period T_p has been selected as a multiple of the pulse length Δ_τ and the pulse amplitude is assumed to be unity. If the first train pulse is denoted by $f_p(t)$ and its Fourier transform by $F_p(\omega)$, then the Fourier transform $F(\omega)$ of the pulse train is given by

$$F(\omega) = \mathscr{F}\{f_p(t) + f_p(t - T_p) + \ldots + f_p[t - (N-1)T_p]\} \tag{5.6-1}$$

$$= F_p(\omega)(1 + e^{-j\omega T_p} + \ldots + e^{-j\omega(N-1)T_p}) \tag{5.6-2}$$

[†]See Papoulis, "The Fourier Integral," McGraw-Hill, New York, 1962.

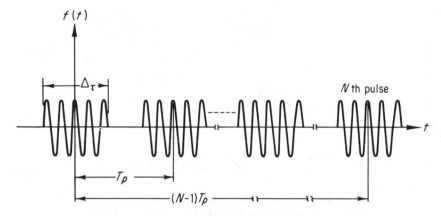

Fig. 5.6-1. Train of N rf pulses.

We obtain Eq. (5.6-2) from (5.6-1) with the aid of the Fourier transform pairs in Table 2.2-1 of Chapter 2. Equation (5.6-2) is a geometric progression (with ratio $e^{-j\omega T_p}$) and can be summed by conventional methods to yield

$$F(\omega) = F_p(\omega) \frac{1 - e^{-j\omega N T_p}}{1 - e^{-j\omega T_p}} \qquad (5.6\text{-}3)$$

or

$$F(\omega) = F_p(\omega) \frac{\sin(\omega N T_p/2)}{\sin(\omega T_p/2)} \exp\left[-j\omega\left(\frac{N-1}{2}\right)T_p\right] \qquad (5.6\text{-}4)$$

A typical plot of $F(\omega)$ is shown in Fig. 5.6-2 for $T_p/\Delta_\tau = 5$ and for carrier frequency $\omega_0 = 2\pi m/T_p$, where m is an integer. Only one-half of the plot about carrier frequency ω_0 is shown.

The width of the overriding envelope (measured between the peak and the first zero of the envelope function) is $2\pi/\Delta_\tau$ and is related to the spectrum of a single pulse; this is evident from Eq. (5.6-4). Note that $F(\omega)$ consists of large spikes, resembling the teeth of a comb, whose centers are separated in frequency by $2\pi/T_p$. The width of an individual tooth (measured between zeros) is of the order of 4π times the reciprocal of the total pulse-train length.

The matched filter for a pulse train has been called a *comb filter*. Its transfer function $H_{MF}(\omega)$ is proportional to $F^*(\omega)$, where $F(\omega)$ is given by Eq. (5.6-4). If the exponential delay term in (5.6-4) is neglected, $F^*(\omega) = F(\omega)$, since the remaining functions are real. The matched filter—as previously indicated—coherently adds all pulses in the train.

Consider first the performance of an idealized approximation to the matched filter, which is commonly referred to as a *uniform comb filter* [5]. Its transfer function is shown in solid lines in Fig. 5.6-3. For reference the envelope of the main lobe of the pulse-train spectrum is shown in dotted lines. The width of a tooth has been chosen as $4\pi/N T_p$. To simplify the analysis

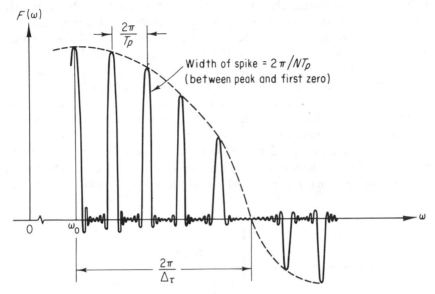

Fig. 5.6-2. Fourier transform of pulse train shown in Fig. 5.6-1 ($T_p/\Delta_\tau = 5$, $\omega_0 = 2\pi m/T_p$ where m is an integer).

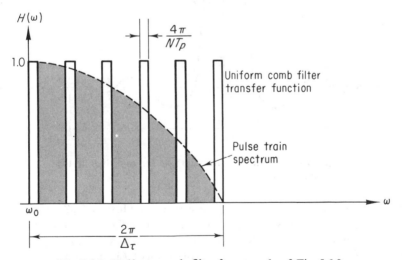

Fig. 5.6-3. Uniform comb filter for example of Fig. 5.6-2.

the input signal is assumed to be processed by a filter with rectangular passband $4\pi/\Delta_\tau$ prior to the comb filter. This eliminates the pulse-train spectrum outside this region. In Section 5.4 [see Eq. (5.4-16)] it was shown that this assumption corresponds to a loss in peak signal-to-noise ratio for each pulse in the train of approximately 1.59 db, with respect to a filter matched exactly to the pulse spectrum. If all pulses are then coherently added, a minimum loss of 1.59 db is encountered for the pulse train, which must be taken into account in the evaluation of the uniform comb filter approximation.

When T_p/Δ_τ is large, there are many teeth in bandwidth $4\pi/\Delta_\tau$ (see Fig. 5.6-3), and end effects and distortions due to the $(\sin Nx)/\sin x$ term in Eq. (5.6-4) are very small; for this case $(\sin Nx)/\sin x$ can be approximated by $\sin Nx/x$, for large N. It is also convenient to consider individual peaks in the pulse-train spectrum as separate signals, each of which is of $(\sin x)/x$ form and processed by a rectangular passband filter whose bandwidth is $4\pi/NT_p$, the width of a uniform comb filter tooth between first zeros (see Fig. 5.6-4). Since the rectangular filter is not optimally selected with

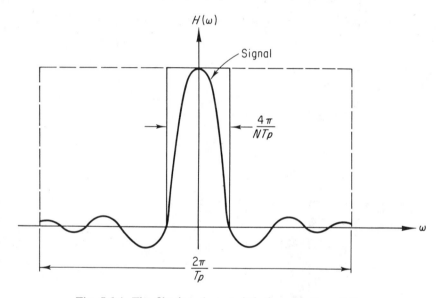

Fig. 5.6-4. The filtering characteristic for each line of the pulse-train spectrum.

respect to a single signal spectrum, there is again a loss of 1.59 db. However, there is also a gain in peak signal-to-noise ratio effected by the bandwidth reduction from $2\pi/T_p$ to $4\pi/NT_p$ about each signal as shown in Fig. 5.6-4. By referring to Eq. (5.4-16), we find the improvement I is given by†

$$I = \frac{\dfrac{8T_p}{2\pi^2\Delta_\tau}\left[\mathrm{Si}\left(\dfrac{\pi\Delta_\tau}{2T_p}\right)\right]^2}{\dfrac{8NT_p}{2\pi^2 \cdot 2\Delta_\tau}\left[\mathrm{Si}\left(\dfrac{\pi\Delta_\tau}{NT_p}\right)\right]^2} = \frac{2}{N}\frac{\left[\mathrm{Si}\left(\dfrac{\pi\Delta_\tau}{2T_p}\right)\right]^2}{\left[\mathrm{Si}\left(\dfrac{\pi\Delta_\tau}{NT_p}\right)\right]^2} \qquad (5.6\text{-}5)$$

†For bandwidths greater than $(4\pi/NT_p)$ the result of Section 5.4 must be modified, since the peak output signal does not occur at $t = 0$ (see Fig. 5.8–2 and discussion).

Note that Eq. (5.4-16) relates the peak signal-to-noise ratio of a rectangular passband filter and a true matched filter for a $(\sin x)/x$ signal spectrum. For large T_p/Δ_τ and N, the sine integrals in (5.6-5) can be approximated by their arguments, resulting in

$$I = \frac{2}{N} \frac{\left[\mathrm{Si}\left(\frac{\pi \Delta_\tau}{2T_p} \right) \right]^2}{\left[\mathrm{Si}\left(\frac{\pi \Delta_\tau}{NT_p} \right) \right]^2} \simeq \frac{N}{2} \tag{5.6-6}$$

When all of the previous gains and losses are combined, the peak signal-to-noise ratio provided by a uniform comb filter, referenced to the peak signal-to-noise ratio out of a filter matched to a single train pulse, is approximately given by (in db)

$$10 \log_{10} \frac{\chi_{\text{out}}}{\chi_1} \simeq 10 \log_{10} N - 6.18 \tag{5.6-7}$$

where the improvement provided by a perfect pulse train matched filter would be $10 \log_{10} N$.

The uniform comb filter has been compared by George and Zamanakos [5] with two other filter approximations, shown in Figs. 5.6-5 and 5.6-6. The first of these, called a *rectangular sine weighted comb filter*, is shown to be 2.1 db better than the uniform comb filter. The second, called a *North type of sine weighted comb filter*, is shown to be 4.3 db better than the uniform comb filter and only 1.9 db poorer than a perfect matched filter.

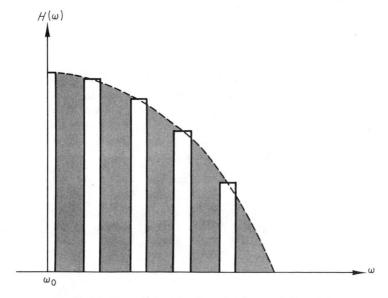

Fig. 5.6-5. Rectangular sin x/x weighted comb filter (*after George and Zamanakos [5], courtesy of IEEE*).

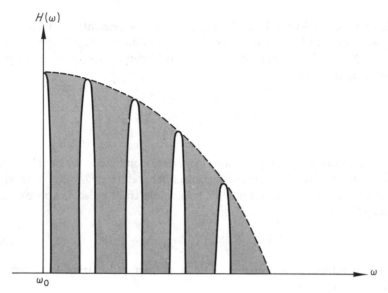

Fig. 5.6-6. North type $-\sin x/x$ weighted comb filter (*after George and Zamanakos* [5], *courtesy of IEEE*).

The performance of a uniform comb filter can be improved by the use of a preliminary rectangular filter with a reduced bandwidth of $2.74\pi/\Delta_\tau$ and a comb filter whose teeth have a bandwidth of $2.74\pi/NT_p$. In this case the improvement in signal-to-noise ratio becomes (in db)

$$10 \log \frac{\chi_{\text{out}}}{\chi_1} = 10 \log_{10} N - 2.05 \qquad (5.6\text{-}8)$$

This is almost equal to that provided by the North type of comb filter.

The filter approximations of Figs. 5.6-3, 5.6-5, and 5.6-6 can be realized by a parallel bank of weighted narrowband filters, each of which is tuned to a discrete frequency with passband characteristics as shown in the figures. For large T_p/Δ_τ (which is the usual case) this type of realization is complex due to the large number of filters required. An alternate approach is to use an active filter like that shown in Fig. 5.6-7.

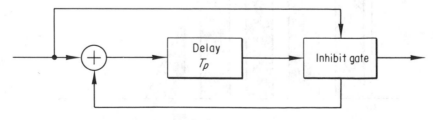

Fig. 5.6-7. Active comb filter.

An impulse applied to the input of the wideband delay line in Fig. 5.6-7 (assumed to have rectangular bandwidth Ω) produces a train of pulses at the output separated by T_p as shown in Fig. 5.6-8. An inhibit gate limits the pulse train to N pulses. The Fourier transform of the impulse response of Fig. 5.6-8 is shown in Fig. 5.6-9. This transfer characteristic is very close to the desired comb filter aproximation.

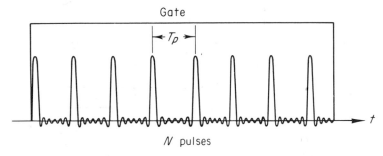

Fig. 5.6-8. Impulse response of delay-line type comb filter.

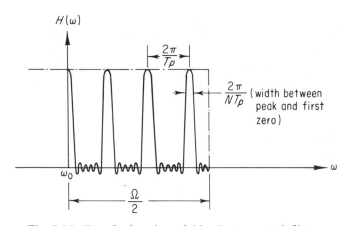

Fig. 5.6-9. Transfer function of delay-line type comb filter.

Another implementation of a comb filter is a sampler followed by a narrow-band filter (see Problem 5.11); this corresponds to a pulse-doppler type of radar processor.

5.7. Example IV: Matched Filter for Digital-Coded Pulse with Rectangular Envelope

Consider next the matched filter solution for signals generated by the delay-line network [6] shown in Fig. 5.7-1. An impulse into the signal generator is

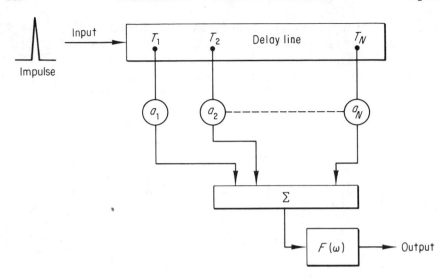

Fig. 5.7-1. Tapped delay-line signal generator.

delayed by an amount T_i, $i = 1, 2, \ldots, N$, at successive taps of the delay line. These delayed impulses are then weighted by a_i, summed and filtered by a network with transfer function $F(\omega)$. The Fourier transform $S_o(\omega)$ of the output signal can be shown to be

$$S_o(\omega) = F(\omega) \sum_{i=1}^{N} a_i e^{-j\omega T_i} \qquad (5.7\text{-}1)$$

In this example the a_i are either $+1$ or -1, $T_i = (i-1)\Delta_\tau$, and $F(\omega) = F_p(\omega)$ is the transform of a monochromatic rectangular pulse $f_p(t)$ of length Δ_τ like that shown in Fig. 5.4-1. With these assumptions, Eq. (5.7-1) becomes

$$S_o(\omega) = F_p(\omega) \sum_{i=1}^{N} a_i e^{-j\omega(i-1)\Delta_\tau} \qquad (5.7\text{-}2)$$

and the inverse Fourier transform $s_o(t)$ is given by

$$s_o(t) = \mathscr{F}^{-1}\{S_o(\omega)\} = \sum_{i=1}^{N} a_i f_p[t - (i-1)\Delta_\tau] \qquad (5.7\text{-}3)$$

where $\mathscr{F}^{-1}\{\quad\}$ denotes an inverse Fourier transform. The signal $s_o(t)$, which is illustrated in Fig. 5.7-2, consists of a series of delayed replicas of $f_p(t)$, each phase shifted by either zero or 180 degrees. From Eq. (5.3-3) the transfer function of the filter matched to $s_o(t)$ is given by

$$H_{\text{MF}}(\omega) = \frac{2k}{N_0} S_o^*(\omega) e^{-j\omega N \Delta_\tau} \qquad (5.7\text{-}4)$$

where a delay of $N\Delta_\tau$ has been added for physical realizability (see Section 5.9) and k is a constant. Substituting (5.7-2) into (5.7-4) yields

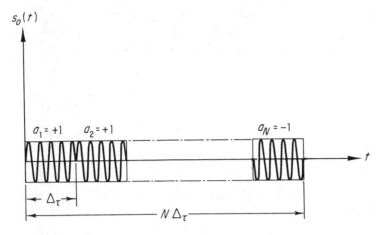

Fig. 5.7-2. Signal generated by tapped delay-line configuration of Fig. 5.7-1.

$$H_{\mathrm{MF}}(\omega) = \frac{2k}{N_0} F_p^*(\omega) \sum_{i=1}^{N} a_i e^{-j\omega(N-i+1)\Delta_\tau} \qquad (5.7\text{-}5)$$

Since $F_p(\omega)$ is real [see Eq. (5.4.-2)], Eq. (5.7-5) can be rewritten as

$$H_{\mathrm{MF}}(\omega) = c F_p(\omega) \sum_{i=1}^{N} a_i e^{-j\omega(N-i+1)\Delta_\tau} \qquad (5.7\text{-}6)$$

where c is a constant equal to $2k/N_0$. A filter with transfer function (5.7-6) is shown in Fig. 5.7-3. The output signal from this filter is given by

$$G(\omega) = c\,|F_p(\omega)|^2 \sum_{i=1}^{N}\sum_{l=1}^{N} a_i a_l e^{-j\omega(N-i+l)\Delta_\tau} \qquad (5.7\text{-}7)$$

The inverse Fourier transform of $G(\omega)$, $g(t)$, is found to be

$$g(t) = c \sum_{i=1}^{N}\sum_{l=1}^{N} a_i a_l r_p[t - (N + l - i)\,\Delta_\tau] \qquad (5.7\text{-}8)$$

where $r_p(t)$ is the inverse transform of $|F_p(\omega)|^2$ and also the autocorrelation function of pulse $f_p(t)$; it is similar to that shown in Fig. 5.4-3. The expression for $g(t)$ is a maximum for $t = N\Delta_\tau$, at which time $i = l$ in the sum of Eq. (5.7-8); this is the time instant when the entire signal $s_o(t)$ is in the matched filter of Fig. 5.7-3 and the Nth term of the signal (last term) coincides with the a_N coefficient of the filter in Fig. 5.7-3. Then,

$$g(N\Delta_\tau) = cNr_p(0) \qquad (5.7\text{-}9)$$

At this time, all subsignals have their coding removed and are delayed appropriately to add up coherently, or in phase, at the output of the summer. The exact shape of the output waveform from the matched filter at other times is determined by the values of a_i in the double sum in Eq. (5.7-8). Some

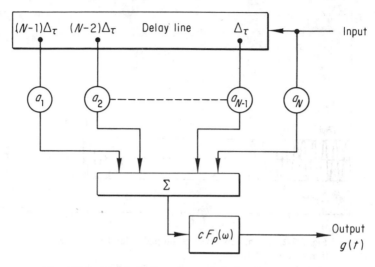

Fig. 5.7-3. Matched filter for coded signal generated by delay-line generator in Fig. 5.7-1.

binary sequences have been found that have the property that all subsidiary peaks surrounding the principal peak never exceed $g(N\Delta_\tau)/N$ [7,8].

If the waveform in Fig. 5.7-2 has amplitude A, then the peak signal-to-noise ratio out of the matched filter, from Eq. (5.3-14), is

$$\chi_p = \frac{2}{N_0}\left(\frac{A^2 N \Delta_\tau}{2}\right) = \frac{A^2 N \Delta_\tau}{N_0} \tag{5.7-10}$$

This is N times the peak that would be obtained with a single subpulse [see Eq. (5.4-8)] since the energy has been increased by just that amount.

5.8. The Stationary Clutter Filter (Urkowitz Filter)

Thus far the examples have treated only matched filters, which by definition maximize signal-to-noise ratio for a known signal in a background of white Gaussian noise. The filter that maximizes signal-to-noise ratio in the presence of nonwhite or colored noise is described by Eq. (5.2-17). This result will now be applied to a problem involving a special form of colored noise known as clutter.

When a target is located in a field of randomly located scatterers, such as vegetation, sea waves, raindrops, or chaff (metalized strips), the radar reflection from these scatterers, referred to as clutter, can be very large and may mask the desired target. The noise in this case includes receiver noise and other sources of thermal noise as well as clutter noise. The clutter reflectors often have random motion with respect to the radar, which increases the difficulty of detecting desired targets. (This problem is discussed more generally in the companion volume.) In this section a limited form of the clutter problem is treated.

The clutter considered here is assumed to be stationary; that is, each individual scatterer has zero relative velocity with respect to the radar. However, the distribution of scatterers relative to each other is random and represents a sample point in a space that includes all possible random configurations. Further, the average cross section of each individual scatterer is assumed small relative to the cross section of the desired target; this implies that reducing the duration of the transmitted pulse enhances the desired target return with respect to the background clutter by reducing the number of overlapping clutter echoes.

The power spectrum of the return from stationary clutter approaches that of the transmitted pulse [9]. This result is not unexpected, since clutter is the sum of many similar signals with arbitrary phases and amplitudes. If $S_i(\omega)$ is the Fourier transform of the transmitted signal, the clutter power spectrum $G_{nn}(\omega)$ may be expressed as

$$G_{nn}(\omega) = \rho |S_i(\omega)|^2 \qquad (5.8\text{-}1)$$

where ρ is a proportionality factor that is related to the number of overlapping echoes. Substituting (5.8-1) into Eq. (5.2-17) yields the optimum transfer function $H_{\text{opt}}(\omega)$:

$$H_{\text{opt}}(\omega) = \frac{ke^{-j\omega t_m}}{\rho S_i(\omega)} \qquad (5.8\text{-}2)$$

For simplicity let $t_m = 0$ and $k = \rho$, so that Eq. (5.8-2) becomes

$$H_{\text{opt}}(\omega) = \frac{1}{S_i(\omega)} \qquad (5.8\text{-}3)$$

The filter described by (5.8-3) is called an inverse filter for obvious reasons. The spectral output $S_o(\omega)$ of the filter is given by

$$S_o(\omega) = S_i(\omega)H_{\text{opt}}(\omega) = 1 \qquad (5.8\text{-}4)$$

and its inverse transform $s_o(t)$ becomes

$$s_o(t) = \delta(t) \qquad (5.8\text{-}5)$$

where $\delta(t)$ is the Dirac delta function. Note that $\delta(t)$ approaches infinity at $t = 0$, corresponding to an infinite output signal-to-noise ratio. In this idealized case the clutter problem is solved by converting the signal reflected from the target into an impulse function, which is easily separated in time from the smaller clutter echoes. However, solution (5.8-5) is not realizable, since the integral of the square of Eq. (5.8-4), representing the output signal energy, is unbounded.

A realizable filter approximation $H_U(\omega)$ of the inverse filter was defined by Urkowitz [10], with the following transfer characteristics:

$$H_U(\omega) = \frac{1}{S_i(\omega)} \quad \text{for} \quad \begin{cases} \omega_0 - \dfrac{\Delta_\omega}{2} < \omega < \omega_0 + \dfrac{\Delta_\omega}{2} \\[2mm] -\omega_0 - \dfrac{\Delta_\omega}{2} < \omega < -\omega_0 + \dfrac{\Delta_\omega}{2} \end{cases} \tag{5.8-6}$$

$$H_U(\omega) = 0 \qquad \text{elsewhere}$$

The output $s_o(t)$ of this filter is given by

$$s_o(t) = \frac{1}{2\pi} \left(\int_{\omega_0 - \Delta\omega/2}^{\omega_0 + \Delta\omega/2} e^{j\omega t}\, d\omega + \int_{-\omega_0 - \Delta\omega/2}^{-\omega_0 + \Delta\omega/2} e^{j\omega t}\, d\omega \right) \tag{5.8-7}$$

which simplifies to

$$s_o(t) = \frac{2}{\pi} \frac{\sin(\Delta_\omega t/2)}{t} \cos \omega_0 t \tag{5.8-8}$$

The maximum signal out of the filter (at $t = 0$) is given by

$$s_{o\text{-max}} = \frac{\Delta_\omega}{\pi} \tag{5.8-9}$$

The average clutter power out of the filter is from Eq. (5.3-11a)

$$\overline{n_o^2(t)} = \frac{1}{2\pi} \int_{-\infty}^{\infty} |H_U(\omega)|^2\, G_{nn}(\omega)\, d\omega = \frac{\rho \Delta_\omega}{\pi} \tag{5.8-10}$$

The peak output signal-to-clutter ratio $[\chi_p]_\text{out}$ is then given by

$$[\chi_p]_\text{out} = \frac{s_{o\text{-max}}^2}{\overline{n_o^2(t)}} = \frac{\Delta_\omega}{\pi \rho} \tag{5.8-11}$$

The improvement in output signal-to-clutter ratio with respect to input signal-to-clutter ratio, assuming a rectangular transmitted pulse like that in Fig. 5.4-1, can be found as follows. The peak input power is A^2. The input clutter power $\overline{n_i^2}$ is given by

$$\overline{n_i^2(t)} = \frac{1}{2\pi} \int_{-\infty}^{\infty} \rho \, |S_i(\omega)|^2\, d\omega \tag{5.8-12a}$$

$$= \rho \int_{-\Delta\tau/2}^{\Delta\tau/2} s_i^2(t)\, dt \tag{5.8-12b}$$

$$\overline{n_i^2(t)} \cong \rho \frac{A^2 \Delta_\tau}{2} \tag{5.8-12c}$$

for large ω_0. Parseval's relation was used in passing from (5.8-12a) to (5.8-12b). The peak input signal-to-clutter ratio is then given by

$$[\chi_p]_\text{in} = \frac{2}{\rho \Delta_\tau} \tag{5.8-13}$$

Equation (5.8-13) shows that the input signal-to-clutter ratio is inversely proportional to transmitted pulse length. The ratio of output to input signal-to-clutter ratios is given by

$$\frac{[\chi_p]_\text{out}}{[\chi_p]_\text{in}} = \frac{\Delta_\omega \Delta_\tau}{2\pi} = \Delta_f \Delta_\tau \tag{5.8-14}$$

where $\Delta_\omega = 2\pi\Delta_f$. Equation (5.8-14) implies that performance can be improved by increasing the transmitted pulse length or the filter bandwidth. However, an increase in pulse length increases the input clutter level; this is exactly compensated for by the increase in ratio (5.8-14). Hence there is no net improvement in performance. However, increasing the filter bandwidth Δ_ω does improve performance.

Next the performance of the Urkowitz filter $H_U(\omega)$, defined in Eq. (5.8-6), is compared with that of a simple rectangular filter with transfer function $H_R(\omega)$; both filters are shown in Fig. 5.8-1. For large ω_0, the output of the

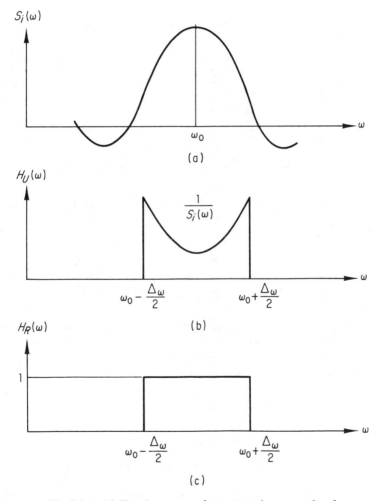

Fig. 5.8-1. (a) Signal spectrum for rectangular narrowband pulse with carrier frequency ω_0; (b) Urkowitz filter; (c) rectangular filter. [*Note:* Only positive spectra are shown.]

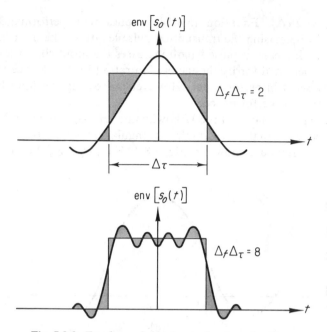

Fig. 5.8-2. Envelope of signal out of rectangular filter.

rectangular filter $s_o(t)$ is given by Eq. (5.4-10), which can be solved to yield

$$s_o(t) = \frac{A}{\pi} \left\{ \text{Si} \left[\frac{\Delta_\omega}{2} \left(t + \frac{\Delta_\tau}{2} \right) \right] - \text{Si} \left[\frac{\Delta_\omega}{2} \left(t - \frac{\Delta_\tau}{2} \right) \right] \right\} \cos \omega_0 t \qquad (5.8\text{-}15)$$

where Si (x) denotes the sine integral defined by Eq. (5.4-13). Sketches of the envelope of $s_o(t)$ (the difference of the two sine integrals) are shown in Fig. 5.8-2 for two different pulse-length filter bandwidth products: $\Delta_f \Delta_\tau = 2$ and $\Delta_f \Delta_\tau = 8$.

Note that in the second case plotted the peak value of the output signal does not occur at $t = 0$. This is due to the sharp cutoff characteristics of the assumed filter, which results in ringing and overshoot.† This phenomenon is not present in bandpass filters with more slowly decaying band edges. For this reason the peak output of the rectangular filter is assumed to occur at $t = 0$, in all cases, so that from (5.8-15)

$$s_{o\text{-max}} = \frac{2A}{\pi} \text{Si} \left(\frac{\Delta_\omega \Delta_\tau}{4} \right) \qquad (5.8\text{-}16)$$

The clutter power out of the rectangular bandpass filter is given by

$$\overline{n_o^2} = \frac{1}{2\pi} \cdot 2 \int_{\omega_0 - \Delta\omega/2}^{\omega_0 + \Delta\omega/2} \rho \, |S_i(\omega)|^2 \, d\omega \qquad (5.8\text{-}17\text{a})$$

†The overshoot is less than 9 per cent.

$$= \frac{\rho A^2}{\pi} \int_{\omega_0 - \Delta\omega/2}^{\omega_0 + \Delta\omega/2} \frac{\sin^2 (\omega - \omega_0) \Delta_\tau/2}{(\omega - \omega_0)^2} \, d\omega \qquad (5.8\text{-}17\text{b})$$

$$= \frac{\rho A^2 \Delta_\tau}{\pi} \int_0^{\Delta\omega\Delta_\tau/4} \frac{\sin^2 x}{x^2} \, dx \qquad (5.8\text{-}17\text{c})$$

$$\overline{n_o^2} = \frac{\rho A^2 \Delta_\tau}{\pi} \left[\mathrm{Si} \left(\frac{\Delta_\omega \Delta_\tau}{2} \right) - \frac{\sin^2 \left(\dfrac{\Delta_\omega \Delta_\tau}{4} \right)}{\Delta_\omega \Delta_\tau/4} \right] \qquad (5.8\text{-}17\text{d})$$

The peak output signal-to-clutter ratio from Eqs. (5.8-16) and (5.8-17d) can be written as

$$[\chi_p]_{\text{out}} = \frac{4}{\pi\rho \, \Delta_\tau} \frac{[\mathrm{Si}\,(x)]^2}{\left[\mathrm{Si}\,(2x) - \dfrac{\sin^2 x}{x} \right]} \qquad (5.8\text{-}18)$$

where $x = \Delta_\omega \Delta_\tau/4$.

The ratio of the peak output signal-to-clutter ratio in (5.8-18) to the input signal-to-clutter ratio given in Eq. (5.8-13) is given by

$$\frac{[\chi_p]_{\text{out}}}{[\chi_p]_{\text{in}}} = \frac{2}{\pi} \frac{[\mathrm{Si}\,(x)]^2}{\left[\mathrm{Si}\,(2x) - \dfrac{\sin^2 x}{x} \right]} \qquad (5.8\text{-}19)$$

Expression (5.8-19) is plotted in Fig. 5.8-3. Equation (5.8-14) for the

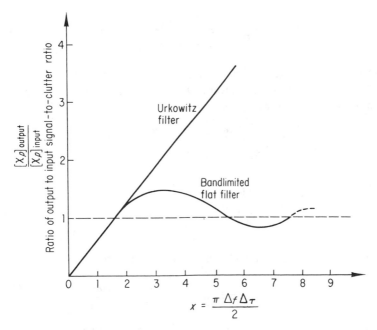

Fig. 5.8-3. Ratio of output to input signal-to-clutter ratios for both Urkowitz filter and rectangular passband filter.

Urkowitz filter is also plotted on Fig. 5.8-3 for comparison purposes. Note that the two filters perform similarly up to $\Delta_\tau \Delta_f = 1.3$. As the filter bandwidth is increased beyond this point, the Urkowitz filter output signal-to-clutter ratio improves while the rectangular filter settles about an asymptotic value of unity (no improvement).

Thus far only clutter has been assumed present. Since no physical system is free of thermal noise, as the filter bandwidth Δ_f is increased to improve performance against clutter, the noise power passed by the filter increases and eventually limits the improvement attainable. In order to account for the effects of thermal noise, the total interference-power density $G_{nn}(\omega)$ can be expressed as

$$G_{nn}(\omega) = \rho \, |\, S_i(\omega)\,|^2 + \frac{N_0}{2} \tag{5.8-20}$$

where $N_0/2$ is the spectral density of white thermal receiver noise. Then the optimum filter transfer function $H_{\text{opt}}(\omega)$, from Eq. (5.2-17), is given by

$$H_{\text{opt}}(\omega) = \frac{k S_i^*(\omega)}{\dfrac{N_0}{2} + \rho \, |\, S_i(\omega)\,|^2} \tag{5.8-21}$$

where again t_m has been set equal to zero. The signal $s_o(t)$ out of the optimum filter is given by

$$s_o(t) = \frac{1}{2\pi} \int_{-\infty}^{\infty} \frac{k\, |\, S_i(\omega)\,|^2}{\dfrac{N_0}{2} + \rho\, |\, S_i(\omega)\,|^2} \, e^{j\omega t} \, d\omega \tag{5.8-22}$$

It can be shown that the maximum value of $s_o(t)$ occurs at $t = 0$ (see Section 5.4). The output interference power $\overline{n_o^2}$ is obtained once again by substituting into Eq. (5.3-11) as follows:

$$\overline{n_o^2} = \frac{1}{2\pi} \int_{-\infty}^{\infty} \frac{k^2\, |\, S_i(\omega)\,|^2}{\left[\dfrac{N_0}{2} + \rho\, |\, S_i(\omega)\,|^2\right]^2} \cdot \left[\frac{N_0}{2} + \rho\, |\, S_i(\omega)\,|^2\right] d\omega \tag{5.8-23}$$

which simplifies to

$$\overline{n_o^2} = \frac{1}{2\pi} k^2 \int_{-\infty}^{\infty} \frac{|\, S_i(\omega)\,|^2}{\dfrac{N_0}{2} + \rho\, |\, S_i(\omega)\,|^2} \, d\omega \tag{5.8-24}$$

The peak output signal-to-interference ratio is then given by

$$[\chi_p]_{\text{out}} = \frac{1}{2\pi} \int_{-\infty}^{\infty} \frac{|\, S_i(\omega)\,|^2}{\dfrac{N_0}{2} + \rho\, |\, S_i(\omega)\,|^2} \, d\omega \tag{5.8-25}$$

To illustrate, consider the linear fm example of Section 5.5 assuming a signal with a large time-bandwidth product. Since $S_i(\omega)$ is approximately constant over the band $\Delta_\omega = 2\pi\,\Delta_f$ and zero outside this band, the input signal energy E may be expressed as

$$E = \frac{1}{\pi} \int_{\omega_0 - \Delta\omega/2}^{\omega_0 + \Delta\omega/2} |S_i(\omega)|^2\, d\omega = 2M^2\Delta_f \qquad (5.8\text{-}26)$$

where $M \cong |S_i(\omega)|$ and Eq. (5.8-25) simplifies to

$$[\chi_p]_{\text{out}} = \frac{2E}{N_0 + \rho\,\dfrac{E}{\Delta_f}} \qquad (5.8\text{-}27)$$

Observe that as the clutter power approaches zero ($\rho \rightarrow 0$), the peak output signal-to-noise ratio approaches the matched filter result, $\chi_{\text{MF}} = 2E/N_0$. As $\rho \rightarrow \infty$, the value of $[\chi_p]_{\text{out}}$ approaches $2\Delta_f/\rho$ and increases with increasing bandwidth, as previously predicted.

Since the results above do not depend on the functional form of the signal as a function of time, the solution applies to any signal with approximately rectangular spectral characteristics. The ratio of clutter power to noise power determines the ultimate performance that can be obtained. If Eq. (5.8-27) is maximized with respect to ρ, for fixed E, the maximum signal-to-interference ratio is obtained when

$$\rho = \frac{N_0 \Delta_f}{E} \qquad (5.8\text{-}28)$$

—that is, when ρ is equal to the ratio of average thermal noise power (rms noise) to signal energy (for a signal with a rectangular spectrum). Solving (5.8-28) for Δ_f shows maximum output signal-to-interference ratio is obtained for Δ_f equal to the ratio of the clutter power ρE to twice the thermal noise spectral density $N_0/2$. The maximum value of $[\chi_p]_{\text{out}}$ is equal to E/N_0 and is just 3 db poorer than that obtained with a matched filter in the presence of white noise alone [see Eq. (5.3-14)].

In the preceding discussion it has been tacitly assumed that clutter noise is a stationary random process. Generally, this approximation is not valid since clutter return is range dependent. However, over reasonably short intervals of time, a nonstationary clutter process can be approximated by a stationary process obtained by analytically extending the clutter return in a selected interval over all time. This approach is also discussed in the companion volume and in Wainstein and Zubakov [11].

5.9. Parameter Estimation and Matched Filtering

In early sections of this chapter the maximization of signal-to-noise ratio was introduced heuristically as a measure of the effectiveness of a receiver in combating noise. The problem considered was that of detecting an exactly known signal in noise. When the structure of the received signal is known but values of one or more signal parameters are unknown, the problem of detecting a signal is closely related to that of parameter estimation. Previously developed ideas can be extended to handle this problem.

For illustrative purposes assume that a signal has an unknown doppler shift due to target motion with respect to the radar, which is one of three discrete frequencies: 0, ω_d, or $2\omega_d$. The zero doppler-shifted signal with spectrum $S_1(\omega)$ is shown in Fig. 5.9-1. Signal spectra $S_2(\omega)$ and $S_3(\omega)$, corre-

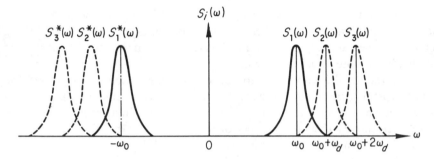

Fig. 5.9-1. Signal spectra for expected doppler shifts of 0, ω_d, and $2\omega_d$.

sponding to doppler shifts ω_d and $2\omega_d$, are approximately related to $S_1(\omega)$ in the right half plane† by

$$S_2(\omega) = S_1(\omega - \omega_d) \tag{5.9-1a}$$

$$S_3(\omega) = S_1(\omega - 2\omega_d) \tag{5.9-1b}$$

These spectra are also shown in Fig. 5.9-1. Since the matched filter transfer function is proportional to the complex conjugate of the signal spectrum [see Eq. (5.3-3)], a separate matched filter can be employed for each expected doppler translation as shown in Fig. 5.9-2. The peak output at time t_m of the filter matched to the true doppler will, by definition, be larger than the output of the remaining two matched filters, which are mismatched to the arriving signal. This criterion can be used to estimate the doppler shift of the arriving signal. Note, however, that owing to the presence of background noise the doppler shift selected can be in error.

By analogy with the result obtained for estimating doppler shift, if the time of arrival of the signal is unknown, strictly a matched filter followed by a sampler is required corresponding to each expected time of arrival; however, this is not necessary in general. For a matched filter impulsive response of duration T the output at any time t is dependent only on the input extending T seconds back into the past. If the sampler is dispensed with, the matched

---

†The spectra are shifted to the left by the same amount in the left half plane.

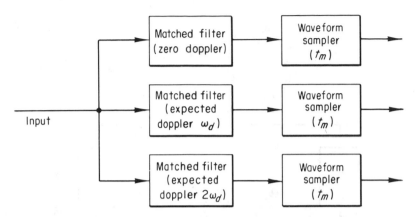

Fig. 5.9-2. Matched filter bank for expected doppler shifts 0, ω_d, and $2\omega_d$.

filter output might resemble that of Fig. 5.9-3 for a simple pulsed radar signal with a large signal-to-noise ratio. Observing the matched filter output on an oscilloscope permits the delay of the peak to be easily estimated. However, since the peak of the waveform envelope may not coincide with a peak in the sine-wave carrier frequency, this can present a problem in a radar with automatic data processing. The problem may be obviated by the simple expedient of extracting the envelope of the waveform. One method for obtaining the envelope is shown in Fig. 5.9-4. In this processor the high-frequency input signal is multiplied by each of two quadrature local oscillator signals. A bandpass filter is introduced in each channel to eliminate the double

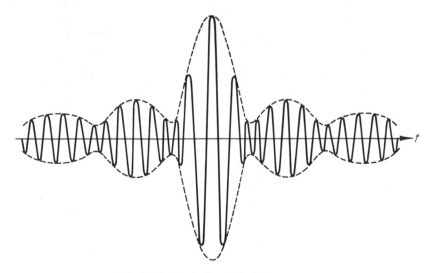

Fig. 5.9-3. Typical matched filter output.

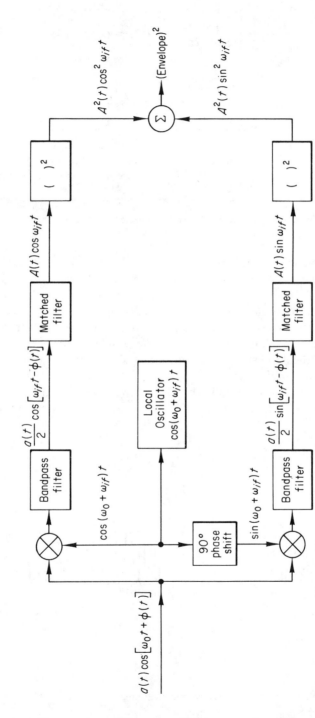

Fig. 5.9-4. Matched filter processor for estimating time delay of the envelope peak.

carrier-frequency term that results from multiplication. The remaining signals are processed by two matched filters and converted to two quadrature signals with envelope $A(t)$. These signals are squared and summed; the sum is equal to the squared envelope of the waveform out of a matched filter. In practice, a much simpler envelope detector is normally employed. The matched filter output is full-wave rectified and followed by a low-pass filter. This technique requires only one channel and a single matched filter, which represents a considerable reduction in equipment complexity.

In summary, by providing a bank of filters matched to all expected signal parameters we may estimate the true parameters by observing the filter that has the largest peak output. In the case of a continuous range of expected doppler frequencies, a number of discretely spaced filters are normally employed. However, for continuous values of expected delays only a single matched filter is required. In this case the output can be observed continuously in order to locate the peak of the output waveform or, when automatic data processing of a received rf waveform is employed, the envelope of the output can be sampled frequently to estimate the echo delay.

5.10. Remarks

As a final note on matched filters, the necessary and sufficient conditions for physical realizability of a linear filter are reviewed; they are

(a) $h(t) = 0$ for $t < 0$. This condition assures that the filter operates only on the past input waveform.

(b) $\int_{-\infty}^{\infty} |h(t)| \, dt < \infty$. This condition assures filter stability; that is, an input of finite duration produces an output that decays to zero as $t \longrightarrow \infty$.

An alternative condition for the physical realizability of linear systems is given by the Paley-Wiener relation [12].

Condition (b) is always satisfied for pulse radar signals. Condition (a) can be satisfied by specifying the matched filter delay such that t_m exceeds Δ_τ (the pulse length). When $\Delta_\tau \longrightarrow \infty$, the matched filter must be approximated [1]. Methods for the synthesis and design of matched filters are discussed in Cook and Bernfeld [13].

PROBLEMS

5.1. Find the peak signal-to-noise ratio out of a matched filter, assuming a narrowband rf signal with triangular envelope and white additive noise of spectral density $N_0/2$ (Fig. 5P-1).

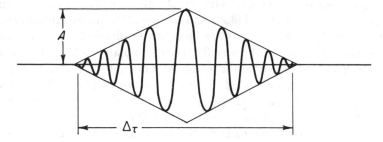

Fig. 5P-1

5.2. Find the peak signal-to-noise ratio out of a matched filter, assuming a narrowband rf signal with an exponentially decaying envelope $Ae^{-\alpha t}$, for $0 < t < \Delta_\tau$, and white background noise of spectral density $N_0/2$ (Fig. 5P-2).

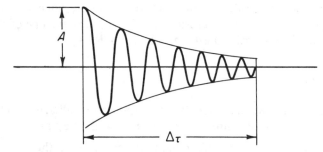

Fig. 5P-2

5.3. Show that Eq. (5.2-14) is a sufficient condition for maximizing signal-to-noise ratio under the stated conditions.

5.4. The output signal from a rectangular filter with a monochromatic rectangular pulse at the input is given by Eq. (5.4-10):

$$s_o(t) = \frac{A\Delta_\tau}{2\pi} \int_{\omega_0 - \Delta\omega/2}^{\omega_0 + \Delta\omega/2} \frac{\sin(\omega - \omega_0)\Delta_\tau/2}{(\omega - \omega_0)\Delta_\tau/2} \cos \omega t \, d\omega$$

Evaluate $s_o(t)$ in terms of the sine integral defined in Eq. (5.4-13).

5.5. The output of a filter matched to a linear fm signal is given by

$$s_o(t) = \int_{t-\Delta\tau/2}^{\Delta\tau/2} \cos\left(\omega_0 \tau + \frac{b\tau^2}{2}\right) \cos\left[\omega_0(\tau - t) + \frac{1}{2}b(\tau - t)^2\right] d\tau$$

for $t > 0$. Introduce a doppler shift ω_d in the first cosine term (signal term). Obtain an expression for the signal output (with mismatched filter) for $t > 0$.

5.6. For the pulse defined by Eq. (5.5-1) derive the Fourier transform and compute the spectrum. Plot the spectrum for $\Delta_f \Delta_\tau = 10, 100,$ and 1000.

5.7. Verify Eq. (5.5-18) in the text with the substitutions indicated. Solve the integral and establish result (5.5-21) with definitions (5.5-22) to (5.5-26).

5.8. Rederive Eq. (5.5-18) with the filter center frequency displaced from the pulse carrier ω_0 by some arbitrary amount; this might represent a doppler-shifted pulse relative to the new filter transfer function. Solve this case for a result similar to that of Eq. (5.5-21) and check the result with reference [4].

5.9. Consider the comb filter example of Section 5.6 and Fig. 5.6-7. Derive an expression for the shape of a typical pulse in the pulse train shown in Fig. 5.6-8. What is the true spectrum of the pulse train shown in Fig. 5.6-9 within bandwidth Ω? Can the uniform comb filter shown in Fig. 5.6-3 be closely approximated by this technique?

5.10. Maximize Eq. (5.8-27) with respect to ρ for fixed E [defined in Eq. (5.8-26)] for the linear fm example of Section 5.5 assuming a large time-bandwidth product.

5.11. The network shown in Fig. 5P-11 illustrates a pulse-doppler type radar processor. The output of the processor can be expressed as

$$e_o(t) = \int_{-\infty}^{\infty} v(x)g(x)h(t - x)\,dx$$

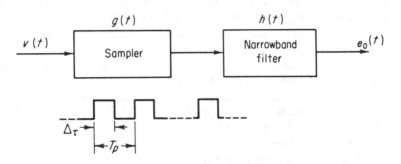

Fig. 5P-11

Let $g(x)$ be a range-gating function, as shown in the diagram, and $h(t)$ a narrowband filter with a rectangular bandpass characteristic and bandwidth $\Delta_\omega = 2\pi/NT_p$.

(a) Treat the product $g(x)h(t - x)$ as a time-varying filter impulse response:

$$\mathscr{H}(t, \tau) = h(\tau)g(t - \tau)$$

and find the transform of $\mathscr{H}(t, \tau)$ with respect to τ.

(b) Show that this transform approximates the matched filter for the pulse train described in Section 5.6.

(c) If $h(t)$ is a North type of comb filter (see Fig. 5.6-6), discuss whether the combination of sampler and comb filter is a better or poorer approximation to the matched filter.

REFERENCES

1. Zadeh, L. A., and Ragazzini, J. R.: Optimum Filters for the Detection of Signals in Noise, *Proc. IRE*, **40**: 1123–1131 (October, 1952).

2. Cook, C. E.: Pulse Compression—Key to More Efficient Radar Transmission, *Proc. IRE*, **48**: (*3*), 310–316 (March, 1960).

3. Cook, C. E.: General Matched-Filter Analysis of Linear FM Pulse Compression, *Proc. IRE*, **49**: (*4*), 831 (April, 1961).

4. DiFranco, J. V.: Closed Form Solution for the Output of a Finite Bandwidth Pulse-Compression Filter, *Proc. IRE*, **49**: (*6*), 1086–1087 (June, 1961).

5. George, S. F., and Zamanakos, A.: Comb Filters for Pulsed Radar Use, *Proc. IRE*, **42**: (*7*), 1159–1165 (July, 1954).

6. Turin, G. L.: An Introduction to Matched Filters, *IRE Trans. on Information Theory*, **IT-6**: (*3*), 311–329 (June, 1960).

7. Storer, J. E., and Turyn, R.: Optimum Finite Code Groups, *Proc. IRE*, **46**: 1649 (September, 1958).

8. Barker, R. H.: Group Synchronizing of Binary Digital Systems, in W. Jackson, ed., "Communication Theory," Academic Press, New York, 1953.

9. Lawson, J. L., and Uhlenbeck, G. E.: "Threshold Signals," MIT Radiation Laboratory Series, Vol. 24, p. 297, McGraw-Hill, New York, 1950.

10. Urkowitz, H.: Filters for Detection of Small Radar Signals in Clutter, *J. Appl. Phys.*, **24**: (*8*), 1024–1031 (August, 1953).

11. Wainstein, L. A., and Zubakov, V. P.: "Extraction of Signals from Noise," Prentice-Hall, Englewood Cliffs, N.J., 1962.

12. Paley, R. E., and Wiener, Norbert: "Fourier Transforms in the Complex Domain," Am. Math. Soc. Colloq. Pub., No. 10, Chap. I, 1934. (See also "Vacuum Tube Amplifiers" by Valley and Wallman.)

13. Cook, C. E., and Bernfeld, M.: "Radar Signals—An Introduction to Theory and Application," Academic Press, New York, 1967.

6

OPTIMUM FILTER FOR
COLORED NOISE†

6.1. Introduction

Equation (5.2-14) is the general solution for a filter with impulse response $h_{opt}(t)$ that maximizes the ratio of peak signal power to average noise level, at $t = t_m$, for a known signal $s_i(t)$. This equation is repeated for convenience:

$$\int_0^T h_{opt}(\tau)\phi_{nn}(t - \tau) \, d\tau = k s_i(t_m - t), \qquad 0 \leqslant t \leqslant T \qquad (6.1\text{-}1)$$

In the derivation of (6.1-1) it was assumed that the optimum filter is linear and the noise is stationary with autocorrelation function $\phi_{nn}(\tau)$.

Integral equation (6.1-1) is an inhomogeneous Fredholm integral equation of the first kind. Three cases may be delineated for solving this equation for $h_{opt}(t)$:

1. When the limits on the integral are doubly infinite in extent ($-\infty$ to $+\infty$), the equation can be solved by Fourier (or Laplace) transform techniques. The transfer function H_{opt} of the optimum filter is given by Eq.

---

†This chapter may be omitted on a first reading.

(5.2-17) and is repeated here:

$$H_{opt}(\omega) = \frac{kS_i^*(\omega)e^{-j\omega t_m}}{G_{nn}(\omega)} \qquad (6.1\text{-}2)$$

where $G_{nn}(\omega)$ is the noise spectral density. Equation (6.1-2) can also be obtained from the Schwarz inequality [1, 2]. The optimum filter in Eq. (6.1-2) is unrealizable, since it operates on the future as well as the past; however, it can be approximated with arbitrary closeness by selecting t_m sufficiently large.

2. When the observation interval is semi-infinite (0 to $+\infty$), Eq. (6.1-1) is known as the Wiener-Hopf integral equation.† A considerable body of literature exists on the solution of this equation using complex-variable theory. This filter operates on the infinite past.

3. When the observation interval is finite, as in Eq. (6.1-1), a continuous solution $h_{opt}(t)$ does not in general exist for a continuous signal $s_i(t)$ unless the kernel $\phi_{nn}(t - \tau)$ has a singularity.‡ However, when the kernel is rational, a closed-form solution of Eq. (6.1-1) can be obtained that involves delta functions and their derivatives. A rational kernel is one whose Fourier transform (spectral density) is of the form $N(\omega^2)/D(\omega^2)$, where $N(\omega^2)$ and $D(\omega^2)$ are polynomials in ω^2.

A rational kernel occurs when white noise has been filtered by a network with transfer function $H(\omega)$ that is constructed of lumped-constant elements (resistors, inductors, and capacitors). The noise spectral density at the filter output is given by $(N_0/2)\,|H(\omega)|^2$ where $N_0/2$ is the noise spectral density. Since $|H(\omega)|^2$ for a lumped filter is a rational function of ω^2, the kernel of the colored-noise process at the filter output is also rational. Most applications encountered by the radar engineer will involve filtered noise satisfying the stated conditions over wide bandwidths. Hence, the rational kernel case has considerable practical significance in applications of the theory.

When the kernel is rational, one method of solution is to express Eq. (6.1-2)—the infinite-limit solution—as a differential equation. The solution of this differential equation is substituted into integral equation (6.1-1), which is then treated as an identity. Values of undetermined constants are selected to satisfy the identity within the interval $(0, T)$. In certain cases the addition of Dirac delta functions and their derivatives to the infinite-limit solution for $h_{opt}(t)$ is required. This technique is demonstrated by the following example.

6.2. Colored *RC* Noise

The *RC* kernel is of the form

$$\phi_{nn}(\tau) = \alpha e^{-\beta|\tau|} \qquad (6.2\text{-}1)$$

---

†See Morse and Feshbach [3], pp. 978–985.
‡See Courant and Hilbert [4], p. 135.

The noise spectral density $G_{nn}(\omega)$ is the Fourier transform of (6.2-1), which is given by

$$G_{nn}(\omega) = \frac{2\alpha\beta}{\beta^2 + \omega^2} \qquad (6.2-2)$$

The autocorrelation function and spectral density of RC noise are shown in Fig. 6.2-1.

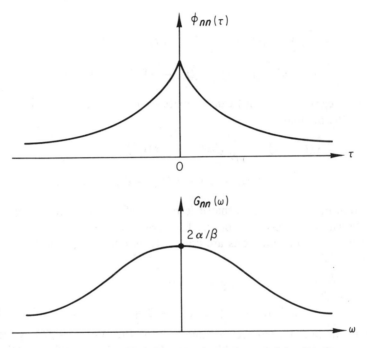

Fig. 6.2-1. Autocorrelation function and spectral density of RC noise.

Substituting Eq. (6.2-2) into Eq. (6.1-2) results in

$$2\alpha\beta H_{\text{opt}}(\omega) = (\beta^2 + \omega^2)kS_i^*(\omega)e^{-j\omega t_m} \qquad (6.2-3)$$

Multiplication of a Fourier transform by $j\omega$ corresponds to differentiation of its inverse transform† (provided the derivative exists); hence, the inverse transform of Eq. (6.2-3) can be written as

$$2\alpha\beta h_{\text{opt}}(t) = k\beta^2 s_i(t_m - t) - k\ddot{s}_i(t_m - t) \qquad (6.2-4)$$

where the dots indicate the order of differentiation. Solving Eq. (6.2-4) for $h_{\text{opt}}(t)$ yields

$$h_{\text{opt}}(t) = \frac{k}{2\alpha\beta} [\beta^2 s_i(t_m - t) - \ddot{s}_i(t_m - t)] \qquad (6.2-5)$$

†See Table 2.2-1.

Next, Eq. (6.2-5) is substituted into Eq. (6.1-1), which is then treated as an identity. This results in the following expression:

$$\frac{1}{2\beta} \int_0^t e^{-\beta(t-\tau)}[\beta^2 s_i(t_m - \tau) - \ddot{s}_i(t_m - \tau)]\, d\tau$$

$$+ \frac{1}{2\beta} \int_t^T e^{-\beta(\tau-t)}[\beta^2 s_i(t_m - \tau) - \ddot{s}_i(t_m - \tau)]\, d\tau = s_i(t_m - t),$$

$$0 \leqslant t \leqslant T \qquad (6.2\text{-}6a)$$

or

$$s_i(t_m - t) = \frac{e^{-\beta t}}{2\beta} \int_0^t e^{\beta\tau}[\beta^2 s_i(t_m - \tau) - \ddot{s}_i(t_m - \tau)]\, d\tau$$

$$+ \frac{e^{\beta t}}{2\beta} \int_t^T e^{-\beta\tau}[\beta^2 s_i(t_m - \tau) - \ddot{s}_i(t_m - \tau)]\, d\tau, \qquad 0 \leqslant t \leqslant T \qquad (6.2\text{-}6b)$$

The integrals in (6.2-6b) can be evaluated by integrating twice by parts, yielding the inequality

$$s_i(t_m - t) \neq s_i(t_m - t) - \frac{e^{-\beta t}}{2\beta}[\beta s_i(t_m) - \dot{s}_i(t_m)]$$

$$- \frac{e^{\beta(t-T)}}{2\beta}[\beta s_i(t_m - T) + \dot{s}_i(t_m - T)], \qquad 0 \leqslant t \leqslant T \qquad (6.2\text{-}7)$$

Inspection of (6.2-7) reveals that it is not satisfied for all values of t within the interval $(0, T)$ because of the last two terms on the right. Suppose, however, two delta functions are added to impulse response $h_{\text{opt}}(t)$ in Eq. (6.2-5) as follows:

$$h_{\text{opt}}(t) = \frac{k}{2\alpha\beta}[\beta^2 s_i(t_m - t) - \ddot{s}_i(t_m - t)]$$

$$+ kA\, \delta(t - \epsilon) + kB\, \delta(t - T + \epsilon), \qquad 0 \leqslant t \leqslant T \qquad (6.2\text{-}8)$$

where the delta functions are contained within the interval $(0, T)$ by the addition of a small quantity ϵ. (This avoids factors of one-half when the delta functions are on the boundaries.) Substituting modified impulse response (6.2-8) into (6.1-1) and for $\epsilon \to 0$, the result is

$$s_i(t_m - t) = s_i(t_m - t) - \frac{e^{-\beta t}}{2\beta}[\beta s_i(t_m) - \dot{s}_i(t_m)]$$

$$- \frac{e^{\beta(t-T)}}{2\beta}[\beta s_i(t_m - T) + \dot{s}_i(t_m - T)] + \alpha A e^{-\beta t} + \alpha B e^{\beta(t-T)},$$

$$0 \leqslant t \leqslant T \qquad (6.2\text{-}9)$$

In order that (6.2-9) be satisfied within the interval $(0, T)$ the values of coefficients A and B must be

$$A = \frac{1}{2\alpha\beta}[\beta s_i(t_m) - \dot{s}_i(t_m)] \qquad (6.2\text{-}10a)$$

$$B = \frac{1}{2\alpha\beta}[\beta s_i(t_m - T) + \dot{s}_i(t_m - T)] \qquad (6.2\text{-}10b)$$

Substituting (6.2-10a) and (6.2-10b) into (6.2-8) yields the impulse response $h_{opt}(t)$:

$$h_{opt}(t) = \frac{k}{2\alpha\beta}\left[\beta^2 s_i(t_m - t) - \ddot{s}_i(t_m - t)\right] + \frac{k}{2\alpha\beta}\left[\beta s_i(t_m) - \dot{s}_i(t_m)\right]\delta(t - \epsilon)$$

$$+ \frac{k}{2\alpha\beta}\left[\beta s_i(t_m - T) + \dot{s}_i(t_m - T)\right]\delta(t - T + \epsilon),$$

$$0 \leqslant t \leqslant T \qquad (6.2\text{-}11)$$

We obtain the derivatives at $t = t_m$ and $t = t_m - T$ by treating $s_i(t)$ as continuous at these two points. To assure realizability t_m must be equal to or greater than T. For $t_m = T$, Eq. (6.2-11) simplifies to

$$h_{opt}(t) = \frac{k}{2\alpha\beta}\{[\beta^2 s_i(T - t) - \ddot{s}_i(T - t)] + \delta(t - \epsilon)[\beta s_i(T) - \dot{s}_i(T)]$$

$$+ \delta(t - T + \epsilon)[\beta s_i(0) + \dot{s}_i(0)]\}, \qquad 0 \leqslant t \leqslant T \qquad (6.2\text{-}12)$$

Equation (6.2-12) is the solution of Eq. (6.1-1) for the RC kernel. The delta functions near $t = 0$ and $t = T$ are the only features that distinguish the solution above from the case in which an infinite observation interval is assumed. The

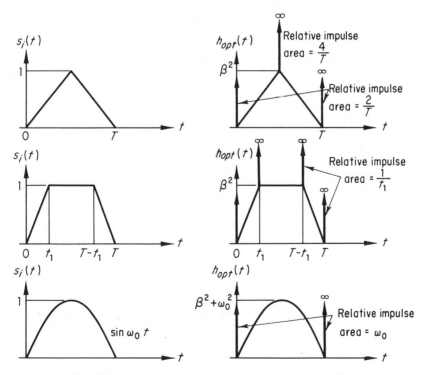

Fig. 6.2-2. Examples of impulse response of optimum filter for filtered RC noise.

Fourier transform of Eq. (6.2-12) gives the transfer function of the optimum filter as

$$H_{\text{opt}}(\omega) = \frac{k}{2\alpha\beta}(\beta^2 + \omega^2)S_i^*(\omega)e^{-j\omega T} + \beta s_i(T) - \dot{s}_i(T)$$
$$+ e^{-j\omega T}[\beta s_i(0) + \dot{s}_i(0)] \qquad (6.2\text{-}13)$$

In Fig. 6.2-2 the optimum filter impulse response of Eq. (6.2-12) is illustrated for a number of elementary (video) waveforms. Note that delta functions occur at discontinuities of $s_i(t)$.

Physical networks normally contain both inductive and capacitive energy-storage elements so that a finite time is required to change energy states. Thus, a physically generated rectangular pulse might resemble the signal shown in Fig. 6.2-3(a). Note that the signal and its derivatives are zero at $t = 0$ and $t = T$. The optimum filter impulse response for such a signal does not contain delta functions, and the solution reduces to that of the case with infinite limits. The impulse response might resemble that shown in Fig. 6.2-3(b).

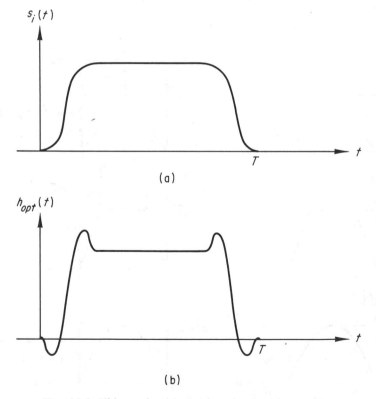

Fig. 6.2-3. Video pulse (a) and its related optimum filter impulse response (b).

The rise and decay intervals are usually small fractions of the total pulse length so that little peak signal-to-noise ratio is sacrificed.

The optimum filter response $e_{\text{opt}}(t)$ to an input $v_i(t)$, assuming that $s(t)$ is continuous at $t = 0$ and T and that $s(t)$ and its derivatives are equal to zero at these boundaries, is given by

$$e_{\text{opt}}(t) = \frac{k}{2\alpha\beta} \int_0^T [\beta^2 s_i(T - \tau) - \ddot{s}_i(T - \tau)]v_i(t - \tau)\,d\tau \qquad (6.2\text{-}14)$$

The peak signal output at $t = T$ is given by

$$e_{\text{opt}}(T) = \frac{k}{2\alpha\beta} \int_0^T [\beta^2 s_i(t) - \ddot{s}_i(t)]v_i(t)\,dt \qquad (6.2\text{-}15)$$

A system with output $e_{\text{opt}}(T)$ is shown in Fig. 6.2-4. This realization, which requires signal multiplication, can be difficult to achieve for signals of large dynamic range, since multiplication is usually accomplished by nonlinear devices. A passive filter implementation is usually preferred; for this to be possible, $S_i(\omega)$ must decay more rapidly than $(1/\omega^2)$, as ω becomes infinitely large.

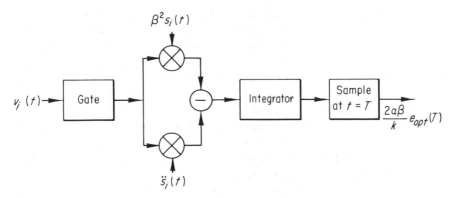

Fig. 6.2-4. Time-domain implementation of the optimum filter for colored *RC* noise.

6.3. Solution of Integral Equation for a Rational Kernel

When the autocorrelation function $\phi_{nn}(\tau)$ is a sum of exponential kernels of the type in Eq. (6.2-1), its Fourier transform $G_{nn}(\omega)$ can be expressed as

$$G_{nn}(\omega) = \frac{N(\omega^2)}{D(\omega^2)} \qquad (6.3\text{-}1)$$

where both $N(\omega^2)$ and $D(\omega^2)$ are polynomials in ω^2 of degrees m and n, respectively, and where $m \leqslant n$. This type of rational kernel characterizes filtered stationary white noise.

The solution for this kernel is obtained in a manner similar to that in Section 6.2. Eq. (6.3-1) is substituted into the solution with infinite limits given by Eq. (6.1-2) which is then reduced to a differential equation by taking the inverse Fourier transform with appropriate identification of terms; for example, multiplication by ω^2 represents the negative second derivative operator, and so on. The result is given by

$$N\left(-\frac{d^2}{dt^2}\right) h_{\text{opt}}(t) = kD\left(-\frac{d^2}{dt^2}\right) s_i(T-t), \qquad 0 \leqslant t \leqslant T \qquad (6.3\text{-}2)$$

with t_m replaced by T for simplicity. The solution to this linear differential equation is treated in many elementary texts† and consists of the sum of a complementary solution to the homogeneous equation

$$N\left(-\frac{d^2}{dt^2}\right) h_{\text{opt}}(t) = 0 \qquad (6.3\text{-}3)$$

and a particular solution that satisfies Eq. (6.3-2). The complementary solution can be expressed as a sum of exponentials of the form

$$\sum_i k_i e^{\alpha_i t} \qquad (6.3\text{-}4)$$

where the α_i are the $2m$ distinct roots of Eq. (6.3-3) written in operator notation ($\mathscr{D} = d/dt$) and the k_i are $2m$ arbitrary constants. The particular solution of Eq. (6.3-2) can generally be found by Fourier or Laplace transform methods and represents a "steady-state" solution. Other methods for obtaining particular solutions are shown in Sokolnikoff [5]; these include the methods of iteration and partial fractions.

The solution of Eq. (6.3-2) and the known kernel are then substituted into Eq. (6.1-1) and the integral is evaluated. The arbitrary constants are selected to satisfy the identity within the interval $(0, T)$. In many cases (for $m \neq n$) delta functions and their derivatives must be added to $h_{\text{opt}}(t)$ at the ends of the interval $(0, T)$ in order to satisfy the identity.

The complete solution, obtained by Zadeh and Ragazzini [6], is given by

$$h_{\text{opt}}(t) = h_{ss}(t) + \sum_{i=0}^{n-m-1} (-1)^i [a_i \overset{(i)}{\delta}(t) + b_i \overset{(i)}{\delta}(T-t)]$$

$$+ \sum_{i=1}^{m} [c_i e^{-h_i t} + d_i e^{-h_i(T-t)}], \qquad 0 \leqslant t \leqslant T \qquad (6.3\text{-}5)$$

where $h_{ss}(t)$ is the solution of Eq. (6.1-1) with infinite limits, $\overset{(i)}{\delta}(t)$ is the ith derivative of the delta function, and $\pm jh_i$ are the zeros of spectral density $G_{nn}(\omega)$. (The h_i's are assumed to be distinct and to have positive real parts.) In [6] it is shown that the $2n$ unknown coefficients a_i, b_i, c_i, and d_i are determined from a set of linear simultaneous equations. When $m = 0$, the exponential terms (complementary solution) do not appear, and when $m = n$ the delta-function terms do not appear.

†See [5], pp. 291–298.

Another solution of the integral equation, when the poles of $G_{nn}(\omega)$ are simple, is given by Laning and Battin [7] in terms of matrix operations and involves the calculation of the inverse of a Vandermonde matrix of order n. Helstrom [8] gives a simpler method, which involves the calculation of the inverse of an $m \times m$ matrix and the use of standard techniques for solving the Wiener-Hopf equation [3]. Because the general solution is laborious and lengthy, it is omitted. Instead, we next treat an example that illustrates the method (and its tediousness).

6.4. A Mixed Kernel (White plus Colored Noise)

When both colored and white noise are present simultaneously, as shown in Fig. 6.4-1, the optimum filter is obtained by a procedure similar to the previous case. Such noise might arise, for example, in a receiver with colored background thermal noise where the antenna input noise is white. The

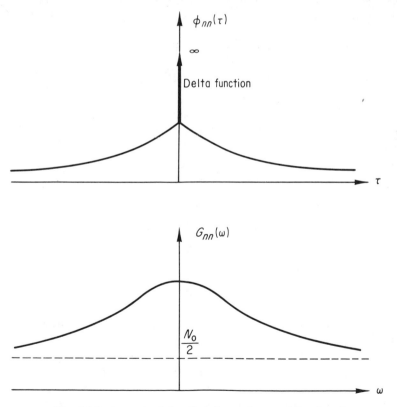

Fig. 6.4-1. Autocorrelation function and spectral density of a mixed kernel including white and colored noise.

autocorrelation function of the noise process shown in Fig. 6.4-1 is given by

$$\phi_{nn}(\tau) = \frac{N_0}{2}\delta(\tau) + \alpha e^{-\beta|\tau|} \qquad (6.4\text{-}1)$$

where $\delta(\tau)$ is the Dirac delta function. When (6.4-1) is substituted into Eq. (6.1-1), the integral equation becomes

$$\frac{N_0}{2}h_{\text{opt}}(t) + \int_0^T \alpha e^{-\beta|t-\tau|}h_{\text{opt}}(\tau)\,d\tau = ks_i(T-t), \qquad 0 \leqslant t \leqslant T \qquad (6.4\text{-}2)$$

Equation (6.4-2) is a Fredholm integral equation of the second kind. A solution generally exists if $(-N_0/2)$ is not an eigenvalue of the homogeneous integral equation (see Courant and Hilbert [4], chap. 3). For real noise processes a negative eigenvalue is impossible, and a solution is assured. In addition, delta functions do not appear in the expression for $h_{\text{opt}}(t)$ as in the general solution of Section 6.3.

This problem can be solved by the technique previously employed. The noise spectral density $G_{nn}(\omega)$ is found from the Fourier transform of $\phi_{nn}(\tau)$ of Eq. (6.4-1):

$$G_{nn}(\omega) = \frac{N_0}{2} + \frac{2\alpha\beta}{\beta^2 + \omega^2} \qquad (6.4\text{-}3a)$$

$$G_{nn}(\omega) = \frac{N_0\omega^2 + N_0\beta^2 + 4\alpha\beta}{2(\omega^2 + \beta^2)} \qquad (6.4\text{-}3b)$$

$G_{nn}(\omega)$ in (6.4-3b) is a rational function of ω^2, distinguishable from previous cases by a spectral density with numerator and denominator of the same degree. Let

$$\mu^2 = \beta^2 + \frac{4\alpha\beta}{N_0} \qquad (6.4\text{-}4)$$

Equation (6.4-3b) can be rewritten as

$$G_{nn}(\omega) = \frac{N_0(\omega^2 + \mu^2)}{2(\omega^2 + \beta^2)} \qquad (6.4\text{-}5)$$

Substituting (6.4-5) into Eq. (6.1-2) yields, for $t_m = T$,

$$(\omega^2 + \mu^2)H_{\text{opt}}(\omega) = \frac{2k}{N_0}(\omega^2 + \beta^2)S_i^*(\omega)e^{-j\omega T} \qquad (6.4\text{-}6)$$

This equation can be converted to a differential equation, as in Sections 6.2 and 6.3, resulting in

$$\ddot{h}_{\text{opt}}(t) - \mu^2 h_{\text{opt}}(t) = \frac{2k}{N_0}[\ddot{s}_i(T-t) - \beta^2 s_i(T-t)] \qquad (6.4\text{-}7)$$

In operator notation the homogeneous equation is

$$(\mathscr{D}^2 - \mu^2)h_{\text{opt}}(t) = 0 \qquad (6.4\text{-}8)$$

so that the complementary solution is given by

$$h_{opt}(t)_c = c_1 e^{\mu t} + c_2 e^{-\mu t} \tag{6.4-9}$$

To obtain the particular solution, let the input signal be

$$s_i(t) = e^{-gt}, \qquad 0 \leqslant t \leqslant T \tag{6.4-10}$$

For $gT \ll 1$, $s_i(t)$ is a nearly rectangular pulse in the interval $(0, T)$. In this case the particular solution is

$$h_{opt}(t)_p = ae^{-g(T-t)} = Ae^{gt} \tag{6.4-11}$$

The total solution is found by combining (6.4-9) and (6.4-11), which yields

$$h_{opt}(t) = c_1 e^{\mu t} + c_2 e^{-\mu t} + Ae^{gt}, \qquad 0 \leqslant t \leqslant T \tag{6.4-12}$$

The constant A is evaluated by substituting particular solution (6.4-11) into differential equation (6.4-7). The result is

$$A = \frac{2}{N_0} \left(\frac{\beta^2 - g^2}{\mu^2 - g^2} \right) e^{-gT} \tag{6.4-13}$$

If signal (6.4-10) and solution (6.4-12) are substituted into integral equation (6.4-2), a number of terms of the form exp (μt), exp $(-\mu t)$, and so on, with undetermined coefficients are obtained. Since the identity must be satisfied at all values of t within the interval 0 to T, these coefficients are equated to the coefficients of like terms on the right side of the equation. After considerable algebraic manipulation, coefficients c_1 and c_2 are found to be

$$c_1 = A \frac{e^{gT}(\mu - \beta)(\mu + \beta)^2(\beta + g) + e^{-\mu T}(\mu + \beta)(\mu - \beta)^2(\beta - g)}{e^{\mu T}(\mu + \beta)^2(\beta^2 - g^2) - e^{-\mu T}(\mu - \beta)^2(\beta^2 - g^2)} \tag{6.4-14}$$

$$c_2 = A \frac{e^{gT}(\mu - \beta)^2(\mu + \beta)(\beta + g) + e^{\mu T}(\mu + \beta)^2(\mu - \beta)(\beta - g)}{e^{\mu T}(\mu + \beta)^2(\beta^2 - g^2) - e^{-\mu T}(\mu - \beta)^2(\beta^2 - g^2)} \tag{6.4-15}$$

Equations (6.4-14) and (6.4-15) complete the description of $h_{opt}(t)$.

The impulsive response corresponding to input signal $s_i(t)$ is shown in Fig. 6.4-2 for various values of α. In each case the smallest value of the impulsive response was normalized to unity. The $\alpha = 0$ case corresponds to white noise alone, for which $h_{opt}(t)$ describes the matched filter impulse response as in Chapter 5. As α increases from zero, $h_{opt}(t)$ approaches the response shown in the middle example of Fig. 6.2-2 with t_1 approaching zero.

In the frequency domain the input noise spectral density is a maximum at $\omega = 0$ and decreases as $\omega \longrightarrow \pm\infty$. As a result the optimum filter weights signal spectral components in the vicinity of $\pm\infty$ more heavily. This high-frequency emphasis of the optimum filter contributes to the spikiness of its impulsive response.

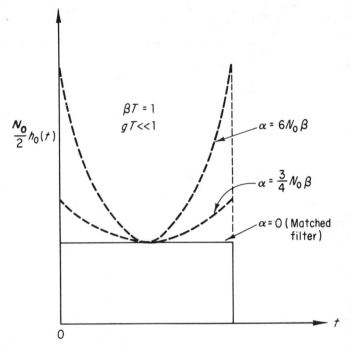

Fig. 6.4-2. Impulsive response of optimum filter for white noise plus *RC* noise.

6.5. Colored Noise with Infinite Observation Interval

When the observation interval extends from $-\infty$ to $+\infty$, the solution for the optimum filter in the presence of colored additive noise can be obtained by another method. The technique consists of "whitening" the input noise in a preliminary filtering operation and following this filter with a conventional matched filter. Thus if the input noise has a spectral density $G_{nn}(\omega)$,

$$G_{nn}(\omega) = \frac{N(\omega^2)}{D(\omega^2)} \qquad (6.5\text{-}1)$$

the prewhitening filter has a transfer function given by

$$|H_W(\omega)|^2 = \frac{N_0}{2} \frac{D(\omega^2)}{N(\omega^2)} \qquad (6.5\text{-}2)$$

The spectral density at the output of the prewhitening filter is $(N_0/2)$. Since $|H_W(\omega)|^2$ is real, it may be factored into two complex conjugate parts as follows:

$$|H_W(\omega)|^2 = \frac{N_0}{2} \frac{D_1(j\omega)D_1^*(j\omega)}{N_1(j\omega)N_1^*(j\omega)} \qquad (6.5\text{-}3)$$

where $D_1 D_1^* = D(\omega^2)$ and $N_1 N_1^* = N(\omega^2)$. From realizability constraints, the prewhitening filter transfer function $H_W(\omega)$ is

$$\sqrt{\frac{N_0}{2}} \left[\frac{D_1(j\omega)}{N_1(j\omega)} \right]$$

The signal out of this filter, $S_i'(\omega)$, is given by

$$S_i'(\omega) = S_i(\omega) H_W(\omega) = \sqrt{\frac{N_0}{2}} \frac{D_1(j\omega)}{N_1(j\omega)} \cdot S_i(\omega) \qquad (6.5\text{-}4)$$

Since the noise is white at this point, the optimum filter corresponding to $S_i'(\omega)$ is a matched filter with transfer function

$$H_{\mathrm{MF}}(\omega) = \sqrt{\frac{2}{N_0}} \cdot k \frac{D_1^*(j\omega)}{N_1^*(j\omega)} \cdot S_i^*(\omega) e^{-j\omega T} \qquad (6.5\text{-}5)$$

When the matched filter is cascaded with the prewhitening filter as shown in Fig. 6.5-1, the overall transfer function of the combination is given by

$$H_{\mathrm{opt}}(\omega) = H_{\mathrm{MF}}(\omega) H_W(\omega) = k \frac{D(\omega^2)}{N(\omega^2)} S_i^*(\omega) e^{-j\omega T} \qquad (6.5\text{-}6a)$$

$$H_{\mathrm{opt}}(\omega) = k \frac{S_i^*(\omega) e^{-j\omega T}}{N(\omega^2)/D(\omega^2)} \qquad (6.5\text{-}6b)$$

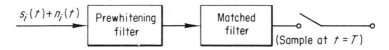

Fig. 6.5-1. Colored-noise optimum filter as a cascade of a prewhitening filter and a matched filter.

which is identical to Eq. (6.1-2) with $G_{nn}(\omega)$ given by (6.5-1) for $t_m = T$.

    As an illustration, consider RC noise where $G_{nn}(\omega)$ is given by Eq. (6.2-2). In this case the prewhitening filter transfer function is

$$H_W(\omega) = \frac{\sqrt{N_0}\,(\beta + j\omega)}{2\sqrt{\alpha\beta}} = K(\beta + j\omega) \qquad (6.5\text{-}7)$$

The matched filter transfer function is given by

$$H_{\mathrm{MF}}(\omega) = \sqrt{\frac{2}{N_0}}\, k \frac{\beta - j\omega}{\sqrt{2\alpha\beta}}\, S_i^*(\omega) e^{-j\omega T} \qquad (6.5\text{-}8)$$

The optimum noise filter is described by

$$H_{\mathrm{opt}}(\omega) = H_{\mathrm{MF}}(\omega) H_W(\omega) = \frac{k}{2\alpha\beta} (\beta^2 + \omega^2)\, S_i^*(\omega) e^{-j\omega T} \qquad (6.5\text{-}9)$$

This result is identical to Eq. (6.2-3), except for a scale factor. As expected, $H_W(\omega)$ is not easily realized.

    The solution for the infinite observation interval can be extended to other situations. Thus if the signal has finite rise and decay times such that the values of the signal and its derivatives are zero at the boundaries ($t = 0$ and $t = T$), then the cases treated in Sections 6.2, 6.3, and 6.4 in which a finite observation time was assumed can be treated in a similar manner.

6.6. Maximum Signal-to-Noise Ratio

The maximum signal-to-noise ratio at the output of the optimum filter can be computed for the infinite-limit case in a straightforward manner. The result is also applicable to the finite observation-interval case when the conditions stated at the end of Section 6.5 are met. From Eq. (6.5-6), the output signal spectrum is

$$S_o(\omega) = S_i(\omega)H_{\text{opt}}(\omega) = k \frac{D(\omega^2)}{N(\omega^2)} |S_i(\omega)|^2 e^{-j\omega T} \qquad (6.6\text{-}1)$$

The output signal time function is given by

$$s_o(t) = \frac{k}{2\pi} \int_{-\infty}^{\infty} \frac{D(\omega^2)}{N(\omega^2)} |S_i(\omega)|^2 e^{j\omega(t-T)} \, d\omega \qquad (6.6\text{-}2)$$

The peak signal output occurs at $t = T$; thus,

$$s_o(T) = \frac{k}{2\pi} \int_{-\infty}^{\infty} \frac{D(\omega^2)}{N(\omega^2)} |S_i(\omega)|^2 \, d\omega \qquad (6.6\text{-}3)$$

The average noise out of the optimum filter is found, in a similar manner, to be

$$\overline{n_o^2(t)} = \frac{1}{2\pi} \int_{-\infty}^{\infty} \left(\frac{N_0}{2}\right)\left(\frac{2k^2}{N_0}\right) \frac{D(\omega^2)}{N(\omega^2)} |S_i(\omega)|^2 \, d\omega \qquad (6.6\text{-}4)$$

When results (6.6-3) and (6.6-4) are combined, the peak signal-to-noise ratio χ_c is given by

$$\chi_c = \frac{[s_o(T)]^2}{\overline{n_o^2(t)}} = \frac{\left[\dfrac{k}{2\pi} \displaystyle\int_{-\infty}^{\infty} \dfrac{D(\omega^2)}{N(\omega^2)} |S_i(\omega)|^2 \, d\omega\right]^2}{\dfrac{k^2}{2\pi} \displaystyle\int_{-\infty}^{\infty} \dfrac{D(\omega^2)}{N(\omega^2)} |S_i(\omega)|^2 \, d\omega} \qquad (6.6\text{-}5)$$

which simplifies to

$$\chi_c = \frac{1}{2\pi} \int_{-\infty}^{\infty} \frac{D(\omega^2)}{N(\omega^2)} |S_i(\omega)|^2 \, d\omega \qquad (6.6\text{-}6)$$

To illustrate this result, consider the example in Section 6.2 where $N(\omega^2) = 2\alpha\beta$ and $D(\omega^2) = \beta^2 + \omega^2$. Substituting into (6.6-6) yields

$$\chi_c = \frac{1}{2\pi(2\alpha\beta)} \int_{-\infty}^{\infty} (\beta^2 + \omega^2) |S_i(\omega)|^2 \, d\omega \qquad (6.6\text{-}7)$$

or

$$\chi_c = \frac{E}{2\alpha\beta}\left[\beta^2 + \frac{\dfrac{1}{2\pi}\displaystyle\int_{-\infty}^{\infty} \omega^2\,|S_i(\omega)|^2\,d\omega}{E}\right]. \tag{6.6-8}$$

where E is the signal energy. Referring to Eqs. (2.5-7) and (2.8-15), note that the second term in the bracket of Eq. (6.6-8) is the normalized second moment $\overline{\omega_s^2}$ of the signal power spectrum. In the case of a video signal whose spectrum is symmetrical about the origin, the normalized second moment is equal to the mean-square frequency and is a measure of signal bandwidth. Equation (6.6-8) for the peak output signal-to-noise ratio then simplifies to

$$\chi_c = \frac{E(\beta^2 + \overline{\omega_s^2})}{2\alpha\beta} \tag{6.6-9}$$

It is apparent that, for fixed β, increasing signal bandwidth results in a higher output signal-to-noise ratio.

PROBLEMS

6.1. (a) For the kernel

$$\phi_{nn}(\tau) = \alpha_1 e^{-\beta_1|\tau|} + \alpha_2 e^{-\beta_2|\tau|}$$

find the spectral density $G_{nn}(\omega)$.

(b) Extend the result of (a) to the general case where

$$\phi_{nn}(\tau) = \sum_i \alpha_i e^{-\beta_i|\tau|}$$

6.2. For the narrowband kernel,

$$\phi_{nn}(\tau) = \alpha e^{-\beta|\tau|}\cos\omega_0\tau$$

find the spectral density $G_{nn}(\omega)$.

6.3. Find the impulsive response of the optimum filter when the autocorrelation function of the background noise is given by Eq. (6.4-1) and the signal is defined by

$$s_i(t) = A(1 - e^{-gt}), \qquad 0 \leqslant t \leqslant T$$

assuming $\beta T = 1$ and $gT \gg 1$. Plot the results showing the effect of increasing α while the white-noise level is maintained constant.

6.4. Find the impulsive response of the optimum filter of problem 6.3 for an infinite observation interval when the input signal is a narrowband signal of carrier frequency ω_0 with a Gaussian shaped envelope given by $A\exp(-at^2)$.

6.5. (a) For the *LRC* filter shown in Fig. 6P-5 find the output noise spectral density with input white noise of spectral density $N_0/2$. The autocorrelation function of the output noise is an example of an *LRC* kernel.

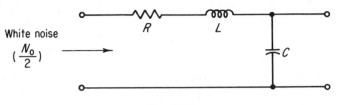

Fig. 6P-5

(b) Find the filter impulsive response that maximizes signal-to-noise ratio by the technique of Section 6.3 and the addition of delta functions and their derivatives.

(c) If the signal and all its derivatives are zero at both ends of the observation interval, what is the result?

(d) For the conditions of part (c) find the peak signal-to-noise ratio at the output.

6.6. (a) To the *LRC* kernel of problem 6.5 add white noise of spectral density $N_0/2$. Find the new spectral density. How does it differ from that of problem 6.5.

(b) For this spectral density find the complementary solution (as in Section 6.4) for the filter impulse response.

(c) For a signal of the form $(Ae^{-gt}, 0 \leqslant t \leqslant T)$ find a particular solution of the differential equation.

(d) Plot the complete optimum filter impulse response as the ratio of white to colored noise is varied, with $gT \gg 1$.

6.7. Derive the general solution of the Fredholm integral equation of the first kind for a rational kernel; that is, establish the validity of result (6.3-5) of Section 6.3.

REFERENCES

1. Van Vleck, J., and Middleton, D.: A Theoretical Comparison of the Visual, Aural, and Meter Reception of Pulsed Signals in the Presence of Noise, *J. Appl. Phys.*, **17**: 940–971 (November, 1946).

2. Dwork, B. M.: Detection of a Pulse Superimposed on Fluctuation Noise, *Proc. IRE*, **38**: 771–774 (July, 1960).

3. Morse, P. M., and Feshbach, H.: "Methods of Theoretical Physics," Vol. 1, McGraw-Hill, New York, 1953.

4. Courant, R., and Hilbert, D.: "Methods of Mathematical Physics," Wiley-Interscience, New York, 1953.

5. Sokolnikoff, I. S., and Sokolnikoff, E. S.: "Higher Mathematics for Engineers and Physicists," McGraw-Hill, New York, 1941.

6. Zadeh, L. A., and Ragazzini, J. R.: Optimum Filters for the Detection of Signals in Noise, *Proc. IRE*, **40**: 1223–1231 (October, 1952).

7. Laning, J. H., and Battin, R. H.: "Random Processes in Automatic Control," McGraw-Hill, New York, 1956.

8. Helstrom, C. W.: Solution of the Detection Integral Equation for Stationary Filtered White Noise, *IEEE Trans. on Information Theory*, **IT-11**: (*3*), 335–339 (July, 1965).

7

THE A POSTERIORI THEORY
OF RECEPTION

7.1. Introduction

Stimulated by Shannon's famous 1948 paper [1], research workers applied information theory to numerous problems in the fields of radar, coding, and communication theory. Among these early investigators were Woodward and Davies in England who wrote a number of papers [2, 3, 4, 5] applying information theory to the problem of optimizing radar reception in the presence of noise.

Using a development different from Shannon's, Woodward and Davies derived Shannon's principal results for discrete and continuous communication channels by postulating only that information is additive; they showed that the receiving process can be described as a transformation from a set of *a priori* probabilities (before reception) to a set of *a posteriori* probabilities (after reception) for every possible transmitted message. They defined an *ideal receiver* as one that calculates the a posteriori probabilities of all possible messages from the received waveform and presents this information to an observer in suitable form.

Since the received waveform usually contains more information than that required to compute the a posteriori probabilities, the ideal receiver must eliminate undesired information. Undesired information in this context refers to extraneous noise interference as well as information not pertinent to the desired output. Once the ideal receiver presents the available information in the form of an a posteriori distribution, an observer can use any suitable criterion to decide which message was actually transmitted. For example, the observer may select the most probable message as the one that was actually transmitted—that is, the message with the greatest a posteriori probability. However, the decision criterion is independent of the ideal receiver and plays no part in its design. This distinguishes *statistical decision theory* from the a posteriori theory of reception.

Woodward and Davies' concept of optimum reception differed from the earlier idea that maximization of output signal-to-noise ratio was the primary function of a receiver, as discussed in previous chapters. This divergence in view is summarized by Woodward in his book [6] in the following words:

> The problem of reception is to gain information from a mixture of wanted signal and unwanted noise, and a considerable literature exists on the subject. Much of it has been concerned with methods of obtaining as large a signal-to-noise ratio as possible on the grounds that noise is what ultimately limits sensitivity and the less there is of it the better. This is a valid attitude as far as it goes, but it does not face up to the problem of extracting information. Sometimes it can be misleading, for there is no general theorem that maximum output signal-to-noise ratio insures maximum gain of information.

In this chapter, the basic ideas of information theory are introduced in sufficient depth to provide a basis for understanding the ideal or a posteriori receiver. We then study the structure of the ideal receiver for various applications. In the case of signals received in a background of band-limited, white, Gaussian noise, the a posteriori receiver has a structure known as a *correlation receiver;* some special forms of the correlation receiver are considered.

7.2. Information Measure—Discrete Case

To define an engineering measure of information, psychological and sociological factors commonly associated with information must be removed from consideration. A suitable point for deriving an information measure is to associate an a priori probability with each of a possible set of messages. The smaller the a priori probability of a message, the greater the magnitude of information conveyed when such a message is received.

Consider the classic example of weather prediction. A forecast predicting snow may or may not contain a significant amount of information, depending on the a priori probability of a snowfall. Thus, if snow is forecast for the month of January in Switzerland, relatively little information has been conveyed since snow is highly probable during that month of the year. Consider, however, an unlikely forecast of snow for the Sahara Desert. Such a prediction would make a tremendous impression on travellers and inhabitants of this region because of its very low a priori probability of occurrence.

To clarify these ideas, let X_1, X_2, . . . , X_n constitute a complete set of possible transmitted signals (messages, events, symbols, and so on). Associate a priori probability $P(X_i)$ with the occurrence of signal X_i. If we choose

$$\sum_{i=1}^{n} P(X_i) = 1$$

one signal in the set is always assumed to be transmitted; however, "no signal" can be included as a member of the set if desired.

Next, let Y_1, Y_2, . . . , Y_m represent the possible received waveforms. If X_i is transmitted and Y_k is received, Y_k may not be identified with X_i because of noise. After waveform Y_k has been received, the state of knowledge of the observer is altered. In the case of a noise-free transmission medium, one signal in the set is unmistakably singled out from all the rest—that is, its probability of occurrence has increased to unity. The probabilities of occurrence of all other signals in the set simultaneously decrease to zero in this case.

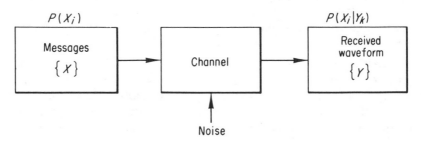

Fig. 7.2-1. Information transfer in a communication channel.

If, however, waveform Y_k is contaminated by noise (see Fig. 7.2-1), there is doubt concerning the exact identity of the transmitted message or signal. This is particularly true when two or more signals in the set are similar to one another. In this more practical case, the situation after reception can be represented by a set of a posteriori probabilities which describe the probabilities that the received waveform Y_k contains signal X_i. The a posteriori probability that signal X_i is present in Y_k is described by the conditional probability $P(X_i \mid Y_k)$—that is, the probability of X_i given Y_k is received.

The information gained about X_i as a result of Y_k being received is related to the change from a priori probability $P(X_i)$ to a posteriori probability $P(X_i \mid Y_k)$. To obtain an expression for information gain, let us associate with each signal X_i prior to transmission a mathematical measure of the *uncertainty*, denoted by $\mathscr{H}(X_i)$, that signal X_i will be sent. $\mathscr{H}(X_i)$ depends on the a priori probability of signal X_i; thus,

$$\mathscr{H}(X_i) = \mathscr{H}[P(X_i)] \tag{7.2-1a}$$
$$= \text{initial or a priori uncertainty of } X_i$$
$$\text{(before reception)} \tag{7.2-1b}$$

Next, denote the *a posteriori uncertainty* that signal X_i is contained in waveform Y_k by $\mathscr{H}(X_i \mid Y_k)$. This uncertainty is a function of the a posteriori probability $P(X_i \mid Y_k)$. Thus,

$$\mathscr{H}(X_i \mid Y_k) = \mathscr{H}[P(X_i \mid Y_k)] \tag{7.2-2a}$$
$$= \text{a posteriori uncertainty that signal } X_i \text{ is}$$
$$\text{contained in waveform } Y_k \tag{7.2-2b}$$

For the mathematical measure of uncertainty to be meaningful, the a posteriori uncertainty $\mathscr{H}(X_i \mid Y_k)$ must be zero when the signal X_i is received in a noise-free channel ($Y_k = X_i$).

Information gained about signal X_i is defined as the reduction in the uncertainty that X_i was transmitted, as a result of receiving waveform Y_k. Denoting this information gain by $\mathscr{I}(X_i \mid Y_k)$, this definition can be expressed as

$$\mathscr{I}(X_i \mid Y_k) = \text{reduction in uncertainty about presence of } X_i$$
$$\text{after } Y_k \text{ is received} \tag{7.2-3a}$$
$$= (\text{a priori uncertainty about } X_i) - (\text{a posteriori}$$
$$\text{uncertainty about } X_i) \tag{7.2-3b}$$
$$= \mathscr{H}(X_i) - \mathscr{H}(X_i \mid Y_k) \tag{7.2-3c}$$

For the noise-free case, $Y_k = X_i$. Since the a posteriori uncertainty $\mathscr{H}(X_i \mid X_i)$ is zero, Eq. (7.2-3c) simplifies to

$$\mathscr{I}(X_i \mid X_i) = \mathscr{H}(X_i) \tag{7.2-4}$$

Equation (7.2-4) states that the information gained about X_i in a noiseless channel is equal to the a priori or initial uncertainty about X_i being transmitted prior to Y_k being received. This result conforms to physical intuition.

In order to relate information gain $\mathscr{I}(X_i \mid Y_k)$ to a priori probability $P(X_i)$ and a posteriori probability $P(X_i \mid Y_k)$, Woodward and Davies [4] postulated that if two statistically independent signals X_i and X_j are transmitted through a noise-free channel, the total information gain about X_i and X_j when treated as a joint or single message should be the sum of the information gains when each is considered as a separate message. This postulate requires that

$$\mathscr{I}(X_i, X_j \,|\, X_i, X_j) = \mathscr{I}(X_i \,|\, X_i) + \mathscr{I}(X_j \,|\, X_j) \tag{7.2-5}$$

With Eq. (7.2-4) applied first to the joint signal (X_i, X_j),

$$\mathscr{I}(X_i, X_j \,|\, X_i, X_j) = \mathscr{H}(X_i, X_j) \tag{7.2-6}$$

Next, applying Eq. (7.2-4) to the individual signals X_i and X_j results in

$$\mathscr{I}(X_i \,|\, X_i) = \mathscr{H}(X_i) \tag{7.2-7}$$

and

$$\mathscr{I}(X_j \,|\, X_j) = \mathscr{H}(X_j) \tag{7.2-8}$$

With Eqs. (7.2-6) through (7.2-8), (7.2-5) becomes

$$\mathscr{H}(X_i, X_j) = \mathscr{H}(X_i) + \mathscr{H}(X_j) \tag{7.2-9}$$

Equation (7.2-9) can be rewritten to indicate the functional dependence of $\mathscr{H}(X_i, X_j)$, $\mathscr{H}(X_i)$, and $\mathscr{H}(X_j)$ on the a priori probabilities $P(X_i, X_j)$, $P(X_i)$, and $P(X_j)$:

$$\mathscr{H}[P(X_i, X_j)] = \mathscr{H}[P(X_i)] + \mathscr{H}[P(X_j)] \tag{7.2-10}$$

where $P(X_i, X_j)$ is the joint a priori probability that both X_i and X_j will be transmitted and $P(X_i)$ and $P(X_j)$ are the separate a priori probabilities of X_i and X_j, respectively. Because of the assumption that X_i and X_j are statistically independent, the joint probability $P(X_i, X_j)$ factors into a product of the individual a priori probabilities,

$$P(X_i, X_j) = P(X_i) \cdot P(X_j) \tag{7.2-11}$$

Substituting Eq. (7.2-11) into (7.2-10) results in

$$\mathscr{H}[P(X_i)P(X_j)] = \mathscr{H}[P(X_i)] + \mathscr{H}[P(X_j)] \tag{7.2-12}$$

Woodward and Davies [4] show that the only nontrivial solution to this functional equation requires $\mathscr{H}[P(X_i)]$ to be of the form

$$\mathscr{H}[P(X_i)] = \pm \mathrm{A} \log_b P(X_i) \tag{7.2-13}$$

By convention, the multiplicative constant A is set equal to unity, the negative sign in Eq. (7.2-13) is selected to make the uncertainty $\mathscr{H}(X_i)$ a positive quantity, and logarithm base b is set equal to two. With these choices, the units of $\mathscr{H}(X_i)$ are called bits (an abbreviation for binary digits).

Result (7.2-13), modified by the convention just stated, and Eq. (7.2-7) permit definitions (7.2-1) and (7.2-2) to be rewritten as

$$\mathscr{H}(X_i) = -\log_2 P(X_i) \quad \text{bits} \tag{7.2-14a}$$

$$= \text{initial uncertainty about } X_i \tag{7.2-14b}$$

$$= \text{information gained about } X_i \text{ in a noise-free}$$
$$\text{channel when } Y_k \text{ is received} \tag{7.2-14c}$$

and

$$\mathscr{H}(X_i \mid Y_k) = -\log_2 P(X_i \mid Y_k) \tag{7.2-15a}$$
$$= \text{a posteriori uncertainty about } X_i$$
$$\text{when } Y_k \text{ is received} \tag{7.2-15b}$$

Substituting Eqs. (7.2-14a) and (7.2-15a) into definition (7.2-3c) yields an expression for information gain in terms of the a priori and a posteriori probabilities:

$$\mathscr{I}(X_i \mid Y_k) = \mathscr{H}(X_i) - \mathscr{H}(X_i \mid Y_k) \tag{7.2-16a}$$
$$= -\log_2 P(X_i) - [-\log_2 P(X_i \mid Y_k)] \tag{7.2-16b}$$
$$= \log_2 \frac{P(X_i \mid Y_k)}{P(X_i)} \text{ bits} \tag{7.2-16c}$$

Woodward and Davies have shown that one need not define information gain as the difference between the a priori and a posteriori uncertainty in order to obtain result (7.2-16c). The same result can be obtained by invoking a second reasonable postulate, which states: if two messages containing the same signal X_i are sent sequentially, the observer may regard the a posteriori probability of X_i, after the first waveform is received, as the a priori probability of X_i before the second message is transmitted; in this case the total information gain about X_i must equal the sum of the information gained from each received waveform separately. The solution of a functional equation that mathematically embodies this postulate, together with result (7.2-13), leads to Eq. (7.2-16c).

Equation (7.2-16c) can be expressed in different form. Suppose both numerator and denominator of Eq. (7.2-16c) are multiplied by $P(Y_k)$. Then, since $P(X_i, Y_k)$ can be expressed as

$$P(X_i, Y_k) = P(X_i \mid Y_k)P(Y_k) \tag{7.2-17}$$

or

$$P(X_i, Y_k) = P(Y_k \mid X_i)P(X_i) \tag{7.2-18}$$

Eq. (7.2-16c) can be rewritten as

$$\mathscr{I}(X_i \mid Y_k) = \log_2 \frac{P(X_i \mid Y_k)P(Y_k)}{P(X_i)P(Y_k)} \tag{7.2-19a}$$
$$= \log_2 \frac{P(X_i, Y_k)}{P(X_i)P(Y_k)} \tag{7.2-19b}$$
$$= \log_2 \frac{P(Y_k \mid X_i)}{P(Y_k)} \tag{7.2-19c}$$

It follows from Eqs. (7.2-19c) and (7.2-16c) that

$$\mathscr{I}(X_i \mid Y_k) = \mathscr{I}(Y_k \mid X_i) \tag{7.2-20}$$

Equation (7.2-20) states that the formulation for information transfer is symmetric—that is, there is as much information about signal X_i given received waveform Y_k as there is information about waveform Y_k given signal X_i. Note further that since both $\mathscr{H}(X_i)$ and $\mathscr{H}(X_i \mid Y_k)$ are positive quantities, Eq. (7.2-16a) yields the inequality

$$\mathscr{I}(X_i \mid Y_k) \leqslant \mathscr{H}(X_i) \tag{7.2-21}$$

Thus Y_k can never provide more information about X_i than that which would be received in a noise-free channel.

7.3. Information Measure—Continuous Case

In the previous discussion, the set of possible signals was assumed finite. This is not true in most practical situations. In radar, for example, an echo from a single target defines an infinite set of signals; members of the set are characterized by such continuous parameters as time delay, rf phase, amplitude, doppler shift, and so on. Hence the set of all possible echo signals forms a continuum in each of several parameters. It can be anticipated that, in passing from a finite discrete signal set to the continuous case, probabilities will be replaced by probability densities in the expressions for uncertainty and information gain.

Consider first the a priori uncertainty $\mathscr{H}(X_i)$ for a discrete signal X_i, $i = 1, 2, \ldots, N$. As N approaches infinity, X_i can be replaced by the continuous variable x. If we divide up the continuum into small intervals or cells Δx wide and approximate $P(X_i)$ by $p(x)\,\Delta x$, where $p(x)$ is a continuous function, the a priori uncertainty that x lies in an interval Δx becomes, from Eq. (7.2-14a),

$$\mathscr{H}(x) = -\log_2 p(x)\,\Delta x \tag{7.3-1}$$

Note that Eq. (7.3-1) states that decreasing Δx increases the uncertainty that x lies within the interval Δx. This result agrees with intuition.

In a similar manner, discrete waveform Y_k can be replaced by a continuous variable y. Then the a posteriori uncertainty of a signal x, given that a waveform y is received, becomes from Eq. (7.2-15a)

$$\mathscr{H}(x \mid y) = -\log_2 p(x \mid y)\Delta x \tag{7.3-2}$$

With information gain defined as the difference between the a priori and a posteriori uncertainties associated with signal x, following reception of waveform y, the following expressions for information gain $\mathscr{I}(x \mid y)$ are obtained:

$$\mathscr{I}(x \mid y) = \mathscr{H}(x) - \mathscr{H}(x \mid y) \tag{7.3-3a}$$

$$= -\log_2 p(x)\Delta x - [-\log_2 p(x \mid y)\,\Delta x] \tag{7.3-3b}$$

$$= \log_2 \frac{p(x \mid y)\Delta x}{p(x)\,\Delta x} \tag{7.3-3c}$$

$$= \log_2 \frac{p(x \mid y)}{p(x)} \tag{7.3-3d}$$

Note that the final expression for information gain in Eq. (7.3-3d) is independent of the interval Δx.

This brief introduction to information theory is concluded with the following observation. If a priori probability density $p(x)$ is known to the observer, the only remaining quantity required to calculate information gain is the a posteriori probability density function $p(x \mid y)$. This result is the basis for the *ideal* or *a posteriori receiver* of Woodward and Davies [4].

7.4. Ideal Receiver

From an information-theory viewpoint, a receiver that computes the a posteriori probability density $p(x \mid y)$ for each possible message x provides the observer with all of the available information contained in the received waveform y. The observer combines this information with a suitable criterion to guess the actual identity of the received signal. If the probability in $p(x \mid y)$ is concentrated around a particular signal x, the guess approaches certainty. When noise is present, however, the probability in $p(x \mid y)$ is usually distributed over many signals.

A receiver that computes the a posteriori density function $p(x \mid y)$ and presents it to the observer has been defined by Woodward and Davies as an *ideal receiver*; it has also come to be known as the *a posteriori receiver*. To describe the functional nature of the ideal or a posteriori receiver, Woodward and Davies utilized the theory of *inverse probability*.

Consider first the definition of *direct probability*. A direct probability describes the chance a particular event will occur due to an assumed cause or hypothesis. Let $H_1, H_2, \ldots, H_N$ represent a complete set of mutually exclusive hypotheses or states of nature; let $E_1, E_2, \ldots, E_M$ denote a set of events or observations. The (direct) probability of event E_j, assuming hypothesis H_i is in effect, can be represented by conditional probability $P(E_j \mid H_i)$.

Very often, however, after an event E_j has occurred, one encounters the problem of guessing which cause or hypothesis in the set H_i, $i = 1, 2, \ldots, N$, was responsible for the event. Since this determination usually cannot be made with certainty, a probability is attached to each hypothesis. These probabilities are called *inverse probabilities*. The (inverse) probability that hypothesis H_i was responsible for observed event E_j can be represented by the conditional probability $P(H_i \mid E_j)$.

Direct probabilities and inverse probabilities can be related by the product law of probability theory (also known as Bayes' rule), which states that the probability of joint event (A, B) may be expressed in either of two forms:

$$P(A, B) = P(A)P(B \mid A) = P(B)P(A \mid B) \tag{7.4-1}$$

From Eq. (7.4-1), inverse probability $P(H_i \mid E_j)$ is related to direct probability $P(E_j \mid H_i)$ by

$$P(H_i \mid E_j) = \frac{P(H_i)P(E_j \mid H_i)}{P(E_j)}, \qquad i = 1, 2, \ldots, N \qquad (7.4\text{-}2)$$

In the context of the theory of inverse probability, the quantities in Eq. (7.4-2) have a special interpretation. Based on our earlier remarks, inverse probability $P(H_i \mid E_j)$ on the left side of Eq. (7.4-2) is the a posteriori probability of hypothesis H_i, $i = 1, 2, \ldots, N$ (after event E_j is observed). The function $P(E_j \mid H_i)$ is known as the likelihood function of E_j; strictly, $P(E_j \mid H_i)$ is interpreted as a function of H_i for an observed event E_j. Further, $P(E_j)$ is not a probability but simply a constant, once event E_j has been observed.

If, instead of a discrete set of hypotheses H_i and events E_j, the variables H and E are continuous, an expression identical to (7.4-2) can be obtained with probabilities replaced by probability densities. Thus,

$$p(H \mid E) = kp(H)p(E \mid H) \qquad (7.4\text{-}3)$$

where k has replaced the constant $1/p(E)$.

When inverse probability is applied to a communication channel, hypothesis H is associated with a possible transmitted signal x and event E with a received waveform y. In this case Eq. (7.4-3) is rewritten as

$$p(x \mid y) = kp(x)p(y \mid x) \qquad (7.4\text{-}4)$$

Equation (7.4-4) states that the probability density of transmitted signal x, after waveform y is received, is proportional to the product of the a priori probability density $p(x)$ and the likelihood function of x, $p(y \mid x)$. Equation (7.4-4) permits the observer to assess the relative probabilities that various signals x were responsible for a received waveform y. The observer may decide, for example, to choose the most probable signal as that signal actually transmitted—that is, the signal x that maximizes $p(x \mid y)$ for a given y. However, whatever criterion the observer uses, the a posteriori distribution $p(x \mid y)$ contains all of the available information in the received message.

7.5. The A Posteriori Distribution in the Presence of White Gaussian Noise

Although it is not yet apparent, Eq. (7.4-4) contains a description of the structure of the ideal receiver; that is, $p(x \mid y)$ describes the operations that must be performed on received waveform $y(t)$ to yield the a posteriori distribution at the receiver output. Let the waveform $u_x(t)$ represent a transmitted message in the set $x = 1, 2, \ldots$, and let $\{n_w(t)\}$ denote an additive white, Gaussian, ergodic random-noise process with zero mean that interferes with the reception of waveform $u_x(t)$. The received waveform $y(t)$ may then be expressed as

$$y(t) = u_x(t) + n_w(t) \qquad (7.5\text{-}1)$$

Since there is a one-to-one correspondence between message x and its associated waveform $u_x(t)$, Eq. (7.4-4) can be replaced by

$$p(x\,|\,y) = kp(x)p(y\,|\,u_x) \qquad (7.5\text{-}2)$$

To obtain $p(x\,|\,y)$ we need only evaluate the likelihood function $p(y\,|\,u_x)$, since the a priori probability density $p(x)$ is presumed known by the observer. The likelihood function $p(y\,|\,u_x)$ describes the effect of the system noise on receiver performance. To calculate $p(y\,|\,u_x)$, note from Eq. (7.5-1) that $p(y - u_x)$, assuming u_x known, is functionally identical to $p_n(n_w)$, the probability density of noise waveform $n_w(t)$. Thus,

$$p(y\,|\,u_x) = p_n(y - u_x) \qquad (7.5\text{-}3)$$

The next task is to obtain an expression for $p_n(n_w)$. For this purpose we employ the sampling theorem discussed in Chapter 2. To apply the theorem, let the set of transmitted waveforms $u_x(t)$ be band-limited. For simplicity, consider first the video case wherein all waveforms $u_x(t)$ are band-limited to $(-\omega_c, \omega_c)$ as shown in Fig. 7.5-1(a). Although the noise process is assumed white, the same bandwidth limitation can be applied to the noise by use of an appropriate band-limiting filter that does not affect the signal; this is shown in Fig. 7.5-1(c) as a distortionless rectangular low-pass filter preceding the ideal receiver. The spectral density of band-limited white noise process $\{n(t)\}$ is shown in Fig. 7.5-1(b).

From the low-frequency sampling theorem [see Eq. (2.4-10)], waveform $n(t)$ can be expressed in sampled form as

$$n(t) = \sum_k n\left(\frac{k}{2f_c}\right) \frac{\sin 2\pi f_c\left(t - \dfrac{k}{2f_c}\right)}{2\pi f_c\left(t - \dfrac{k}{2f_c}\right)} \qquad (7.5\text{-}4)$$

where $\omega_c = 2\pi f_c$. Since the noise is assumed to be Gaussian with zero mean, a noise sample $n_k = n(k/2f_c)$ is a Gaussian random variable with probability density function

$$p(n_k) = \frac{1}{\sqrt{2\pi\sigma_k^2}} \exp\left(-\frac{n_k^2}{2\sigma_k^2}\right) \qquad (7.5\text{-}5)$$

where σ_k^2 is the variance of sample n_k [see Eq. (3.13-1)]. Since the noise random process is assumed stationary, $\sigma_k^2 = \sigma^2$, for all k. If the magnitude of the white-noise power spectral density is $N_0/2$ watts/Hz, as shown in Fig. 7.5-1(b), then it follows from Eq. (4.6-29) that

$$\sigma_k^2 = \sigma^2 = \left(\frac{1}{2\pi}\right)\left(\frac{N_0}{2}\right)(2\omega_c) = f_c N_0 \qquad (7.5\text{-}6)$$

It follows from sampling theorem representation (7.5-4) that the probability density function of a sample waveform of a band-limited random process is simply the joint probability density function of the set of sample

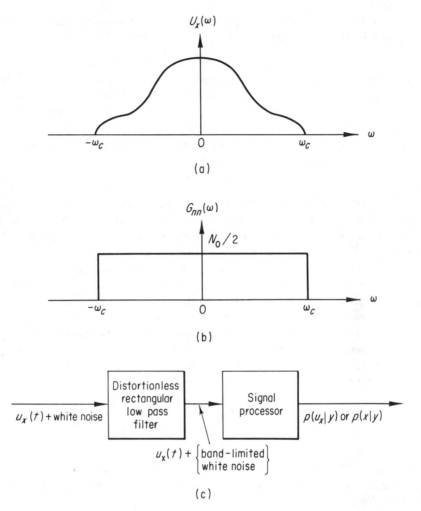

Fig. 7.5-1. (a) Signal spectrum.   **(b)** Noise spectral density out of low-pass filter.   **(c)** Functional block diagram of ideal receiver.

values that uniquely describe the waveform. In the case of a received wave-form T seconds long, (approximately) $T/(1/2f_c)$, or $2f_cT$, samples are required to describe each of the waveforms $y(t)$, $n(t)$, and $u_x(t)$. These waveforms can be represented in vector notation by $\mathbf{y}$, $\mathbf{n}$, and $\mathbf{u}_x$, each with $2f_cT$ coordinates, as described in Chapter 2. In this notation the probability density function $p_n(\mathbf{n})$ of noise waveform $n(t)$ may be expressed as the joint probability density functions of the set of noise waveform samples $n_1, n_2, \ldots, n_{2f_cT}$ as follows:

$$p_n(\mathbf{n}) = p_n(n_1, n_2, \ldots, n_{2f_cT}) \qquad (7.5\text{-}7)$$

It will now be shown that noise samples n_k are uncorrelated random variables. To demonstrate this result, it is sufficient to show that the expected value of $n_k \cdot n_l$ is zero for all $k \neq l$, since $\overline{n_k}$ is zero for all k. The Wiener-Khintchine theorem, which is discussed in Section 4.6, relates the autocorrelation function $\phi_{nn}(\tau)$ of a wide-sense stationary random process $\{n(t)\}$ to the power spectral density $G_{nn}(\omega)$ of the process, as follows:

$$\phi_{nn}(\tau) = \frac{1}{2\pi} \int_{-\infty}^{\infty} G_{nn}(\omega) e^{j\omega\tau} \, d\omega \qquad (7.5\text{-}8)$$

In this example, $G_{nn}(\omega)$ has the constant value of $N_0/2$ over the interval $(-\omega_c, \omega_c)$, as shown in Fig. 7.5-1(b). Substituting $N_0/2$ into Eq. (7.5-8) yields

$$\phi_{nn}(\tau) = \frac{1}{2\pi} \int_{-\omega_c}^{\omega_c} \frac{N_0}{2} e^{j\omega\tau} \, d\omega = \frac{N_0 \sin \omega_c \tau}{2\pi\tau} \qquad (7.5\text{-}9)$$

Since noise samples n_k and n_l are separated by a time interval $(k - l)/2f_c$, $\overline{n_k n_l}$ can be evaluated using Eq. (7.5-9) as

$$\overline{n_k n_l} = \phi_{nn}\left(\frac{k - l}{2f_c}\right) \qquad (7.5\text{-}10\text{a})$$

$$= \left(\frac{N_0}{2\pi}\right)\left(\frac{2f_c}{k - l}\right) \sin (k - l)\pi \qquad (7.5\text{-}10\text{b})$$

$$= 0, \qquad \text{for } k \neq l \qquad (7.5\text{-}10\text{c})$$

This demonstrates that random variables n_k and n_l are uncorrelated for all $k \neq l$. Note that for $k = l$, Eq. (7.5-10b) becomes

$$\overline{n_k^2} = \sigma^2 = N_0 f_c \qquad (7.5\text{-}11)$$

which is identical to result (7.5-6).

Since random variables n_k are uncorrelated and Gaussian, they are statistically independent. Thus, Eq. (7.5-7) can be replaced by

$$p_n(\mathbf{n}) = p(n_1) \cdot p(n_2) \cdot \ldots \cdot p(n_{2f_c T}) \qquad (7.5\text{-}12)$$

where probability density function $p(n_k)$ for each sample n_k is given by Eq. (7.5-5) with $N_0 f_c$ substituted for σ_k^2:

$$p(n_k) = \frac{1}{\sqrt{2\pi f_c N_0}} \exp\left(-\frac{n_k^2}{2f_c N_0}\right) \qquad (7.5\text{-}13)$$

Substituting Eq. (7.5-13) into (7.5-12) results in

$$p_n(\mathbf{n}) = \frac{1}{(2\pi f_c N_0)^{f_c T}} \exp\left(-\frac{1}{2f_c N_0} \sum_{k=1}^{2f_c T} n_k^2\right) \qquad (7.5\text{-}14)$$

Let the samples of waveforms $\mathbf{y}$ and $\mathbf{u}_x$ be denoted respectively by y_k and u_{x-k}, $k = 1, 2, \ldots, 2f_c T$. Then from Eq. (7.5-3) $p(\mathbf{y} \mid \mathbf{u}_x)$ is obtained by replacing n_k in Eq. (7.5-14) by $(y_k - u_{x-k})$—that is,

$$p(\mathbf{y} \mid \mathbf{u}_x) = p_n(\mathbf{n}) \mid_{n_k = y_k - u_{x-k}} \qquad (7.5\text{-}15\text{a})$$

Combining Eqs. (7.5-14) and (7.5-15a) yields

$$p(\mathbf{y}\,|\,\mathbf{u}_x) = \frac{1}{(2\pi f_c N_0)^{f_c T}} \exp\left[-\frac{1}{2f_c N_0} \sum_{k=1}^{2f_c T} (y_k - u_{x-k})^2\right] \qquad (7.5\text{-}15b)$$

The sum in the exponent of Eq. (7.5-15b) can be replaced by an integral by means of the following (approximate) relation demonstrated in Chapter 2 for a real band-limited waveform $f(t)$ with samples f_k [see Eq. (2.4-18a)]:

$$\int_0^T f^2(t)\, dt \cong \frac{1}{2f_c} \sum_{k=1}^{2f_c T} f_k^2 \qquad (7.5\text{-}16)$$

From Eq. (7.5-16), Eq. (7.5-15b) can be rewritten as

$$p(\mathbf{y}\,|\,\mathbf{u}_x) = k_1 \exp\left\{-\frac{1}{N_0} \int_0^T [y(t) - u_x(t)]^2\, dt\right\} \qquad (7.5\text{-}17)$$

where

$$k_1 = \frac{1}{(2\pi f_c N_0)^{f_c T}} \qquad (7.5\text{-}18)$$

When Eq. (7.5-17) is substituted into (7.5-2), the a posteriori probability distribution $p(x\,|\,\mathbf{y})$ becomes

$$p(x\,|\,\mathbf{y}) = k_2 p(x) \exp\left\{-\frac{1}{N_0} \int_0^T [y(t) - u_x(t)]^2\, dt\right\} \qquad (7.5\text{-}19)$$

Equation (7.5-19) is a general result that applies to band-limited signals in additive white Gaussian noise.

As an example, suppose that all messages x have equal a priori probabilities of being transmitted. Then $p(x)$ is a constant. Assume further that the observer selects the message x that maximizes $p(x\,|\,\mathbf{y})$—that is, the most likely signal $u_x(t)$ based on the observation $y(t)$. This corresponds to a waveform choice that minimizes the integrand in Eq. (7.5-19). Note that this is equivalent to selecting the signal $u_x(t)$ that has the least mean-square departure from the received waveform $y(t)$.

7.6. Sufficient Receiver (Correlation Receiver)

When the integrand in Eq. (7.5-19) is expanded, the expression becomes

$$p(x\,|\,\mathbf{y}) = k_2 p(x) \exp\left\{-\frac{1}{N_0}\left[\int_0^T y^2(t)\, dt + \int_0^T u_x^2(t)\, dt\right]\right\}$$
$$\cdot \exp\left[\frac{2}{N_0} \int_0^T y(t)u_x(t)\, dt\right] \qquad (7.6\text{-}1)$$

Since the received waveform $y(t)$ is known, the integral $\int_0^T y^2(t)\, dt$ in the first exponential term is a constant. Further, if the energy in each transmitted waveform $u_x(t)$ is chosen the same for all messages x, then the integral

$\int_0^T u_x^2(t)\, dt$ is also a constant and is independent of x. Observe also that the noise power density $N_0/2$ can be measured prior to reception of $y(t)$. Therefore, the entire first exponential term in Eq. (7.6-1) can be included in a single constant of proportionality, as follows:

$$p(x\,|\,\mathbf{y}) = k_3 p(x)\exp\left[\frac{2}{N_0}\int_0^T y(t)u_x(t)\,dt\right] \qquad (7.6\text{-}2)$$

Since the a priori probability $p(x)$ is assumed known to the observer and is independent of the received waveform $y(t)$, the only significant operation in Eq. (7.6-2) on received waveform $y(t)$ is described by the term

$$q(x) = \frac{2}{N_0}\int_0^T y(t)u_x(t)\,dt \qquad (7.6\text{-}3)$$

The function $q(x)$ is called a *sufficient statistic*. It contains information derived from received waveform $\mathbf{y}$ just sufficient to permit calculation of the complete a posteriori probability density $p(x\,|\,\mathbf{y})$ using Eq. (7.6-2).

In many situations insufficient information concerning the a priori probability density $p(x)$ is available. In these cases, an a priori distribution may be assumed in order to realize an ideal receiver. Alternatively, suppose the observer is provided with a sufficient statistic; he can then weigh the a priori probability of the various possible messages x in accordance with his own judgment. While this procedure sounds somewhat chaotic, since different observers may render dissimilar decisions, in practice this usually will not be the case; qualified observers placed in the same situation will develop, after a period of time, similar a priori probability weightings of the data as a result of common experience on the same "job." Previously, a receiver that computes $p(x\,|\,\mathbf{y})$ was defined as the ideal receiver. A receiver that computes and presents a sufficient statistic at its output instead of the actual a posteriori distribution is called a *sufficient receiver*.

Consider expression (7.6-3) for the sufficient statistic $q(x)$ in greater detail. The functional form of $q(x)$ is a cross correlation† of the received waveform $\mathbf{y}$ with signal waveform $\mathbf{u}_x$ (at a particular instant) for each possible message x. If it is assumed that received waveform $\mathbf{y}$ contains signal waveform $\mathbf{u}_k$ plus noise—that is,

$$y(t) = u_k(t) + n(t) \qquad (7.6\text{-}4)$$

then

$$q(x) = \frac{2}{N_0}\int_0^T u_k(t)u_x(t)\,dt + \frac{2}{N_0}\int_0^T n(t)u_x(t)\,dt \qquad (7.6\text{-}5)$$

When $\mathbf{u}_x = \mathbf{u}_k$, the first integral in Eq. (7.6-5) is a maximum and equal to

---

†The (finite time) cross-correlation function between two real functions with finite energy, $f_1(t)$ and $f_2(t)$, is defined by $\int_{-\infty}^{\infty} f_1(t)f_2(t+\tau)\,dt$; for complex functions, by $\int_{-\infty}^{\infty} f_1^*(t)f_2(t+\tau)\,dt$.

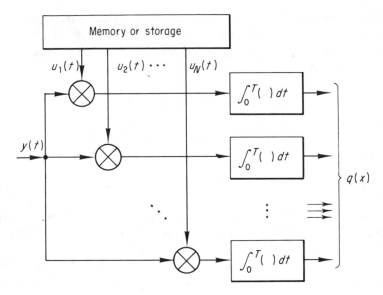

Fig. 7.6-1. Active correlation receiver.

the energy in $u_k(t)$. The second integral represents the uncertainty contributed by noise to the value of $q(x)$ for each waveform $\mathbf{u}_x$. Thus, if the observer selects a message x, corresponding to the peak value of $q(x)$, as the true message, the result can be in error as a result of a noise contribution, which shifts the location of the peak by an amount related to the signal energy and noise spectral density. A specific example is considered in the next section.

A receiver that computes $q(x)$ cross-correlates the received waveform $\mathbf{y}$ with stored replicas of each possible waveform $\mathbf{u}_x$; hence it is known as a *correlation receiver*. It can be implemented using several different configurations. One is shown in Fig. 7.6-1. The incoming signal is divided among N channels (after preamplification). Each channel corresponds to a different stored waveform or message. For example, in the kth channel the received waveform $\mathbf{y}$ is multiplied by a stored replica of waveform $\mathbf{u}_k$ and the result is integrated over the period T; this period corresponds to the duration of $u_k(t)$. (It is assumed implicitly in this model that the time of arrival of signal $\mathbf{u}_k$ is known. This situation applies, for example, to a synchronized communication channel.) The set of outputs shown in Fig. 7.6-1 collectively represent sufficient statistic $q(x)$. If the a priori probability for each signal $\mathbf{u}_k$ is constant and each has the same energy, the channel with the greatest output identifies the most probable message; that is, the most probable message is the one with the greatest cross correlation.

Another implementation of Eq. (7.6-5) is shown in Fig. 7.6-2. In this case, the functions of waveform storage, multiplication, and integration in Fig.

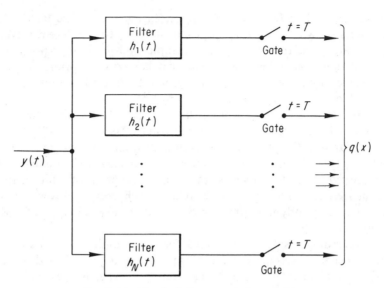

Fig. 7.6-2. Matched filter correlation receiver.

7.6-1 are replaced by a filter and a sampling gate. To understand this receiver, recall that the output of a linear filter can be obtained by convolving the input signal with filter impulse response $h(t)$. If the filter is realizable, its output $e(t)$ due to an input waveform $y(t)$ is given by [see Eq. (2.2-16) and the discussion that follows]

$$e(t) = \int_0^\infty h(\tau)y(t - \tau)\,d\tau, \qquad t \geqslant 0 \qquad (7.6\text{-}6)$$

Let impulse response $h_k(t)$ in Fig. 7.6-2 be related to message waveform $u_k(t)$ by

$$h_k(t) = \begin{cases} u_k(T - t), & 0 \leqslant t \leqslant T \\ 0, & \text{otherwise} \end{cases} \qquad (7.6\text{-}7)$$

Substituting Eq. (7.6-7) into (7.6-6) results in output

$$e_k(t) = \int_0^T u_k(T - \tau)y(t - \tau)\,d\tau, \qquad t \geqslant 0 \qquad (7.6\text{-}8)$$

With $(t - \tau)$ replaced by s, Eq. (7.6-8) becomes

$$e_k(t) = \int_0^T u_k(s)y(t - T + s)\,ds \qquad (7.6\text{-}9)$$

The filter output, sampled at time $t = T$ by the sampling gate, is

$$e_k(T) = \int_0^T u_k(s)y(s)\,ds \qquad (7.6\text{-}10)$$

which is identical to Eq. (7.6-3) for $q(x_k)$, except for a scale factor. Thus, if the filters shown in Fig. 7.6-2 are each matched to a different waveform $\mathbf{u}_x$ and sampled at time $t = T$, their sampled outputs correspond to the entire function $q(x)$. When each waveform has equal a priori probability and equal energy, the filter with the greatest sampled output amplitude indicates the most probable transmitted message.

The reader may have recognized that Eq. (7.6-7) describes the impulse response of a filter matched to signal waveform $u_k(t)$ [see Eq. (5.3-2)]. Matched filters appeared in Chapter 5 in connection with maximizing signal-to-noise ratio for signals in white Gaussian noise. The appearance of matched filters in the present context also stems from the assumption of white Gaussian noise interference, although a different approach was employed. Matched filters occur frequently in optimum receivers that have this type of additive interference.

The preceding results were derived for (band-limited) video signals. A similar procedure is employed for obtaining the a posteriori distribution $p(x \mid \mathbf{y})$ of a set of (band-limited) high-frequency signals $s_x(t)$ with carrier frequency ω_0. As before,

$$y(t) = s_x(t) + n(t) \qquad (7.6\text{-}11)$$

where $y(t)$, $s_x(t)$, and $n(t)$ are each band-limited to $\pm\Delta_\omega/2$ radians per second about the carrier frequency ω_0. For $n(t)$ described in terms of an in-phase component $n_I(t)$ and quadrature component $n_Q(t)$:

$$n(t) = n_I(t) \cos \omega_0 t - n_Q(t) \sin \omega_0 t \qquad (7.6\text{-}12)$$

a sampling-theorem representation of waveform $n(t)$ [see Eq. (2.6-8)] is given by

$$n(t) = \mathbf{n} = \sum_{k=1}^{T\Delta_f} \left[n_I\left(\frac{k}{\Delta_f}\right) \cos \omega_0 t - n_Q\left(\frac{k}{\Delta_f}\right) \sin \omega_0 t \right] \frac{\sin \pi\Delta_f\left(t - \dfrac{k}{\Delta_f}\right)}{\pi\Delta_f\left(t - \dfrac{k}{\Delta_f}\right)} \qquad (7.6\text{-}13)$$

where $\Delta_\omega = 2\pi\Delta_f$.

Note that waveform $n(t)$, or vector $\mathbf{n}$, is described in Eq. (7.6-13) by $T\Delta_f$ in-phase waveform samples and $T\Delta_f$ quadrature waveform samples: $n_I(k/\Delta_f) = n_{I\text{-}k}$ and $n_Q(k/\Delta_f) = n_{Q\text{-}k}$, respectively, at time instants $1/\Delta_f$ apart. Hence,

$$p_n(\mathbf{n}) = p_n(n_{I\text{-}1}, n_{I\text{-}2}, \ldots, n_{I\text{-}T\Delta_f}, n_{Q\text{-}1}, n_{Q\text{-}2}, \ldots, n_{Q\text{-}T\Delta_f}) \qquad (7.6\text{-}14)$$

When the background noise is a white Gaussian random process, it can be shown that samples $n_{I\text{-}k}$ and $n_{Q\text{-}k}$ are uncorrelated Gaussian random variables for all k and hence are statistically independent. Therefore, Eq. (7.6-14) can be rewritten as

$$p_n(\mathbf{n}) = p(n_{I\text{-}1}) \cdot p(n_{I\text{-}2}) \cdot \ldots \cdot p(n_{I\text{-}T\Delta_f}) \cdot p(n_{Q\text{-}1}) \cdot p(n_{Q\text{-}2}) \cdot \ldots \cdot p(n_{Q\text{-}T\Delta_f}) \qquad (7.6\text{-}15)$$

Since each sample is a Gaussian random variable with variance $N_0 \Delta_f$, Eq. (7.6-15) becomes

$$p_n(\mathbf{n}) = \frac{1}{(2\pi N_0 \Delta_f)^{T\Delta_f}} \exp\left[-\frac{1}{2N_0\Delta_f} \sum_{k=1}^{T\Delta_f} (n_{I-k}^2 + n_{Q-k}^2) \right] \qquad (7.6\text{-}16)$$

It follows from Eq. (2.6-19) for $f(t) = g(t)$ that (7.6-16) can be rewritten as

$$p_n(\mathbf{n}) \cong k_1 \exp\left[-\frac{1}{N_0} \int_0^T n^2(t)\, dt \right] \qquad (7.6\text{-}17)$$

With the sampling-theorem representations of waveforms $y(t)$ and $s_x(t)$ denoted by vectors $\mathbf{y}$ and $\mathbf{s}_x$, the conditional probability density $p(\mathbf{y} \,|\, \mathbf{s}_x)$ can be related to $p_n(\mathbf{n})$, as in the low-frequency case, by replacing $n(t)$ in (7.6-17) by $[y(t) - s_x(t)]$. The result is

$$p(\mathbf{y}\,|\,\mathbf{s}_x) = k_1 \exp\left\{ -\frac{1}{N_0} \int_0^T [y(t) - s_x(t)]^2\, dt \right\} \qquad (7.6\text{-}18)$$

Hence, the a posteriori probability of message x, $p(x \,|\, \mathbf{y})$, is given by

$$p(x\,|\,\mathbf{y}) = k_2\, p(x) \exp\left\{ -\frac{1}{N_0} \int_0^T [y(t) - s_x(t)]^2\, dt \right\} \qquad (7.6\text{-}19)$$

Note that Eq. (7.6-19) is identical to (7.5-19) except for a minor change in notation. The argument from this point on parallels that for the video-frequency case. The sufficient statistic $q(x)$ is once again given by

$$q(x) = \frac{1}{N_0} \int_0^T y(t) s_x(t)\, dt \qquad (7.6\text{-}20)$$

To compute $q(x)$, the received waveform $y(t)$ is cross-correlated with stored replicas of known waveforms $s_x(t)$; a sufficient receiver for $q(x)$ can be implemented by an active correlator bank as shown in Fig. 7.6-1 or by a set of filters matched to the known waveforms $s_x(t)$ followed by a sampling gate as shown in Fig. 7.6-2.

7.7. Estimating Radar Range of a Stationary Target with Known Cross Section

In radar, typical questions to be answered are: *Is a target present? If it is present, what is its range? If it is moving, what is its velocity?* Determining a target's presence or absence is the detection problem; measurement of the parameters of a target is the *estimation* problem. In this and subsequent sections, the a posteriori theory of reception will be applied to radar detection and estimation problems. Since a posteriori theory is better suited to parameter estimation than detection, we shall consider first the problem of range estimation.

For illustrative purposes, we shall determine the structure of an ideal a posteriori receiver that provides estimates of the radar range of a stationary point target with known radar cross section. Elapsed time τ between the transmission of a radar signal and the reception of an echo from a target is related to target range R by

$$\tau = \frac{2R}{c} \qquad (7.7\text{-}1)$$

where c is the velocity of light; hence, estimating the radar range of a target is equivalent to estimating elapsed time τ.

The ideal radar receiver calculates the a posteriori probability density $p(\tau \mid \mathbf{y})$ of target echo delay from the received waveform $\mathbf{y}$. If $p(\tau)$ describes the a priori probability density that the target echo is received τ seconds after transmission, the a posteriori probability density of target echo delay is then given by

$$p(\tau \mid \mathbf{y}) = k_1 p(\tau) p(\mathbf{y} \mid \tau) \qquad (7.7\text{-}2)$$

For simplicity assume that $p(\tau)$ is uniform over the interval $(0, T)$, where T is much greater than waveform duration T_s. In this case Eq. (7.7-2) becomes

$$p(\tau \mid \mathbf{y}) = k_2 p(\mathbf{y} \mid \tau) \qquad (7.7\text{-}3)$$

The radar problem is an interesting variation of the general a posteriori reception problem. Instead of an ensemble of transmitted waveforms representing different messages, a radar (usually) transmits one waveform $s(t)$. Since range information is desired, the set of delayed waveforms $s(t - \tau)$ represents the message set where delay τ is a continuous random variable within the interval $(0, T)$. Message waveform $s(t - \tau)$ is assumed completely known except for delay τ; hence, Eq. (7.7-3) can be rewritten as

$$p(\tau \mid \mathbf{y}) = k_2 p[\mathbf{y} \mid s(t - \tau)] \qquad (7.7\text{-}4)$$

To apply earlier results, let $s(t)$ be a band-limited waveform. It follows from Eq. (7.6-18) that

$$p(\tau \mid \mathbf{y}) = k_2 \exp \left\{ -\frac{1}{N_0} \int_0^T [y(t) - s(t - \tau)]^2 \, dt \right\} \qquad (7.7\text{-}5)$$

Expanding the integrand in Eq. (7.7-5) yields

$$p(\tau \mid \mathbf{y}) = k_2 \exp \left[-\frac{1}{N_0} \int_0^T y^2(t) \, dt - \frac{1}{N_0} \int_0^T s^2(t - \tau) \, dt \right]$$
$$\cdot \exp \left[\frac{2}{N_0} \int_0^T y(t) s(t - \tau) \, dt \right] \qquad (7.7\text{-}6)$$

Only those terms in (7.7-6) which contain τ explicitly are of interest; hence, the first integral may be treated as a constant. In addition, for $T \gg T_s$, the second integral is a constant equal to the signal energy since $s(t - \tau)$ is (almost always) wholly within the interval $(0, T)$. Thus the last exponential in Eq. (7.7-6) is the only term that is an explicit function of τ. Hence Eq. (7.7-6) simplifies to

$$p(\tau \mid \mathbf{y}) = k_3 \exp \left[\frac{2}{N_0} \int_0^T y(t)s(t - \tau)\, dt \right] \qquad (7.7\text{-}7)$$

A sufficient receiver calculates the sufficient statistic,

$$q(\tau) = \frac{2}{N_0} \int_0^T y(t)s(t - \tau)\, dt \qquad (7.7\text{-}8)$$

for all possible values of τ. This can be accomplished by the receiver shown in Fig. 7.6-1 or 7.6-2. However, a passive matched filter implementation can be realized that is considerably simpler than the rather complex realization shown in Fig. 7.6-2. Instead of an ensemble of matched filters, each followed by a sampling gate, only a single matched filter is required. To demonstrate this result, recall that signal waveform duration T_s is assumed to be much shorter than observation interval T. The limits on Eq. (7.7-8) can be modified to indicate that signal $s(t - \tau)$, for $s(t)$ defined over interval $(0, T_s)$, truncates the integrand to the interval $(\tau, \tau + T_s)$:

$$q(\tau) = \frac{2}{N_0} \int_\tau^{\tau + T_s} y(t)s(t - \tau)\, dt \qquad (7.7\text{-}9)$$

If variable of integration t is replaced by $(T_s + \tau - t_1)$, Eq. (7.7-9) becomes

$$q(\tau) = \frac{2}{N_0} \int_0^{T_s} y(T_s + \tau - t_1)s(T_s - t_1)\, dt_1 \qquad (7.7\text{-}10)$$

By defining

$$h(t_1) = \frac{2}{N_0} s(T_s - t_1), \qquad 0 \leqslant t_1 \leqslant T_s \qquad (7.7\text{-}11)$$

we can rewrite Eq. (7.7-10) as

$$q(\tau) = \int_0^{T_s} y(T_s + \tau - t_1)h(t_1)\, dt_1 \qquad (7.7\text{-}12)$$

Equation (7.7-12) represents the output of a filter with impulsive response $h(t_1)$ evaluated at time $(\tau + T_s)$. It follows that if the filter output as a function of time is denoted by $e_h(t)$, then

$$q(\tau) = e_h(\tau + T_s) \qquad (7.7\text{-}13)$$

Note that Eq. (7.7-11) describes $h(t)$ as the impulse response of a filter matched to waveform $s(t)$ (with delay T_s for filter realizability). Thus, passing $y(t)$ through a filter matched to the transmitted waveform $s(t)$ generates an output signal that is identical to the sufficient statistic $q(\tau)$ for all τ, except for the addition of a small delay T_s. This demonstrates that one filter can match all expected delays, eliminating the need for the filter bank of Fig. 7.6-2.

The sufficient statistic $q(\tau)$ is generally not useful as an estimator of target range without additional processing. To see why, we describe the transmitted and received signals employing the complex representation discussed in Section 2.5. A narrowband signal $s(t)$ with amplitude modulation $a(t)$, phase modulation $\theta(t)$, and carrier frequency $\omega_0 = 2\pi f_0$ can be written in the following ways:

$$s(t) = a(t) \cos [\omega_0 t + \theta(t)] \tag{7.7-14a}$$
$$= \text{Re} \, [a(t)e^{j\theta(t)} e^{j\omega_0 t}]$$
$$= \text{Re} \, [\tilde{s}(t)e^{j\omega_0 t}] \tag{7.7-14b}$$
$$= \text{Re} \, [\psi(t)] \tag{7.7-14c}$$

$\tilde{s}(t)$ is called the complex envelope of complex signal $\psi(t)$ and contains information concerning both the amplitude and phase modulation of real signal $s(t)$. Similarly, we can write

$$y(t) = \text{Re} \, [\tilde{y}(t)e^{j\omega_0 t}] \tag{7.7-15a}$$
$$= \text{Re} \, [\gamma(t)] \tag{7.7-15b}$$

and

$$n(t) = \text{Re} \, [\tilde{n}(t)e^{j\omega_0 t}] \tag{7.7-16a}$$
$$= \text{Re} \, [\nu(t)] \tag{7.7-16b}$$

where $\tilde{y}(t)$ and $\tilde{n}(t)$ are the complex envelopes of real signals $y(t)$ and $n(t)$, respectively.

The expression for the a posteriori distribution $p(\tau|\mathbf{y})$ in complex notation is

$$p(\tau|\mathbf{y}) = k_1 p(\tau) \exp \left[-\frac{1}{2N_0} \int_0^T |\gamma(t) - \psi(t - \tau)|^2 \, dt \right] \tag{7.7-17a}$$

$$= k_1 p(\tau) \exp \left[-\frac{1}{2N_0} \int_0^T |\gamma(t)|^2 \, dt - \frac{1}{2N_0} \int_0^T |\psi(t - \tau)|^2 \, dt \right]$$
$$\cdot \exp \left\{ \text{Re} \left[\frac{1}{N_0} \int_0^T \gamma^*(t)\psi(t - \tau) \, dt \right] \right\} \tag{7.7-17b}$$

$$= k_2 p(\tau) \exp \left\{ \text{Re} \left[\frac{1}{N_0} \int_0^T \gamma^*(t)\psi(t - \tau) \, dt \right] \right\} \tag{7.7-17c}$$

$$= k_2 p(\tau) \exp \left\{ \text{Re} \left[\frac{1}{N_0} e^{-j\omega_0 \tau} \int_0^T \tilde{y}^*(t)\tilde{s}(t - \tau) \, dt \right] \right\} \tag{7.7-17d}$$

The factor of $\frac{1}{2}$ preceding the integral in Eq. (7.7-17a) is required for energy normalization, as discussed in Chapter 2. Equation (7.7-17d) is obtained by substituting definitions (7.7-14b) and (7.7-15b) into (7.7-17c). Note that the exponent in (7.7-17d) is the sufficient statistic $q(\tau)$:

$$q(\tau) = \text{Re} \left\{ \left[\frac{1}{N_0} \int_0^T \tilde{y}^*(t)\tilde{s}(t - \tau) \, dt \right] e^{-j\omega_0 \tau} \right\} \tag{7.7-18}$$

Comparison of (7.7-18) with (7.7-14b) suggests that the complex envelope of $q(\tau)$, denoted by $\tilde{q}(\tau)$, be defined by

$$\tilde{q}(\tau) = \frac{1}{N_0} \int_0^T \tilde{y}^*(t)\tilde{s}(t - \tau) \, dt \tag{7.7-19}$$

so that the sufficient statistic $q(\tau)$ can be written in the following ways:

$$q(\tau) = \text{Re} \left[\tilde{q}(\tau) e^{-j\omega_0 \tau} \right] \qquad (7.7\text{-}20\text{a})$$

$$= \text{Re} \left[| \tilde{q}(\tau) | e^{-j[\omega_0 \tau - \arg \tilde{q}(\tau)]} \right] \qquad (7.7\text{-}20\text{b})$$

$$= | \tilde{q}(\tau) | \cos \left[\omega_0 \tau - \arg \tilde{q}(\tau) \right] \qquad (7.7\text{-}20\text{c})$$

Substituting Eq. (7.7-20c) back into (7.7-17d) results in

$$p(\tau \,|\, \mathbf{y}) = k_2 p(\tau) e^{| \tilde{q}(\tau) | \cos[\omega_0 \tau - \arg \tilde{q}(\tau)]} \qquad (7.7\text{-}21)$$

The fluctuation of the exponent in Eq. (7.7-21) at the carrier-frequency rate is represented by the fine structure in the a posteriori distribution $p(\tau \,|\, \mathbf{y})$ shown in Fig. 7.7-1. The fluctuation is most apparent in the vicinity of τ

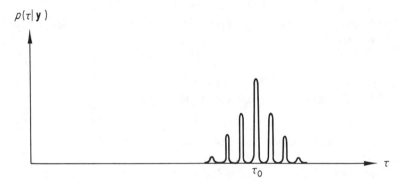

Fig. 7.7-1. Fine structure in a posteriori probability density function of range delay.

close to the true target time delay τ_0, where $| \tilde{q}(\tau) |$ is near its maximum value. This fine structure is due to the sensitive dependence of target echo phase $\omega_0 \tau$ upon range delay at microwave frequencies; that is, small changes in target range result in large phase changes. In search radars, phase information is generally not used for estimating target range because of the inherent high ambiguity in $p(\tau \,|\, \mathbf{y})$ shown in Fig. 7.7-1. (The rate of change of rf phase, however, is used frequently to measure target velocity.) To eliminate this ambiguity, rf phase $\omega_0 \tau$ in (7.7-21) is assumed to be a random variable ϕ with a uniform probability density function over the interval $(0, 2\pi)$. This assumption reflects maximum ignorance of the target echo initial phase. Then the a posteriori distribution of Eq. (7.7-21) can be rewritten as a conditional joint probability density in the variables τ and ϕ. Thus,

$$p(\tau, \phi \,|\, \mathbf{y}) = k p(\tau, \phi) e^{| \tilde{q}(\tau) | \cos[\phi - \arg \tilde{q}(\tau)]} \qquad (7.7\text{-}22)$$

To obtain the marginal probability density function $p(\tau \,|\, \mathbf{y})$ from $p(\tau, \phi \,|\, \mathbf{y})$, Eq. (7.7-22) is integrated with respect to ϕ. For this purpose τ and ϕ are assumed to be independent random variables. The statistical independence of τ and ϕ is a reasonable assumption at microwave frequencies where, as previously mentioned, ϕ fluctuates rapidly for practically nonmeasurable changes in τ. Thus,

$$p(\tau, \phi) = p(\tau) \cdot p(\phi) \qquad (7.7\text{-}23a)$$

$$= \frac{1}{2\pi} p(\tau) \qquad (7.7\text{-}23b)$$

The marginal probability density function $p(\tau \,|\, \mathbf{y})$ is computed to be

$$p(\tau \,|\, \mathbf{y}) = \int_0^{2\pi} p(\tau, \phi \,|\, \mathbf{y}) \, d\phi \qquad (7.7\text{-}24a)$$

$$= k p(\tau) \left[\frac{1}{2\pi} \int_0^{2\pi} e^{|q(\tau)| \cos[\phi - \arg \tilde{q}(\tau)]} \, d\phi \right] \qquad (7.7\text{-}24b)$$

The bracketed term in Eq. (7.7-24b) is a standard tabulated integral† which leads to the result

$$p(\tau \,|\, \mathbf{y}) = k p(\tau) I_0(|\,\tilde{q}\,|) \qquad (7.7\text{-}25)$$

where $I_0(x)$ is the modified Bessel function of order zero and argument x. From Eq. (7.7-20c) note that the envelope of q is

$$\text{env}\{q\} = |\,\tilde{q}\,| \qquad (7.7\text{-}26)$$

Hence, when fine-structure phase information is not used for estimating range, it is not $q(\tau)$ that is the sufficient statistic but the envelope of $q(\tau)$. A sufficient receiver that eliminates ambiguous fine-structure range information consists of a filter matched to the transmitted waveform followed by an envelope detector, as shown in Fig. 7.7-2.

The accuracy with which target range can be estimated by the sufficient receiver of Fig. 7.7-2 is considered next. For this purpose the incoming waveform is split into signal and noise components. If the true value of the echo time delay is denoted by τ_0, the received waveform $\gamma(t)$ is given by

$$\gamma(t) = \psi(t - \tau_0) + \nu(t) \qquad (7.7\text{-}27)$$

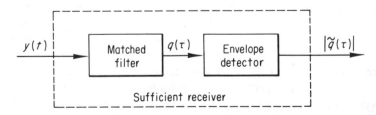

Fig. 7.7-2. Sufficient receiver for estimating target range using a narrowband radar waveform.

$$\dagger \frac{1}{2\pi} \int_0^{2\pi} e^{x \cos (\phi + \theta)} \, d\phi = I_0(x)$$

or, in complex envelope notation, by

$$\bar{y}(t) = \bar{s}(t - \tau_0)e^{j\omega\tau_0} + \tilde{n}(t) \qquad (7.7\text{-}28)$$

When Eq. (7.7-28) is substituted into (7.7-19), $|\tilde{q}|$ becomes

$$|\tilde{q}(\tau)| = \left| \frac{1}{N_0} \int_0^T \bar{s}^*(t - \tau_0)e^{-j\omega\tau_0}\bar{s}(t - \tau)\, dt + \frac{1}{N_0} \int_0^T \tilde{n}^*(t)\bar{s}(t - \tau)\, dt \right| \qquad (7.7\text{-}29)$$

Since $e^{-j\omega\tau_0}$ is independent of t and is of unit magnitude, it can be factored out of Eq. (7.7-29); however, this introduces a factor $e^{j\omega\tau_0}$ in the second integral in Eq. (7.7-29). Because $\tilde{n}^*(t)$ is a random-noise waveform, multiplication of the waveform by the exponential $e^{j\omega\tau_0}$ has no effect on the statistical properties of the noise. Thus, by redefining $\tilde{n}^*(t)$, we can rewrite Eq. (7.7-29) as

$$|\tilde{q}(\tau)| = \left| \frac{1}{N_0} \int_0^T \bar{s}^*(t - \tau_0)\bar{s}(t - \tau)\, dt + \frac{1}{N_0} \int_0^T \tilde{n}^*(t)\bar{s}(t - \tau)\, dt \right| \qquad (7.7\text{-}30)$$

Next, define

$$\tilde{g}(\tau) = \frac{1}{N_0} \int_0^T \bar{s}^*(t - \tau_0)\bar{s}(t - \tau)\, dt \qquad (7.7\text{-}31)$$

and

$$\tilde{h}(\tau) = \frac{1}{N_0} \int_0^T \tilde{n}^*(t)\bar{s}(t - \tau)\, dt \qquad (7.7\text{-}32)$$

Then, Eq. (7.7-30) can be written as

$$|\tilde{q}(\tau)| = |\tilde{g}(\tau) + \tilde{h}(\tau)| \qquad (7.7\text{-}33)$$

It is clear from Eqs. (7.7-31) through (7.7-33) that $\tilde{g}(\tau)$ is a low-frequency complex function that represents the signal portion of $\tilde{q}$ and $\tilde{h}(\tau)$ is a low-frequency complex function that represents the noise contribution to $\tilde{q}$.

In this treatment of range estimation, it has been assumed that target existence has already been established. Since reliable detection requires a peak signal-to-noise ratio in excess of 13 db (20 to 1), estimating target range on a single radar observation is meaningful only when the received signal-to-noise ratio is at least this great. This can also be seen in Fig. 7.7-3, which illustrates the a posteriori distribution $p(\tau \mid \mathbf{y})$ for a target in the vicinity of τ_0 for several values of signal-to-noise ratio. It is clear from the figure that there is a threshold below which detection and estimation are not reliable.

The accuracy with which τ can be estimated is related to the width of the peak in the a posteriori probability density $p(\tau \mid \mathbf{y})$ which, for large signal-to-noise ratios, occurs in the vicinity of true target echo delay τ_0. In this region the contribution of the signal component $\tilde{g}(\tau)$ to $\tilde{q}(\tau)$ is much larger than the noise component $\tilde{h}(\tau)$. With $\tilde{q}(\tau)$ approximated by $\tilde{g}(\tau)$ in the vicinity of $\tau = \tau_0$, Eq. (7.7-25) is approximately given by

$$p(\tau \mid \mathbf{y})|_{\tau \sim \tau_0} = k\, p(\tau) I_0(|\tilde{g}(\tau)|) \qquad (7.7\text{-}34)$$

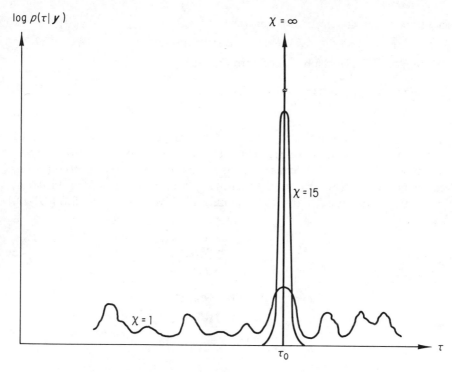

Fig. 7.7-3. A posteriori probability density function as function of signal-to-noise ratio.

The function $I_0(x)$ behaves in the vicinity of large values of its argument†
like e^x, so that Eq. (7.7-34) can be further approximated by

$$p(\tau \mid \mathbf{y})|_{\tau \sim \tau_0} = k p(\tau) e^{|\tilde{g}(\tau)|} \qquad (7.7\text{-}35)$$

Inspection of Eq. (7.7-31) indicates that the maximum value of $|\tilde{g}(\tau)|$
occurs when $\tau = \tau_0$. Let $\tilde{g}(\tau)$ be expanded in a Taylor series about $\tau = \tau_0$:

$$\tilde{g}(\tau) = \tilde{g}(\tau_0) + \frac{\dot{\tilde{g}}(\tau_0)}{1!}(\tau - \tau_0) + \frac{\ddot{\tilde{g}}(\tau_0)}{2!}(\tau - \tau_0)^2 + \ldots \qquad (7.7\text{-}36)$$

where the number of dots denotes the order of the derivative with respect
to τ. The term $\tilde{g}(\tau_0)$ is found (neglecting end effects) from Eq. (7.7-31) with
$\tau = \tau_0$:

$$\tilde{g}(\tau_0) = \frac{1}{N_0} \int_0^T |\tilde{s}(t - \tau_0)|^2 \, dt \qquad (7.7\text{-}37a)$$

$$= \frac{2E}{N_0} \qquad (7.7\text{-}37b)$$

†The asymptotic expansion of $I_0(x)$ for $x \gg 1$ is given by

$$I_0(x) \cong \frac{e^x}{\sqrt{2\pi x}}\left(1 + \frac{1}{8x} + \ldots\right)$$

where E is the energy in the real signal $s(t)$—that is,

$$\int_0^T s^2(t)\,dt = E \tag{7.7-38}$$

To obtain Eq. (7.7-37b) we used Eq. (2.8-7), which states that the energy in complex envelope $\tilde{s}(t)$ is twice that in real signal $s(t)$. The term $\dot{\tilde{g}}(\tau_0)$ is evaluated by differentiating Eq. (7.7-31) with respect to τ and then replacing τ by τ_0. Since

$$\frac{d\tilde{s}(t-\tau)}{d\tau} = -\frac{d\tilde{s}(t-\tau)}{dt} = -\dot{\tilde{s}}(t-\tau) \tag{7.7-39}$$

$\dot{\tilde{g}}(\tau_0)$ is given by

$$\dot{\tilde{g}}(\tau_0) = -\frac{1}{N_0}\int_0^T \tilde{s}^*(t-\tau_0)\dot{\tilde{s}}(t-\tau_0)\,dt \tag{7.7-40a}$$

If $\tilde{s}(t-\tau_0)$ is assumed to be entirely contained in the search interval $(0, T)$, the limits in Eq. (7.7-40a) can be extended to plus and minus infinity, respectively, without altering the value of $\dot{\tilde{g}}(\tau_0)$. If, in addition, $(t-\tau_0)$ is replaced by v, (7.7-40a) becomes

$$\dot{\tilde{g}}(\tau_0) = -\frac{1}{N_0}\int_{-\infty}^{\infty} \tilde{s}^*(v)\dot{\tilde{s}}(v)\,dv, \qquad 0 < \tau_0 \leqslant T \tag{7.7-40b}$$

From relation (2.8-13), Eq. (7.7-40b) reduces to

$$\dot{\tilde{g}}(\tau_0) = 0 \tag{7.7-41}$$

The expression for $\ddot{\tilde{g}}(\tau_0)$ is obtained in a similar manner; thus,

$$\ddot{\tilde{g}}(\tau_0) = \frac{1}{N_0}\int_0^T \tilde{s}^*(t-\tau_0)\ddot{\tilde{s}}(t-\tau_0)\,dt \tag{7.7-42a}$$

$$= \frac{1}{N_0}\int_{-\infty}^{\infty} \tilde{s}^*(t)\ddot{\tilde{s}}(t)\,dt \tag{7.7-42b}$$

From Eq. (2.8-21) Eq. (7.7-42b) can be rewritten as

$$\ddot{\tilde{g}}(\tau_0) = -\left(\frac{2E}{N_0}\right)\overline{(\omega-\omega_0)^2} \tag{7.7-43}$$

where $\overline{(\omega-\omega_0)^2}$ is interpreted as the mean-square bandwidth of signal $s(t)$. Denoting the mean-square bandwidth by Δ_ω^2:

$$\Delta_\omega^2 = \overline{(\omega-\omega_0)^2} \tag{7.7-44}$$

we can rewrite (7.7-43) as

$$\ddot{\tilde{g}}(\tau_0) = -\Delta_\omega^2 \frac{2E}{N_0} \tag{7.7-45}$$

With (7.7-37b), (7.7-41), and (7.7-45) substituted back into Eq. (7.7-36), the Taylor series expansion of $\tilde{g}(\tau)$, in the vicinity of τ_0, is given by

$$\tilde{g}(\tau) = \frac{2E}{N_0} - \frac{\Delta_\omega^2}{2}\left(\frac{2E}{N_0}\right)(\tau-\tau_0)^2 + \cdots \tag{7.7-46}$$

Replacing $\tilde{g}(\tau)$ in Eq. (7.7-35) by the first two terms of expression (7.7-46) yields

$$p(\tau\,|\,\mathbf{y})|_{\tau\sim\tau_0} = k_1 p(\tau)\exp\left[-\frac{\Delta_\omega^2}{2}\left(\frac{2E}{N_0}\right)(\tau-\tau_0)^2\right] \qquad (7.7\text{-}47)$$

where exp $(2E/N_0)$ has been absorbed in the constant. In the vicinity of τ_0 the a priori probability $p(\tau)$ is generally a slowly varying function compared to the exponential term in Eq. (7.7-47) and can be ignored. Therefore, it may be concluded that $p(\tau\,|\,\mathbf{y})$ is approximately Gaussian near τ_0 with variance σ_τ^2 given by

$$\sigma_\tau^2 = \frac{1}{\Delta_\omega^2(2E/N_0)} \qquad (7.7\text{-}48)$$

Equation (7.7-48) states that the rms error in estimating echo time delay is inversely proportional to rms signal bandwidth Δ_ω and the square root of $2E/N_0$, which is equal to the peak signal-to noise ratio out of the matched filter in Fig. 7.7-2 (see Chapter 5). Result (7.7-48) is used to predict radar range accuracy for large values of $2E/N_0$. For a simple monochromatic rf pulse with no phase modulation, pulse duration is inversely proportional to rms bandwidth Δ_ω. Hence, in this simple case, rms error is directly proportional to the transmitted signal pulse duration.

Some additional comments are appropriate at this point. Because of the approximations made in deriving Eq. (7.7-47), the a posteriori distribution peaks at true time delay τ_0. If this were always true, the observer would need only to locate the peak of this distribution and no error would be made in estimating target range. But in deriving Eq. (7.7-47) the noise contribution was ignored; this component of $|\tilde{q}(\tau)|$ displaces the peak by a random amount. The variance of this displacement is shown in Woodward and Davies [2] to be approximately equal to the variance σ_τ^2.

The complex envelope function $\tilde{g}(\tau)$ plays an extremely important role in radar waveform design. It was previously demonstrated that the waveform out of a sufficient receiver that consists of a filter matched to the transmitted signal followed by an envelope detector is given by $|\tilde{q}(\tau)|$. Equation (7.7-33) indicates that $|\tilde{q}(\tau)|$ in the vicinity of a large target is approximately equal to $|\tilde{g}(\tau)|$. In order to direct attention to the shape of the output waveform, τ_0 is set equal to zero in (7.7-31) and the scale factor $1/N_0$ is dropped; the modified function, denoted by $|\tilde{e}_{\mathrm{MF}}(\tau)|$, is given by

$$|\tilde{e}_{\mathrm{MF}}(\tau)| = \left|\int_0^T \tilde{s}^*(t)\tilde{s}(t-\tau)\,dt\right| \qquad (7.7\text{-}49)$$

When an echo is received from a target whose radar range is changing at a constant velocity v, it is shown in Appendix A that, for a transmitted narrowband signal with complex envelope $\tilde{s}(t)$ and carrier frequency ω_0, the complex envelope of the received signal is (approximately) $\tilde{s}(t)e^{-j\omega_d t}$.

Frequency ω_d is called the doppler shift of the signal and is related to radial target velocity v by

$$\omega_d = \pm \frac{2v}{c} \omega_0 \qquad (7.7\text{-}50)$$

where c is again the velocity of propagation of the radar signal through the transmission medium. For a target approaching the radar, the doppler shift in Eq. (7.7-50) is positive; otherwise, it is negative. [Radial target velocity corresponds to the addition of ω_d to carrier frequency ω_0 in the real narrowband signal description given in (7.7-14a).]

By retracing the steps leading to (7.7-49), it can be shown that for a moving target Eq. (7.7-49) is replaced by

$$|\tilde{e}_{\mathrm{MF}}(\tau, \omega_d)| = |\int \tilde{s}^*(t)\tilde{s}(t - \tau)e^{j\omega_d t}\,dt| \qquad (7.7\text{-}51)$$

Note that τ in Eq. (7.7-51) can be positive or negative, since the nominal target echo time delay τ_0 was set equal to zero. For a radar signal of duration T_s, it is easy to show that $|\tilde{e}_{\mathrm{MF}}(\tau, \omega_d)|$ is nonzero only for τ in the interval $(-T_s, T_s)$. With the trivial substitution of $-\tau$ for τ and replacing the integrand in Eq. (7.7-51) by its complex conjugate, we obtain an expression called the *radar ambiguity function*. This function, denoted by $\chi(\tau, \omega_d)$, is given by

$$\chi(\tau, \omega_d) = \left|\int_{-\infty}^{\infty} \tilde{s}(t)\tilde{s}^*(t + \tau)e^{-j\omega_d t}\,dt\right| \qquad (7.7\text{-}52)$$

The radar ambiguity function has become a major criterion of radar performance and design since it was first introduced by Woodward [6].

7.8. A Posteriori Detection

The design of an optimum receiver for detecting a target echo can be approached in many ways. One approach, suggested by a posteriori theory, is to design a receiver whose output yields an estimate of the target echo amplitude. The probability of a target's existence can be related to the probability that the estimated amplitude will be larger than some minimum value (chosen for reliable detection).† Thus, if echo amplitude is treated as a random variable, the joint a posteriori probability density $p(\tau, A\,|\,\mathbf{y})$ becomes for unknown range delay τ and echo amplitude A,

$$p(\tau, A\,|\,\mathbf{y}) = k p(\tau, A) p(\mathbf{y}\,|\,A, \tau) \qquad (7.8\text{-}1a)$$
$$= k p(\tau, A) p[\mathbf{y}\,|\,s(t - \tau)] \qquad (7.8\text{-}1b)$$

The delayed echo $s(t - \tau)$ can be written as $A s_0(t - \tau)$ in order to show

---

†This approach is closely related to maximum likelihood detection. See Helstrom [7], pp. 240–244.

explicitly the functional dependence of the received signal on both A and τ. Hence, Eq. (7.8-1b) can be rewritten as

$$p(\tau, A \mid \mathbf{y}) = k p(\tau, A) p[\mathbf{y} \mid A s_0(t - \tau)] \tag{7.8-2}$$

If it is assumed that $s(t)$ is a band-limited signal, earlier results can be used [see Eq. (7.6-19)] to obtain the following expression for $p[\mathbf{y} \mid A s_0(t - \tau)]$:

$$p[\mathbf{y} \mid A s_0(t - \tau)] = \exp \left\{ -\frac{1}{N_0} \int_0^T [y(t) - A s_0(t - \tau)]^2 \, dt \right\} \tag{7.8-3}$$

Substituting Eq. (7.8-3) into (7.8-2) and expanding the exponential term results in

$$p(\tau, A \mid \mathbf{y}) = k p(\tau, A) \exp \left[-\frac{1}{N_0} \int_0^T y^2(t) \, dt \right] \exp \left[-\frac{A^2}{N_0} \int_0^T s_0^2(t - \tau) \, dt \right]$$

$$\cdot \exp \left[\frac{2A}{N_0} \int_0^T y(t) s_0(t - \tau) \, dt \right] \tag{7.8-4}$$

Equation (7.8-4) can be simplified. Note that since $y(t)$ is a known received waveform, the first exponential term in Eq. (7.8-4) can be absorbed into the constant multiplier. Next, let $s_0(t)$ be normalized to unit energy:

$$\int_0^T s_0^2(t) \, dt = 1 \tag{7.8-5}$$

Finally, define a new quantity $q_0(\tau)$ by analogy with the previous definition of $q(\tau)$ in Eq. (7.7-8), as follows:

$$q_0(\tau) = \frac{2}{N_0} \int_0^T y(t) s_0(t - \tau) \, dt \tag{7.8-6}$$

It follows from Eqs. (7.7-8) and (7.8-6) that

$$q(\tau) = A q_0(\tau) \tag{7.8-7}$$

With Eqs. (7.8-5) through (7.8-7), Eq. (7.8-4) can be replaced by

$$p(\tau, A \mid \mathbf{y}) = k_1 p(\tau, A) e^{-A^2/N_0} e^{A q_0(\tau)} \tag{7.8-8}$$

Consider first the problem of estimating target range when amplitude A is a random variable. The marginal a posteriori distribution $p(\tau \mid \mathbf{y})$ is found by integrating $p(\tau, A \mid \mathbf{y})$ with respect to A. For this purpose we write

$$p(\tau, A) = p(\tau) p(A \mid \tau) \tag{7.8-9}$$

Substituting Eq. (7.8-9) into (7.8-8) yields for $p(\tau \mid \mathbf{y})$

$$p(\tau \mid \mathbf{y}) = k_1 p(\tau) \int_A p(A \mid \tau) e^{-A^2/N_0 + A q_0(\tau)} \, dA \tag{7.8-10}$$

If it is assumed that a priori probabilities are known to the observer, Eq. (7.8-10) indicates that (after integration) a knowledge of $q_0(\tau)$ is sufficient to compute $p(\tau \mid \mathbf{y})$. Thus, a sufficient receiver for range estimation computes the sufficient statistic $q_0(\tau)$ when echo amplitude is unknown. Since $q_0(\tau)$

differs from $q(\tau)$ by a scale factor, the structure of a sufficient receiver for unknown signal amplitude is essentially unchanged from a sufficient receiver for known signal amplitude.

Returning to the detection problem, we compute the marginal a posteriori probability density function of target amplitude A by integrating $p(\tau, A \mid \mathbf{y})$, given in Eq. (7.8-8), with respect to τ. Thus

$$p(A \mid \mathbf{y}) = k_1 p(A) e^{-A^2/N_0} \int_0^T p(\tau \mid A) e^{A q_0(\tau)} \, d\tau \qquad (7.8\text{-}11)$$

It is apparent from Eq. (7.8-11) that the computation of $p(A \mid \mathbf{y})$ still requires the computation of $q_0(\tau)$. Further, received waveform $y(t)$ appears only in the function $q_0(\tau)$. Thus $q_0(\tau)$ is still a necessary statistic, and the structure of the optimum receiver for estimating target amplitude requires a structure for estimating range as a preliminary step.

To interpret the remaining operations described in Eq. (7.8-11), assume that the a priori distribution $p(\tau \mid A)$ is independent of A; with Eq. (7.8-7), Eq. (7.8-11) can be rewritten as

$$p(A \mid \mathbf{y}) = k_1 p(A) e^{-A^2/N_0} \int_0^T p(\tau) e^{q(\tau)} \, d\tau \qquad (7.8\text{-}12)$$

The quantity under the integral sign in Eq. (7.8-12) is the unnormalized a posteriori distribution for target echo delay. Hence, to estimate amplitude, it is necessary to compute the area under the unnormalized a posteriori distribution over the entire search interval. Note, however, that for a single target with a large radar cross section, the area under the unnormalized distribution is essentially a measure of the area under a single spike, whose peak value is related to target echo amplitude (see Fig. 7.7-3). For example, a target return with a 13-db signal-to-noise ratio produces a spike approximately e^{20} in relative amplitude in the unnormalized a posteriori distribution.

The preceding development describes a roundabout method for arriving at an optimum detection receiver. The a posteriori theory can be related more closely to practice in the following way. The starting point of a posteriori theory, Eq. (7.4-2), relates inverse probabilities and direct probabilities; it is repeated here in slightly different notation:

$$P(H_i \mid \mathbf{y}) = \frac{P(H_i) P(\mathbf{y} \mid H_i)}{P(\mathbf{y})} \qquad (7.8\text{-}13)$$

where the H_i denote mutually exclusive hypotheses and the observed events $\mathbf{y}$ are initially assumed to be discrete. It is assumed further that only two hypotheses are possible: (1) signal present, which is denoted by hypothesis "S," and (2) signal absent, denoted by hypothesis "0." Let $P(S)$ denote the a priori probability that signal is present and $P(0)$ the a priori probability that signal is absent. Then Eq. (7.8-13) can be written as

$$P(S \mid \mathbf{y}) = \frac{P(S) P(\mathbf{y} \mid S)}{P(\mathbf{y})} \qquad (7.8\text{-}14)$$

and

$$P(0\,|\,\mathbf{y}) = \frac{P(0)P(\mathbf{y}\,|\,0)}{P(\mathbf{y})} \qquad (7.8\text{-}15)$$

The event $\mathbf{y}$ has two characterizations, depending on whether signal is present or not; therefore, the probability of y is given by

$$P(\mathbf{y}) = P(S)P(\mathbf{y}\,|\,S) + P(0)P(\mathbf{y}\,|\,0) \qquad (7.8\text{-}16)$$

Substituting Eq. (7.8-16) into Eqs. (7.8-14) and (7.8-15) results in

$$P(S\,|\,\mathbf{y}) = \frac{P(S)P(\mathbf{y}\,|\,S)}{P(S)P(\mathbf{y}\,|\,S) + P(0)P(\mathbf{y}\,|\,0)} \qquad (7.8\text{-}17)$$

and

$$P(0\,|\,\mathbf{y}) = \frac{P(0)P(\mathbf{y}\,|\,0)}{P(S)P(\mathbf{y}\,|\,S) + P(0)P(\mathbf{y}\,|\,0)} \qquad (7.8\text{-}18)$$

When the numerator and denominator of Eqs. (7.8-17) and (7.8-18) are divided by the product, $P(S)P(\mathbf{y}\,|\,0)$, they can be rewritten as

$$P(S\,|\,\mathbf{y}) = \frac{\dfrac{P(\mathbf{y}\,|\,S)}{P(\mathbf{y}\,|\,0)}}{\dfrac{P(\mathbf{y}\,|\,S)}{P(\mathbf{y}\,|\,0)} + \dfrac{P(0)}{P(S)}} = \frac{\ell(\mathbf{y})}{\ell(\mathbf{y}) + \dfrac{P(0)}{P(S)}} \qquad (7.8\text{-}19)$$

and

$$P(0\,|\,\mathbf{y}) = \frac{\dfrac{P(0)}{P(S)}}{\dfrac{P(\mathbf{y}\,|\,S)}{P(\mathbf{y}\,|\,0)} + \dfrac{P(0)}{P(S)}} = \frac{\dfrac{P(0)}{P(S)}}{\ell(\mathbf{y}) + \dfrac{P(0)}{P(S)}} \qquad (7.8\text{-}20)$$

where $\ell\,(\mathbf{y})$ is defined by

$$\ell(\mathbf{y}) = \frac{P(\mathbf{y}\,|\,S)}{P(\mathbf{y}\,|\,0)} \qquad (7.8\text{-}21)$$

The quantity $\ell(\mathbf{y})$ is called the likelihood ratio, since it is the ratio of two likelihood functions. When $\mathbf{y}$ is a continuous random variable, Eq. (7.8-21) is replaced by

$$\ell(\mathbf{y}) = \frac{p(\mathbf{y}\,|\,S)}{p(\mathbf{y}\,|\,0)} \qquad (7.8\text{-}22)$$

A possible detection criterion is to decide signal is present when the a posteriori probability $P(S\,|\,\mathbf{y})$ exceeds $P(0\,|\,\mathbf{y})$. Since

$$P(S\,|\,\mathbf{y}) + P(0\,|\,\mathbf{y}) = 1 \qquad (7.8\text{-}23)$$

this is equivalent to deciding a signal is present when $P(S\,|\,\mathbf{y})$ is greater than 0.5. Observe that the a posteriori probabilities of signal and no signal in Eqs. (7.8-19) and (7.8-20) are monotonic functions of the likelihood ratio; the probability of signal increases when the likelihood ratio increases while the probability of no signal simultaneously decreases. If a receiver is constructed that calculates the likelihood ratio for each waveform **y**, the existence probability of a target can be calculated from (7.8-19) with a knowledge of the ratio of the a priori probabilities of no signal and signal. Likelihood-ratio receivers are discussed more fully later in connection with statistical decision theory.

7.9. Band-Limited Signals in Colored Noise

The results in previous sections have been derived for additive white Gaussian noise. In this section necessary modifications are investigated for nonwhite noise. Consider first low-pass signals band-limited to $(-\omega_c, \omega_c)$. Extension to high-frequency narrowband signals is straightforward.

As in the case of white noise, Eq. (7.5-2), which relates the a posteriori density $p(x \mid \mathbf{y})$ of a set of messages x with associated waveforms $u_x(t)$ to the a priori distribution $p(x)$ and the likelihood function $p(\mathbf{y} \mid x)$, is the starting point:

$$p(x \mid \mathbf{y}) = kp(x)p(\mathbf{y} \mid x) \qquad (7.9\text{-}1a)$$

$$= kp(x)p(\mathbf{y} \mid \mathbf{u}_x) \qquad (7.9\text{-}1b)$$

Vector notations $\mathbf{y}$, $\mathbf{u}_x$ and $\mathbf{n}$ are respectively employed for band-limited waveforms† $y(t)$, $u_x(t)$ and $n(t)$, each of which is described over an interval T by $2f_cT$ waveform samples spaced $1/2f_c$ apart (where $\omega_c = 2\pi f_c$). The probability density of a band-limited random waveform is described by the joint probability density of its waveform samples. Thus,

$$p(\mathbf{y} \mid \mathbf{u}_x) = p(y_1, y_2, \ldots, y_{2f_cT} \mid \mathbf{u}_x) \qquad (7.9\text{-}2)$$

where $y_k = y(k/2f_c)$ is the kth waveform sample. If the noise waveform samples are $n_k = n(k/2f_c)$, the probability density of **n** is described by the joint density

$$p_n(\mathbf{n}) = p_n(n_1, n_2, \ldots, n_{2f_cT}) \qquad (7.9\text{-}3)$$

With Eqs. (7.5-3), (7.5-7), and (7.5-15a), Eq. (7.9-2) becomes

$$p(\mathbf{y} \mid \mathbf{u}_x) = p_n[(y_1 - u_{x\text{-}1}), (y_2 - u_{x\text{-}2}), \ldots, (y_{2f_cT} - u_{x\text{-}2f_cT})] \quad (7.9\text{-}4)$$

Equation (7.9-4) states that the conditional probability density $p(\mathbf{y} \mid \mathbf{u}_x)$ is equal to $p_n(\mathbf{n})$ with n_k replaced by $(y_k - u_{x\text{-}k})$ for $k = 1, 2, \ldots, 2f_cT$.

To obtain $p(\mathbf{y} \mid \mathbf{u}_x)$ we need to know $p_n(n_1, n_2, \ldots, n_{2f_cT})$. At this point the present development deviates from that of Section 7.5. In the case of white noise, Gaussian samples $n(k/2f_c)$, $k = 1, 2, \ldots, 2f_cT$, are uncorrelated and hence statistically independent. As a result $p_n(n_1, n_2, \ldots, n_{2f_cT})$ can be

---

†An ideal bandpass filter in the interval $(-\omega_c, \omega_c)$ is assumed at the input to the receiver, as discussed in Section 7.5 [see Fig. 7.5-1(b)].

factored into a product of density functions. When the noise is colored,† the noise samples are no longer statistically independent; in this case $p_n(n_1, n_2, \ldots, n_{2f_cT})$ is a multidimensional Gaussian density function described by Eq. (3.14-7) with zero mean:

$$p_n(n_1, n_2, \ldots, n_{2f_cT}) = \frac{1}{(2\pi)^{f_cT} |\Lambda|^{1/2}} \exp\left(-\frac{1}{2}\mathbf{n}^T\Lambda^{-1}\mathbf{n}\right) \qquad (7.9\text{-}5)$$

As noted in section 3.14, $\mathbf{n}^T$ is the transpose of matrix $\mathbf{n}$, defined by

$$\mathbf{n} = \begin{bmatrix} n_1 \\ n_2 \\ \cdot \\ \cdot \\ \cdot \\ n_{2f_cT} \end{bmatrix} \qquad (7.9\text{-}6)$$

Matrix Λ^{-1} is the inverse of matrix Λ, $|\Lambda|$ denotes the determinant of Λ, and Λ is the covariance matrix of the noise samples with $2f_cT$ rows and $2f_cT$ columns. Element λ_{ik} of the covariance matrix denotes the covariance between waveform samples n_i and n_k, that are separated in time by $(i - k)/2f_c$ seconds; λ_{ik} is related to the colored-noise autocorrelation function $\phi_{nn}(\tau)$ by

$$\lambda_{ik} = \phi_{nn}\left(\frac{i-k}{2f_c}\right) \qquad (7.9\text{-}7)$$

where $\phi_{nn}(\tau)$ is (from the Wiener-Khintchine theorem) the Fourier transform of the colored-noise spectral density $G_{nn}(\omega)$.

It follows from Eqs. (7.9-1) through (7.9-5) that

$$p(x\,|\,\mathbf{y}) = k_1 p(x) \exp\left[-\tfrac{1}{2}(\mathbf{y} - \mathbf{u}_x)^T\Lambda^{-1}(\mathbf{y} - \mathbf{u}_x)\right] \qquad (7.9\text{-}8)$$

If the indicated matrix multiplications are performed, Eq. (7.9-8) becomes

$$p(x\,|\,\mathbf{y}) = k_1 p(x) \exp\left\{-\tfrac{1}{2}\left[(\mathbf{y}^T\Lambda^{-1}\mathbf{y}) - (\mathbf{y}^T\Lambda^{-1}\mathbf{u}_x) - (\mathbf{u}_x^T\Lambda^{-1}\mathbf{y}) + (\mathbf{u}_x^T\Lambda^{-1}\mathbf{u}_x)\right]\right\}$$
$$(7.9\text{-}9)$$

Each of the four bracketed matrices in (7.9-9) is of quadratic form; after matrix multiplication, each term reduces to a scalar quantity. The first bracket in Eq. (7.9-9) is a known constant upon reception of waveform $\mathbf{y}$. The last bracket in Eq. (7.9-9) also reduces to a known constant since $\mathbf{u}_x$ is exactly known (a priori) for each x. The remaining second and third bracketed matrices are transposes of one another, since Λ is symmetric for a wide-sense stationary random process [see Eq. (4.4-9)]. Hence a sufficient a posteriori receiver calculates either term—$(\mathbf{y}^T\Lambda^{-1}\mathbf{u}_x)$, for example. With this statistic and the a priori density function, the a posteriori distribution $p(x\,|\,\mathbf{y})$ can then be computed from (7.9-9). Based upon this distribution, the observer

†See Section 4.7 for description of colored noise.

renders a decision concerning the identity of the transmitted signal in accordance with a predetermined criterion; for example, select that x which maximizes $p(x \mid \mathbf{y})$.

Let the sufficient statistic $(\mathbf{y}^T \Lambda^{-1} \mathbf{u}_x)$ be denoted once again by $q(x)$. Note that if the noise is white, the covariance matrix Λ is given by

$$\Lambda = \sigma^2 \mathbf{I} = f_c N_0 \mathbf{I} \qquad (7.9\text{-}10)$$

where $\mathbf{I}$ is the identity matrix and σ^2 is the variance of the noise process at each sampling instant. For this case the sufficient statistic $q(x)$ becomes $(1/f_c N_0)\mathbf{y}^T \mathbf{u}_x$ which can also be expressed as

$$q(x) = \left(\frac{1}{f_c N_0}\right) \sum_{k=1}^{2f_c T} y_k u_{x-k}$$

With (approximate) sampling-theorem relation (7.5-16), the sufficient statistic can be rewritten as

$$q(x) = \frac{2}{N_0} \int_0^T y(t) u_x(t)\, dt$$

which is identical to the result in Eq. (7.6-3).

When the noise is colored, a similar approach is adopted. The inverse covariance matrix Λ^{-1} is factored into a constant times the product of a matrix $\mathbf{W}$ and its transpose as follows:

$$\Lambda^{-1} = \frac{1}{\sigma^2} \mathbf{W}^T \mathbf{W} \qquad (7.9\text{-}11)$$

Such factorization is always possible for a positive definite matrix;† also every covariance matrix (and its inverse) is positive definite.‡ The matrix $\mathbf{W}$ has $2f_c T$ rows and columns, as Λ^{-1} does. Substituting Eq. (7.9-11) into the matrix expression $(\mathbf{y}^T \Lambda^{-1} \mathbf{u}_x)$ results in

$$q(x) = \mathbf{y}^T \Lambda^{-1} \mathbf{u}_x = \frac{1}{\sigma^2} (\mathbf{y}^T \mathbf{W}^T \mathbf{W} \mathbf{u}_x) \qquad (7.9\text{-}12a)$$

$$= \frac{1}{\sigma^2} (\mathbf{W}\mathbf{y})^T (\mathbf{W}\mathbf{u}_x) \qquad (7.9\text{-}12b)$$

If we let

$$\mathbf{Y} = \begin{bmatrix} Y_1 \\ Y_2 \\ \cdot \\ \cdot \\ \cdot \\ Y_{2f_c T} \end{bmatrix} = \mathbf{W}\mathbf{y} \qquad (7.9\text{-}13)$$

†See Guillemin [8], pp. 150–152.
‡See Middleton [9], p. 20.

and

$$\mathbf{U}_x = \begin{bmatrix} U_{x-1} \\ U_{x-2} \\ \cdot \\ \cdot \\ \cdot \\ U_{x-2f_cT} \end{bmatrix} = \mathbf{W}\mathbf{u}_x \qquad (7.9\text{-}14)$$

then the sufficient statistic can be rewritten as

$$q(x) = \frac{1}{\sigma^2}\,\mathbf{Y}^T\mathbf{U}_x \qquad (7.9\text{-}15)$$

Note that $\mathbf{Y}$ and $\mathbf{U}_x$ are each column matrices with elements that can be interpreted as sample values of waveforms $Y(t)$ and $U_x(t)$; the premultiplication of column matrices $\mathbf{y}$ and $\mathbf{u}_x$ by $\mathbf{W}$ corresponds to a weighting of the waveform samples of $y(t)$ and $u_x(t)$ which converts them to output waveforms $Y(t)$ and $U_x(t)$. Hence, $\mathbf{W}$ can be interpreted as the time-domain description of a filter with transfer function $H_W(\omega)$.

The sufficient statistic $(1/\sigma^2)\mathbf{Y}^T\mathbf{U}_x$ is proportional to the cross correlation[†] between waveforms $Y(t)$ and $U_x(t)$. One implementation of a sufficient receiver that yields $q(x)$ at its output is shown in Fig. 7.9-1 (for comparison, see Fig. 7.6-1).

To determine the nature of filtering operation $H_W(\omega)$ corresponding to matrix $\mathbf{W}$, consider its effect on a colored noise waveform $n(t)$. Denoting the filter output by $N(t)$, or $\mathbf{N}$, we may write, as in Eqs. (7.9-13) and (7.9-14),

$$\mathbf{N} = \mathbf{W}\mathbf{n} \qquad (7.9\text{-}16)$$

From (7.9-16) the covariance matrix of samples N_k of waveform $N(t)$ can be computed as

$$\overline{\mathbf{N}\mathbf{N}^T} = \overline{(\mathbf{W}\mathbf{n})(\mathbf{W}\mathbf{n})^T} \qquad (7.9\text{-}17\text{a})$$
$$= \mathbf{W}\overline{\mathbf{n}\mathbf{n}^T}\mathbf{W}^T \qquad (7.9\text{-}17\text{b})$$
$$= \mathbf{W}\mathbf{\Lambda}\mathbf{W}^T \qquad (7.9\text{-}17\text{c})$$

Solving (7.9-11) for $\mathbf{\Lambda}$ and substituting into (7.9-17c) yields the following result:

$$\overline{\mathbf{N}\mathbf{N}^T} = \sigma^2\mathbf{W}(\mathbf{W}^T\mathbf{W})^{-1}\mathbf{W}^T \qquad (7.9\text{-}18\text{a})$$
$$= \sigma^2(\mathbf{W}\mathbf{W}^{-1})(\mathbf{W}^{T\text{-}1}\mathbf{W}^T) \qquad (7.9\text{-}18\text{b})$$
$$= \sigma^2\mathbf{I} \qquad (7.9\text{-}18\text{c})$$

Equation (7.9-18c) indicates that waveform samples N_k are statistically independent and each has variance σ^2; hence the N_k are samples of a white Gaussian random process. Thus the premultiplication of $\mathbf{n}$ by matrix $\mathbf{W}$ can

†See footnote on p. 215.

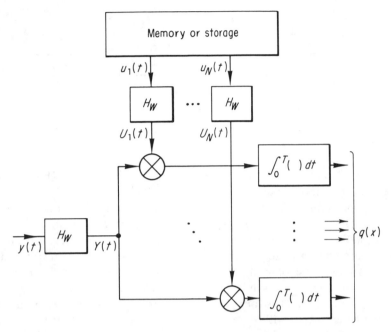

Fig. 7.9-1. Cross-correlation receiver for signals in colored noise.

be interpreted as a filtering operation that whitens the colored input noise (cf. Section 6.5). It follows that if the power spectral density of the input colored noise is described by $G_{nn}(\omega)$ within the frequency interval $(-\omega_c, \omega_c)$, then, to whiten the input noise, the filter transfer function $H_W(\omega)$ must satisfy

$$|H_W(\omega)|^2 = \frac{1}{G_{nn}(\omega)}, \qquad -\omega_c \leqslant \omega \leqslant \omega_c \qquad (7.9\text{-}19)$$

The filter $H_W(\omega)$ is frequently called a prewhitening filter.

The input signal is also altered by filter $H_W(\omega)$. If $u_k(t)$ is the transmitted signal, then it is altered to $U_k(t)$ by the prewhitening filter (see Fig. 7.9-1). It was previously shown that a sufficient receiver for known signals in white Gaussian noise cross-correlates stored replicas of the signal with the input waveform. This is illustrated in Fig. 7.9-1 by the cross correlation of $Y(t)$ with each $U_k(t)$.

When the preceding ideas are applied to a radar that estimates target range in background colored noise, a sufficient receiver can be implemented in a single channel by extending earlier results as shown in Fig. 7.9-2. Block $|H_W(\omega)|$ in Fig. 7.9-2 is the prewhitening filter. The matched filter shown is matched to waveform $\mathbf{U} = \mathbf{Wu}$, where $\mathbf{u}$ is the sampled transmitted wave-

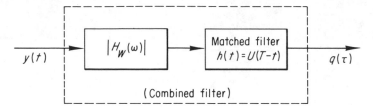

Fig. 7.9-2. Passive filter realization for range estimation in colored noise.

form. Both filters usually can be combined into a single filter whose output is the sufficient statistic $q(\tau)$.

In the preceding discussion the existence of a prewhitening filter was assumed. The realizability of such filters warrants comment. A random process $\{n(t)\}$ is *deterministic* if $n(t)$ can be predicted (by a linear operation) for all future and past times from knowledge of any finite waveform sample of $n(t)$. An example of a deterministic random process is described by $\cos(\omega_0 t + \theta)$, where θ is a random variable. Deterministic random processes are not satisfactory models for system noise, since a signal immersed in such noise can be extracted without error. It is shown in Doob[†] that a necessary and sufficient condition for a random process with spectral density $G_{nn}(\omega)$ to be *nondeterministic* is that

$$\int_{-\infty}^{\infty} \left| \frac{\log G_{nn}(\omega)}{1 + \omega^2} \right| d\omega < \infty \tag{7.9-20}$$

For comparative purposes, a necessary and sufficient condition for physical realizability of a filter with a square integrable transfer function $H(\omega)$ is given by the Paley-Wiener criterion:

$$\int_{-\infty}^{\infty} \left| \frac{\log |H(\omega)|}{1 + \omega^2} \right| d\omega < \infty \tag{7.9-21}$$

The similarity between Eqs. (7.9-20) and (7.9-21) is apparent. Suppose white noise of spectral density $N_0/2$ is passed through a realizable filter with transfer function $H(\omega)$; the output colored noise process spectral density $G_{nn}(\omega)$ is given by $(N_0/2)\,|H(\omega)|^2$. To determine whether this process is nondeterministic, $G_{nn}(\omega)$ is substituted into Eq. (7.9-20) as follows:

$$\int_{-\infty}^{\infty} \left| \frac{\log G_{nn}(\omega)}{1 + \omega^2} \right| d\omega = \int_{-\infty}^{\infty} \left| \frac{\log N_0/2\,|H(\omega)|^2}{1 + \omega^2} \right| d\omega \tag{7.9-22a}$$

$$= \int_{-\infty}^{\infty} \left| \frac{\log N_0/2}{1 + \omega^2} + \frac{2\log |H(\omega)|}{1 + \omega^2} \right| d\omega \tag{7.9-22b}$$

$$\leqslant \int_{-\infty}^{\infty} \left| \frac{\log N_0/2}{1 + \omega^2} \right| d\omega + 2\int_{-\infty}^{\infty} \left| \frac{\log |H(\omega)|}{1 + \omega^2} \right| d\omega \tag{7.9-22c}$$

†See Doob [10], p. 584.

The first integral in Eq. (7.9-22c) is finite. The second integral is finite due to realizability condition (7.9-21). Hence (7.9-20) is satisfied; it follows that the colored noise process generated by passing white noise through a realizable filter is nondeterministic.

Consider next the inverse problem of converting colored nondeterministic noise with finite power into white noise by a realizable filter transformation. Superficially this is impossible since an idealized white-noise model has constant spectral density over an infinite frequency interval and hence infinite power. Physical realizations of white noise processes have spectral densities that are flat over a wide frequency interval but eventually fall off exponentially. In applications, a noise spectral density that is flat over an interval which encompasses the significant signal bandwidth is often a satisfactory approximation to white noise; this approach is adopted here.

A filter $H_W(\omega)$ with a colored-noise input of spectral density $G_{nn}(\omega)$ produces an output noise spectral density given by $|H_W(\omega)|^2 G_{nn}(\omega)$. If $G_{nn}(\omega)$ is nonzero over the significant signal spectral region of interest, we require that the spectral density out of the prewhitening filter be flat over this interval. Beyond this frequency band, the filter characteristic is shaped to satisfy the requirements of square integrability and the Paley-Wiener criterion. This results in a realizable filter and an output noise spectral density that is white over the significant region of the signal spectrum. In this respect a prewhitening filter can be made physically realizable.

Equation (7.9-20) describes a significant constraint on nondeterministic processes. Thus if a random process possesses a spectral density that is zero over any continuous frequency interval, Eq. (7.9-20) is not satisfied; hence the process is deterministic. Thus a band-limited process is deterministic. It follows that the sampling-theorem representation of a nondeterministic random process is only an approximation to reality, as previously stated. In practice, however, the sampling-theorem representation has been found to be a good approximation of both deterministic and nondeterministic signals that are essentially band-limited.

7.10. Arbitrary Signals in Colored Noise

The problem of reception in colored noise will now be treated more rigorously using a procedure that does not depend on the sampling theorem. Instead, a Karhunen-Loéve orthonormal expansion is employed that characterizes arbitrary random processes, whether band-limited or not. Unfortunately the results obtained by this procedure often cannot be interpreted as simply as in the sampling-theorem approach.

In Section 4.10 it was shown that a finite-duration sample of a Gaussian random process $\{n(t)\}$ can be described by an orthonormal expansion with uncorrelated Gaussian coefficients. Thus, waveform $n(t)$ in the observation interval $(0, T)$ can be represented by

$$n(t) = \sum_{k=1}^{\infty} n_k \psi_k(t), \qquad 0 \leqslant t \leqslant T \tag{7.10-1}$$

where the coefficients n_k are related to waveform $n(t)$ by

$$n_k = \int_0^T n(t) \psi_k(t) \, dt \tag{7.10-2}$$

Coefficients n_k do not correspond to waveform time samples in this development. [From (7.10-2) note that n_k can be generated by passing $n(t)$ through a filter with impulsive response $\psi_k(T - t)$ and sampling the output at time T(see Section 7.6).] The functions $\psi_k(t)$ form a complete orthonormal set over the interval $(0, T)$; they are solutions to integral equation

$$\int_0^T \phi_{nn}(t - s) \psi_k(s) \, ds = \lambda_k \psi_k(t), \qquad 0 \leqslant t, s \leqslant T \tag{7.10-3}$$

where $\phi_{nn}(\tau)$ is the autocorrelation function of the (stationary) noise random process. Corresponding to each eigenfunction ψ_k in Eq. (7.10-3) there is an eigenvalue λ_k which is a real positive number equal to the variance of random variable n_k.

A difficulty of the present method is that an infinite set $\{n_k\}$ may be required to describe $n(t)$ by Eq. (7.10-1). To circumvent this problem, the coefficients n_k are usually ordered so that the random variable with the largest variance is designated as n_1 (and its variance as λ_1); the next largest is designated n_2 (with variance λ_2); and so on. Then $n(t)$ is approximated by the first m coefficients (with $m \to \infty$) in Eq. (7.10-1). Since each n_k is a statistically independent Gaussian random variable, the distribution $p_n(n_1, n_2, \ldots, n_m)$ is given by

$$p_n(n_1, n_2, \ldots, n_m) = p(n_1) \cdot p(n_2) \cdot \ldots \cdot p(n_m) \tag{7.10-4a}$$

$$= \prod_{k=1}^{m} \frac{1}{(2\pi\lambda_k)^{1/2}} \exp\left(-\frac{n_k^2}{2\lambda_k}\right) \tag{7.10-4b}$$

From Eqs. (7.5-3) and (7.10-4b) $p(\mathbf{y} \mid \mathbf{u}_x)$ is given by

$$p(\mathbf{y} \mid \mathbf{u}_x) = \prod_{k=1}^{m} \frac{1}{(2\pi\lambda_k)^{1/2}} \exp\left[-\frac{(y_k - u_{x-k})^2}{2\lambda_k}\right] \tag{7.10-5a}$$

$$= \frac{1}{(2\pi)^{m/2}\left(\prod_{k=1}^{m} \lambda_k\right)^{1/2}} \exp\left[-\sum_{k=1}^{m} \frac{(y_k - u_{x-k})^2}{2\lambda_k}\right] \tag{7.10-5b}$$

where, analogous to Eqs. (7.10-1) and (7.10-2),

$$y(t) = \sum_{k=1}^{\infty} y_k \psi_k(t) \tag{7.10-6}$$

$$y_k = \int_0^T y(t) \psi_k(t) \, dt \tag{7.10-7}$$

and

$$u_x(t) = \sum_{k=1}^{\infty} u_{x-k}\psi_k(t) \qquad (7.10\text{-}8)$$

$$u_{x-k} = \int_0^T u_x(t)\psi_k(t)\, dt \qquad (7.10\text{-}9)$$

Except for the uncertainty in m and the unequal variances, Eq. (7.10-5b) is similar to Eq. (7.5-15) for a band-limited signal in white Gaussian noise. The exponent in Eq. (7.10-5b) can be expanded into three terms:

$$\left\langle \sum_{k=1}^{m} \frac{y_k^2}{2\lambda_k}, \quad \sum_{k=1}^{m} \frac{u_{x-k}^2}{2\lambda_k}, \quad \sum_{k=1}^{m} \frac{y_k u_{x-k}}{\lambda_k} \right\rangle$$

The last term is the sufficient statistic $q(x)$. It has the appearance of a cross correlation between a waveform with components y_k and a signal with components u_{x-k}/λ_k. Let

$$U_{x-k} = \frac{u_{x-k}}{\lambda_k}, \qquad k = 1, 2, 3, \ldots \qquad (7.10\text{-}10)$$

and define

$$U_x(t) = \sum_{k=1}^{\infty} U_{x-k}\psi_k(t), \qquad 0 \leqslant t \leqslant T \qquad (7.10\text{-}11)$$

Then sufficient statistic $q(x)$ can be expressed as

$$q(x) = \sum_{k=1}^{m} \frac{y_k u_{x-k}}{\lambda_k} = \sum_{k=1}^{m} y_k U_{x-k} \qquad (7.10\text{-}12)$$

The sum in Eq. (7.10-12) can be transformed into an integral by the following identity. Let $f(t)$ and $g(t)$ be two functions defined over the interval $(0, T_1)$ with expansions of the form given in Eq. (7.10-1). Then,

$$\int_0^{T_1} f(t)g(t)\, dt = \int_0^{T_1} \sum_{i=1}^{\infty} f_i\psi_i(t) \sum_{k=1}^{\infty} g_k\psi_k(t)\, dt \qquad (7.10\text{-}13\text{a})$$

$$= \sum_{i,k}^{\infty} f_i g_k \int_0^{T_1} \psi_i(t)\psi_k(t)\, dt \qquad (7.10\text{-}13\text{b})$$

Since the $\psi_k(t)$ are orthonormal, the desired result follows:

$$\int_0^{T_1} f(t)g(t)\, dt = \sum_{k=1}^{\infty} f_k g_k \qquad (7.10\text{-}14)$$

From identity (7.10-14), and for m approaching infinity, (7.10-12) can be rewritten as

$$q(x) = \lim_{m\to\infty} \sum_{k=1}^{m} y_k U_{x-k} = \int_0^T y(t) U_x(t)\, dt \qquad (7.10\text{-}15)$$

when the limit exists. It follows from Eq. (7.10-15) that sufficient statistic $q(x)$ can be obtained by passing the received waveform through a filter $h_x(t)$ matched to $U_x(t)$—that is,

$$h_x(t) = U_x(T - t) \qquad (7.10\text{-}16)$$

and sampling the output at time $t = T$.

To appreciate the significance of filter $h_x(t)$ defined in (7.10-16), observe that $U_x(t)$ is the solution of the integral equation

$$u_x(t) = \int_0^T \phi_{nn}(t - s) U_x(s)\, ds, \qquad 0 \leqslant s, t \leqslant T \qquad (7.10\text{-}17)$$

This can be shown by replacing $u_x(t)$ and $U_x(t)$ in Eq. (7.10-17) by their respective expansions, Eqs. (7.10-8) and (7.10-11); the result is

$$\sum_{k=1}^{\infty} u_{x-k} \psi_k(t) = \int_0^T \phi_{nn}(t - s) \sum_{k=1}^{\infty} U_{x-k} \psi_k(s)\, ds \qquad (7.10\text{-}18a)$$

$$= \sum_{k=1}^{\infty} U_{x-k} \int_0^T \phi_{nn}(t - s) \psi_k(s)\, ds \qquad (7.10\text{-}18b)$$

$$= \sum_{k=1}^{\infty} U_{x-k} \lambda_k \psi_k(t) \qquad (7.10\text{-}18c)$$

It follows from Eq. (7.10-18c) that

$$U_{x-k} = \frac{u_{x-k}}{\lambda_k}, \qquad k = 1, 2, \ldots$$

which is identical to expression (7.10-10) and completes the proof.

Integral equation (7.10-17) is similar to integral equation (5.2-14), whose solution yields the impulse response of a filter that maximizes the peak signal power to average noise level for a known signal immersed in stationary colored noise. This integral equation, called an inhomogeneous Fredholm integral equation of the first kind, occurs frequently in noise problems. A number of special cases were treated in Chapters 5 and 6.

To relate present results to those of the previous section, let interval $(0, T)$ approach infinity so that (7.10-17) can be replaced by

$$u_x(t) = \int_{-\infty}^{\infty} \phi_{nn}(t - s) U_x(s)\, ds \qquad (7.10\text{-}19)$$

Equation (7.10-19) is the familiar convolution integral; its Fourier transform is given by

$$u_x(\omega) = G_{nn}(\omega) U_x(\omega) \qquad (7.10\text{-}20)$$

where $u_x(\omega)$ and $U_x(\omega)$ describe the transforms of $u_x(t)$ and $U_x(t)$, respectively, and $G_{nn}(\omega)$ is the spectral density of the stationary colored-noise process. Solving Eq. (7.10-20) for $U_x(\omega)$ yields

$$U_x(\omega) = \frac{u_x(\omega)}{G_{nn}(\omega)} \qquad (7.10\text{-}21)$$

If we denote the Fourier transform of $h_x(t)$ by $H_x(\omega)$, then from Eqs. (7.10-16) and (7.10-21) $H_x(\omega)$ is given by

$$H_x(\omega) = \frac{u_x^*(\omega) e^{-j\omega T}}{G_{nn}(\omega)} \qquad (7.10\text{-}22a)$$

$$= \left(\frac{1}{\sqrt{G_{nn}(\omega)}} \right) \left[\frac{u_x^*(\omega) e^{-j\omega T}}{\sqrt{G_{nn}(\omega)}} \right] \qquad (7.10\text{-}22b)$$

Equation (7.10-22a) is similar to Eq. (5.2-17), which describes the transfer function of a filter that maximizes signal-to-noise ratio; Eq. (7.10-22b) shows that filter $H_x(\omega)$ can be interpreted as the cascade of a prewhitening filter (with a transfer function whose magnitude is described by the first bracket) and a filter matched to a modification of the transmitted signal. This result is discussed in Chapter 6. A passive filter implementation of an a posteriori sufficient receiver for optimum reception in colored noise is shown in Fig. 7.10-1.

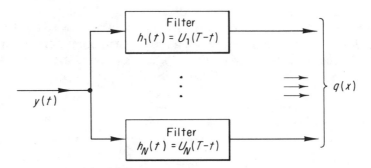

Fig. 7.10-1. Passive filter realization for sufficient receiver (colored noise).

7.11. Signals Perturbed by a Random Channel

It was implicitly assumed in previous discussions that the propagating medium is distortionless and time-invariant. In some applications the transmission medium is time-varying; examples are scatter communication, sonar, and radar astronomy. In both scatter communication and sonar, the transmitted and received signals propagate over many constantly changing paths. As a result, the composite received signal waveform, which is the sum of many multipath signals, becomes random. Similarly, in radio astronomy the reflected signal from a planet, such as Venus, consists of echoes from widely distributed scattering areas on the planetary surface. Owing to the rotation of the earth and the observed planet, path lengths traversed by various reflections change continuously. This results in a composite reflected signal which is random and is similar to that obtained in multipath propagation.

The most comprehensive treatment of optimum receivers for signals distorted by random time-varying channels and corrupted by additive Gaussian noise is due to Kailath [11], who assumed a Gaussian channel. The problem can be formulated as follows: a waveform $u_x(t)$, representing message x in set $\{x\}$, is transmitted through a linear time-variant Gaussian channel

A of finite memory; the resulting composite waveform is $z_x(t)$. Further, independent Gaussian noise $n(t)$ is added to waveform $z_x(t)$ prior to reception; the resulting received waveform is $y(t)$. It is desired to determine the structure of a sufficient receiver whose output, together with the known a priori probability density $p(x)$, provides the a posteriori probability density $p(x|\mathbf{y})$ to an observer.

To simplify the interpretation of the optimum receiver structure the sampling-theorem is employed for waveform representation in the channel model developed by Kailath [12]; the characteristics of the linear time-variant channel are simulated† by a discrete tapped delay line with time-varying gains $a_i(t)$, as shown in Fig. 7.11-1. The taps are spaced one time unit apart, where the time unit is established from system input (or output) total bandwidth constraints or, when the propagation paths are separable, tap locations can be arranged to coincide with the actual paths.

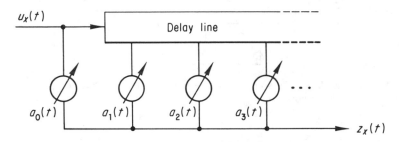

Fig. 7.11-1. Tapped delay-line representation of linear time-varying channel (*after Kailath* [11], *courtesy of IEEE*).

If the input and channel time functions are sampled at the tap spacing rate, it follows from Fig. 7.11-1 that the sampled waveform output $z(m \Delta t) = z(m)$ is given by

$$z(m) = \sum_{k=0}^{m} a_k(m)u_x(m - k) \qquad (7.11\text{-}1)$$

where m and k are integers. Equation (7.11-1) is the discrete analogue of the convolution theorem for a time-varying filter with impulse response $h(\tau, t)$. (See Section 2.2.) Equation (7.11-1) can be rewritten in matrix notation as

$$\mathbf{z}_x = \mathbf{A}\mathbf{u}_x \qquad (7.11\text{-}2)$$

where $\mathbf{z}_x$ and $\mathbf{u}_x$ are column matrices of the sampled values of $z_x(t)$ and $u_x(t)$, and $\mathbf{A}$ is a rectangular matrix representation of the channel. For example, for a three-tap channel,

---

†Kailath shows that an arbitrary linear time-variant filter can be represented as closely as desired by a discrete filter of the type shown in Fig. 7.11-1.

$$
\begin{bmatrix} z_x(0) \\ z_x(1) \\ z_x(2) \\ z_x(3) \end{bmatrix} = \begin{bmatrix} a_0(0) & 0 & 0 \\ a_1(1) & a_0(1) & 0 \\ a_2(2) & a_1(2) & a_0(2) \\ 0 & a_2(3) & a_1(3) \\ 0 & 0 & a_2(4) \end{bmatrix} \begin{bmatrix} u_x(0) \\ u_x(1) \\ u_x(2) \end{bmatrix} \tag{7.11-3}
$$

A channel covariance matrix Λ_A can be defined; thus, for $a_i(t)$ with zero mean, Λ_A is given by†

$$
\Lambda_A = \begin{bmatrix} \overline{a_0^2(0)} & \overline{a_0(0)a_0(1)} & \ldots & \overline{a_0(0)a_1(1)} & \ldots \\ \overline{a_0(1)a_0(0)} & \overline{a_0^2(1)} & \ldots & & \\ \vdots & & & & \\ \vdots & & & & \\ \overline{a_1(1)a_0(0)} & \overline{a_1(1)a_0(1)} & \ldots & \overline{a_1^2(1)} & \ldots \\ \vdots & & & & \\ \vdots & & & & \end{bmatrix} \tag{7.11-4}
$$

The channel covariance matrix Λ_A is symmetric and almost always nonsingular. An exception is the covariance matrix of a random time invariant channel, which can be described by one tap; in such situations Kailath [12] has shown that by redefining matrices, the modified channel covariance matrix can be made nonsingular. Henceforth, Λ_A is assumed to be nonsingular. It can be shown that Λ_{z_x} is also symmetric and nonsingular.

The received waveform $\mathbf{y}$ is the sum of signal $\mathbf{z}_x$ and an additive Gaussian noise waveform $\mathbf{n}$:

$$
\mathbf{y} = \mathbf{z}_x + \mathbf{n} \tag{7.11-5}
$$

If channel $\mathbf{A}$ contains a random component $\mathbf{A}^r$ plus a mean component $\bar{\mathbf{A}}$, Eq. (7.11-5) can be rewritten as

$$
\mathbf{y} = \mathbf{A}^r \mathbf{u}_x + \bar{\mathbf{A}} \mathbf{u}_x + \mathbf{n} \tag{7.11-6}
$$

Both $\mathbf{z}_x$ and $\mathbf{n}$ are Gaussian and their statistics are assumed to be known a priori. In particular, a zero mean noise process $\{n(t)\}$ is assumed with nonsingular covariance matrix Λ_n; further, $\mathbf{z}_x$ and $\mathbf{n}$ are statistically independent. It follows that $\mathbf{y}$ is a sampled Gaussian random waveform with covariance matrix $\Lambda_{y|x}$ given by

$$
\Lambda_{y|x} = \Lambda_{z_x} + \Lambda_n \tag{7.11-7}
$$

The likelihood function $p(\mathbf{y} \mid \mathbf{u}_x)$ (see Sections 7.4 and 7.5) can be written as

$$
p(\mathbf{y} \mid \mathbf{u}_x) = \frac{1}{(2\pi)^{N/2} \mid \Lambda_{y|x} \mid^{1/2}} \exp\left[-\frac{1}{2} (\mathbf{y} - \bar{\mathbf{z}}_x)^T \Lambda_{y|x}^{-1} (\mathbf{y} - \mathbf{z}_x) \right] \tag{7.11-8}
$$

---

†In [12] it is shown how covariance matrix Λ_{z_x} can be computed from covariance matrix Λ_A for a given $\mathbf{u}_x$; however, it is not needed in this discussion.

where N is the number of waveform samples in vector $\mathbf{y}$. The a posteriori distribution $p(x\,|\,\mathbf{y})$ is given by [see Eq. (7.5-2)]

$$p(x\,|\,\mathbf{y}) = k_1 p(x)p(\mathbf{y}\,|\,\mathbf{u}_x) \qquad (7.11\text{-}9a)$$

$$= k_2 p(x) \exp\left[-\tfrac{1}{2}(\mathbf{y} - \bar{\mathbf{z}}_x)^T \Lambda_{y|x}^{-1}(\mathbf{y} - \bar{\mathbf{z}}_x)\right] \qquad (7.11\text{-}9b)$$

A sufficient statistic of the a posteriori distribution $p(x\,|\,\mathbf{y})$ is

$$q(x) = (\mathbf{y} - \bar{\mathbf{z}}_x)^T \Lambda_{y|x}^{-1}(\mathbf{y} - \bar{\mathbf{z}}_x) \qquad (7.11\text{-}10)$$

Depending on additional assumptions about the channel and the additive noise, various optimum receiver structures are obtained. A number of examples are now considered.

(a) The case for $\mathbf{A}^r = 0$ corresponds to a deterministic channel; for noise that is white and stationary, $\Lambda_n = \sigma^2 \mathbf{I}_n$ [see Eq. (7.9-10)], where σ^2 is the variance of the (band-limited) noise process and $\mathbf{I}$ is the identity matrix. In this case, Eq. (7.11-10) becomes

$$q(x) = \frac{1}{\sigma^2}(\mathbf{y} - \bar{\mathbf{z}}_x)^T(\mathbf{y} - \bar{\mathbf{z}}_x) \qquad (7.11\text{-}11a)$$

$$= \frac{1}{\sigma^2}(\mathbf{y}^T\mathbf{y} + \bar{\mathbf{z}}_x^T\bar{\mathbf{z}}_x - 2\bar{\mathbf{z}}_x^T\mathbf{y}) \qquad (7.11\text{-}11b)$$

The first term in (7.11-11b) is independent of x and the second is exactly known for given x; hence $q(x)$ can be simplified to (see Section 7.9)

$$q'(x) = \bar{\mathbf{z}}_x^T\mathbf{y} \qquad (7.11\text{-}12)$$

By methods analogous to those discussed in Sections 7.6 and 7.9, receiver elements that compute $q'(x)$ for waveform $u_x(t)$ are shown in Figs. 7.11-2(a) and 7.11-2(b). To obtain the entire function $q'(x)$, additional receiver elements are required for the remainder of set $\{x\}$. The receiver in Fig. 7.11-2(a) is an active correlator; a matched filter implementation is shown in Fig. 7.11-2(b). These results are similar to those obtained in Section 7.6, Figs. 7.6-1 and 7.6-2.

When the noise is colored, covariance matrix $\Lambda_{y|x}$ can be factored into the product of a matrix $\mathbf{W}$ and its transpose [see Eq. (7.9-11) and related discussion]. In this case (7.11-10) can be rewritten as

$$q(x) = (\mathbf{y} - \bar{\mathbf{z}}_x)^T\mathbf{W}^T\mathbf{W}(\mathbf{y} - \bar{\mathbf{z}}_x) \qquad (7.11\text{-}13a)$$

$$= (\mathbf{W}\mathbf{y})^T(\mathbf{W}\mathbf{y}) + (\mathbf{W}\bar{\mathbf{z}}_x)^T(\mathbf{W}\bar{\mathbf{z}}_x) - 2(\mathbf{W}\bar{\mathbf{z}}_x)^T(\mathbf{W}\mathbf{y}) \qquad (7.11\text{-}13b)$$

The sufficient statistic is given by the last term in (7.11-13b):

$$q'(x) = (\mathbf{W}\bar{\mathbf{z}}_x)^T(\mathbf{W}\mathbf{y}) \qquad (7.11\text{-}14)$$

An active correlator element that computes $q'(x)$ for one waveform $u_x(t)$ is shown in Fig. 7.11-3. The matrix $\mathbf{W}$ may be interpreted, as discussed earlier, as a prewhitening filter with impulse response $h_W(t)$ that whitens the noise component of $\mathbf{y}$. Since the signal component in $\mathbf{y}$ is also altered, the known reference signal must be modified prior to cross correlation.

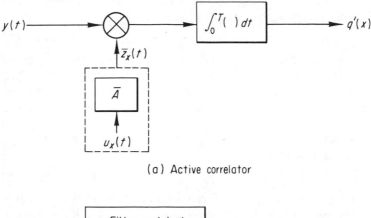

(a) Active correlator

(b) Matched filter

Fig. 7.11-2. Receiver elements for a deterministic channel with white additive Gaussian noise.

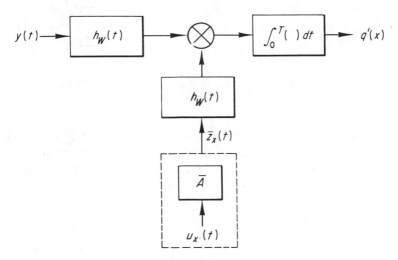

Fig. 7.11-3. Receiver element for deterministic channel with additive colored Gaussian noise.

(b) Consider next a purely random channel—that is, $\bar{\mathbf{A}} = 0$. Since $\mathbf{A}^r$ is unknown, it cannot be used to calculate $\mathbf{z}_x$ as in the previous example [see Fig. 7.11-2(a)]. For this case, Kailath [11] shows that the optimum receiver estimates $\mathbf{z}_x$ and cross-correlates this estimate with $\mathbf{y}$. Let $\mathbf{H}_x$ denote a linear filter (not necessarily realizable) that operates on $\mathbf{y}$ to yield a minimum variance estimate of $\mathbf{z}_x$, denoted by $\hat{\mathbf{z}}_x$; that is,

$$\hat{\mathbf{z}}_x = \mathbf{H}_x \mathbf{y} \tag{7.11-15}$$

The matrix multiplication $\mathbf{H}_x \mathbf{y}$ in (7.11-15) requires the number of columns in matrix $\mathbf{H}_x$ to equal the number of rows in $\mathbf{z}_x$, assuming both $\mathbf{y}$ and $\mathbf{z}_x$ have the same duration. In order that $\hat{\mathbf{z}}_x$ and $\mathbf{z}_x$ also have the same duration, the number of rows in $\mathbf{H}_x$ must equal the number of rows in $\mathbf{z}_x$. Hence $\mathbf{H}_x$ is square.

When (7.11-15) is substituted into (7.11-12), the sufficient statistic for additive white noise becomes

$$q'(x) = \mathbf{y}^T \mathbf{H}_x \mathbf{y} \tag{7.11-16}$$

One element of a sufficient receiver that computes $q'(x)$ is shown in Fig. 7.11-4. As expected, the filter $\mathbf{H}_x$ yields the estimate $\hat{\mathbf{z}}_x$; this estimate is then cross-correlated with input $\mathbf{y}$. For colored noise, the result is modified only slightly; if (7.11-15) is substituted into (7.11-14), the sufficient statistic becomes

$$q'(x) = (\mathbf{W}\mathbf{H}_x \mathbf{y})^T (\mathbf{W}\mathbf{y}) \tag{7.11-17}$$

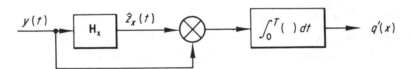

Fig. 7.11-4. Receiver element for purely random channel with additive white Gaussian noise.

An element of the sufficient receiver is shown in Fig. 7.11-5. The action of prewhitening filter $h_W(t)$ is similar to that described in the previous example.

An expression for estimating filter $\mathbf{H}_x$ can be obtained by minimizing the average mean-square difference between $\hat{\mathbf{z}}_x$ and $\mathbf{z}_x$ (see reference [11]). If

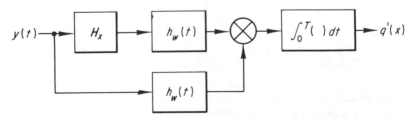

Fig. 7.11-5. Receiver element for purely random channel with additive colored Gaussian noise.

this procedure is followed, the result is

$$\mathbf{H}_x = (\mathbf{I} + \Lambda_n \Lambda_{z_x}^{-1})^{-1} \qquad (7.11\text{-}18)$$

Both Λ_n and $\Lambda_{z_x}^{-1}$ are symmetric; however, their product is not symmetric unless they commute. Hence, $\mathbf{H}_x$ is not symmetric in general. For additive white Gaussian noise, (7.11-18) becomes

$$\mathbf{H}_x = (\mathbf{I} + \sigma^2 \Lambda_{z_x}^{-1})^{-1} \qquad (7.11\text{-}19)$$

which *is* symmetric. It is shown in Kailath [11] that a filter defined by a symmetric matrix is unrealizable, since it operates on future as well as past data. However, this filter can be replaced by a realizable filter that is obtained from $\mathbf{H}_x$ by setting all matrix elements above the main diagonal equal to zero and doubling all those below the main diagonal.

The above results can be extended to Gaussian channels that contain both a deterministic component and a random component (see problem 7.9). Further, in weak-signal cases, approximate solutions can be obtained for non-Gaussian-channel statistics; these solutions are obtained from Taylor series expansions of the probability density function of signal plus noise.

7.12. Summary

In the a posteriori theory of reception, a posteriori probabilities and probability densities provide a complete description of the results of the receiving process. An ideal receiver, by definition, calculates and presents the a posteriori distribution of a desired quantity or quantities to an observer. Owing to the dependence of the a posteriori distribution on a priori information, which may not always be available, the concept of a sufficient receiver was introduced. A sufficient receiver calculates and presents to the observer a sufficient statistic; this statistic contains the essential information in the received waveform and permits the a posteriori distribution to be computed when the a priori statistics are known. When reception occurs in additive Gaussian noise, the a posteriori theory leads to sufficient receivers that employ matched filters. However, the theory has been found to be useful in many situations where the optimum receiver structure is not obvious.

PROBLEMS

7.1. When message X_1 is transmitted through a communication channel, the probability that Y_1 is received is 0.75 and Y_2 is 0.25. When X_2 is transmitted, the probability of Y_1 is 0.4 and Y_2 is 0.6. This can be represented by the accompanying diagram (Fig. 7P-1). If a priori probability $P(X_1) = 0.7$ and

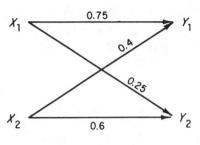

Fig. 7P-1

$P(X_2) = 0.3$, compute the uncertainties $\mathscr{H}(X_1)$ and $\mathscr{H}(X_2)$ and the a posteriori uncertainties $\mathscr{H}(X_i | Y_k)$, $i, k = 1, 2$. What are the information gains $\mathscr{I}(X_i | Y_k)$, $i, k = 1, 2$, through the channel?

7.2. In applications of information theory, *average information transfer* is often of considerable interest. Information can be averaged with respect to the recipient, the sender, or both. The expressions for each case are

$$\mathscr{I}_{X_i} = \sum_k \mathscr{I}(Y_k | X_i) P(Y_k | X_i)$$

$$\mathscr{I}_{Y_k} = \sum_i \mathscr{I}(X_i | Y_k) P(X_i | Y_k)$$

$$\mathscr{I} = E[\mathscr{I}_{X_i}] = E[\mathscr{I}_{Y_k}] = \sum_{i, k} \mathscr{I}(X_i | Y_k) P(X_i, Y_k)$$

Compute these quantities for the example in problem 7.1.

7.3. A communication channel transmits a message in the set $\{X_1, X_2, X_3, X_4\}$ with equal a priori probability. Suppose that half of the messages, in a long sequence of messages, are erroneously received; each error is distributed equally among the other three messages. Find the average information per channel message.

7.4. Show that the radar ambiguity function [see Eq. (7.7-52)] can be expressed as

$$\chi(\tau, \omega_d) = \frac{1}{2\pi} \left| \int \tilde{S}^*(\omega) \tilde{S}(\omega + \omega_d) e^{-j\omega\tau} \, d\omega \right|$$

where $\tilde{S}(\omega)$ is the Fourier transform of $s(t)$.

7.5. The radar ambiguity function [see Eq. (7.7-52)] is frequently represented as a two-dimensional surface over a plane with axes τ and ω_d. Calculate the radar ambiguity function for a rectangular rf pulse. What are the dimensions of the main lobe of this surface along the τ and ω_d axes about the origin?

7.6. The exact doppler effect resulting from reflection from a moving target changes transmitted signal $s(t)$ to $s(at)$, where

$$a = \frac{c \pm v}{c \mp v}$$

Find the exact and approximate doppler effect on the Fourier transform of $s(t)$. (See Appendix A.)

7.7. For a two-component sampling-theorem representation of a received waveform, (y_1, y_2), and a signal, (u_{x-1}, u_{x-2}), and a 2×2 covariance matrix Λ, calculate sufficient statistic $q(x) = \mathbf{y}^T \Lambda^{-1} \mathbf{u}_x$. Draw the block diagram of a sufficient receiver that implements each indicated operation. What is the significance of the processor?

7.8. Verify (7.10-20) by carrying out the Fourier transformation of Eq. (7.10-19). With $h_x(t) = U_x(T - t)$ verify the solution obtained in Eq. (7.10-22a), using the relations in Chapter 2.

7.9. For a channel like that discussed in Section 7.11, but with a deterministic (specular) component as well as a random component, develop the optimum receiver structure assuming additive white Gaussian noise. Interpret the structure as a combination of the deterministic and random channel cases.

REFERENCES

1. Shannon, C. E.: A Mathematical Theory of Communication, *Bell System Tech. J.*, **27**: 379–423, 623–656 (July and October, 1948).

2. Woodward, P. M., and Davies, I. L.: A Theory of Radar Information, *Phil. Mag.*, **41**: 1001–1017 (October, 1950).

3. Woodward, P. M.: Information Theory and the Design of Radar Receivers, *Proc. IRE*, **39**: 1521–1524 (December, 1951).

4. Woodward, P. M., and Davies, I. L.: Information Theory and Inverse Probability in Telecommunications, *Proc. IEE*, **99**: (*Pt. III*), 37–44 (March, 1952).

5. Davies, I. L.: On Determining the Presence of Signals in Noise, *Proc. IEE*, **99**: (*Pt. III*), 45–51 (March, 1952).

6. Woodward, P. M.: "Probability and Information Theory with Applications to Radar," Pergamon, Oxford, England, 1953.

7. Helstrom, C. W.: "Statistical Theory of Signal Detection," Pergamon, Oxford, England, 1960.

8. Guillemin, E. A.: "Mathematics of Circuit Analysis," Wiley, New York, 1949.

9. Middleton, D.: "An Introduction to Statistical Communication Theory," McGraw-Hill, New York, 1960.

10. Doob, J. L.: "Stochastic Processes," Wiley, New York, 1953.

11. Kailath, T.: Correlation Detection of Signals Perturbed by a Random Channel, *IRE Trans. on Information Theory*, **IT-6**: (*3*), 361–366 (June, 1960).

12. Kailath, T.: Sampling Models for Linear Time-Variant Filters, Res. Lab. of Electronics Tech. Report 352, Mass. Inst. of Technology, Cambridge, Mass. (May, 1959).

8

STATISTICAL
DECISION THEORY

8.1. Introduction

In the preceding chapters two different criteria were discussed for optimizing
the performance of a communication or radar system. It was shown in Chapter 5 that maximizing output signal-to-noise ratio leads to a matched filter
receiver implementation for a signal immersed in white Gaussian noise.
This criterion stems from the intuitive notion that noise is an undesirable
interfering signal whose strength, relative to the desired signal, should be
minimized for best performance. In Chapter 7 it was shown that an ideal a
posteriori receiver, defined as one that preserves all desired information in
the received waveform, also can be realized with a matched filter for a signal
immersed in white Gaussian noise. It is reassuring that such different criteria
result in similar optimum receiver structures; however, it should not be
inferred that both criteria will always produce the same result in other
applications.

Another approach to the optimum reception problem that has received considerable attention is based on the methods of *statistical inference*. The general theory of statistical inference by means of *hypothesis testing* was pioneered in the 1930's by Neyman and Pearson [1]. Subsequently, hypothesis testing was expanded by Wald [2, 3] and others [4–8] into *statistical decision theory*.

Neyman-Pearson hypothesis testing was first applied to the problem of signal detection at the Radiation Laboratory during the Second World War [5]. The application of decision theory to both detection and parameter estimation received broad stimulation and organization from the work of Middleton and Van Meter [6, 7, 8] during the 1950's. A complete exposition of this subject appears in Middleton [8]. Because of its breadth and generality, Middleton's treatment of statistical decision theory is followed in this text.

Decision theory implicitly recognizes that, in the real world, decisions have to be made with imperfect, noise-contaminated data. How such decisions can be made in an optimal fashion is its principal concern. Decision theory has been criticized by exponents of the a posteriori theory because information is lost in the decision process. The advocates of decision theory reply that an optimum decision receiver preserves all of the available information up to the point where a decision is made; further, information lost in the decision process is a penalty that must be paid when information is ultimately to be acted upon. Despite this philosophical difference, both theories have many similarities in practice.

In decision theory the various possible events that can occur are characterized as *hypotheses*. For example, the presence or absence of a signal in a noisy waveform may be viewed as two alternative mutually exclusive hypotheses. The object of statistical decision theory is to formulate a *decision rule* that operates on the received data to decide which hypothesis, among all possible hypotheses, best describes the truth. Obviously, the term *best* must be suitably defined. In applications, the best decision rule depends strongly on the criterion of optimality chosen by the system designer. Experience, judgment, and an understanding of the problem are required if one is to make a good choice.

8.2. Elements of a Statistical Decision Problem

The basic elements of a statistical decision problem are: (1) a set of hypotheses that characterize the possible true states of nature; (2) a test in which data are obtained from which we wish to infer the truth; (3) a decision rule that operates on the data to decide in an optimal fashion which hypothesis in fact best describes the true state of nature; (4) a criterion of optimality that reflects the cost of correct and incorrect decisions.

The following simple example, unrelated to communication theory, demonstrates the relationships between the basic elements of a statistical decision problem. An enterprising racehorse owner wishes to maximize his winnings at the racetrack (criterion of optimality) by means of an optimum betting procedure (decision rule). A statistician friend finds a strong correlation between previous wins and losses of the owner's horse and the clocked time in which the horse runs a measured mile the day before a race (the test). This correlation in shown in the graphs of Fig. 8.2-1, which describe the relative frequency of occurrence of various clocked times T that were measured the day before the horse won a race, and the relative frequency of values of T that were measured prior to the horse's losing a race. These two curves are the (conditional) probability density of T given that a win occurs next day and the (conditional) probability density of T given that a loss occurs next day.

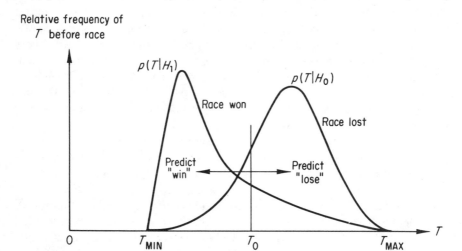

Fig. 8.2-1. Measured statistics.

Because of the many (statistical) factors that affect a race, knowledge of T (the test) does not yield errorless prediction of the result. The decision rule in this case is simply to choose an optimum threshold value of T, called T_0; a "win" is predicted when the measured T before a race is smaller than T_0 and a "lose" is predicted when T is greater than T_0, as shown in Fig. 8.2-1.

The value of T_0 to be chosen depends on the costs assigned to correct and incorrect decisions. Assume that in an average race the odds are such that a \$2 bet returns \$15 if the horse wins. Hence a correct prediction of a "win" results in a net gain of \$13, or, from a different point of view, a loss of $-\$13$. However, if a "win" is predicted and the horse loses, there is a \$2 loss. If a "lose" is predicted and no bet is placed, the owner suffers no loss; but, if a "lose" is predicted and the horse wins, the owner suffers a loss of $ 13 since he did not place a bet. These losses can be presented in the form of a *cost* or *loss* matrix (sometimes called a payoff matrix) as shown in Table 8.2-1.

TABLE 8.2-1

Loss or Cost Matrix
(Loss = −Gain)

	$T \leqslant T_0$: *Predict win*	$T > T_0$: *Predict lose*
Horse wins	−$13	+13
Horse loses	+2	0

Proceeding with the problem, let

$$\begin{cases} H_0 & \text{be the hypothesis that the horse loses} \\ H_1 & \text{be the hypothesis that the horse wins} \end{cases} \quad (8.2\text{-}1)$$

and let C_{ij} represent the owner's loss when H_i is predicted and H_j actually occurs. Thus the cost matrix is given by

$$\mathbf{C} = \begin{bmatrix} C_{11} & C_{01} \\ C_{10} & C_{00} \end{bmatrix} \quad (8.2\text{-}2)$$

where

C_{11} = cost or loss when H_1 (win) is chosen and H_1 occurs

= −$13 (gain) (8.2-3a)

C_{01} = loss when H_0 (lose) is chosen and H_1 (win) occurs

= $13 (loss) (8.2-3b)

C_{10} = loss when H_1(win) is chosen and H_0 (lose) occurs

= $2 (loss) (8.2-3c)

C_{00} = loss when H_0 (lose) is chosen and H_0 occurs

= $0 (8.2-3d)

To find the value of threshold T_0, the expected or average loss (cost) per race is computed as a function of T_0 and then T_0 is varied to minimize the loss. The expected loss L is the average value of the four losses, where each is weighted according to its probability of occurrence. Thus,

$$L(T_0) = \text{average loss} = \sum_{\substack{i=0,1 \\ j=0,1}} C_{ij} P(H_i \text{ is chosen and } H_j \text{ occurs}) \quad (8.2\text{-}4)$$

where $L(T_0)$ is written to emphasize the dependence of the loss function on threshold T_0. Now, by the rules of probability,

$$P(H_1 \text{ is chosen and } H_1 \text{ occurs}) = P(T \leqslant T_0 | H_1) \cdot P(H_1) \quad (8.2\text{-}5)$$

$$P(H_0 \text{ is chosen and } H_1 \text{ occurs}) = P(T > T_0 | H_1) \cdot P(H_1) \quad (8.2\text{-}6)$$

$$P(H_0 \text{ is chosen and } H_0 \text{ occurs}) = P(T > T_0 | H_0) \cdot P(H_0) \qquad (8.2\text{-}7)$$

$$P(H_1 \text{ is chosen and } H_0 \text{ occurs}) = P(T \leqslant T_0 | H_0) \cdot P(H_0) \qquad (8.2\text{-}8)$$

Looking into past performance, we find that the horse won 10 per cent of all the races in which he was entered. Assuming this trend will continue, let

$$P(H_1) = 0.1 \qquad (8.2\text{-}9)$$

$$P(H_0) = 1 - P(H_1) = 0.9 \qquad (8.2\text{-}10)$$

Further, from the definitions of probability, and from the probability density plots of $p(T|H_0)$ and $p(T|H_1)$ in Fig. 8.2-1,

$$P(T \leqslant T_0 | H_1) = \int_{T_{\min}}^{T_0} p(T|H_1)\, dT \qquad (8.2\text{-}11)$$

$$P(T > T_0 | H_1) = \int_{T_0}^{T_{\max}} p(T|H_1)\, dT \qquad (8.2\text{-}12\text{a})$$

$$= 1 - P(T \leqslant T_0 | H_1) \qquad (8.2\text{-}12\text{b})$$

$$P(T \leqslant T_0 | H_0) = \int_{T_{\min}}^{T_0} p(T|H_0)\, dT \qquad (8.2\text{-}13)$$

$$P(T > T_0 | H_0) = \int_{T_0}^{T_{\max}} p(T|H_0)\, dT \qquad (8.2\text{-}14\text{a})$$

$$= 1 - P(T \leqslant T_0 | H_0) \qquad (8.2\text{-}14\text{b})$$

Equations (8.2-12b) and (8.2-14b) follow from the area constraints on probability density functions, namely

$$\int_{T_{\min}}^{T_{\max}} p(T|H_1)\, dT = 1 \qquad (8.2\text{-}15)$$

and

$$\int_{T_{\min}}^{T_{\max}} p(T|H_0)\, dT = 1 \qquad (8.2\text{-}16)$$

Substituting Eqs. (8.2-3) and (8.2-5) through (8.2-14) into Eq. (8.2-4) results in

$$L(T_0) = C_{11} P(T \leqslant T_0 | H_1) P(H_1) + C_{10} P(T \leqslant T_0 | H_0) P(H_0)$$
$$+ C_{01} P(T > T_0 | H_1) P(H_1) + C_{00} P(T > T_0 | H_0) P(H_0) \qquad (8.2\text{-}17\text{a})$$

$$L(T_0) = C_{11} P(T \leqslant T_0 | H_1) P(H_1) + C_{10} P(T \leqslant T_0 | H_0) P(H_0)$$
$$+ C_{01} [1 - P(T \leqslant T_0 | H_1)] P(H_1) \qquad (8.2\text{-}17\text{b})$$
$$+ C_{00} [1 - P(T \leqslant T_0 | H_0)] P(H_0)$$

$$L(T_0) = C_{01} P(H_1) + C_{00} P(H_0) - (C_{01} - C_{11}) P(T \leqslant T_0 | H_1) P(H_1)$$
$$+ (C_{10} - C_{00}) P(T \leqslant T_0 | H_0) P(H_0) \qquad (8.2\text{-}17\text{c})$$

$$L(T_0) = C_{01} P(H_1) + C_{00} P(H_0) - (C_{01} - C_{11}) P(H_1) \int_{T_{\min}}^{T_0} p(T|H_1)\, dT$$
$$+ (C_{10} - C_{00}) P(H_0) \int_{T_{\min}}^{T_0} p(T|H_0)\, dT \qquad (8.2\text{-}17\text{d})$$

To find the value of T_0 that minimizes the average loss, (8.2-17d) is differentiated with respect to T_0 and set equal to zero; the result is

$$\frac{dL(T_0)}{dT_0} = -(C_{01} - C_{11})P(H_1)p(T_0\,|\,H_1) + (C_{10} - C_{00})P(H_0)p(T_0\,|\,H_0)$$

$$= 0 \qquad\qquad (8.2\text{-}18)$$

Equation (8.2-18) can be rearranged as follows:

$$\frac{p(T_0\,|\,H_1)}{p(T_0\,|\,H_0)} = \frac{(C_{10} - C_{00})P(H_0)}{(C_{01} - C_{11})p(H_1)} \qquad\qquad (8.2\text{-}19)$$

Equation (8.2-19) can be evaluated by substituting the numbers for C_{ij} and $P(H_i)$ given in Eqs. (8.2-3), (8.2-9), and (8.2-10):

$$\frac{p(T_0\,|\,H_1)}{p(T_0\,|\,H_0)} = \frac{(2 - 0)(0.9)}{(13 + 13)(0.1)} = 0.69 \qquad\qquad (8.2\text{-}20)$$

Equation (8.2-20) states that the optimum threshold is chosen at a point T_0 where $p(T_0\,|\,H_1)$ is approximately 70 per cent of the value $p(T_0\,|\,H_0)$. This is illustrated in Fig. 8.2-1. If the test statistic T preceding a race is less than or equal to the value of T_0 so chosen, a "win" is predicted; if T is greater than this value, a "lose" is predicted. Note that because of the large loss associated with a "lose" prediction when the horse in fact wins, the threshold is biased to favor "win" predictions; this can easily be shown by substituting a smaller value for C_{01} in Eq. (8.2-20).

In the example above, the optimality criterion selected was the minimization of (average) loss to the racehorse owner; this is equivalent to maximizing his gain. Decision rules that minimize average loss are called *Bayes rules* and are considered in greater detail later. The radar problem, formulated as a problem in statistical decision theory, is considered next.

8.3. Reception as a Decision Problem

The fundamental objectives of a radar system are to detect the presence of objects and estimate their positions and/or motions in space relative to the radar. The latter is accomplished by analyzing the modifications of the radar signal as a result of electromagnetic reflection from the object and is usually called parameter estimation. Both detection and parameter estimation can be formulated as problems in statistical decision theory.

A. Detection

A radar echo is invariably immersed in additive noise, and possibly also in clutter return. Since noise and clutter are random phenomena, a decision concerning the presence or absence of a target echo is statistical in nature.

In order to minimize the number of incorrect decisions, it seems reasonable to take advantage of a priori information concerning the echo signal structure and the statistical properties of the noise—and clutter, if present. For the present, noise is assumed to be the principal obstacle to reliable target echo detection.

The detection of a signal in noise can be formulated as a problem in hypothesis testing in the following way. A finite-duration sample of a noisy waveform is received that may, or may not, contain a signal. The hypothesis that the received waveform *does not* contain a signal is to be tested against the hypothesis that the received waveform *does* contain a signal. The first hypothesis, denoted by H_0, is often called the *null* hypothesis while the second, denoted by H_1, is referred to as the *alternative* hypothesis. If the signal to be detected is deterministic—that is, its structure is completely known—then H_1 is called a *simple* alternative. In radar this situation almost never occurs, since echo amplitude and phase are usually unknown. When the signal to be detected is a member of a finite or infinite set of signals, H_1 is called a *composite* alternative. In this case if we decide H_1 is true, we can conclude only that one member of the signal class is present whose identity is not revealed by the test.

It is convenient now to introduce vector-space notation. The class of possible signals (echoes) can be represented as points **s** in signal space Ω. Each point in the space represents a waveform with a particular combination of signal parameter values such as amplitude, phase, doppler, and so on. When possible, a probability of occurrence is assigned to each combination of signal parameters. This information is contained in a joint a priori probability density function $\sigma(\mathbf{s})$ over all the points **s** in signal space Ω.

In a similar way a noise space can be defined whose points **n** describe all possible waveform realizations of the noise process within the observation interval. From the statistical and spectral properties of the noise, an a priori joint probability density $p(\mathbf{n})$ can be deduced that describes the frequency of occurrence of waveforms in this space.

Next an observation space, Γ, is defined whose points **v** represent all possible joint combinations of signal and noise waveforms within the observation interval. The frequency of occurrence of members of this space can also be described by an a priori probability density function, which is written as a conditional probability $p(\mathbf{v}|\mathbf{s})$ to show explicitly the dependence of the observed waveform **v** on signal **s**. For this purpose it is convenient to include the null hypothesis $\mathbf{s} = 0$ as a point in signal space Ω.

An essential feature of statistical decision theory is the decision rule used to arrive at a decision. Note that the decision rule depends only on the observed waveform **v** and not on signal **s**. A decision rule that leads to decision d as a result of observation **v** is denoted by $D(d|\mathbf{v})$. Decision rule $D(d|\mathbf{v})$ mathematically describes the conditional probability of deciding d, having observed **v**; that is, for a particular waveform **v** there is only a probability (in the most general case) that a decision $d_1 = $ "yes" or $d_2 = $ "no" will be made. Such a decision rule is called a *randomized decision rule* and its implementation requires a chance mechanism as part of the receiver structure. In practical applications, the decision rule usually reduces to a *nonrandom decision rule* where a probability of 0 or 1 is assigned to $D(d_1 | \mathbf{v})$ and $D(d_2 | \mathbf{v})$ for each observation **v**. In this case a chance mechanism is not required in the receiver.

The set of possible decisions **d** in a statistical decision problem can be described as points in a decision space Δ. If the interpretation of a decision rule $D(\mathbf{d} | \mathbf{v})$ as a probability (or probability density in the event of a continuum of possible decisions) is retained, then $D(\mathbf{d} | \mathbf{v})$ describes the probability (density) of each point in decision space for every possible waveform **v**. In a signal detection problem, decision space contains only two points: signal present and signal absent.

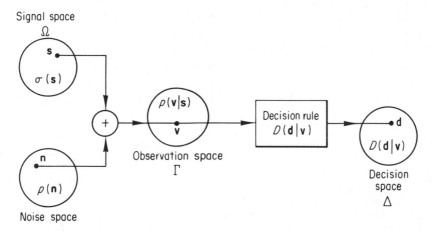

Fig. 8.3-1. Reception as a decision problem.

In Fig. 8.3-1 the general decision problem is shown in terms of the various spaces just defined. Observe that a decision rule may be interpreted as an operator that maps points in observation space into points in decision space with a preassigned probability $D(\mathbf{d} | \mathbf{v})$. The essence of the decision problem is to choose decision rules that accomplish this mapping in an optimum way with respect to a particular criterion of excellence. The mathematical operations embodied in the decision rule define the operations to be performed by an "optimum" decision receiver on the received waveform **v** in order to render a decision **d** in accordance with the selected criterion of excellence.

B. Parameter estimation

Some attributes of a radar target can be deduced from modifications of the reflected radar waveform. These modifications are conveniently characterized by unknown signal parameters of an otherwise deterministic echo signal structure. Theoretically, were it not for the presence of noise, the values of these parameters can be measured to any desired degree of precision.

Parameter estimation is formulated as a problem in statistical decision theory by extending the ideas introduced in the discussion of radar detection. In detection, observation space Γ is mapped into two points in decision space Δ by means of decision rule $D(\mathbf{d} \,|\, \mathbf{v})$, namely *signal present* and *signal absent*. If decision space Δ is enlarged to include a selected subset of points in signal space Ω, Fig. 8.3-1 also describes the parameter-estimation problem in terms of decision theory. In fact the set of points in decision space may contain the entire set of points in signal space; in this case the dimensionality of signal space is identical with that of decision space. Often, however, less precision is required, in which case the dimensionality of decision space is smaller than that of signal space. Figure 8.3-2 illustrates two possible situations. In one, decision space and signal space have the same (discrete) dimensionality. The second, indicated by a dotted line, shows a decision space with a smaller dimensionality. A similar situation exists when signal space is of infinite dimension.

In summary, parameter estimation divides observation space Γ into

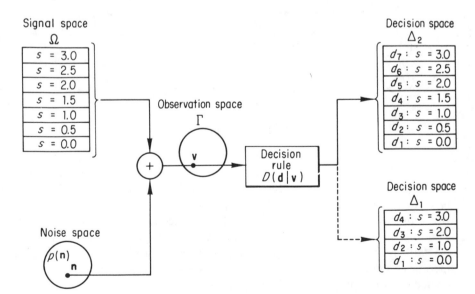

Fig. 8.3-2. Example of parameter estimation as a decision problem.

subsets of points that are mapped by a decision rule into signal points in decision space Δ; that is, decision d_i is assigned to observed waveform $\mathbf{v}$, in accordance with decision rule $D(d_i | \mathbf{v})$, when $\mathbf{v}$ is a member of the ith subset of points in Γ. As before, the optimum decision rule is determined by the selected criterion of optimality.

C. Similarity between detection and parameter estimation

In both detection and parameter estimation, the decision rule maps the space of observations Γ onto the space of decisions Δ. In simple detection the space of decisions contains only two points, while in parameter estimation the dimensionality of decision space is usually greater. Note that if all the points in decision space corresponding to "signal present" ($\mathbf{s} \neq 0$) are grouped together, a parameter-estimation receiver can be used to indicate the existence of a signal and therefore is also a solution to the detection problem. However, a decision receiver that is optimum for parameter estimation may not be optimum for detection; hence, it is necessary to treat detection and parameter estimation as separate statistical decision problems.

8.4. Definition of Loss Function

In order to select an optimum decision rule in a statistical decision problem, we evaluate the relative performance of each possible decision rule, selecting the rule that yields the best performance. Clearly a method of evaluating performance is required. For this purpose Wald introduced the notion of a simple *cost or loss function*, which associates a quantitative cost $C(\mathbf{s}, \mathbf{d})$ with each point $\mathbf{s}$ in signal space Ω and each point $\mathbf{d}$ in decision space Δ. (The cost function was introduced as the owner's loss in the example of Section 8.2.) The cost function describes the loss incurred by a receiving system that results in a decision $\mathbf{d}$ when the input signal is $\mathbf{s}$. In the case of a correct decision, the loss or cost function may be interpreted as a gain.

A substantial theory has been developed for problems in which *average loss* is used as a measure of comparative system performance. This choice is motivated by the fact that average loss is representative of system performance evaluated over all possible modes of behavior. A decision rule that describes a receiver with the least average loss is called a *Bayes rule* and the receiver a *Bayes receiver*. Other performance criteria lead to different decision rules— for example, minimax, Neyman-Pearson, and others. These are discussed later in this chapter.

For reasons that will become clear, it is convenient to define two loss functions. The *conditional loss* $L_c(D | \mathbf{s})$ is a useful measure of loss when the input signal is known, or when the input signal is not known and the a priori probability density $\sigma(\mathbf{s})$ over signal space Ω is also unknown. If the a priori probability density $\sigma(\mathbf{s})$ is known, a more complete performance loss rating is provided by the *average loss* $L(D, \sigma)$.

The conditional loss $L_c(D\,|\,\mathbf{s})$ is defined as the mathematical expectation of the loss with respect to all possible decisions $\mathbf{d}$ for a given $\mathbf{s}$ and decision rule D. Thus,

$$L_c(D\,|\,\mathbf{s}) = E_{\mathbf{d}|\mathbf{s}}[C(\mathbf{s}, \mathbf{d})] \qquad (8.4\text{-}1a)$$

$$= \int_\Delta C(\mathbf{s}, \mathbf{d})p(\mathbf{d}\,|\,\mathbf{s})\,d\mathbf{d} \qquad (8.4\text{-}1b)$$

In words, the conditional loss is the sum of costs associated with all possible decisions weighted by their probability of occurrence, assuming that $\mathbf{s}$ is the true state of nature. The conditional probability density of deciding $\mathbf{d}$ given $\mathbf{s}$, $p(\mathbf{d}\,|\,\mathbf{s})$, can also be expressed as

$$p\,(\mathbf{d}\,|\,\mathbf{s}) = \int_\Gamma p(\mathbf{d}, \mathbf{v}\,|\,\mathbf{s})\,d\mathbf{v} \qquad (8.4\text{-}2)$$

The form of Eq. (8.4-2) indicates that $p(\mathbf{d}\,|\,\mathbf{s})$ can be considered a (conditional) marginal density function that can be derived from the (conditional) joint probability density function $p(\mathbf{d}, \mathbf{v}\,|\,\mathbf{s})$. By means of the product rule,

$$p(\mathbf{d}, \mathbf{v}\,|\,\mathbf{s}) = D(\mathbf{d}\,|\,\mathbf{v}, \mathbf{s})p(\mathbf{v}\,|\,\mathbf{s}) \qquad (8.4\text{-}3)$$

(8.4-2) can be rewritten as

$$p(\mathbf{d}\,|\,\mathbf{s}) = \int_\Gamma D(\mathbf{d}\,|\,\mathbf{v}, \mathbf{s})p(\mathbf{v}\,|\,\mathbf{s})d\mathbf{v} \qquad (8.4\text{-}4a)$$

$$= \int_\Gamma D(\mathbf{d}\,|\,\mathbf{v})p(\mathbf{v}\,|\,\mathbf{s})\,d\mathbf{v} \qquad (8.4\text{-}4b)$$

The transition from Eq. (8.4-4a) to (8.4-4b) used the result

$$D(\mathbf{d}\,|\,\mathbf{v}, \mathbf{s}) = D(\mathbf{d}\,|\,\mathbf{v}) \qquad (8.4\text{-}5)$$

which follows from the fact that decision rule $D(\mathbf{d}\,|\,\mathbf{v})$ is a function only of data $\mathbf{v}$, as noted earlier, and not $\mathbf{s}$. Inserting (8.4-4b) into (8.4-1) finally results in the conditional loss

$$L_c(D\,|\,\mathbf{s}) = \int_\Gamma p(\mathbf{v}\,|\,\mathbf{s})\,d\mathbf{v} \int_\Delta C(\mathbf{s}, \mathbf{d})\,D(\mathbf{d}\,|\,\mathbf{v})\,d\mathbf{d} \qquad (8.4\text{-}6)$$

When the input signal is not known but a priori probability density function $\sigma(\mathbf{s})$ is known, the average loss $L(D, \sigma)$, which is defined as the mathematical expectation of the conditional loss with respect to the input signal statistics, may be written as

$$L(D, \sigma) = E_\mathbf{s}[L_c(D\,|\,\mathbf{s})] \qquad (8.4\text{-}7a)$$

$$= \int_\Omega \sigma(\mathbf{s})\,d\mathbf{s} \int_\Gamma p(\mathbf{v}\,|\,\mathbf{s})d\mathbf{v} \int_\Delta C(\mathbf{s}, \mathbf{d})\,D(\mathbf{d}\,|\,\mathbf{v})\,d\mathbf{d} \qquad (8.4\text{-}7b)$$

Alternatively, the average loss can be defined as the sum of costs associated with decisions $\mathbf{d}$ and inputs $\mathbf{s}$ weighted according to their joint probability of occurrence. Thus,

$$L(D, \sigma) = E_{\mathbf{d}, \mathbf{s}}[C(\mathbf{s}, \mathbf{d})] \qquad (8.4\text{-}8a)$$

$$= \int_\Omega \int_\Delta C(\mathbf{s}, \mathbf{d}) p(\mathbf{d}, \mathbf{s}) d\mathbf{d}\, d\mathbf{s} \qquad (8.4\text{-}8b)$$

$$= \int_\Omega \sigma(\mathbf{s})\, d\mathbf{s} \int_\Delta C(\mathbf{s}, \mathbf{d}) p(\mathbf{d}\,|\,\mathbf{s})\, d\mathbf{d} \qquad (8.4\text{-}8c)$$

The inner integral in (8.4-8c) is the conditional loss defined in Eq. (8.4-1b); therefore $L(D, \sigma)$ can also be written as

$$L(D, \sigma) = \int_\Omega L_c(D\,|\,\mathbf{s}) \sigma(\mathbf{s})\, d\mathbf{s} \qquad (8.4\text{-}9)$$

which is a restatement of relation (8.4-7a).

In summary, the average loss function $L(D, \sigma)$ provides a measure for evaluating the performance of different systems when complete a priori statistics concerning the signal and noise are available. The loss function and Bayes criterion are applied next to the binary detection problem.

8.5. Binary Detection

In binary detection a decision is rendered between one of two possible outcomes: *noise alone* or *signal plus noise*. The decision permits only the existence of a target to be established; target parameters such as range, velocity, and so on are not obtained without further processing. Binary detection is a useful introduction to radar detection theory. More realistic solutions of the radar problem are treated later in this chapter and in those which follow.

Let H_0 denote the null hypothesis that noise alone is present and H_1 the composite alternative hypothesis that signal plus noise is present. This may be stated concisely as

$$\begin{aligned} H_0: & \quad \mathbf{s} \in \Omega_0 \\ H_1: & \quad \mathbf{s} \in \Omega_1 \end{aligned} \qquad (8.5\text{-}1)$$

where Ω_0 and Ω_1 are nonoverlapping regions of signal space. It follows from the definitions of H_0 and H_1 that Ω_0 contains the single point $\mathbf{s} = 0$ and Ω_1 contains all points, $\mathbf{s} \neq 0$.

An expression for the a priori probability density $\sigma(\mathbf{s})$ defined over signal space can be found as follows. Let $\mathfrak{p}$ and $\mathfrak{q}$ be the a priori probabilities of signal present and signal absent, respectively. Then the probability density function $\sigma(\mathbf{s})$ can be expressed as

$$\sigma(\mathbf{s}) = \mathfrak{q}\delta(\mathbf{s} - 0) + \mathfrak{p}w(\mathbf{s}) \qquad (8.5\text{-}2)$$

where the Dirac delta function $\delta(\mathbf{s} - 0)$ describes the discrete probability distribution of **s** over Ω_0 and $w(\mathbf{s})$ describes the probability density of **s** over space Ω_1. In accordance with the above,

$$\int_{\Omega_1} w(\mathbf{s})d\mathbf{s} = 1 \tag{8.5-3}$$

When Eq. (8.5-2) is substituted into (8.4-7b), the expression for the average loss $L(D, \sigma)$ may be rewritten as

$$L(D, \sigma) = \mathfrak{q} \int_\Gamma p(\mathbf{v}\,|\,0)\,d\mathbf{v} \int_\Delta C(0, \mathbf{d})D(\mathbf{d}\,|\,\mathbf{v})\,d\mathbf{d}$$
$$+ \mathfrak{p} \int_{\Omega_1} w(\mathbf{s})\,d\mathbf{s} \int_\Gamma p(\mathbf{v}\,|\,\mathbf{s})d\mathbf{v} \int_\Delta C(\mathbf{s}, \mathbf{d})D(\mathbf{d}\,|\,\mathbf{v})\,d\mathbf{d} \tag{8.5-4}$$

Equation (8.5-4) can be simplified with the definition

$$E_\mathbf{s}[p(\mathbf{v}\,|\,\mathbf{s})] = \overline{p(\mathbf{v}\,|\,\mathbf{s})}_\mathbf{s} = \int_{\Omega_1} w(\mathbf{s})p(\mathbf{v}\,|\,\mathbf{s})\,d\mathbf{s} \tag{8.5-5}$$

as follows:

$$L(D, \sigma) = \mathfrak{q} \int_\Gamma p(\mathbf{v}\,|\,0)\,d\mathbf{v} \int_\Delta C(0, \mathbf{d})D(\mathbf{d}\,|\,\mathbf{v})\,d\mathbf{d}$$
$$+ \mathfrak{p} \int_\Gamma \overline{p(\mathbf{v}\,|\,\mathbf{s})}_\mathbf{s}\,d\mathbf{v} \int_\Delta C(\mathbf{s}, \mathbf{d})D(\mathbf{d}\,|\,\mathbf{v})\,d\mathbf{d} \tag{8.5-6}$$

Let cost assignments $C(\mathbf{s}, \mathbf{d})$ and $C(0, \mathbf{d})$ be made as shown in Table 8.5-1 where C_α and $C_{\bar\beta}$ denote costs of errors; C_α is the penalty or loss associated with deciding signal is present when, in fact, there is no signal; $C_{\bar\beta}$ is the loss associated with deciding no signal when, in fact, there is a signal. The

TABLE 8.5-1

Cost Matrix for Binary Detection

		Signal S	
		$\mathbf{s} = 0$	$\mathbf{s} \neq 0$
Decision **d**	d_0	$C_{1-\alpha}$	$C_{\bar\beta}$
	d_1	C_α	$C_{1-\bar\beta}$

notation reflects the definitions of α and $\bar\beta$ given in Eqs. (8.5-10) and (8.5-11), respectively; α is the false-alarm probability and $\bar\beta$ the average missed-detection probability. The quantities $C_{1-\alpha}$ and $C_{1-\bar\beta}$ represent the costs of correct decisions—that is,

$$C_{1-\alpha} = C(\mathbf{s} \in \Omega_0; d_0) \tag{8.5-7a}$$
$$C_{1-\bar\beta} = C(\mathbf{s} \in \Omega_1; d_1) \tag{8.5-7b}$$

These costs can be carried through the remaining derivations; however, since no penalty is usually associated with correct decisions, it is convenient to set the costs of correct decisions equal to zero (see problem 8.4):

$$C_{1-\alpha} = C_{1-\bar{\beta}} = 0 \tag{8.5-8}$$

Substituting the cost matrix of Table 8.5-1 and Eq. (8.5-8) into Eq. (8.5-6), the loss equation becomes

$$L(D, \sigma) = \mathfrak{q}C_\alpha \int_\Gamma D(d_1 \,|\, \mathbf{v}) p(\mathbf{v} \,|\, 0) \, d\mathbf{v}$$

$$+ \,\mathfrak{p}C_{\bar{\beta}} \int_\Gamma D(d_0 \,|\, \mathbf{v}) \,\overline{p(\mathbf{v} \,|\, \mathbf{s})}_\mathbf{s} \, d\mathbf{v} \tag{8.5-9}$$

Equation (8.5-9) can be written in another form of theoretical interest. If α denotes the probability of deciding a signal is present when there is no signal (commonly called a type I error or a false alarm in radar terminology) and $\bar{\beta}$ denotes the probability of deciding that signal is absent when a signal is present (commonly called a type II error or a missed detection in radar), then it follows from previous remarks that

$$\alpha = \int_\Gamma p(\mathbf{v} \,|\, 0) D(d_1 \,|\, \mathbf{v}) \, d\mathbf{v} \tag{8.5-10}$$

$$\bar{\beta} = \int_\Gamma \overline{p(\mathbf{v} \,|\, \mathbf{s})}_\mathbf{s} \, D(d_0 \,|\, \mathbf{v}) \, d\mathbf{v} = \left(\int_\Gamma p(\mathbf{v} \,|\, \mathbf{s}) D(d_0 \,|\, \mathbf{v}) \, d\mathbf{v} \right)_\mathbf{s}$$

$$= \overline{\beta(\mathbf{s})}_\mathbf{s} \tag{8.5-11}$$

Note that $\bar{\beta}$ by definition is the type II error probability averaged with respect to the a priori distribution of signal. Substituting (8.5-10) and (8.5-11) into (8.5-9) results in

$$L(D, \sigma) = \mathfrak{q}\alpha C_\alpha + \mathfrak{p}\bar{\beta}C_{\bar{\beta}} \tag{8.5-12}$$

Equation (8.5-12) relates average loss $L(D, \sigma)$ to the a priori probability of signal $\mathfrak{p} = 1 - \mathfrak{q}$, the probabilities of type I and type II errors, α and $\bar{\beta}$, and the costs of type I and type II errors, C_α and $C_{\bar{\beta}}$, respectively.

8.6. Bayes Decision Rule

Bayes decision rule D_B results from the minimization of $L(D, \sigma)$. Since binary decision space Δ contains only the two points d_0 (no signal) and d_1 (signal plus noise), decision rule $D_B(\mathbf{d}, \mathbf{v})$ satisfies the probability normalizing relation

$$D_B(d_0 \,|\, \mathbf{v}) + D_B(d_1 \,|\, \mathbf{v}) = 1 \tag{8.6-1}$$

Substituting Eq. (8.6-1) into (8.5-9) and eliminating $D_B(d_1 \,|\, \mathbf{v})$ yields

$$L(D, \sigma) = \mathfrak{q}C_\alpha + \int_\Gamma D_B(d_0 \,|\, \mathbf{v})[\mathfrak{p}C_{\bar{\beta}}\overline{p(\mathbf{v} \,|\, \mathbf{s})}_\mathbf{s} - \mathfrak{q}C_\alpha p(\mathbf{v} \,|\, 0)] \, d\mathbf{v} \tag{8.6-2}$$

Note that $D_B(d_0 \,|\, \mathbf{v})$ is positive and less than unity; also $\mathfrak{p}$, $\mathfrak{q}$, C_α, $C_{\bar{\beta}}$ are positive quantities. Then, to minimize $L(D, \sigma)$, choose

$$\begin{cases} D_B(d_0 \mid \mathbf{v}) = 1 \\ D_B(d_1 \mid \mathbf{v}) = 0 \end{cases} \tag{8.6-3}$$

—that is, decide signal is absent when

$$\mathfrak{p} C_{\bar{\beta}} \overline{p(\mathbf{v} \mid \mathbf{s})}_s < \mathfrak{q} C_\alpha p(\mathbf{v} \mid 0) \tag{8.6-4}$$

and choose

$$\begin{cases} D_B(d_0 \mid \mathbf{v}) = 0 \\ D_B(d_1 \mid \mathbf{v}) = 1 \end{cases} \tag{8.6-5}$$

—that is, decide signal present when

$$\mathfrak{p} C_{\bar{\beta}} \overline{p(\mathbf{v} \mid \mathbf{s})}_s \geqslant \mathfrak{q} C_\alpha p(\mathbf{v} \mid 0) \tag{8.6-6}$$

Note that the Bayes decision rule that results from the minimization of $L(D, \sigma)$ is *nonrandom*. Inequalities (8.6-4) and (8.6-6) can be rewritten in terms of a concisely defined function $\ell(\mathbf{v})$, called the *generalized likelihood ratio*,

$$\ell(\mathbf{v}) = \frac{\mathfrak{p} \overline{p(\mathbf{v} \mid \mathbf{s})}_s}{\mathfrak{q} p(\mathbf{v} \mid 0)} \tag{8.6-7}$$

With this definition, the Bayes decision rule reduces to

$$\begin{cases} \text{decide } d_1 \text{ when } \ell(\mathbf{v}) \geqslant \mathcal{K} \text{ (signal present)} \\ \text{decide } d_0 \text{ when } \ell(\mathbf{v}) < \mathcal{K} \text{ (signal absent)} \end{cases} \tag{8.6-8}$$

where

$$\mathcal{K} = \frac{C_\alpha}{C_{\bar{\beta}}} \tag{8.6-9}$$

Equation (8.6-8) specifies a test strategy in terms of likelihood ratio $\ell(\mathbf{v})$, which is a function of data $\mathbf{v}$, and threshold $\mathcal{K}$, which is a function of error cost assignments. The Bayes decision rule divides observation space Γ into two regions Γ' and Γ'' which are separated by the boundary $\ell(\mathbf{v}) = \mathcal{K}$, as shown in Fig. 8.6-1. The acceptance region Γ' for hypothesis H_0 ($\mathbf{s} \in \Omega_0$) contains all $\mathbf{v}$ for which $\ell(\mathbf{v}) < \mathcal{K}$; the rejection region Γ'' for hypothesis H_0 contains all $\mathbf{v}$ for which $\ell(\mathbf{v}) \geqslant \mathcal{K}$. The rejection region for hypothesis H_0 is, of course, the acceptance region for hypothesis H_1 ($\mathbf{s} \in \Omega_1$).

When $\mathfrak{p}$, $\mathfrak{q}$, $w(\mathbf{s})$, $p(\mathbf{v} \mid \mathbf{s})$, and cost assignments C_α and $C_{\bar{\beta}}$ are known, the Bayes strategy requires that the generalized likelihood ratio be computed for received data $\mathbf{v}$ and the result compared with a threshold $\mathcal{K}$, defined in (8.6-9). In general, the computation of the likelihood ratio is a complex nonlinear operation on data $\mathbf{v}$. In radar, approximations for the important cases of threshold signals and very large signals permit physical interpretations of receiver structure.

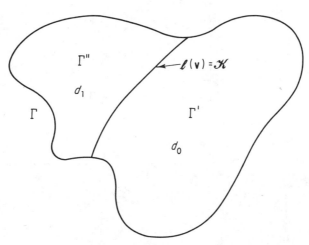

Fig. 8.6-1. Decision regions for binary detection.

8.7. Examples of Bayes Detectors

A few examples are now presented to illustrate the Bayes detection strategy.

Example 1: Signal **s** is assumed to be a positive constant A and data **v** consist of three observations v_1, v_2, and v_3. Under hypothesis H_1, each of the data points v_i consists of signal A plus a noise sample n_i, $i = 1, 2, 3$. The noise variates n_i are assumed to be statistically independent Gaussian random variables with zero mean $(\overline{n_i} = 0)$ and variance σ^2. For $\mathfrak{p} = \mathfrak{q} = \frac{1}{2}$, the generalized likelihood ratio $\ell(\mathbf{v})$ can be written as

$$\ell(\mathbf{v}) = \frac{\dfrac{1}{\sqrt{2\pi\sigma^2}} \exp\left[-\dfrac{\sum\limits_{i=1}^{3} (v_i - A)^2}{2\sigma^2}\right]}{\dfrac{1}{\sqrt{2\pi\sigma^2}} \exp\left(-\dfrac{\sum\limits_{i=1}^{3} v_i^2}{2\sigma^2}\right)} \tag{8.7-1}$$

The boundary that divides observation space into two parts is defined by $\ell(\mathbf{v}) = \mathscr{K}$. From (8.7-1)

$$\mathscr{K} = \exp\left[-\frac{3A^2}{2\sigma^2} + \frac{A\sum\limits_{i=1}^{3} v_i}{\sigma^2}\right] \tag{8.7-2}$$

Replacing $\mathscr{K}$ by $C_\alpha/C_{\bar{\beta}}$ and taking logarithms, we can rewrite Eq. (8.7-2) as

$$\sum_{i=1}^{3} v_i = \frac{3A}{2} + \frac{\sigma^2}{A} \ln \frac{C_\alpha}{C_{\bar{\beta}}} = M \tag{8.7-3}$$

where M is a constant. Decision surface (8.7-3) describes a plane, as shown in Fig. 8.7-1. When data vector **v** occurs in the region (Γ'') above the plane, the

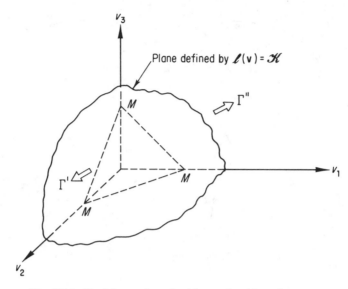

Fig. 8.7-1. Decision regions for binary detection of a constant (dc) signal in Gaussian noise.

decision is *signal present* (**s** = *A*); below the plane (Γ'), the decision is *signal absent* (**s** = 0). If the dimensionality of **v** is increased from three to *N*, the boundary surface in observation space becomes a hyperplane.

The quantity $\sum v_i$ in this example is called the test statistic. This quantity describes all of the operations that must be performed on the raw data to conduct a Bayes test for minimizing average loss $L(D, \sigma)$. Thus the Bayes receiver adds data points v_i, $i = 1, 2, 3$, and compares the sum with a fixed threshold. Several realizations of an actual Bayes processor can be synthesized for performing a test that detects a positive mean *A* in Gaussian noise.

The result above is valid only for $A > 0$. If the signal is a negative constant (that is, **s** = −*A*), the optimum test strategy becomes

$$\text{decide } \textit{signal present} \text{ when } \quad -\sum_{i=1}^{3} v_i \geqslant M \qquad (8.7\text{-}4a)$$

$$\text{decide } \textit{signal absent} \text{ when } \quad -\sum_{i=1}^{3} v_i < M \qquad (8.7\text{-}4b)$$

Equations (8.7-4) describe a boundary decision surface that is a plane parallel to that in Fig. 8.7-1 and equally displaced from the origin in the opposite direction. In this case a decision of *signal present* is made when **v** describes a point below the plane defined by Eqs. (8.7-4), and a *signal absent* decision is made when **v** is above the surface.

For the case in which the mean can be either positive or negative, both decision surfaces are required. A decision of *signal present* is made when the point **v** is either above the plane defined by Eq. (8.7-3) or below the plane defined by Eq.

(8.7-4). A *signal absent* decision is made whenever the data point **v** is sandwiched between the two planes.

Example 2: Consider next a signal consisting of three samples, s_1, s_2, and s_3 where each sample is a statistically independent Gaussian random variable with zero mean and a variance of σ^2. This might represent a noise-jammer signal received in a background of additive system noise. For this case, with $\mathfrak{p} = \mathfrak{q} = \frac{1}{2}$, the likelihood ratio becomes

$$\ell(\mathbf{v}) = \overline{\exp\left[-\frac{1}{2\sigma^2}\sum_{i=1}^{3}(-2v_is_i + s_i^2)\right]}_{\mathbf{s}} \qquad (8.7\text{-}5)$$

Since the s_i are statistically independent, $\ell(\mathbf{v})$ can be averaged separately with respect to each random variable s_i in Eq. (8.7-5). Using the result,

$$\frac{1}{\sqrt{2\pi\sigma^2}}\int_{-\infty}^{\infty} e^{-s_i^2/2\sigma^2} \cdot e^{-(s_i^2 - 2v_is_i)/2\sigma^2}\, ds_i = \frac{1}{\sqrt{2}}\, e^{v_i^2/4\sigma^2} \qquad (8.7\text{-}6)$$

we find the decision surface for this example is given by

$$\sum_{i=1}^{3} v_i^2 = 4\sigma^2 \ln(2\sqrt{2}\,\mathscr{K}) = \mathscr{K}_1 \qquad (8.7\text{-}7)$$

The decision surface boundary described by (8.7-7) is a sphere, as shown in Fig. 8.7-2. When data **v** falls within the sphere, a *signal absent* decision d_0 is made; when **v** falls outside of the sphere, a *signal present* decision d_1 is made. If the dimensionality of **s** is greater than three, the decision surface becomes a hyper-sphere in observation space.

In this example, the test statistic is

$$\sum_{i=1}^{3} v_i^2$$

It describes the computations to be performed by the optimum Bayes receiver. The

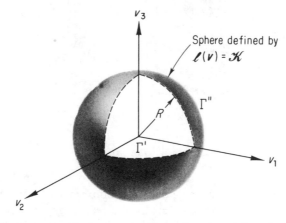

Fig. 8.7-2. Decision regions for binary detection of Gaussian signal in Gaussian noise.

quantity v_i^2 may be interpreted as the power in each of the data points; the total power

$$\sum_{i=1}^{3} v_i^2$$

is then compared with a threshold that depends on the cost assignments. When the variance for each signal sample differs, a solution similar to (8.7-7) is obtained but with a different threshold value.

8.8. Error Probabilities

Expressions for type I and type II error probabilities α and $\bar{\beta}$, respectively, are given by Eqs. (8.5-10) and (8.5-11). These expressions apply, in general, to both Bayes and non-Bayes decision rules and are not restricted to optimum systems. In statistical terminology α, the probability of rejecting H_0 when in fact it is true, is called the *level* or *size* of the test; $1 - \bar{\beta}$, the probability of rejecting H_0 when in fact it is false, is called the *power* of the test. In radar $1 - \bar{\beta}$ is the probability of target detection.

Since observation space Γ consists of nonoverlapping regions Γ' and Γ'', we can rewrite Eqs. (8.5-10) and (8.5-11) for a Bayes decision rule receiver as

$$\alpha = \int_{\Gamma'} p(\mathbf{v}\,|\,0)\,D_B(d_1\,|\,\mathbf{v})\,d\mathbf{v} + \int_{\Gamma''} p(\mathbf{v}\,|\,0)D_B(d_1\,|\,\mathbf{v})\,d\mathbf{v} \qquad (8.8\text{-}1)$$

$$\bar{\beta} = \int_{\Gamma'} \overline{p(\mathbf{v}\,|\,\mathbf{s})_s}\,D_B(d_0\,|\,\mathbf{v})\,d\mathbf{v} + \int_{\Gamma''} \overline{p(\mathbf{v}\,|\,\mathbf{s})_s}\,D_B(d_0\,|\,\mathbf{v})\,d\mathbf{v} \qquad (8.8\text{-}2)$$

Note that from earlier remarks concerning Bayes decision rules (8.6-3) and (8.6-5)

$$\left.\begin{array}{l} D_B(d_1\,|\,\mathbf{v}) = 0 \\ D_B(d_0\,|\,\mathbf{v}) = 1 \end{array}\right\} \quad \text{for } \mathbf{v} \text{ in } \Gamma' \qquad (8.8\text{-}3)$$

and

$$\left.\begin{array}{l} D_B(d_0\,|\,\mathbf{v}) = 0 \\ D_B(d_1\,|\,\mathbf{v}) = 1 \end{array}\right\} \quad \text{for } \mathbf{v} \text{ in } \Gamma'' \qquad (8.8\text{-}4)$$

so that Eqs. (8.8-1) and (8.8-2) simplify to

$$\alpha = \int_{\Gamma''} p(\mathbf{v}\,|\,0)\,d\mathbf{v} \qquad (8.8\text{-}5)$$

$$\bar{\beta} = \int_{\Gamma'} \overline{p(\mathbf{v}\,|\,\mathbf{s})_s}\,d\mathbf{v} \qquad (8.8\text{-}6)$$

To illustrate, consider a simple example in which signal space contains a single member $\mathbf{s} = s$ and a single observation v is made. In this case observation space Γ may be represented by the real line $-\infty \leqslant v < \infty$. Probability densities $p(v\,|\,0)$ and $p(v\,|\,s)$ are graphed with real line v as abscissa, as shown in Fig. 8.8-1. Partitioning of observation space into two parts is

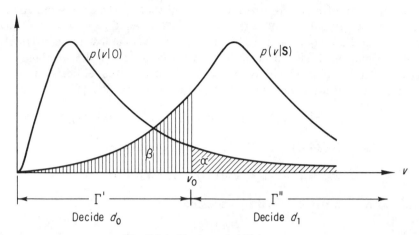

Fig. 8.8-1. Error probabilities.

equivalent to partitioning the real line $-\infty \leqslant v < \infty$ by a point v_0, which is obtained by solving $\ell(v) = \mathscr{K}$ for $v = v_0$. It follows from these remarks and Eqs. (8.8-5) and (8.8-6) that α and $\bar{\beta}$ are given for this example by

$$\alpha = \int_{v_0}^{\infty} p(v\,|\,0)\,dv \qquad (8.8\text{-}7)$$

and

$$\bar{\beta} = \beta = \int_{-\infty}^{v_0} p(v\,|\,s)\,dv \qquad (8.8\text{-}8)$$

Equations (8.8-7) and (8.8-8) state that the type I error or false-alarm probability is the area under probability density curve $p(v\,|\,0)$ over the interval in v for which *signal present* is decided; and the type II error or false-dismissal probability is the area under probability density curve $p(v\,|\,s)$ over the interval in v for which signal absent is decided. This result is analogous to that in Example 1 of Section 8.7 where α can be evaluated by a volume integral of $p(\mathbf{v}\,|\,0)$ above the plane defined by $\ell(\mathbf{v}) = \mathscr{K}$ (signal present) and the $\bar{\beta}$ error can be obtained by the integral of $\overline{p(\mathbf{v}\,|\,s)}_s$ below the plane (signal absent). Similarly, in Example 2 of Section 8.7 α can be obtained by integrating $p(\mathbf{v}\,|\,0)$ outside the sphere (signal present), while $\bar{\beta}$ can be computed by integrating $\overline{p(\mathbf{v}\,|\,s)}_s$ within the sphere (signal absent).

8.9. The Ideal Observer

Another criterion can be adopted for optimizing the binary detection strategy, namely, minimize the total probability of error P_{te}, defined by

$$P_{te} = \mathsf{q}\alpha + \mathsf{p}\bar{\beta} \qquad (8.9\text{-}1)$$

A comparison of Eq. (8.9-1) with (8.5-12) shows that the expression for P_{te} is the same as that for average loss $L(D, \sigma)$ when $C_\alpha = C_{\bar{\beta}} = 1$. Hence, minimizing P_{te} is equivalent to minimizing the average loss for a particular cost assignment. This optimization criterion is called the *ideal-observer* criterion and originated with Siegert [11]. The ideal-observer decision rule is found by substituting Eqs. (8.5-10) and (8.5-11) into (8.9-1):

$$P_{te} = \mathfrak{q} \int_\Gamma p(\mathbf{v}\,|\,0)D(d_1\,|\,\mathbf{v})\,d\mathbf{v} + \mathfrak{p} \int_\Gamma \overline{p(\mathbf{v}\,|\,\mathbf{s})_\mathbf{s}}\,D(d_0\,|\,\mathbf{v})\,d\mathbf{v} \qquad (8.9\text{-}2)$$

With constraint (8.6-1), Eq. (8.9-2) can be rewritten as

$$P_{te} = \int_\Gamma D(d_0\,|\,\mathbf{v})[\mathfrak{p}\overline{p(\mathbf{v}\,|\,\mathbf{s})_\mathbf{s}} - \mathfrak{q}p(\mathbf{v}\,|\,0)]\,d\mathbf{v} + \mathfrak{q} \qquad (8.9\text{-}3)$$

This expression is minimized by choosing

$$\begin{cases} D_I(d_0\,|\,\mathbf{v}) = 1 \\ D_I(d_1\,|\,\mathbf{v}) = 0 \end{cases} \qquad (8.9\text{-}4)$$

—that is, decide no signal when

$$\ell(\mathbf{v}) = \frac{\mathfrak{p}\overline{p(\mathbf{v}\,|\,\mathbf{s})_\mathbf{s}}}{\mathfrak{q}p(\mathbf{v}\,|\,0)} < 1 \qquad (8.9\text{-}5)$$

and by choosing

$$\begin{cases} D_I(d_1\,|\,\mathbf{v}) = 1 \\ D_I(d_0\,|\,\mathbf{v}) = 0 \end{cases} \qquad (8.9\text{-}6)$$

—that is, decide signal is present when

$$\ell(\mathbf{v}) = \frac{\mathfrak{p}\overline{p(\mathbf{v}\,|\,\mathbf{s})_\mathbf{s}}}{\mathfrak{q}p(\mathbf{v}\,|\,0)} \geqslant 1 \qquad (8.9\text{-}7)$$

Comparing this result with that in (Eq. (8.6-8) shows that the ideal-observer strategy is equivalent to a Bayes strategy with threshold $\mathscr{K} = 1$; hence, the Bayes class of optimum detectors includes the ideal observer as a sub-class. A threshold of unity corresponds to cost assignments $C_\alpha = C_{\bar{\beta}} = 1$, as noted above, so that the ideal observer is assessed an equal penalty for either type of mistake. For this reason, the ideal-observer strategy is seldom used in radar, since the cost of a missed radar detection is generally higher than the cost of a false alarm. (This problem is discussed in greater detail in Chapter 17.)

8.10. The Neyman-Pearson Criterion

The Neyman and Pearson theory of hypothesis testing antedates the development of statistical decision theory. It does not require knowledge of a priori signal statistics, nor does it require an explicit assignment of cost functions.

Neyman and Pearson define an optimum test as one that minimizes the probability of certain errors. In a test of hypothesis H, two types of errors can be made: H may be rejected when it is true, or it may be accepted when it is false. It seems reasonable that an optimum test should minimize the probability of committing both types of errors—that is, the test should have a small probability of rejecting H when it is true and a large probability of rejecting H when it is false. A test with a probability ϵ of rejecting H when it is true is called a *test of level* ϵ. The Neyman-Pearson criterion asserts that among all tests of level ϵ, the "best" test is one that has the greatest probability of rejecting H when it is false.

When applied to radar, the Neyman-Pearson test is a test between two alternative hypotheses, H_0 and H_1, only one of which is true: H_0 is usually called the null hypothesis and H_1, the alternative hypothesis; H_0 refers to the hypothesis that only noise is present, while H_1 denotes signal plus noise. If H_1 is decided when H_0 is true, the error is called a type I error (false alarm). If H_0 is decided when H_1 is true, the error is called a type II error (missed target detection or false dismissal). The Neyman-Pearson criterion requires that, for a fixed false-alarm probability α, a test be found that minimizes the missed target-detection probability β or, equivalently, that maximizes the probability of target detection $(1 - \beta)$.

In general, hypothesis H_1 can be a simple or composite hypothesis. In the classical Neyman-Pearson test, hypothesis H_1 is assumed to be simple— that is, the signal consists of a single known value, $\mathbf{s} = s_1$. The simple alternative hypothesis does not apply to radar, since the target echo is generally a function of many variables. When signal space consists of more than one element, H_1 is a composite hypothesis. In this case the probability of a type II error is a function of the signal parameters. For this situation the classical Neyman-Pearson strategy must be modified. A simple modification is considered in this section and others in Sections 8.13, 8.14, and 8.15.

One extension [3, 8] of the Neyman-Pearson test strategy, when H_1 is a composite hypothesis, is to minimize the *total* type II error probability *that has been averaged with respect to the a priori probability density of signal*†— that is, minimize $\mathfrak{p} \bar{\beta}$ subject to a total fixed type I error probability $\mathfrak{q}\alpha$. This extension is referred to, for convenience, as the *modified* Neyman-Pearson criterion. As before, $\mathfrak{p}$ is the a priori probability of signal present, $\mathfrak{q} = 1 - \mathfrak{p}$ is the a priori probability of signal absent, and $\bar{\beta}$ is given by Eq. (8.5-11). Following the method of Lagrange, the best decision rule D_{NP}, in the modified Neyman-Pearson sense, minimizes

$$L_{\mathrm{NP}} = \mathfrak{p}\bar{\beta} + \lambda \mathfrak{q}\alpha \qquad (8.10\text{-}1)$$

where λ is the Lagrange multiplier that is undetermined at this point. The notation L_{NP} is adopted since expression (8.10-1) is the same as expression

†In this case a priori statistics are required.

(8.5-12) for average loss with $C_\alpha = 1$ and $C_{\bar{\beta}} = \lambda$. Substituting Eqs. (8.5-10) and (8.5-11) into (8.10-1) yields

$$L_{\text{NP}} = \mathfrak{p} \int_\Gamma \overline{p(\mathbf{v}\,|\,\mathbf{s})}_\mathbf{s} D(d_0\,|\,\mathbf{v})\,d\mathbf{v} + \lambda\mathfrak{q} \int_\Gamma p(\mathbf{v}\,|\,0)D(d_1\,|\,\mathbf{v})d\mathbf{v} \quad (8.10\text{-}2)$$

With Eq. (8.6-1), (8.10-2) becomes

$$L_{\text{NP}} = \int_\Gamma D(d_0\,|\,\mathbf{v})[\mathfrak{p}\overline{p(\mathbf{v}\,|\,\mathbf{s})}_\mathbf{s} - \lambda\mathfrak{q}p(\mathbf{v}\,|\,0)]\,d\mathbf{v} + \lambda\mathfrak{q} \quad (8.10\text{-}3)$$

This expression is minimized by choosing

$$\begin{cases} D_{\text{NP}}(d_0\,|\,\mathbf{v}) = 1 \\ D_{\text{NP}}(d_1\,|\,\mathbf{v}) = 0 \end{cases} \quad (8.10\text{-}4)$$

—that is, decide no signal when

$$\ell(\mathbf{v}) = \frac{\mathfrak{p}\overline{p(\mathbf{v}\,|\,\mathbf{s})}_\mathbf{s}}{\mathfrak{q}p(\mathbf{v}\,|\,0)} < \lambda \quad (8.10\text{-}5)$$

and choosing

$$\begin{cases} D_{\text{NP}}(d_0\,|\,\mathbf{v}) = 0 \\ D_{\text{NP}}(d_1\,|\,\mathbf{v}) = 1 \end{cases} \quad (8.10\text{-}6)$$

(signal present) when

$$\ell(\mathbf{v}) \geqslant \lambda \quad (8.10\text{-}7)$$

Comparing this result with that of Eq. (8.6-8), we may note that the *modified* Neyman-Pearson strategy corresponds to a Bayes test strategy with threshold $\mathscr{K} = \lambda$. The choice of λ is not arbitrary but depends on the specification of α, since its value

$$\alpha = \int_\Gamma p(\mathbf{v}\,|\,0)\,D_{\text{NP}}(d_1\,|\,\mathbf{v})\,d\mathbf{v} \quad (8.10\text{-}8)$$

is determined by the surface $\ell(\mathbf{v}) = \lambda$ separating the regions Γ' and Γ'' in observation space. This strategy is frequently employed in radar problems.

8.11. The Minimax Approach

To apply Bayes' criterion for minimizing average loss, it is necessary to know the statistics of the noise process, as well as $p(\mathbf{v}\,|\,\mathbf{s})$ and a priori signal statistics $\sigma(\mathbf{s})$. In many practical cases probability density function $\sigma(\mathbf{s})$ is not known and it is not feasible to obtain experimental data to establish $\sigma(\mathbf{s})$. As a result Bayes' criterion cannot be applied. Another criterion that may be reasonable in such situations is the *minimax criterion*.

As an illustrative example, consider a situation in which signal space Ω contains four points, denoted by $s_1, s_2, s_3,$ and s_4. It follows from definition

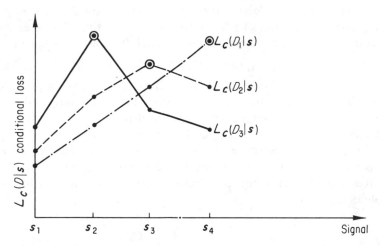

Fig. 8.11-1. Conditional loss function for discrete s as a function of decision rule D_i.

(8.4-1b) that there is a conditional loss $L_c\,(D\,|\,s_i)$, $i = 1, 2, 3, 4$, associated with each member of signal space; the values of conditional loss are dependent on decision rule D. Figure 8.11-1 shows three sets of conditional losses corresponding to three different decision rules D_1, D_2, and D_3. The maximum value of conditional loss is circled for each signal. Note that decision rule D_2 results in a peak or maximum conditional loss that is less than the maximum losses resulting from decision rules D_1 and D_3. A decision rule that minimizes the maximum conditional loss is called a minimax rule. If the set of admissible decision rules contains only rules D_1, D_2, and D_3, then rule D_2 is the minimax rule.

A minimax decision rule D_M results in a maximum conditional loss equal to, or less than, that resulting from any other admissible decision rule D:

$$\max_{\mathbf{s}} L_c(D_M\,|\,\mathbf{s}) \leqslant \max_{\mathbf{s}} L_c(D\,|\,\mathbf{s}) \qquad \text{for all } D \qquad (8.11\text{-}1)$$

or

$$\max_{\mathbf{s}} L_c(D_M\,|\,\mathbf{s}) = \max_{\mathbf{s}} \min_{D} L_c(D\,|\,\mathbf{s}) \qquad (8.11\text{-}2)$$

For very general conditions, which are almost always met in practical problems, Wald [3] has shown that

$$\max_{\mathbf{s}} \min_{D} L_c(D\,|\,\mathbf{s}) = \min_{D} \max_{\mathbf{s}} L_c(D\,|\,\mathbf{s}) \qquad (8.11\text{-}3)$$

from which the origin of the name minimax is apparent.

Minimax rules have a number of interesting properties. For example, it can be shown [3] that a minimax decision rule D_M is a Bayes rule relative to a least-favorable a priori distribution $\sigma_{lf}(\mathbf{s})$. Further, the Bayes average loss $L_M(D_M, \sigma_{lf})$, corresponding to D_M and $\sigma_{lf}(\mathbf{s})$, is larger than the Bayes average loss corresponding to any other a priori signal distribution—that is,

$$L_M(D_M, \sigma_{lf}) \geqslant L_B(D_B, \sigma) \qquad \text{for all } \sigma(\mathbf{s}) \tag{8.11-4}$$

where L_B is the Bayes loss resulting from Bayes rule D_B and a priori signal distribution $\sigma(\mathbf{s})$. Thus, the minimax average loss is the largest Bayes loss when all a priori distributions $\sigma(\mathbf{s})$ are considered.

To illustrate, consider once again Example 1 in Section 8.7 where a test was obtained for the presence of a positive mean A in Gaussian noise with variance σ^2. When only one observation v is available, Eq. (8.7-3), which describes the boundary between decision regions in observation space, simplifies to

$$v_0 = \frac{A}{2} + \frac{\sigma^2}{A} \ln \frac{\mathfrak{q} C_\alpha}{\mathfrak{p} C_\beta} \tag{8.11-5}$$

For known A, σ^2, and specified cost assignments C_α and C_β, v_0 is a function only of $\mathfrak{p}$ (since $\mathfrak{q} = 1 - \mathfrak{p}$). The error probabilities can then be expressed as

$$\alpha = \int_{v_0(\mathfrak{p})}^{\infty} p(v \,|\, 0) \, dv \tag{8.11-6}$$

and

$$\beta = \int_{-\infty}^{v_0(\mathfrak{p})} p(v \,|\, A) \, dv \tag{8.11-7}$$

from which both α and β can be found as functions of $\mathfrak{p}$.

From Eq. (8.5-12), the Bayes average loss is given by

$$L_B(\sigma) = (1 - \mathfrak{p}) \alpha(\mathfrak{p}) C_\alpha + \mathfrak{p} \beta(\mathfrak{p}) C_\beta \tag{8.11-8}$$

This loss can be computed for various values of a priori probability $\mathfrak{p}$ of signal present and plotted as shown in Fig. 8.11-2. The maximum loss is obtained by differentiating Eq. (8.11-8) with respect to $\mathfrak{p}$ and setting the result equal to zero; this yields

$$\alpha(\mathfrak{p}) C_\alpha = \beta(\mathfrak{p}) C_\beta \tag{8.11-9}$$

Transcendental equation (8.11-9) can be solved for $\mathfrak{p} = \mathfrak{p}_M$, at which the maximum loss occurs. When $\mathfrak{p}$ is equal to $\mathfrak{p}_M$, the Bayes loss is equal to the minimax loss. This follows since the minimax solution corresponds to the Bayes strategy for the worst a priori signal statistics. The minimax criterion in effect compensates for ignorance of the true state of nature by assuming the worst state of nature.

To summarize, a Bayes decision rule takes into consideration all of the a priori statistics relating to both signal and noise. When signal statistics are unavailable, a minimax decision rule sometimes offers a reasonable alternative. A minimax rule is a Bayes rule relative to a least favorable distribution;

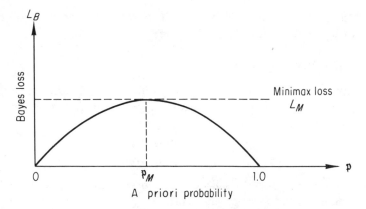

Fig. 8.11-2. Bayes loss and minimax solution.

the minimax average loss is the maximum of all Bayes losses. For this reason it is often criticized as being too conservative. The reasonableness of a minimax solution must be examined in each application.

8.12. Review

Nonrandom decision rules have been found conforming to a number of definitions of *best* or *optimum*. Let us review briefly the results.

A. Bayes solution

A Bayes rule minimizes the average loss $L(D, \sigma)$ defined in Eq. (8.4-7b). The Bayes solution is

$$\text{Decide:} \quad d_1 \; (\textit{signal present}) \qquad \text{when } \ell(\mathbf{v}) \geqslant \mathscr{K}$$
$$\text{Decide:} \quad d_0 \; (\textit{no signal}) \qquad \text{when } \ell(\mathbf{v}) < \mathscr{K} \qquad (8.12\text{-}1)$$

where $\mathscr{K} = C_\alpha / C_{\bar{\beta}}$ and $\ell(\mathbf{v})$ is the *generalized likelihood ratio* defined by

$$\ell(\mathbf{v}) = \frac{\mathfrak{p}\,\overline{p(\mathbf{v} \mid \mathbf{s})_\mathbf{a}}}{q\,p(\mathbf{v} \mid 0)} \qquad (8.12\text{-}2)$$

Equation $\ell(\mathbf{v}) = \mathscr{K}$ defines the boundary in observation space that separates *signal present* from *signal absent* decisions.

B. Ideal observer

The ideal observer decision rule minimizes the total (average) probability of error, P_{te}, defined in Eq. (8.9-1). The decision rule is

$$\text{Decide:} \quad d_1 \qquad \text{when } \ell(\mathbf{v}) \geqslant 1$$
$$\text{Decide:} \quad d_0 \qquad \text{when } \ell(\mathbf{v}) < 1 \qquad (8.12\text{-}3)$$

This solution corresponds to a Bayes system in which the costs of errors are assigned equal weight $(C_\alpha = C_{\bar\beta})$.

C. Modified Neyman-Pearson

In this case the decision rule minimizes the *total* type II error probability $\mathfrak{p}\bar\beta$ subject to the constraint of a fixed total type I error probability $\mathfrak{q}\alpha$. The solution is

$$\text{Decide:} \quad d_1 \quad \text{when } \ell(\mathbf{v}) \geqslant \lambda$$
$$\text{Decide:} \quad d_0 \quad \text{when } \ell(\mathbf{v}) < \lambda \tag{8.12-4}$$

This solution corresponds to a Bayes system with $\mathscr{K} = \lambda$. The threshold λ is determined by the specification of $\mathfrak{q}\alpha$.

D. Minimax approach

A minimax decision corresponds to a Bayes rule for a priori signal statistics $\sigma_{if}(\mathbf{s})$ that maximize Bayes loss. This solution represents a conservative approach, since the worst a priori state of nature is assumed. The rule is

$$\text{Decide:} \quad d_1 \quad \text{when } \ell(\mathbf{v}) \geqslant \mathscr{K}$$
$$\text{Decide:} \quad d_0 \quad \text{when } \ell(\mathbf{v}) < \mathscr{K} \tag{8.12-5}$$

where the generalized likelihood ratio $\ell(\mathbf{v})$ is obtained for a priori probability density $\sigma_{if}(\mathbf{s})$.

Of significance is the fact that all the preceding tests lead to a comparison of the likelihood ratio $\ell(\mathbf{v})$ with different threshold values. A receiver that computes $\ell(\mathbf{v})$ for all possible $\mathbf{v}$ and compares this value with an adjustable threshold can realize any of the preceding tests. It is shown in the next two chapters that the structure of such a receiver, in the presence of white Gaussian noise, consists of a matched filter with additional processing, depending on the specific radar model assumed.

8.13. Bayes Solutions for Complex Cost Functions

In general binary detection, signal $\mathbf{s}$ is a function of a number of parameters $\boldsymbol{\theta}$. Such parameters often include: amplitude $(\theta_1 = A)$, time delay $(\theta_2 = \tau)$, initial phase $(\theta_3 = \phi)$, and so on. In radar, the signal parameters provide information about various target parameters such as range, range-rate, acceleration, azimuth and elevation angle, angular rate and acceleration, and so forth. The signal statistics are described by the a priori (existence) probability $\mathfrak{p}$ and the a priori probability density $w(\boldsymbol{\theta})$. The present discussion differs from that in Section 8.5 [see Eq. (8.5-8)], in that the cost of a correct decision, $C_{1-\bar\beta}$ is not chosen equal to zero; instead $C_{1-\bar\beta}$ is assumed to be a function of signal parameters $\boldsymbol{\theta}$.

In the cost matrix of Table 8.13-1, $C_{1-\overline{\beta}}(\boldsymbol{\theta})$ is the cost of a correctly detected signal. Substituting this matrix and Eq. (8.5-2), with $w(\mathbf{s})$ replaced by $w(\boldsymbol{\theta})$, into Eq. (8.5-4) yields the average loss function,

$$L(D, \sigma) = \mathfrak{q}C_\alpha \int_\Gamma D(d_1 \mid \mathbf{v}) p(\mathbf{v} \mid 0) \, d\mathbf{v}$$

$$+ \mathfrak{p}C_{\overline{\beta}} \int_\theta w(\boldsymbol{\theta}) \, d\boldsymbol{\theta} \int_\Gamma D(d_0 \mid \mathbf{v}) p[\mathbf{v} \mid \mathbf{s}(\boldsymbol{\theta})] \, d\mathbf{v}$$

$$+ \mathfrak{p} \int_\theta w(\boldsymbol{\theta}) \, d\boldsymbol{\theta} \int_\Gamma C_{1-\overline{\beta}}(\boldsymbol{\theta}) D(d_1 \mid \mathbf{v}) p[\mathbf{v} \mid \mathbf{s}(\boldsymbol{\theta})] \, d\mathbf{v} \qquad (8.13\text{-}1)$$

[When $C_{1-\overline{\beta}}(\boldsymbol{\theta})$ in (8.13-1) is set equal to a constant independent of parameters $\boldsymbol{\theta}$, it can be shown that minimizing Eq. (8.13-1) leads again to a Bayes solution, similar to that discussed in Section 8.6, in terms of the generalized likelihood ratio of Eq. (8.6-7).] With Eq. (8.6-1), expression (8.13-1) reduces to

$$L(D, \sigma) = \mathfrak{p}C_{\overline{\beta}} + \int_\Gamma D(d_1 \mid \mathbf{v}) \left\{ \mathfrak{p} \int_\theta C_{1-\overline{\beta}}(\boldsymbol{\theta}) w(\boldsymbol{\theta}) p[\mathbf{v} \mid \mathbf{s}(\boldsymbol{\theta})] \, d\boldsymbol{\theta} \right.$$

$$\left. - \mathfrak{p}C_{\overline{\beta}} \overline{p[\mathbf{v} \mid \mathbf{s}(\boldsymbol{\theta})]_\theta} + \mathfrak{q}C_\alpha p(\mathbf{v} \mid 0) \right\} d\mathbf{v} \qquad (8.13\text{-}2)$$

Minimizing (8.13-2) yields the Bayes decision rule D_B:

$$\left. \begin{array}{l} D_B(d_1 \mid \mathbf{v}) = 1 \\ D_B(d_0 \mid \mathbf{v}) = 0 \end{array} \right\} \quad \text{(signal present)} \qquad (8.13\text{-}3)$$

when

$$\mathfrak{p} \left\{ C_{\overline{\beta}} \overline{p[\mathbf{v} \mid \mathbf{s}(\boldsymbol{\theta})]_\theta} - \int_\theta C_{1-\overline{\beta}}(\boldsymbol{\theta}) w(\boldsymbol{\theta}) p[\mathbf{v} \mid \mathbf{s}(\boldsymbol{\theta})] d\boldsymbol{\theta} \right\} \geqslant \mathfrak{q}C_\alpha p(\mathbf{v} \mid 0) \qquad (8.13\text{-}4)$$

TABLE 8.13-1

Cost Matrix for Complex Cost Functions

		Signal s	
		s = 0	s ≠ 0
Decision **d**	d_0	0	$C_{\overline{\beta}}$
	d_1	C_α	$C_{1-\overline{\beta}}(\boldsymbol{\theta})$

Otherwise, decide *signal absent*. This inequality can be rewritten as

$$\frac{\mathfrak{p} \overline{p[\mathbf{v} \mid \mathbf{s}(\boldsymbol{\theta})]_\theta}}{\mathfrak{q} p(\mathbf{v} \mid 0)} - \frac{\mathfrak{p} \int_\theta C_{1-\overline{\beta}}(\boldsymbol{\theta}) w(\boldsymbol{\theta}) p[\mathbf{v} \mid \mathbf{s}(\boldsymbol{\theta})] d\boldsymbol{\theta}}{\mathfrak{q} C_{\overline{\beta}} p(\mathbf{v} \mid 0)} \geqslant \frac{C_\alpha}{C_{\overline{\beta}}} \qquad (8.13\text{-}5)$$

The first term in (8.13-5) is the generalized likelihood ratio $\ell(\mathbf{v})$ defined in (8.6-7). Expression (8.13-5) is similar to Eq. (8.6-8) with the addition of a second term that depends on cost assignment $C_{1-\overline{\beta}}(\boldsymbol{\theta})$.

In Chapter 17 we consider the solution of a problem of this type for a signal parameter $\theta_1 = \tau$, where τ is the expected time of signal arrival,† assuming two different cost matrix assignments. In one case, $C_{1-\bar{\beta}}(\boldsymbol{\theta})$ is set equal to zero; this yields a solution that depends on the generalized likelihood ratio and results in a Bayes receiver that averages the output of a matched filter with respect to the a priori probability density of τ.

In the second case, the cost function $C_{1-\bar{\beta}}(\boldsymbol{\theta})$ is chosen to be a step function with penalty C_m when the detection occurs at an arrival time other than the true value. The Bayes strategy for this case is a threshold test for each of a set of discrete expected arrival times. The threshold is determined by both the cost assignments and the a priori probability density of expected arrival time τ. This strategy corresponds to the use of separate *range bin* tests—which is intuitively reasonable.

8.14. *Preferred* Neyman-Pearson Strategy

In many situations the previous approach cannot be used because a priori statistics are lacking or a reasonable basis for choosing cost penalties $C_i(\boldsymbol{\theta})$ is not available. An alternate approach is provided by the *preferred* Neyman-Pearson strategy, which differs from the *modified* Neyman-Pearson strategy discussed earlier. The preferred Neyman-Pearson strategy is to find a decision surface that separates the acceptance and rejection regions (with respect to hypothesis H_0) such that the type II error probability $\beta(\boldsymbol{\theta})$ is minimized for a fixed value of α (the *level* of the test); or equivalently, the probability of detection (the *power* of the test) is maximized. Since type II error probability $\beta(\boldsymbol{\theta})$ is in general a function of signal parameters $\boldsymbol{\theta}$, the solution differs for each set of parameters. In special cases, the test is the same for all admissible values of $\boldsymbol{\theta}$; such a test is called *uniformly most powerful*. Unfortunately, uniformly most powerful tests do not occur very often. The following two examples are illustrative.

Example 1: In the first example of Section 8.7 where a test for a positive mean A in Gaussian noise was obtained, the test strategy consists of a likelihood-ratio test with a planar boundary surface given by

$$\sum_{i=1}^{3} v_i = M$$

The region in observation space Γ'' for which signal is said to be present consists of all points **v** in the volume above the plane shown in Fig. 8.7-1.

Suppose instead that mean A is positive but otherwise unknown. Since test level α is established by integrating $p(\mathbf{v}\,|\,0)$ over Γ'' [see Eq. (8.8-5)], the specification of test level α implicitly defines the value of M and hence the position of the boundary surface. This result is independent of parameter A; hence the region Γ''' is the same for all values of A and the test is a uniformly most powerful one.

---

†Time of arrival is approximated by a discrete set.

Example 2: Suppose mean A is unknown with respect to both magnitude and sign. Recall that when mean A in the example of Section 8.7 is negative, space Γ''' consists of the region below the plane

$$\sum_{i=1}^{3} v_i = -M$$

Hence space Γ''' is not the same for positive and negative values of A so that the test in this case is not uniformly most powerful.

When a uniformly most powerful solution cannot be found, other criteria can be employed. For example, the class of tests may be reduced by considering only those with some additional desirable characteristic. A uniformly most powerful test may then exist within the reduced class. An example is the class of *unbiased* tests, discussed in Lehman [13], whose power functions† have a minimum at $\mathbf{s} = 0$. These tests are beyond the scope of the present discussion.

8.15. *Intuitive* Substitute

When a uniformly most powerful test does not exist, an alternate *intuitive* strategy is to average the power of the test—that is, the probability of detection $P_d(\boldsymbol{\theta})$—with respect to both the a priori probability density function governing the signal parameters $\boldsymbol{\theta}_1$ whose statistics are known and a least favorable a priori density function governing the parameters $\boldsymbol{\theta}_2$ whose statistics are unknown; a test is then sought that maximizes the averaged detection probability. This approach is related to the *modified* Neyman-Pearson strategy discussed previously, in which $\overline{P_d(\boldsymbol{\theta})_{\theta_1}}$ is maximized for a fixed level α, and to the minimax strategy, where an averaging is performed with respect to least favorable a priori statistics $\sigma_{lf}(\boldsymbol{\theta}_2)$. This is a conservative philosophy, since, on an average, the value of P_d obtained is the worst that can be expected.

For some radar parameters, solutions obtained with the intuitive substitute approach yield good results; in other cases poor results are obtained. For example, consider the following radar parameters: amplitude A, delay τ, doppler ω_d, and initial phase ϕ. Statistical information is usually available concerning signal amplitude; this is expressed by describing the target model as *Rayleigh*, *one dominant plus Rayleigh*, and so on (see Chapter 1). Averaging P_d with respect to the appropriate amplitude probability density generally leads to a satisfactory result (see Chapters 9 and 11). However, for both delay and doppler, the intuitive approach provides unsatisfactory results. In particular, averaging over the regions of uncertainty of delay and doppler leads to an unsatisfactory test in a multiple-target environment. (This problem is considered in Chapter 17.)

---

†The power function of a test is a plot of the power of the test as a function of parameters θ over the space of random parameters.

In the case of starting phase ϕ, the intuitive averaging approach does provide satisfactory performance. A priori information concerning ϕ is usually unavailable; hence a least favorable distribution is employed. A uniform probability density function can be shown to be the least favorable since, for any other distribution, a system can be designed to reduce errors by emphasizing values of ϕ that have high probability. Averaging phase leads to an optimum receiver structure in which a matched filter is followed by an envelope detector (see Chapter 9). When compared to an optimum receiver for which ϕ is assumed to be known, it is shown in Chapter 9 that the loss in detectability for ϕ with a uniform probability density is small (less than 1 db) in the region of primary concern to the radar designer, namely, high signal-to-noise ratio. Thus, the intuitive approach in this case results in a receiver that is both simple to implement and only slightly poorer than that provided by a receiver based on complete knowledge of phase.

8.16. Fixed and Sequential Testing

In the discussion of statistical decision theory it has been tacitly assumed that a decision is rendered after a fixed observation interval in which data are collected. The observations made during this interval may consist, in general, of discrete or (sampled) continuous input waveforms. In some systems the observation interval is not fixed but is of variable length and is dependent on the input data. This might be advantageous in an application where it is desirable to keep the observation interval as short as possible. For example, when a large radar echo signal is received from a nearby target, it may be desirable to take advantage of this circumstance to shorten the observation interval. This is especially true if the target represents a possible threat and quick retaliatory action is required.

A test procedure for a variable-length observation period has been developed by Wald [9] and is known as a *sequential test*. A similar concept was considered by Neyman and Pearson [10] in 1933 as an extension of their theory of hypothesis testing. They defined three possible decisions: accept H, reject H, and no decision. In Wald's method of sequential testing, it is decided whether to make a decision based upon the data already taken or to continue taking more data following each measurement. Thus the length of the observation interval depends on the quality of the available data. Although it is theoretically possible for a test to continue indefinitely, it has been shown [9] that on the average the observation interval is shorter in a sequential test than in a fixed test. Furthermore, in practice a sequential test is usually truncated after some predetermined number of observations.

More work has been done in applying decision theory to nonsequential testing than to sequential testing. However, as improved signal-processing techniques become available, sequential testing procedures will undoubtedly assume greater significance. Sequential testing and its application to the radar detection problem are discussed in Chapter 16.

8.17. Concluding Remarks

The application of statistical decision theory to problems in communications and radar is being actively pursued. Despite the power of the method, certain limitations restrict its range of application. These limitations result from requirements on the system model that can never be completely satisfied in practice.

One limitation is concerned with cost assignments. Assignments are usually made by the system designer and therefore are subject to individual bias. Fortunately, the structure of the optimum system is often fairly insensitive to variations in cost assignment. For example, the structure of a Bayes receiver for simple radar binary detection is independent of the magnitude of the preassigned costs.† The sensitivity of solutions to cost assignments in other problems should be investigated in order to assess their effect on system complexity and performance under variable conditions. In practice, the number of useful cost functions is usually limited by both reasonableness and computability.

A more important limitation stems from the need for a priori information concerning both the signal and the noise processes. If such information is not available, the theory cannot be applied completely. It may be recalled that a priori probabilities also played an important role in the formulation of a posteriori theory. If a priori information is available about the noise process but signal statistics are unavailable, a solution may still be possible by invoking other criteria such as minimax. In some cases, an adaptive system in which the system decision rule varies as "learning" takes place may be desirable. System performance generally degrades as the uncertainty about a priori information increases.

The choice of a criterion of optimality has already been discussed. The selection of an appropriate criterion is influenced by the system designers' judgment, by the a priori data, and by possible external system constraints. Although other optimization criteria have been suggested, none has a theory as well-developed as Bayes and minimax. In addition, it is often difficult in practice to realize the optimum system without some approximation. More elegant optimality criteria may only further complicate analysis without yielding substantially improved performance.

In the following chapters, statistical decision theory is applied to various problems in detection. Specific results are obtained for the structure of the

---

†This is not true for complex cost assignments, however, as previously noted.

optimum receiver and expected performance. In several important cases performance is presented in graphical form as an aid to system design.

PROBLEMS

8.1. In the problem of Section 8.2, assume the racehorse owner is not penalized when a bet is not placed—that is, $C_{01} = 0$. With the other quantities as given, find the likelihood ratio and interpret the result.

8.2. Given a priori probability of signal $\mathfrak{p}$, a priori probability of no signal $\mathfrak{q}$, cost matrix

$$C = \begin{bmatrix} 0 & C_\beta \\ C_\alpha & 0 \end{bmatrix}$$

and

$$p(v \mid 0) = \frac{1}{\sqrt{2\pi\sigma^2}} \exp\left(-\frac{v^2}{2\sigma^2}\right)$$

$$\overline{p(v \mid s)}_s = \frac{1}{\sqrt{2\pi\sigma^2}} \exp\left[-\frac{(v - A)^2}{2\sigma^2}\right]$$

(a) find threshold v_0 from $\ell(v_0) = C_\alpha / C_\beta$;
(b) find v_0 in terms of $\mathfrak{p}$ for $A = 1$, $\sigma = 1$, $C_\alpha = 1$, and $C_\beta = 2$;
(c) compute α and β for $\mathfrak{p}$ ranging in value from 0.1 to 0.9;
(d) plot α and β as functions of $\mathfrak{p}$;
(e) compute the Bayes loss $L_B = \mathfrak{p}\beta C_\beta + \mathfrak{q}\alpha C_\alpha$ and plot as a function of $\mathfrak{p}$;
(f) find the minimax loss from (e) and determine the value of $\mathfrak{p}$ for which the minimax loss occurs.

8.3. Using the results of problem 8.2 plot the expected loss as a function of $\mathfrak{p}$ for a Bayes receiver that is optimized for $\mathfrak{p} = 0.25$.
(a) Explain the deviation between the Bayes loss curve of problem 8.2 and the loss plotted for this problem.
(b) Why is minimax optimum from the loss standpoint when $\mathfrak{p}$ is unknown?

8.4. Find threshold $\mathscr{K}$ in Section 8.6, when $C_{1-\alpha}$ and $C_{1-\beta}$ are not equal to zero.

8.5. Let H_0 be the simple hypothesis that an observation v has a rectangular distribution:

$$p_0(v) = \begin{cases} \frac{1}{4}, & -2 \leqslant v \leqslant 2 \\ 0, & \text{otherwise} \end{cases}$$

and let H_1 be the simple hypothesis that v has an exponential distribution:

$$p_1(v) = \frac{1}{2} \exp\left(-|v|\right)$$

Let $\mathfrak{p}$ be the a priori probability that H_0 is true.
(a) Find the Bayes test for hypothesis H_0 assuming equal costs for type I and type II errors ($C_\alpha = C_\beta$, $C_{1-\alpha} = C_{1-\beta} = 0$).
(b) For $\mathfrak{p} = \frac{1}{4}$, find the total probability of making an error.

8.6. Repeat problem 8.5 when two independent observations v_1 and v_2 are available with the same statistics as in 8.5.

8.7. Suppose that hypothesis H_1 in problem 8.5 is composite such that v has an exponential distribution with unknown parameter a:

$$p_1(v) = \frac{a}{2} \exp\left(-a\,|v|\right)$$

For $a > 0$, determine whether a uniformly most powerful Neyman-Pearson test exists.

8.8. (a) For the composite hypothesis case discussed in Section 8.13 derive the Bayes strategy for a cost matrix given by

$$\mathbf{d} \begin{array}{c} \\ d_0 \\ d_1 \end{array} \begin{array}{cc} \mathbf{s} = 0 & \mathbf{s}(\theta) \neq 0 \\ \begin{bmatrix} 0 & C_\beta(\theta) \\ C_\alpha & C_{1-\beta}(\theta) \end{bmatrix} \end{array}$$

(b) Verify the result of Eq. (8.13-4) when $C_\beta(\theta) = C_{\bar{\beta}}$.

(c) Verify the result with $C_\beta(\boldsymbol{\theta}) = C_{\bar{\beta}}$ and $C_{1-\beta}(\boldsymbol{\theta}) = 0$.

8.9. Let signal $s(\theta)$ be a function of a single discrete parameter θ which takes on values in the set $(\theta_1, \theta_2, \ldots, \theta_N)$. If the a priori probability of θ_i is given by $\mathfrak{p}_i$, minimize the total type II error $\sum_i \mathfrak{p}_i \beta_i$ subject to a fixed type I error and the constraint on $\sum_i \mathfrak{p}_i$. Maximize this expression with respect to $\mathfrak{p}_i$ for $i = 1, 2, \ldots, N$ and interpret the result. (See Chap. 5 of reference [14].)

8.10. Given the probability densities of the waveform envelopes of narrowband noise and signal-plus-noise, respectively, for N independent samples r_i:

$$p_n(\mathbf{r}) = \prod_{i=1}^{N} \frac{r_i}{\sigma^2} \exp\left(-\frac{r_i^2}{2\sigma^2}\right)$$

$$p_{s+n}(\mathbf{r}) = \prod_{i=1}^{N} \frac{r_i}{\sigma^2} \exp\left(-\frac{r_i^2 + A^2}{2\sigma^2}\right) I_0\left(\frac{r_i A}{\sigma^2}\right)$$

(a) find the likelihood ratio;

(b) for the small-signal case, $A/\sigma \ll 1$, find the approximate test statistic by a series expansion of the likelihood ratio;

(c) interpret the expression above in terms of the structure of the threshold detector.

REFERENCES

1. Neyman, J., and Pearson, E.: On the Problem of the Most Efficient Tests of Statistical Hypotheses, *Phil. Trans. Roy. Soc.*, **A231**: 289 (1933).

2. Wald, A.: Contributions to the Theory of Statistical Estimation and Testing of Hypotheses, *Ann. Math. Stat.*, **10**: 299 (1939).

3. Wald, A.: "Statistical Decision Functions," Wiley, New York, 1950.

4. Blackwell, D., and Girshick, M.A.: "Theory of Games and Statistical Decisions," Wiley, New York, 1954.

5. Lawson, J. L., and Uhlenbeck, G. E.: "Threshold Signals," McGraw-Hill, New York, 1950.

6. Van Meter, D., and Middleton, D.: Modern Statistical Approaches to Reception in Communication Theory, *IRE Trans.*, **PGIT-4**: 119–145 (Sept., 1954).

7. Middleton, D., and Van Meter, D.: Detection and Extraction of Signals in Noise from the Point of View of Statistical Decision Theory, *J. Soc. Ind. Appl. Math.*, **3**: 192 (1955); **4**: 86 (1956).

8. Middleton, D.: "An Introduction to Statistical Communication Theory," McGraw-Hill, New York, 1960.

9. Wald, A.: "Sequential Analysis," Wiley, New York, 1947.

10. Neyman, J., and Pearson, E. S.: The Testing of Statistical Hypotheses in Relation to Probability a Priori, *Proc. Cambridge Phil. Soc.*, **29** (1933).

11. Lawson, J. L., and Uhlenbeck, G. E.: "Threshold Signals," Preface and Sec. 7.5, McGraw-Hill, New York, 1950.

12. Cramer, H.: "Mathematical Methods of Statistics," Princeton University Press, Princeton, N.J., 1946.

13. Lehman, E. L.: "Testing Statistical Hypotheses," Wiley, New York, 1959.

14. Helstrom, C.: "Statistical Theory of Signal Detection," Pergamon, Oxford, England, 1960.

PART IV

OPTIMUM RADAR
DETECTION

In Part IV the statistical decision-theory concepts developed in Chapter 8 are applied to the radar detection problem. The theory is used to determine the structure and performance of the optimum detection receiver for a number of practically important cases. In Chapter 9 detection based on a single observation is studied. Decision theory is applied to derive the optimum receiver structure for an exactly known signal, a signal of unknown phase, and finally a signal of unknown phase and amplitude.

In Chapters 10 and 11 detection based on multiple observations is considered; these chapters include the principal results of Marcum [1] and Swerling [2]. In Chapter 10 we consider the optimum detection receiver for detecting nonfluctuating incoherent pulse trains and evaluate receiver performance. Extensive graphical results are presented based on computer calculations carried out by Fehlner [3]. The four Swerling [2] models for fluctuating incoherent pulse trains are studied in Chapter 11; these include scan-to-scan and pulse-to-pulse amplitude fluctuations. The optimum receiver structure and performance are evaluated for these cases and results are presented graphically, based again on the computer results of Fehlner [3].

In Chapter 12 various forms of the radar equation and the fundamental quantities entering into it are considered from a system viewpoint. Typical examples are treated to illustrate the application of the radar equation and the results of Chapters 9 through 11 to the search radar problem. The concept of *cumulative* detection probability is introduced in Chapter 13. A procedure is described there for minimizing the power-aperture product of a search radar.

REFERENCES

1. Marcum, J. I.: A Statistical Theory of Target Detection by Pulsed Radar, Rand Research Memo RM-754 (December, 1947). Also, *IRE Trans. on Information Theory*, **IT-6:** (*2*), 59–144 (April, 1960).

2. Swerling, P.: Probability of Detection for Fluctuating Targets, Rand Research Memo RM-1217 (March, 1954). Also, *IRE Trans. on Information Theory*, **IT-6:** (*2*), 269–308 (April, 1960).

3. Fehlner, L. F.: Target Detection by a Pulsed Radar, Report TB451, Applied Physics Laboratory, Johns Hopkins University, Silver Spring, Maryland (July, 1962).

DETECTION
BASED ON A
SINGLE OBSERVATION

9.1. Introduction

The detection of signals on a single observation is sometimes called single-hit detection, coherent detection, or detection subject to predetection integration. The latter designations indicate that the rf structure of the radar signal within the observation interval is deterministic, or exactly known, except possibly for initial phase, time of occurrence, amplitude, and doppler.

The case of an exactly known signal immersed in white Gaussian noise is considered first since considerable understanding and insight are gained thereby. The results are then generalized to encompass signals of unknown phase and unknown amplitude. Signals of unknown arrival time and doppler require special consideration and are discussed below and at greater length in Chapter 17. In the latter part of the chapter earlier results are generalized to signals in colored noise.

Unknown target parameters that are frequently encountered in radar detection problems include range delay, doppler (velocity), angle of arrival, amplitude, and initial rf phase. For a signal of unknown delay, doppler, and angle of arrival it is shown in Chapter 17 that the optimum *Bayes* detection criterion, relative to a specific cost matrix assignment for a search radar, corresponds to a separate likelihood-ratio test in every range-angle-doppler resolution cell in the search volume. The dimensions of each resolution cell are approximately: the antenna beam width in angle, $1/\Delta_f$ in range, and $1/\Delta_\tau$ in doppler, where Δ_f and Δ_τ are the signal bandwidth and duration, respectively. Since the same test is performed in each resolution cell, it is sufficient to investigate the likelihood-ratio test in a typical cell; for simplicity, the cell chosen is the first range bin, in an arbitrary antenna beam, and zero doppler. The length of the observation-interval is equal to the signal-waveform duration. The observation-interval for simple uncoded pulses is approximately equal to the range resolution-cell dimension $1/\Delta_f$, as shown in Fig. 9.1-1. For coded pulses the waveform duration is usually greater than $1/\Delta_f$; hence the observation interval might resemble that shown in Fig. 9.1-2.

With respect to unknown signal amplitude, the detection criterion is also *Bayes*. The Bayes criterion requires that we average the likelihood ratio with respect to the a priori signal-amplitude statistics (see Chapter 8). For this purpose a statistical model of target return must be known, or assumed; for example, the signal amplitude may be classified as nonfluctuating or fluctuating. In the case of fluctuating signals, several density functions have been found to correspond to radar observations of aircraft and satellites.

The optimality criterion adopted for unknown initial phase is the *intuitive substitute* discussed in Section 8.15. This criterion requires that the likelihood ratio be averaged with respect to a least favorable distribution for initial phase; a uniform probability density function satisfies this require-

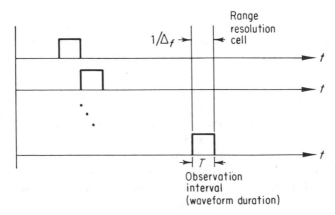

Fig. 9.1-1. Range resolution cell size and related observation interval for a simple rectangular pulse.

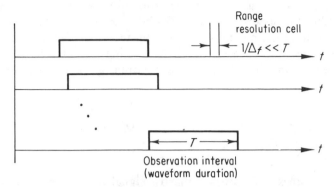

Fig. 9.1-2. Range resolution cell size and related observation interval for coded rectangular pulses with $\Delta_f \gg 1/T$.

ment since it reflects maximum ignorance of initial phase. The loss in detectability resulting from use of this criterion is relatively small for high signal-to-noise ratios.

9.2. Optimum Detection of an Exactly Known Signal

For simplicity we consider first optimum detection of an exactly known signal. It was shown in Chapter 8 that the Bayes detection strategy for an exactly known signal is a likelihood-ratio test. The optimum decision rule specifies that the likelihood ratio, which is a function of data $v(t)$ observed in interval T, be compared with a threshold $\mathcal{K}$. Knowledge of the form of the likelihood ratio permits specification of a test statistic that involves all of the essential operations on the data $v(t)$. Since the test statistic also describes the structure of the optimum Bayes detection receiver, the performance of the receiver can be evaluated by considering the properties of the test statistic. Depending on the absence or presence of signal, waveform $v(t)$ consists of either a sample function of a zero mean white Gaussian noise process, or of an exactly known signal of duration T added to the noise.

$$v(t) = s(t) + n(t) \quad (1)$$
$$or = \{ n(t) \quad (0)$$

A. Structure of optimum receiver

To derive an expression for the likelihood ratio, it is convenient to employ the sampling-theorem waveform representation. Although the Karhunen-Loéve expansion is more elegant, the sampling theorem leads to expressions that are more easily interpreted for obtaining optimum receiver structure. The signal is assumed to be an exactly known band-limited waveform with spectrum confined to the interval $(-\omega_c, \omega_c)$. (Note that an rf signal is not excluded by this description, since ω_c can be very large.) For analytic convenience the received waveform $v(t)$ is assumed to be prefiltered by an ideal low-pass filter that is distortionless within $(-\omega_c, \omega_c)$ and zero outside the interval; this filter has no effect on band-limited input signal $s(t)$ but results in a band-limited noise process $\{n(t)\}$. With this assumption input waveform $v(t)$ can be described by the sampling theorem given in Chapter 2; thus, for signal present,

$$v_i = s_i + n_i, \qquad i = 1, 2, \ldots, 2f_cT \tag{9.2-1}$$

where $v_i = v(i/2f_c)$, $s_i = s(i/2f_c)$, and $n_i = n(i/2f_c)$. Equation (9.2-1) can be written in vector notation as

$$\mathbf{v} = \mathbf{s} + \mathbf{n} \tag{9.2-2}$$

When hypothesis H_1 is true—that is, signal plus noise is present—the conditional joint probability density function governing the samples v_i, given s_i, can be obtained by the method discussed in Section 7.5. [See also Eq. (3.14-9).] The density function is

$$p(\mathbf{v}\,|\,\mathbf{s}) = \frac{1}{(2\pi)^{N/2}(N_0 f_c)^N} \exp\left[-\frac{\sum\limits_{i=1}^{N}(v_i - s_i)^2}{2f_cN_0}\right] \tag{9.2-3}$$

where $N = 2f_cT$. In similar manner, when hypothesis H_0 is true (noise alone), the conditional joint probability density function $p(\mathbf{v}\,|\,0)$ is given by

$$p(\mathbf{v}\,|\,0) = \frac{1}{(2\pi)^{N/2}(f_cN_0)^N} \exp\left(-\frac{\sum\limits_{i=1}^{N} v_i^2}{2f_cN_0}\right) \tag{9.2-4}$$

The likelihood-ratio expression [see Eq. (8.6-7)], when there are no random signal parameters, is given by

$$\ell(\mathbf{v}) = \frac{\mathfrak{p}\,p(\mathbf{v}\,|\,\mathbf{s})}{\mathfrak{q}\,p(\mathbf{v}\,|\,0)} \tag{9.2-5}$$

A comparison of $\ell(\mathbf{v})$ with threshold $\mathscr{K}$, which depends on the cost assignments for a particular problem, is equivalent to a comparison of $\ell'(\mathbf{v})$,

$$\ell'(\mathbf{v}) = \frac{p(\mathbf{v}\,|\,\mathbf{s})}{p(\mathbf{v}\,|\,0)} \tag{9.2-6}$$

with a modified threshold $\mathscr{K}'$, where the a priori probabilities $\mathfrak{p}$ and $\mathfrak{q}$ have been absorbed in threshold $\mathscr{K}'$:

$$\mathscr{K}' = \frac{\mathfrak{q}}{\mathfrak{p}}\,\mathscr{K} \tag{9.2-7}$$

Cost assignments are made in accordance with the specific strategy employed —ideal observer, minimax, and so on.

From (9.2-4) through (9.2-6),

$$\ell'(\mathbf{v}) = \frac{\exp\left\{-\sum\limits_{i=1}^{2f_cT}[v_i - s_i]^2/2f_cN_0\right\}}{\exp\left(-\sum\limits_{i=1}^{2f_cT} v_i^2/2f_cN_0\right)} \tag{9.2-8}$$

With approximate relation (2.4-18), Eq. (9.2-8) can be rewritten as

$$\ell'(\mathbf{v}) = \frac{\exp\left\{-\dfrac{1}{N_0}\displaystyle\int_0^T [v(t) - s(t)]^2\, dt\right\}}{\exp\left(-\dfrac{1}{N_0}\displaystyle\int_0^T v^2(t)\, dt\right)} \tag{9.2-9a}$$

$$\ell'(\mathbf{v}) = \exp\left[-\frac{1}{N_0}\int_0^T s^2(t)\, dt + \frac{2}{N_0}\int_0^T v(t)\, s(t)\, dt\right] \tag{9.2-9b}$$

The first integral in Eq. (9.2-9b) is equal to signal energy E; thus,

$$\ell'(\mathbf{v}) = \exp\left[-\frac{E}{N_0} + \frac{2}{N_0}\int_0^T v(t)s(t)\, dt\right] \tag{9.2-10}$$

Since the exponential in (9.2-10) is a monotonic function of its exponent, an equivalent threshold test is given by comparing

$$\ln \ell'(\mathbf{v}) = -\frac{E}{N_0} + \frac{2}{N_0}\int_0^T v(t)s(t)\, dt \tag{9.2-11}$$

with a suitably modified threshold. Hence, the threshold test is

$$\frac{2}{N_0}\int_0^T v(t)s(t)\, dt \begin{cases} \geqslant \dfrac{E}{N_0} + \ln \mathscr{K}' & (\textit{signal present}) \\[2mm] < \dfrac{E}{N_0} + \ln \mathscr{K}' & (\textit{signal absent}) \end{cases} \tag{9.2-12}$$

The quantity on the left side of Eq. (9.2-12) is called the test statistic; it describes the essential operations that must be performed by the optimum Bayes detector on the data $v(t)$. This test statistic, previously discussed in Section 7.6, represents the cross correlation of $v(t)$ and $s(t)$. As in the development of Chapter 7, the optimum Bayes detector for an exactly known signal is implemented by the processors shown in Figs. 9.2-1 and 9.2-2, which are active and passive realizations of a cross correlator. Note that in Fig. 9.2-1 the active correlator realization requires a multiplier. In Fig. 9.2-2 a passive matched filter is employed to realize the cross correlator, as discussed in Chapter 7. The output of the correlator is sampled

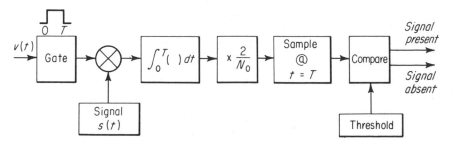

Fig. 9.2-1. Active realization of Bayes detector for an exactly known signal.

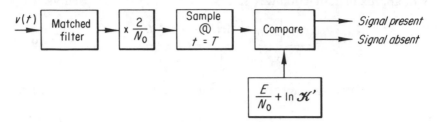

Fig. 9.2-2. Matched filter realization of Bayes detector for an exactly known signal.

at $t = T$ and the sampled value is compared with the specified threshold; a decision is then rendered on the presence or absence of the signal.

B. Performance of optimum receiver

To evaluate the performance of the optimum receiver for an exactly known signal, let

$$y(T) = \frac{2}{N_0} \int_0^T v(t)s(t)\, dt \qquad (9.2\text{-}13)$$

With relation (2.4-17), $y(T)$ can be rewritten as

$$y(T) = \frac{1}{f_c N_0} \sum_{i=1}^{2f_c T} v_i s_i \qquad (9.2\text{-}14)$$

The statistics of $y(T)$ are governed by input noise process $\{n(t)\}$. Since $\{n(t)\}$ [and therefore $\{v(t)\}$] is a Gaussian noise process, waveform samples v_i are Gaussian random variables. Hence, $y(T)$ is a sum of Gaussian random variables and is also a Gaussian random variable.

When hypothesis H_0 is true—that is, noise alone is present—(9.2-14) becomes

$$y(T) = \frac{1}{f_c N_0} \sum_{i=1}^{2f_c T} n_i s_i \qquad (9.2\text{-}15)$$

Since $\overline{n_i} = 0$ for all i, it follows from (9.2-15) that the mean value of $y(T)$ is

$$\overline{y(T)} = \frac{1}{f_c N_0} \sum_{i=1}^{2f_c T} \overline{n_i} s_i = 0 \qquad (9.2\text{-}16)$$

Waveform samples n_i are statistically independent for all i (see Section 7.5); hence the variance of $y(T)$ is

$$\overline{y^2(T)} = \frac{1}{(f_c N_0)^2} \sum_{i=1}^{2f_c T} \sum_{j=1}^{2f_c T} \overline{n_i n_j} s_i s_j \qquad (9.2\text{-}17a)$$

$$\overline{y^2(T)} = \frac{1}{(f_c N_0)^2} \sum_{i=1}^{2f_c T} \overline{n_i^2} s_i^2 \qquad (9.2\text{-}17b)$$

The quantity $\overline{n_i^2}$ is equal to $f_c N_0$ for a zero-mean process [see Eq. (7.5-6)], so that Eq. (9.2-17b) can be rewritten as

$$\overline{y^2(T)} = \frac{1}{f_c N_0} \sum_{i=1}^{2f_c T} s_i^2 \qquad (9.2\text{-}18)$$

or, from approximate sampling relation (2.4-17), as

$$\overline{y^2(T)} = \frac{2}{N_0} \int_0^T s^2(t)\, dt = \frac{2E}{N_0} \qquad (9.2\text{-}19)$$

where E is the signal energy. The quantity $2E/N_0$ appears very often in radar theory; it is convenient to define

$$\mathscr{R} = \frac{2E}{N_0} \qquad (9.2\text{-}20)$$

Then,

$$\overline{y^2(T)} = \mathscr{R} \qquad \text{variance of } y(T) \qquad (9.2\text{-}21)$$

Thus test statistic $y(T)$ is a Gaussian random variable with zero mean and variance equal to $\mathscr{R}$ for hypothesis H_0. The probability of a type I error, or probability of false alarm P_{fa}, which is the probability that $y(T)$ exceeds a threshold y_b when noise alone is present, is given by

$$P_{fa} = \frac{1}{\sqrt{2\pi\mathscr{R}}} \int_{y_b}^{\infty} e^{-y^2/2\mathscr{R}}\, dy \qquad (9.2\text{-}22)$$

Eq. (9.2-22) can be rewritten as

$$P_{fa} = \frac{1}{\sqrt{2\pi}} \int_{z_b}^{\infty} e^{-z^2/2}\, dz \qquad (9.2\text{-}23)$$

where

$$z_b = \frac{y_b}{\sqrt{\mathscr{R}}} = \frac{y_b}{\sqrt{2E/N_0}} \qquad (9.2\text{-}24)$$

When hypothesis H_1 is true (signal present), $y(T)$ is given by

$$y(T) = \frac{2}{N_0} \int_0^T [n(t) + s(t)]s(t)\, dt \qquad (9.2\text{-}25a)$$

$$= \frac{1}{f_c N_0} \sum_{i=1}^{2f_c T} (n_i + s_i)s_i \qquad (9.2\text{-}25b)$$

Analogous to the earlier development, the mean value of $y(T)$ is

$$\overline{y(T)} = \frac{1}{f_c N_0} \sum_{i=1}^{2f_c T} s_i^2 = \frac{2}{N_0} \int_0^T s^2(t)\, dt \qquad (9.2\text{-}26a)$$

$$\overline{y(T)} = \frac{2E}{N_0} = \mathscr{R} \qquad (9.2\text{-}26b)$$

and the variance of $y(T)$ is

$$\text{var}\{y(T)\} = \overline{y^2(T)} - [\overline{y(T)}]^2 \qquad (9.2\text{-}27a)$$

$$\text{var}\{y(T)\} = \frac{1}{(f_c N_0)^2} \sum_{i=1}^{2f_c T} \sum_{j=1}^{2f_c T} \overline{(n_i + s_i)(n_j + s_j)}\, s_i s_j - \mathscr{R}^2 \qquad (9.2\text{-}27b)$$

$$\text{var}\{y(T)\} = \frac{2E}{N_0} + \frac{4E^2}{N_0^2} - \mathscr{R}^2 = \mathscr{R} \qquad (9.2\text{-}27\text{c})$$

Thus, for hypothesis H_1, $y(T)$ is a Gaussian random variable with mean $\mathscr{R}$ and variance $\mathscr{R}$. The probability of detection P_d is equal to $1 - \beta$, where β is the type II error probability; P_d is computed as

$$P_d = \frac{1}{\sqrt{2\pi\mathscr{R}}} \int_{y_b}^{\infty} e^{-(y-\mathscr{R})^2/2\mathscr{R}} \, dy \qquad (9.2\text{-}28\text{a})$$

$$P_d = \frac{1}{\sqrt{2\pi}} \int_{z_b}^{\infty} e^{-z^2/2} \, dz \qquad (9.2\text{-}28\text{b})$$

where

$$z_b = \frac{y_b - \mathscr{R}}{\sqrt{\mathscr{R}}} \qquad (9.2\text{-}29)$$

Equation (9.2-23) for false-alarm probability P_{fa} and (9.2-28b) for probability of detection P_d both depend on the evaluation of an integral of the form

$$\frac{1}{\sqrt{2\pi}} \int_{z_b}^{\infty} e^{-z^2/2} \, dz = 1 - \frac{1}{\sqrt{2\pi}} \int_{-\infty}^{z_b} e^{-z^2/2} \, dz \qquad (9.2\text{-}30)$$

The integral on the right of Eq. (9.2-30) is a tabulated function (see reference [2]), from which both P_d and P_{fa} can be evaluated for selected values of threshold y_b and $\mathscr{R}$. In Fig. 9.2-3 P_d is plotted versus $\mathscr{R}$ with P_{fa} as a parameter for values of greatest interest in radar ($\mathscr{R} > 5$ db). A sharp performance threshold is indicated by the steep ascent of the curves in the region of 5–15 db.

A word of caution is in order at this point concerning the use of the plotted curves in Fig. 9.2-3 and other performance graphs that appear later in this text. It was shown in Chapter 5 that $2E/N_0$ is equal to the ratio of maximum instantaneous signal power to average noise power out of a matched filter [see Eq. (5.3-14)]. Unfortunately, instantaneous signal power is not consistently interpreted in the literature. In Fig. 9.2-4, the solid line is the envelope of the peak instantaneous signal power of a narrowband rf waveform. The dotted line indicates the envelope of the average instantaneous power of this waveform where the averaging is over an rf cycle. A simple computation shows that

$$\left(\begin{array}{c}\text{peak instantaneous}\\ \text{signal power}\end{array}\right) = 2 \left(\begin{array}{c}\text{average instantaneous}\\ \text{signal power}\end{array}\right) \qquad (9.2\text{-}31)$$

Corresponding to peak and average instantaneous signal power, we can write

$$(S/N)_{\text{peak}} = \frac{\left(\begin{array}{c}\text{peak instantaneous}\\ \text{signal power}\end{array}\right)}{(\text{average noise power})} \qquad (9.2\text{-}32)$$

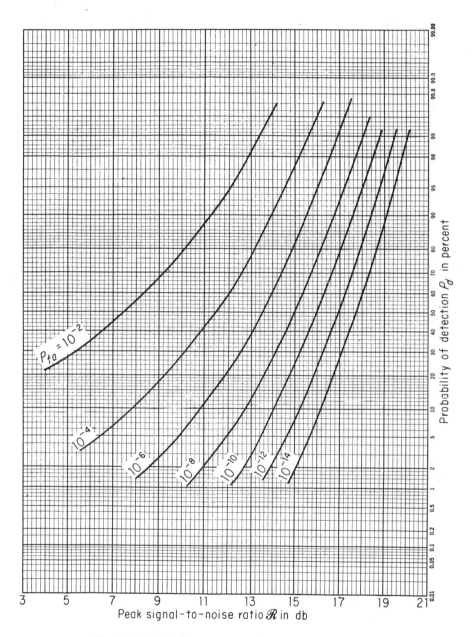

$P_{fa} = 10^{-2}$

10^{-4}

10^{-6}

10^{-8}

10^{-10}

10^{-12}

10^{-14}

Peak signal-to-noise ratio $\mathcal{R}$ in db

Probability of detection P_d in percent

Fig. 9.2-3. Detection characteristic for an exactly known signal.

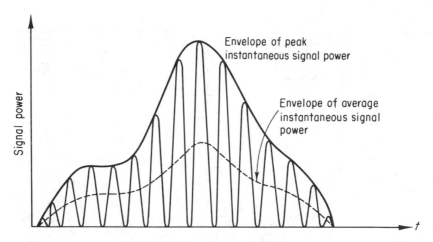

Fig. 9.2-4. Instantaneous signal power.

and

$$(S/N)_{\text{average}} = \frac{\left(\begin{array}{c}\text{average instantaneous}\\ \text{signal power}\end{array}\right)}{(\text{average noise power})} \tag{9.2-33}$$

It follows from Eqs. (9.2-20), (9.2-31) and the development leading to Eq. (5.3-14) that

$$\mathscr{R} = \frac{2E}{N_0} = [(S/N)_{\text{peak}}]_{\text{max}} = 2[(S/N)_{\text{average}}]_{\text{max}} \tag{9.2-34}$$

In many references, performance graphs of the type shown in Fig. 9.2-3 are plotted with respect to (S/N) without further qualification. The parameter $\mathscr{R}$ is employed in this text to avoid the confusion associated with such usage. Because of the factor of two which appears in Eq. (9.2-34), the reader may find a 3 db discrepancy between the performance plots in this text and those found elsewhere.

9.3. Optimum Detection of a Signal of Unknown Phase

A. Structure of optimum receiver

Consider a narrowband signal of duration T given by

$$s(t) = Aa(t - \tau_0) \cos\left[\omega_0(t - \tau_0) + \theta(t - \tau_0) + \phi\right] \tag{9.3-1}$$

where A is signal amplitude, $a(t)$ is an envelope function of duration T, $\theta(t)$ is the signal phase modulation, τ_0 is the target echo delay, and ϕ is the initial phase. As mentioned previously, initial phase ϕ is generally assumed to be a random variable since it is not known to within a fraction of 2π

radians. This is illustrated by the fact that 2π radians at a frequency of 1000 MHz corresponds to an incremental time delay of one nanosecond (10^{-9} seconds); for a signal bandwidth of 10 MHz, range resolution is about 100 nanoseconds and accuracy may be of the order of 10 nanoseconds (10^{-8} second). Thus the range measurement precision corresponds roughly to 20π radians of the rf signal. Owing to this large uncertainty, initial rf phase is considered a random variable with a probability density function that is uniform in the interval $(0, 2\pi)$. As discussed in Chapter 8, a uniform probability density function assumes the least a priori information about ϕ and corresponds to a least favorable state of nature. By virtue of this assumption the *intuitive substitute* criterion discussed in Chapter 8 is employed with respect to phase.

Without loss in generality, τ_0 can be set equal to zero†; for this case Eq. (9.3-1) becomes

$$s(t) = Aa(t) \cos [\omega_0 t + \theta(t) + \phi] \tag{9.3-2}$$

Equation (9.3-2) can also be written as

$$s(t) = \mathring{s}_I(t) \cos \phi - \mathring{s}_Q(t) \sin \phi \tag{9.3-3}$$

where

$$\mathring{s}_I(t) = Aa(t) \cos [\omega_0 t + \theta(t)] \tag{9.3-4a}$$

$$\mathring{s}_Q(t) = Aa(t) \sin [\omega_0 t + \theta(t)] \tag{9.3-4b}$$

The quantities $\mathring{s}_I(t)$ and $\mathring{s}_Q(t)$ are called the rf quadrature components of narrowband signal $s(t)$.

The generalized likelihood ratio in this case is given by

$$\overline{\ell(\mathbf{v})}_\phi = \frac{\mathfrak{p}\,\overline{p(\mathbf{v} \mid \phi)}_\phi}{\mathfrak{q}\,p(\mathbf{v} \mid 0)} \tag{9.3-5a}$$

$$= \frac{\mathfrak{p}}{\mathfrak{q}} \, E_\phi[\ell'(\mathbf{v} \mid \phi)] \tag{9.3-5b}$$

where the averaging with respect to ϕ assumes a uniform density function. Substituting Eq. (9.3-3) into Eq. (9.2-10) yields the following expression for $\ell'(\mathbf{v} \mid \phi)$:

$$\ell'(\mathbf{v} \mid \phi) = \exp \left\{ -\frac{E}{N_0} + \frac{2}{N_0} \left[\int_0^T v(t)\, \mathring{s}_I(t)\, dt \right] \cos \phi \right.$$

$$\left. - \frac{2}{N_0} \left[\int_0^T v(t)\, \mathring{s}_Q(t)\, dt \right] \sin \phi \right\} \tag{9.3-6}$$

We define

$$\mathring{y}_I(T) = \frac{2}{N_0 \sqrt{\mathscr{R}}} \int_0^T v(t) \mathring{s}_I(t)\, dt \tag{9.3-7a}$$

$$\mathring{y}_Q(T) = \frac{2}{N_0 \sqrt{\mathscr{R}}} \int_0^T v(t) \mathring{s}_Q(t)\, dt \tag{9.3-7b}$$

---

†This corresponds to a test for the presence of signal in the first range bin, as mentioned in Section 9.1.

where $\mathscr{R} = 2E/N_0$. Note that $\mathring{y}_I(T)$ may be interpreted as the output at time T of a filter matched to $(2/N_0\sqrt{\mathscr{R}})\mathring{s}_I(t)$ and $\mathring{y}_Q(T)$ as the output of a filter matched to $(2/N_0\sqrt{\mathscr{R}})\mathring{s}_Q(t)$ at time T. It follows from definitions (9.3-4a), (9.3-4b), (9.3-7a), and (9.3-7b) that $\mathring{y}_I(T)$ and $\mathring{y}_Q(T)$ may also be interpreted as the sampled values at time T of the quadrature rf components of a signal out of a filter matched to waveform $(2/N_0\sqrt{\mathscr{R}})s(t)$. With definitions (9.3-7), Eq. (9.3-6) can be rewritten as

$$\ell'(\mathbf{v} \mid \phi) = e^{-\mathscr{R}/2} \exp\{\sqrt{\mathscr{R}}\,[\mathring{y}_I(T)\cos\phi - \mathring{y}_Q(T)\sin\phi]\} \qquad (9.3\text{-}8)$$

The exponent in Eq. (9.3-8) is proportional to the difference between the projections on the real axis of orthogonal† vectors $\mathring{y}_I$ and $\mathring{y}_Q$, as shown in Fig. 9.3-1. Equation (9.3-8) can also be written as

$$\ell'(\mathbf{v} \mid \phi) = e^{-\mathscr{R}/2} \exp\{\sqrt{\mathscr{R}[\mathring{y}_I^2(T) + \mathring{y}_Q^2(T)]}\cos(\phi + \alpha)\} \qquad (9.3\text{-}9)$$

where $\alpha = \tan^{-1}\mathring{y}_Q(T)/\mathring{y}_I(T)$. With the definition

$$r(T) = \sqrt{\mathring{y}_I^2(T) + \mathring{y}_Q^2(T)} \qquad (9.3\text{-}10)$$

Eq. (9.3-9) becomes

$$\ell'(\mathbf{v} \mid \phi) = e^{-\mathscr{R}/2} \exp\{\sqrt{\mathscr{R}}\,r(T)\cos(\phi + \alpha)\} \qquad (9.3\text{-}11)$$

It follows from previous remarks and definition (9.3-10) that $r(T)$ is the envelope of the signal out of a filter matched to waveform $(2/N_0\sqrt{\mathscr{R}})s(t)$, sampled at time T.

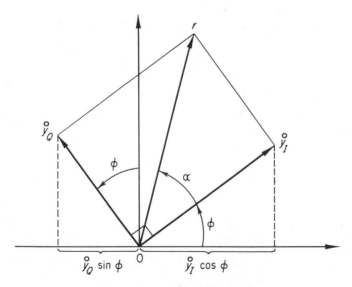

Fig. 9.3-1. Relationship between quadrature components.

†The orthogonal relationship is approximately true for a narrowband signal.

Substituting (9.3-11) into Eq. (9.3-5b) and averaging the result with respect to ϕ† yields

$$\overline{\ell'(\mathbf{v}\,|\,\phi)}_\phi = e^{-\mathscr{R}/2} \int_\phi e^{\sqrt{\mathscr{R}}\,r(T)\cos(\phi+\alpha)}\, p(\phi)\, d\phi \qquad (9.3\text{-}12\text{a})$$

$$= e^{-\mathscr{R}/2}\frac{1}{2\pi} \int_0^{2\pi} e^{\sqrt{\mathscr{R}}\,r(T)\cos(\phi+\alpha)}\, d\phi \qquad (9.3\text{-}12\text{b})$$

Since‡

$$\frac{1}{2\pi}\int_0^{2\pi} e^{a\cos(\theta+\xi)}\, d\theta = I_0(a) \qquad (9.3\text{-}13)$$

where $I_0(x)$ is the modified Bessel function of the first kind and order zero, Eq. (9.3-12b) becomes

$$\overline{\ell'(\mathbf{v}\,|\,\phi)}_\phi = e^{-\mathscr{R}/2}I_0[\sqrt{\mathscr{R}}\,r(T)] \qquad (9.3\text{-}14)$$

The test strategy for a signal of unknown phase is then given by

$$I_0[\sqrt{\mathscr{R}}\,r(T)] \begin{cases} \geq e^{\mathscr{R}/2}\mathscr{K}' & (\textit{signal present}) \\ < e^{\mathscr{R}/2}\mathscr{K}' & (\textit{signal absent}) \end{cases} \qquad (9.3\text{-}15)$$

where $\mathscr{K}'$ is defined in Eq. (9.2-7). The threshold test described in (9.3-15) can be implemented as shown in Fig. 9.3-2.

The receiver implementation shown in Fig. 9.3-2 can be simplified. The function $I_0(x)$ is a monotonic function of x, as shown in Fig. 9.3-3; hence the $I_0(\ )$ computation can be eliminated by properly adjusting the threshold. Then the test statistic is simply sampled envelope function $r(T)$ which describes all of the essential operations on the data v. The quantity r can

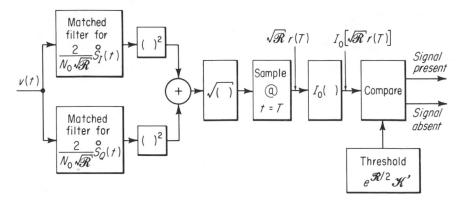

Fig. 9.3-2. Optimum detector for signal of unknown phase.

† $p(\phi) = \dfrac{1}{2\pi}$ in the interval $(0, 2\pi)$.

‡ See reference [2], p. 376.

be obtained by envelope detecting the output of a single matched filter as shown in Fig. 9.3-4 and sampling at time T. For narrowband signals the receiver implementations of Figs. 9.3-2 and 9.3-4 are practically identical with respect to performance.

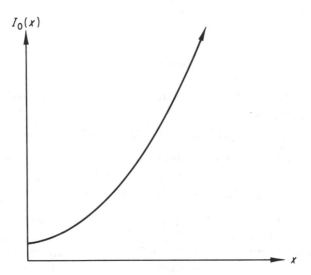

Fig. 9.3-3. Plot of function $I_0(x)$.

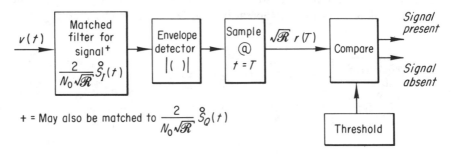

Fig. 9.3-4. Simplified optimum detector for signal of unknown phase.

B. Performance of the optimum receiver

To evaluate the performance of the optimum detector for signals of unknown phase the distribution of test statistic $r(T)$ is derived for hypotheses signal present and signal absent. For this purpose the spectrum of narrowband signal $s(t)$ is assumed to be contained in the interval $(-\omega_c, \omega_c)$, where $\omega_c = 2\pi f_c$. Then we can rewrite Eqs. (9.3-7a) and (9.3-7b) in sampling-theorem notation as [see Eqs. (9.2-13) and (9.2-14)]

$$\mathring{y}_I(T) = \frac{1}{f_c N_0 \sqrt{\mathscr{R}}} \sum_{i=1}^{2f_c T} v_i \mathring{s}_{I-i} \tag{9.3-16a}$$

$$\mathring{y}_Q(T) = \frac{1}{f_c N_0 \sqrt{\mathscr{R}}} \sum_{i=1}^{2f_c T} v_i \mathring{s}_{Q-i} \tag{9.3-16b}$$

As in Section 9.2 both $\mathring{y}_I(T)$ and $\mathring{y}_Q(T)$ are Gaussian random variables when the additive noise process is Gaussian.

For hypothesis H_0(noise alone), the mean and variance of $\mathring{y}_I(T)$ and $\mathring{y}_Q(T)$ can be found as follows:

$$\overline{\mathring{y}_I(T)} = \frac{1}{f_c N_0 \sqrt{\mathscr{R}}} \sum_{i=1}^{2f_c T} \overline{n_i} \mathring{s}_{I-i} = 0 \tag{9.3-17a}$$

$$\overline{\mathring{y}_Q(T)} = \frac{1}{f_c N_0 \sqrt{\mathscr{R}}} \sum_{i=1}^{2f_c T} \overline{n_i} \mathring{s}_{Q-i} = 0 \tag{9.3-17b}$$

assuming a zero mean noise process. Also,

$$\overline{\mathring{y}_I^2(T)} = \left(\frac{1}{f_c N_0 \sqrt{\mathscr{R}}}\right)^2 \sum_{i=1}^{2f_c T} \sum_{j=1}^{2f_c T} \overline{n_i n_j} \mathring{s}_{I-i} \mathring{s}_{I-j} \tag{9.3-18a}$$

$$= \left(\frac{1}{f_c N_0 \sqrt{\mathscr{R}}}\right)^2 \sum_{i=1}^{2f_c T} \overline{n_i^2} \mathring{s}_{I-i}^2 \tag{9.3-18b}$$

$$= \frac{1}{f_c N_0 \mathscr{R}} \sum_{i=1}^{2f_c T} \mathring{s}_{I-i}^2 \cong \frac{2}{N_0 \mathscr{R}} \int_0^T \mathring{s}_I^2(t)\, dt \tag{9.3-18c}$$

$$= \frac{2E}{N_0 \mathscr{R}} = 1 \tag{9.3-18d}$$

Likewise,

$$\overline{\mathring{y}_Q^2(T)} = 1 \tag{9.3-19}$$

since

$$\int_0^T \mathring{s}_I^2(t)\, dt = \int_0^T \mathring{s}_Q^2(t)\, dt = E \tag{9.3-20}$$

The covariance between Gaussian random variables $\mathring{y}_I$ and $\mathring{y}_Q$ is given by

$$\overline{\mathring{y}_I \mathring{y}_Q} = \left(\frac{1}{f_c N_0 \sqrt{\mathscr{R}}}\right)^2 \sum_{i=1}^{2f_c T} \sum_{j=1}^{2f_c T} \overline{n_i n_j}\, \mathring{s}_{I-i} \mathring{s}_{Q-j} \tag{9.3-21a}$$

$$= \left(\frac{1}{f_c N_0 \sqrt{\mathscr{R}}}\right)^2 \sum_{i=1}^{2f_c T} \overline{n_i^2}\, \mathring{s}_{I-i} \mathring{s}_{Q-i} \tag{9.3-21b}$$

$$= \frac{1}{f_c N_0 \mathscr{R}} \sum_{i=1}^{2f_c T} \mathring{s}_{I-i} \mathring{s}_{Q-i} \tag{9.3-21c}$$

$$= \frac{2A^2}{N_0 \mathscr{R}} \int_0^T a^2(t) \sin\left[\omega_0 t + \theta(t)\right] \cos\left[\omega_0 t + \theta(t)\right] dt \tag{9.3-21d}$$

$$\overline{\mathring{y}_I \mathring{y}_Q} = \frac{A^2}{N_0 \mathscr{R}} \int_0^T a^2(t) \sin\left[2\omega_0 t + 2\theta(t)\right] dt \sim \frac{1}{2\omega_0 T} \tag{9.3-21e}$$

Since $\omega_0 \gg 1/2T$ for a narrowband signal, the covariance between $\dot{y}_I$ and $\dot{y}_Q$ is close to zero; hence components $\dot{y}_I$ and $\dot{y}_Q$ are essentially uncorrelated and therefore statistically independent. Then, from Eqs. (9.3-17) through (9.3-19), the joint density function of $\dot{y}_I$ and $\dot{y}_Q$, for hypothesis H_0, is given by

$$p(\dot{y}_I, \dot{y}_Q) = p(\dot{y}_I)p(\dot{y}_Q) = \frac{1}{2\pi} \exp\left(-\frac{\dot{y}_I^2 + \dot{y}_Q^2}{2}\right) \qquad (9.3\text{-}22)$$

It follows from the definitions of r and α that [see Eq. (9.3-10) and Fig. 9.3-1]

$$\dot{y}_I = r \cos \alpha \qquad (9.3\text{-}23a)$$
$$\dot{y}_Q = r \sin \alpha \qquad (9.3\text{-}23b)$$

With transformation (9.3-23) the joint probability density $p(r, \alpha)$ can be derived from $p(\dot{y}_I, \dot{y}_Q)$ in (9.3-22) and is given by†

$$p(r, \alpha) = \frac{r}{2\pi} \exp\left(-\frac{r^2}{2}\right) \qquad (9.3\text{-}24)$$

Marginal probability density function $p(r)$ is obtained by integrating $p(r, \alpha)$ over the domain of random variable α; thus,

$$p(r) = \int_0^{2\pi} p(r, \alpha)\, d\alpha = \int_0^{2\pi} \frac{r}{2\pi} e^{-r^2/2}\, d\alpha \qquad (9.3\text{-}25a)$$

$$p(r) = r \exp\left(-\frac{r^2}{2}\right) \qquad (9.3\text{-}25b)$$

This function is called the Rayleigh density function and is plotted in Fig. 9.3-5.

From the definition of false-alarm probability,

$$P_{fa} = \int_{r_b}^{\infty} r e^{-r^2/2}\, dr \qquad (9.3\text{-}26a)$$

$$P_{fa} = \exp\left(-\frac{r_b^2}{2}\right) \qquad (9.3\text{-}26b)$$

Eq. (9.3-26b) can be solved for the threshold r_b required to obtain a specified false-alarm probability as follows:

$$r_b = \sqrt{2 \ln \frac{1}{P_{fa}}} \qquad (9.3\text{-}27)$$

Consider next the case when signal plus noise is present—that is, when hypothesis H_1 is true. Then, since

$$v(t) = \dot{s}_I(t) \cos \phi - \dot{s}_Q(t) \sin \phi + n(t) \qquad (9.3\text{-}28)$$

it follows from (9.3-16a) and (9.3-16b) that

†See Section 3.10 for a discussion of transformations of random variables.

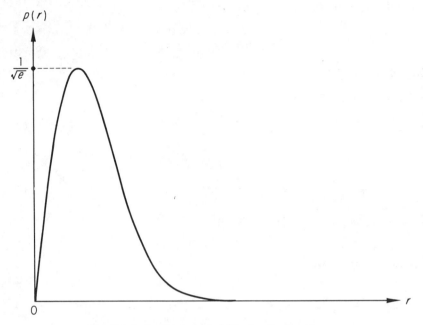

Fig. 9.3-5. Rayleigh probability density function.

$$\mathring{y}_I(T) = \frac{1}{f_c N_0 \sqrt{\mathscr{R}}} \sum_{i=1}^{2f_c T} [(\mathring{s}_{I-i} \cos \phi - \mathring{s}_{Q-i} \sin \phi)\mathring{s}_{I-i} + n_i \mathring{s}_{I-i}]$$

$$= \frac{1}{f_c N_0 \sqrt{\mathscr{R}}} \sum_{i=1}^{2f_c T} (\mathring{s}_{I-i}^2 \cos \phi - \mathring{s}_{Q-i}\mathring{s}_{I-i} \sin \phi + n_i \mathring{s}_{I-i}) \qquad (9.3\text{-}29a)$$

and

$$\mathring{y}_Q(T) = \frac{1}{f_c N_0 \sqrt{\mathscr{R}}} \sum_{i=1}^{2f_c T} [(\mathring{s}_{I-i} \cos \phi - \mathring{s}_{Q-i} \sin \phi)\mathring{s}_{Q-i} + n_i \mathring{s}_{Q-i}]$$

$$= \frac{1}{f_c N_0 \sqrt{\mathscr{R}}} \sum_{i=1}^{2f_c T} (\mathring{s}_{I-i}\mathring{s}_{Q-i} \sin \phi - \mathring{s}_{Q-i}^2 \sin \phi + n_i \mathring{s}_{Q-i}) \qquad (9.3\text{-}29b)$$

From Eqs. (9.3-21c) through (9.3-21e), the second term of the sum in (9.3-29a) and the first term of the sum in (9.3-29b) are approximately zero. With the sampling-theorem relations, Eq. (9.3-20), and the definition of $\mathscr{R}$, Eqs. (9.3-29) can be written more simply as

$$\mathring{y}_I(T) = \sqrt{\mathscr{R}} \cos \phi + \frac{1}{f_c N_0 \sqrt{\mathscr{R}}} \sum_{i=1}^{2f_c T} n_i \mathring{s}_{I-i} \qquad (9.3\text{-}30a)$$

$$\mathring{y}_Q(T) = -\sqrt{\mathscr{R}} \sin \phi + \frac{1}{f_c N_0 \sqrt{\mathscr{R}}} \sum_{i=1}^{2f_c T} n_i \mathring{s}_{Q-i} \qquad (9.3\text{-}30b)$$

It follows from Eqs. (9.3-30) that $\mathring{y}_I$ and $\mathring{y}_Q$ are again Gaussian random variables with mean values

$$\overline{\dot{y}_I(T)} = \sqrt{\mathscr{R}} \cos \phi \tag{9.3-31a}$$

$$\overline{\dot{y}_Q(T)} = -\sqrt{\mathscr{R}} \sin \phi \tag{9.3-31b}$$

In similar manner,

$$\text{var}(\dot{y}_I) = \overline{\dot{y}_I^2(T)} - [\overline{\dot{y}_I(T)}]^2 \tag{9.3-32a}$$

$$= \mathscr{R} \cos^2 \phi + \left(\frac{1}{f_c N_0 \sqrt{\mathscr{R}}}\right)^2 \sum_{i=1}^{2f_c T} \sum_{j=1}^{2f_c T} \overline{n_i n_j} \, \mathring{s}_{I-i} \mathring{s}_{I-j}$$

$$+ \frac{\cos \phi}{f_c N_0} \sum_{i=1}^{2f_c T} \overline{n_i} \, \mathring{s}_{I-i} - \mathscr{R} \cos^2 \phi \tag{9.3-32b}$$

$$= \left(\frac{1}{f_c N_0 \sqrt{\mathscr{R}}}\right)^2 \sum_{i=1}^{2f_c T} \sum_{j=1}^{2f_c T} \overline{n_i n_j} \, \mathring{s}_{I-i} \mathring{s}_{I-j} \tag{9.3-32c}$$

Equation (9.3-32c) reduces to [see Eqs. (9.3-18)]

$$\text{var}(\dot{y}_I) = 1 \tag{9.3-33}$$

It can also be shown that

$$\text{var}(\dot{y}_Q) = 1 \tag{9.3-34}$$

By steps analogous to those in Eqs. (9.3-21a) through (9.3-21e), it can be shown that the covariance between $\dot{y}_I$ and $\dot{y}_Q$ tends to zero for narrowband signals.

The preceding results permit the joint density function of $\dot{y}_I$ and $\dot{y}_Q$, for hypothesis H_1, to be written as

$$p(\dot{y}_I, \dot{y}_Q \,|\, \phi)$$
$$= \frac{1}{2\pi} \exp\left[-\frac{(\dot{y}_I - \sqrt{\mathscr{R}} \cos \phi)^2 + (\dot{y}_Q + \sqrt{\mathscr{R}} \sin \phi)^2}{2}\right] \tag{9.3-35a}$$

$$p(\dot{y}_I, \dot{y}_Q \,|\, \phi)$$
$$= \frac{1}{2\pi} \exp\left[-\frac{\dot{y}_I^2 + \dot{y}_Q^2 + \mathscr{R} - 2\sqrt{\mathscr{R}}(\dot{y}_I \cos \phi - \dot{y}_Q \sin \phi)}{2}\right] \tag{9.3-35b}$$

The probability density function $p(r, \alpha \,|\, \phi)$ can be obtained from (9.3-35b) by means of transformation (9.3-23) and is given by

$$p(r, \alpha \,|\, \phi) = \frac{r}{2\pi} \exp\left[-\frac{r^2 + \mathscr{R} - 2\sqrt{\mathscr{R}} \, r \cos(\phi + \alpha)}{2}\right] \tag{9.3-36}$$

By integrating Eq. (9.3-36) with respect to α, the conditional marginal density function $p(r \,|\, \phi)$ is obtained; thus,

$$p(r \,|\, \phi) = re^{-(r^2 + \mathscr{R})/2} \frac{1}{2\pi} \int_0^{2\pi} e^{r\sqrt{\mathscr{R}} \cos(\phi + \alpha)} \, d\alpha \tag{9.3-37a}$$

$$= re^{-(r^2 + \mathscr{R})/2} I_0(r\sqrt{\mathscr{R}}) \tag{9.3-37b}$$

where Eq. (9.3-13) was used to evaluate the integral in (9.3-37a). Probability density function $p(r)$ is obtained by averaging $p(r \,|\, \phi)$ with respect to

ϕ assuming a uniform probability density; thus

$$p(r) = \int_0^{2\pi} \frac{1}{2\pi} p(r \mid \phi) \, d\phi = re^{-(r^2+\mathcal{R})/2} I_0(r\sqrt{\mathcal{R}}) \qquad (9.3\text{-}38)$$

The probability density function in Eq. (9.3-38) is called the *Rician* or *modified Rayleigh* density function. A plot of the Rician density function is shown in Fig. 9.3-6 for a relatively small and large value of $\mathcal{R}$.

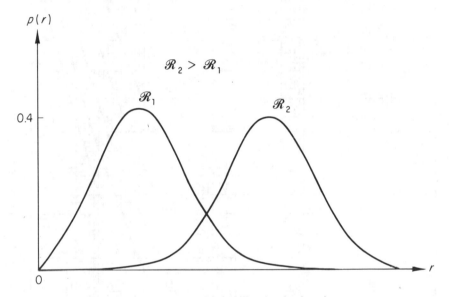

Fig. 9.3-6. Rician probability density function.

The probability of detection may now be computed as in Section 9.2 with density function (9.3-38):

$$P_d = \int_{r_b}^{\infty} re^{-(r^2+\mathcal{R})/2} I_0(r\sqrt{\mathcal{R}}) \, dr \qquad (9.3\text{-}39)$$

Replacing r_b in Eq. (9.3-39) by Eq. (9.3-27) yields

$$P_d = \int_{\sqrt{2\ln(1/P_{fa})}}^{\infty} re^{-(r^2+\mathcal{R})/2} I_0(r\sqrt{\mathcal{R}}) \, dr \qquad (9.3\text{-}40)$$

The integral in Eq. (9.3-39) is called the Q function by Marcum [3]:

$$Q(r_b, \sqrt{\mathcal{R}}) = \int_{r_b}^{\infty} re^{-(r^2+\mathcal{R})/2} I_0(r\sqrt{\mathcal{R}}) \, dr \qquad (9.3\text{-}41)$$

This function is related to a tabulated function called the incomplete Toronto function by

$$Q(r_b, \sqrt{\mathcal{R}}) = 1 - T_{r_b/\sqrt{2}}\left(1, 0, \sqrt{\frac{\mathcal{R}}{2}}\right) \qquad (9.3\text{-}42)$$

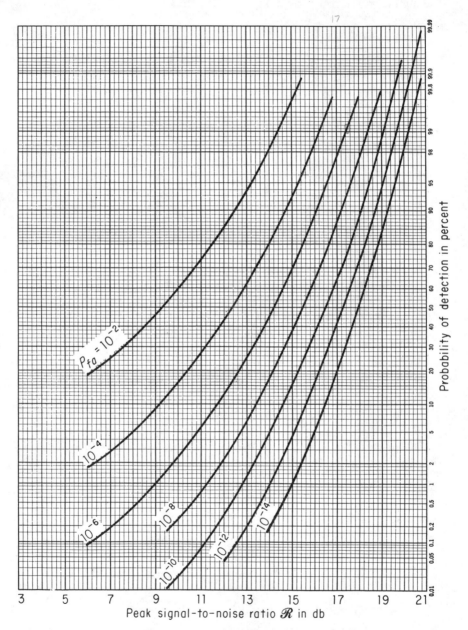

Fig. 9.3-7. Detection characteristic for signal of unknown phase.

Hence the probability of detection can also be expressed as

$$P_d = 1 - T_{\sqrt{\ln(1/P_{fa})}}\left(1, 0, \sqrt{\frac{\mathscr{R}}{2}}\right)$$

(9.3-43)

From plots of the Q function and incomplete Toronto function in Marcum [3], a set of detection curves for signals of unknown phase, similar to those in Fig. 9.2-3, is plotted in Fig. 9.3-7. (See discussion at end of Section 9.2 concerning units of $\mathscr{R}$.) A study of Fig. 9.3-7 and Fig. 9.2-3 reveals that an increase in signal-to-noise ratio $\mathscr{R}$ of approximately 0.8 db is required for a signal of unknown phase compared to a signal of known phase at a P_d of 50 per cent and a P_{fa} of approximately 10^{-6}. The difference in $\mathscr{R}$ decreases for smaller values of P_{fa}. For a (larger) P_{fa} of 10^{-2} the loss is of the order of 1.8 db at 50 per cent P_d. As $\mathscr{R}$ increases, the gap between corresponding curves diminishes and the performance of the envelope detector approaches the performance of the optimum receiver for an exactly known signal.

9.4. Optimum Detection of a Rayleigh Fluctuating Signal

A. Structure of the optimum receiver

A signal of unknown amplitude and phase occurs frequently in practice. As in the previous case, the strategy for unknown phase is to average the likelihood ratio employing a least favorable a priori distribution. With respect to amplitude, realistic statistical models are available. For example, echoes from targets that consist of an aggregate of many random scatterers follow a Rayleigh probability distribution, as discussed in Chapter 1. Radar echoes from many types of aircraft satisfy this distribution. When a Rayleigh a priori distribution is assumed for amplitude and a uniform distribution is assumed for phase, a generalized likelihood-ratio expression for the optimum test strategy, as discussed in Chapter 8 and Section 9.1, yields a comparison test of the envelope r defined in Eq. (9.3-10) with a threshold (see reference [4] for a derivation of this result). This test is the same as that for the previous case in which amplitude was assumed known. Hence the structure of the optimum receiver is the same as those in Figs. 9.3-2 and 9.3-4.

This result can be generalized. It will now be shown that the structure of the optimum receiver is independent of the form of the amplitude probability density function. To show this, let

$$\mathring{s}_I(t) = A\Sigma_I(t)$$

(9.4-1a)

$$\mathring{s}_Q(t) = A\Sigma_Q(t)$$

(9.4-1b)

and let $\mathscr{E}$ be defined by

$$\mathscr{E} = \int_0^T \Sigma_I^2(t)\, dt = \int_0^T \Sigma_Q^2(t)\, dt$$

(9.4-2)

It follows from (9.4-1) and (9.3-20) that

$$E = A^2 \mathscr{E} \qquad (9.4\text{-}3)$$

It is convenient to normalize $\mathscr{E}$ (and therefore Σ_I and Σ_Q) as follows:

$$\frac{2\mathscr{E}}{N_0} = 1 \qquad (9.4\text{-}4)$$

To evaluate $\overline{\ell'(\mathbf{v} \mid \phi, A)_{\phi, A}}$, Eq. (9.3-14) is a suitable starting point; with the preceding definitions Eq. (9.3-14) becomes for this case

$$\overline{\ell'(\mathbf{v} \mid \phi, A)_\phi} = e^{-A^2/2} I_0[Ar(T)] \qquad (9.4\text{-}5)$$

and

$$\overline{\ell'(\mathbf{v} \mid \phi, A)_{\phi, A}} = \int_A e^{-A^2/2} I_0(Ar) p(A)\, dA = f(r) \qquad (9.4\text{-}6)$$

Note that $f(r)$ in (9.4-6) is a monotonic function of r; this result is independent of the sign and distribution of A since $I_0(x)$ is an even function of its argument and $p(A)$ is always positive. Therefore sampled envelope r is a sufficient statistic; this completes the proof.

Choosing a value for the probability of false alarm fixes the value of threshold r_b [see (9.3-27)]. The set of points in Γ'' for which test statistic r exceeds r_b, when signal plus noise is present, is independent of the signal-amplitude statistics. Hence, as discussed in Section 8.14, this corresponds to a *uniformly most powerful* test with respect to amplitude. This general result applies to the discussion in the next section also, where a different statistical model is considered for signal amplitude A.

B. Performance of optimum receiver

The optimum detector for a Rayleigh fluctuating signal of unknown phase is an envelope detector as shown in Figs. 9.3-2 and 9.3-4. An expression for false-alarm probability has already been established for this detector in Section 9.3; only the probability of detection remains to be calculated. The probability of detection in this case is defined as the averaged detection probability, with respect to amplitude statistics. A direct method for obtaining this result is to average the detection probability for a signal of unknown phase [see Eq. (9.3-39)] with respect to the Rayleigh density function. The Rayleigh density function is given by

$$p(A) = \frac{A}{A_0^2} \exp\left(-\frac{A^2}{2A_0^2}\right), \qquad A \geqslant 0 \qquad (9.4\text{-}7)$$

where A_0 is the peak of the density function, or most probable value of A. Note that from Eqs. (9.4-3) and (9.4-4)

$$\mathscr{R} = \frac{2E}{N_0} = A^2 \qquad (9.4\text{-}8)$$

The probability of detection is obtained by substituting (9.4-8) into (9.3-39) and averaging the result with respect to (9.4-7) as follows:

$$P_d = \int_0^\infty \int_{r_b}^\infty \frac{A}{A_0^2} e^{-A^2/2A_0^2} r e^{-(r^2+A^2)/2} I_0(Ar) \, dr \, dA \qquad (9.4\text{-}9)$$

which can also be written in the form

$$P_d = \int_{r_b}^\infty r e^{-r^2/2} \, dr \int_0^\infty \frac{A}{A_0^2} e^{-(1+A_0^2)A^2/2A_0^2} I_0(Ar) \, dA \qquad (9.4\text{-}10)$$

The integral in (9.4-10) with respect to A is a special case of the following tabulated integral (see reference [5]):

$$\int_0^\infty t^{\mu-1} I_\nu(\alpha t) e^{-p^2 t^2} \, dt$$

$$= \frac{\Gamma\left(\dfrac{\mu+\nu}{2}\right)\left(\dfrac{\alpha}{2p}\right)^\nu}{2p^\mu \Gamma(\nu+1)} e^{\alpha^2/4p^2} {}_1F_1\left(\frac{\nu-\mu}{2}+1, \nu+1; -\frac{\alpha^2}{4p^2}\right) \qquad (9.4\text{-}11)$$

where $\Gamma(x)$ is the gamma function and ${}_1F_1(a, c; z)$ is the confluent hypergeometric function. Upon substituting $t = A$, $\mu = 2$, $\nu = 0$, $\alpha = r$, and $p = \sqrt{A_0^2 + 1/2A_0^2}$ into Eq. (9.4-11), Eq. (9.4-10) becomes

$$P_d = \int_{r_b}^\infty r e^{-r^2/2} \left[\frac{1}{1+A_0^2} e^{A_0^2 r^2/2(1+A_0^2)} {}_1F_1\left(0, 1; -\frac{A_0^2 r^2}{2(1+A_0^2)}\right)\right] dr \qquad (9.4\text{-}12)$$

The power series expansion of the confluent hypergeometric function is given by

$${}_1F_1(a, c; z) = 1 + \frac{az}{c} + \frac{a(a+1)z^2}{2!c(c+1)} + \cdots \qquad (9.4\text{-}13)$$

From (9.4-13), ${}_1F_1(0, 1; z) = 1$ so that (9.4-12) simplifies to

$$P_d = \frac{1}{1+A_0^2} \int_{r_b}^\infty r \exp\left[-\frac{r^2}{2(1+A_0^2)}\right] dr \qquad (9.4\text{-}14\text{a})$$

$$P_d = \exp\left[-\frac{r_b^2}{2(1+A_0^2)}\right] \qquad (9.4\text{-}14\text{b})$$

Replacing threshold r_b by Eq. (9.3-27) yields

$$P_d = \exp\left(\frac{\ln P_{fa}}{1+A_0^2}\right) \qquad (9.4\text{-}15\text{a})$$

$$P_d = (P_{fa})^{1/(1+A_0^2)} \qquad (9.4\text{-}15\text{b})$$

It follows from (9.4-8) that

$$\overline{\mathscr{R}} = \overline{A^2} = \int_0^\infty \frac{A^3}{A_0^2} \exp\left(-\frac{A^2}{2A_0^2}\right) dA \qquad (9.4\text{-}16\text{a})$$

$$\overline{\mathscr{R}} = 2A_0^2 \qquad (9.4\text{-}16\text{b})$$

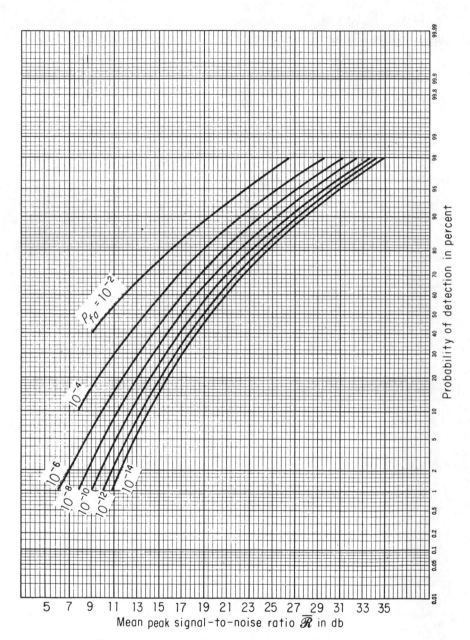

Fig. 9.4-1. Optimum detection characteristic for a Rayleigh fluctuating target (unknown phase).

Using Eq. (9.4-16b), Eq. (9.4-15b) can be expressed in terms of (average) signal-to-noise ratio $\bar{\mathscr{R}}$ as follows:

$$P_d = (P_{fa})^{1/(1+\bar{\mathscr{R}}/2)} \qquad (9.4\text{-}17)$$

or

$$P_{fa} = (P_d)^{1+\bar{\mathscr{R}}/2} \qquad (9.4\text{-}18)$$

This relation, which predicts radar detection performance for a Rayleigh fluctuating target with unknown initial phase, is plotted in Fig. 9.4-1 (see discussion at end of Section 9.2 concerning units of $\mathscr{R}$).

9.5. Optimum Detection of a One-Dominant-Plus-Rayleigh Fluctuating Signal; Structure and Performance

For a signal of unknown amplitude and phase that is characterized by a one-dominant-plus-Rayleigh density function with respect to amplitude, the optimum receiver performs, as above, an envelope comparison test (since it is uniformly most powerful with respect to amplitude). We need only calculate the probability of detection with respect to this new distribution.

The one-dominant-plus-Rayleigh density function is given by

$$p(A) = \frac{9A^3}{2A_0^4} \exp\left(-\frac{3A^2}{2A_0^2}\right), \qquad A \geqslant 0 \qquad (9.5\text{-}1)$$

This density function describes the return from a target consisting of one large dominant reflector plus many small random scatterers. Substituting (9.4-8) into (9.3-39) and averaging the result with respect to amplitude A using (9.5-1) yields

$$P_d = \int_0^\infty \int_{r_b}^\infty \frac{9A^3}{2A_0^4} e^{-3A^2/2A_0^2} r e^{-(r^2+A^2)/2} I_0(Ar)\, dr\, dA \qquad (9.5\text{-}2)$$

Equation (9.5-2) can be rewritten as

$$P_d = \int_{r_b}^\infty r e^{-r^2/2}\, dr \int_0^\infty \frac{9A^3}{2A_0^4} \exp\left[-\frac{(A_0^2+3)A^2}{2A_0^2}\right] I_0(Ar)\, dA \qquad (9.5\text{-}3)$$

The integral with respect to A can be evaluated using Eq. (9.4-11) with $\mu = 4$, $t = A$, $\nu = 0$, $\alpha = r$, and $p = \sqrt{(A_0^2+3)/2A_0^2}$, to yield

$$P_d = \frac{9}{(A_0^2+3)^2} \int_{r_b}^\infty r\left(1 + \frac{A_0^2 r^2}{2(3+A_0^2)}\right) e^{-3r^2/2(3+A_0^2)}\, dr \qquad (9.5\text{-}4)$$

To evaluate the integral in (9.5-4) let

$$v = \frac{\sqrt{3}\, r}{\sqrt{3 + A_0^2}} \qquad (9.5\text{-}5)$$

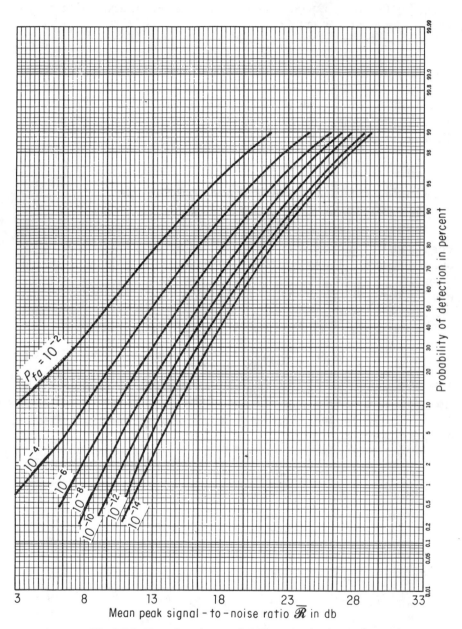

Fig. 9.5-1. Detection characteristic for a one-dominant-plus-Rayleigh fluctuating target.

Then (9.5-4) can be rewritten as

$$P_d = \frac{3}{3 + A_0^2} \int_{\sqrt{3}\, r_b / \sqrt{3 + A_0^2}}^{\infty} v \left(1 + \frac{A_0^2 v^2}{6} \right) e^{-v^2/2} \, dv \tag{9.5-6a}$$

$$P_d = \exp\left[-\frac{3 r_b^2}{2(3 + A_0^2)} \right] \left(\frac{3}{3 + A_0^2} \right) \left[1 + \frac{A_0^2}{3} + \frac{r_b^2}{2\left(1 + \frac{3}{A_0^2}\right)} \right] \tag{9.5-6b}$$

With respect to a one-dominant-plus-Rayleigh probability density function, average signal-to-noise ratio $\overline{\mathscr{R}}$ is calculated as

$$\overline{\mathscr{R}} = \int_0^\infty \frac{9 A^5}{2 A_0^4} \exp\left(-\frac{3 A^2}{2 A_0^2} \right) dA \tag{9.5-7a}$$

$$\overline{\mathscr{R}} = \tfrac{4}{3} A_0^2 \tag{9.5-7b}$$

Substituting (9.5-7b) into Eq. (9.5-6b) yields

$$P_d = \frac{1}{1 + \frac{\overline{\mathscr{R}}}{4}} \left(1 + \frac{\overline{\mathscr{R}}}{4} - \frac{\ln P_{fa}}{1 + \frac{4}{\overline{\mathscr{R}}}} \right) \exp\left(\frac{\ln P_{fa}}{1 + \frac{\overline{\mathscr{R}}}{4}} \right) \tag{9.5-8a}$$

$$P_d = \frac{1}{1 + \frac{4}{\overline{\mathscr{R}}}} \left(1 + \frac{4}{\overline{\mathscr{R}}} - \frac{\ln P_{fa}}{1 + \frac{\overline{\mathscr{R}}}{4}} \right) \exp\left(\frac{\ln P_{fa}}{1 + \frac{\overline{\mathscr{R}}}{4}} \right) \tag{9.5-8b}$$

Equation (9.5-8b) is plotted in Fig. 9.5-1 (see discussion at end of Section 9.2 concerning units of $\mathscr{R}$). With these curves radar performance of the optimum detection system for a *one-dominant-plus-Rayleigh* target of unknown phase can be predicted.

9.6. Analytic Approximations

In practical applications it is frequently desirable to be able to relate P_{fa}, P_d, and peak signal-to-noise ratio $\mathscr{R}$ by approximate expressions that yield quick estimates of system performance parameters. In this section some exact and approximate expressions are derived that permit rapid evaluation using two reference graphs.

A. Exactly known signal

To simplify notation it is convenient to introduce some new definitions. Let

$$b = \frac{1}{\sqrt{2\pi}} \int_T^\infty e^{-t^2/2} \, dt \tag{9.6-1}$$

Denoting the integral in (9.6-1) by $\Phi(T)$, we can rewrite (9.6-1) as

$$b = \Phi(T) \tag{9.6-2}$$

and its inverse as

$$\Phi^{-1}(b) = T \tag{9.6-3}$$

With these definitions the expression for false-alarm probability given by Eqs. (9.2-23) and (9.2-24) can be rewritten as follows:

$$y_b = \sqrt{\mathscr{R}} \, \Phi^{-1}(P_{fa}) \tag{9.6-4}$$

In similar manner Eqs. (9.2-28b) and (9.2-29) can be solved for threshold y_b to yield

$$y_b = \mathscr{R} + \sqrt{\mathscr{R}} \, \Phi^{-1}(P_d) \tag{9.6-5}$$

These equations can be solved for $\mathscr{R}$ to give

$$\mathscr{R} = [\Phi^{-1}(P_{fa}) - \Phi^{-1}(P_d)]^2 \tag{9.6-6}$$

which is an exact expression. Functions $\Phi^{-1}(P_{fa})$ and $\Phi^{-1}(P_d)$ are plotted in Figs. 9.6-1 and 9.6-2. With these curves and Eq. (9.6-6) a rapid evaluation of $\mathscr{R}$ can be made for selected values of P_{fa} and P_d.

B. Signal of unknown phase

It has been shown by Rice [10] and Marcum [3] that expression (9.3-39) for P_d can be expanded in an asymptotic series as follows:

$$P_d = \frac{1}{2} - \frac{1}{2}\left[\frac{1}{\sqrt{2\pi}}\int_{-\beta+\alpha}^{\beta-\alpha} e^{-t^2/2}\,dt\right]$$
$$+ \frac{1}{2\alpha\sqrt{2\pi}}e^{-(\beta-\alpha)^2/2}\left[1 - \frac{\beta-\alpha}{4\alpha} + \frac{1+(\beta-\alpha)^2}{8\alpha^2} + \cdots\right] \tag{9.6-7}$$

where

$$\alpha = \sqrt{\mathscr{R}} \tag{9.6-8}$$

and [see Eq. (9.3-27)]

$$\beta = \sqrt{2\ln\frac{1}{P_{fa}}} \tag{9.6-9}$$

A first-order approximation of P_d is obtained by retaining only the first two terms in Eq. (9.6-7):

$$P_d \cong \frac{1}{2} - \frac{1}{2}\left[\frac{1}{\sqrt{2\pi}}\int_{-\beta+\alpha}^{\beta-\alpha} e^{-t^2/2}\,dt\right] \tag{9.6-10a}$$

$$P_d \cong \frac{1}{\sqrt{2\pi}}\int_{\beta-\alpha}^{\infty} e^{-t^2/2}\,dt \tag{9.6-10b}$$

From Eqs. (9.6-2), (9.6-8), and (9.6-9), Eq. (9.6-10b) can be rewritten as

$$P_d \cong \Phi\left(\sqrt{2\ln\frac{1}{P_{fa}}} - \sqrt{\mathscr{R}}\right) \tag{9.6-11}$$

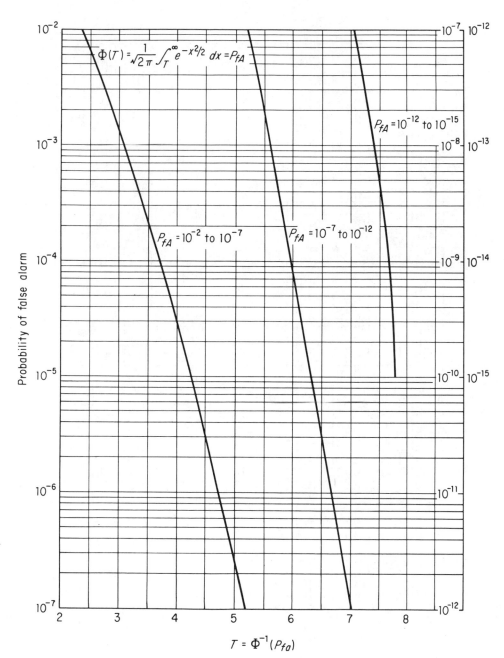

Fig. 9.6-1. P_{fa} vs. $\Phi^{-1}(P_{fa})$.

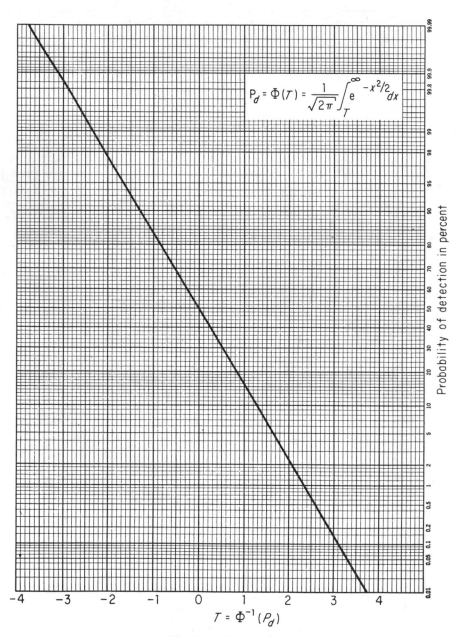

$$P_d = \Phi(T) = \frac{1}{\sqrt{2\pi}} \int_T^\infty e^{-x^2/2} dx$$

Probability of detection in percent

$$T = \Phi^{-1}(P_d)$$

Fig. 9.6-2. P_d vs. $\Phi^{-1}(P_d)$.

Solving for $\mathscr{R}$ results in

$$\mathscr{R} \cong \left[\sqrt{2 \ln \frac{1}{P_{fa}}} - \Phi^{-1}(P_d) \right]^2 \qquad (9.6\text{-}12)$$

With Eq. (9.6-12) and Fig. 9.6-2 peak signal-to-noise ratio $\mathscr{R}$ can be estimated for any given P_d and P_{fa}. Approximation (9.6-12) is most accurate for large $\mathscr{R}$ and small P_{fa}.

C. Fluctuating signal

For a Rayleigh signal no approximation is required, since exact solution (9.4-17) or (9.4-18) can be solved for $\bar{\mathscr{R}}$ to yield

$$\bar{\mathscr{R}} = 2 \left(\frac{\ln P_{fa}}{\ln P_d} - 1 \right) \qquad (9.6\text{-}13)$$

Equation (9.6-13) permits a rapid estimate of average signal-to-noise ratio $\bar{\mathscr{R}}$ for given P_{fa} and P_d.

The one-dominant-plus-Rayleigh model results cannot be significantly simplified. For this case the value of $\bar{\mathscr{R}}$ can be computed from exact expression (9.5-8b) when P_{fa} and P_d are given.

9.7. Likelihood Ratio for Colored Noise

The single-hit detection problem is now treated employing the Karhunen-Loéve representation for both signal and noise waveforms. Signal $s(t)$ is assumed to be imbedded in a zero-mean narrowband stationary Gaussian noise process $\{n(t)\}$ with arbitrary spectral density (colored noise). For convenience, narrowband waveforms are specified in complex notation (see Section 2.5) as follows:

$$s(t) = \text{Re}\,[\tilde{s}(t)e^{j\omega_0 t}], \qquad 0 \leqslant t \leqslant T \qquad (9.7\text{-}1)$$

$$n(t) = \text{Re}\,[\tilde{n}(t)e^{j\omega_0 t}], \qquad 0 \leqslant t \leqslant T \qquad (9.7\text{-}2)$$

$$v(t) = \text{Re}\,[\tilde{v}(t)e^{j\omega_0 t}], \qquad 0 \leqslant t \leqslant T \qquad (9.7\text{-}3)$$

where $\tilde{s}(t)$, $\tilde{n}(t)$, and $\tilde{v}(t)$ are respectively the complex envelopes of input signal, noise, and the received waveform, and where ω_0 is a known carrier frequency. With these definitions hypothesis H_0 (noise alone) corresponds to

$$\tilde{v}(t) = \tilde{n}(t), \qquad 0 \leqslant t \leqslant T \qquad (9.7\text{-}4)$$

and hypothesis H_1 (signal plus noise) corresponds to

$$\tilde{v}(t) = \tilde{s}(t) + \tilde{n}(t), \qquad 0 \leqslant t \leqslant T \qquad (9.7\text{-}5)$$

In order that complex waveforms $\tilde{v}(t)$, $\tilde{s}(t)$, and $\tilde{n}(t)$ may be represented by Karhunen-Loéve expansions, the results of Section 4.10 must be extended to complex processes. Analogous to (9.7-1) through (9.7-3) define the complex envelope $\tilde{\phi}_{nn}(\tau)$ of autocorrelation function $\phi_{nn}(\tau)$ of a narrow-band process $\{n(t)\}$ by

$$\phi_{nn}(\tau) = \text{Re}\left[\tilde{\phi}_{nn}(\tau)e^{j\omega_0\tau}\right] \tag{9.7-6}$$

The complex autocorrelation function $\tilde{\phi}_{nn}(\tau)$ can be related to complex envelope $\tilde{n}(t)$ as follows. From the definition of $\phi_{nn}(\tau)$ and Eq. (9.7-2),

$$\phi_{nn}(\tau) = \overline{n(t+\tau)n(t)} \tag{9.7-7a}$$

$$\phi_{nn}(\tau) = \overline{\{\text{Re}\left[\tilde{n}(t+\tau)e^{j\omega_0(t+\tau)}\right]\}\left\{\dfrac{\tilde{n}(t)e^{j\omega_0 t} + \tilde{n}*e^{-j\omega_0 t}}{2}\right\}} \tag{9.7-7b}$$

$$\phi_{nn}(\tau) = \text{Re}\left[\dfrac{\overline{\tilde{n}(t+\tau)\tilde{n}(t)}e^{j\omega_0(2t+\tau)}}{2}\right] + \text{Re}\left[\dfrac{\overline{\tilde{n}(t+\tau)\tilde{n}*(t)}e^{j\omega_0\tau}}{2}\right] \tag{9.7-7c}$$

For a stationary process the first term in Eq. (9.7-7c) is zero since it can be shown that [6]

$$\overline{\tilde{n}(t+\tau)\tilde{n}(t)} = 0 \tag{9.7-8}$$

Equation (9.7-7c) then simplifies to

$$\phi_{nn}(\tau) = \text{Re}\left[\dfrac{\overline{\tilde{n}(t+\tau)\tilde{n}*(t)}}{2}e^{j\omega_0\tau}\right] \tag{9.7-9}$$

A comparison of Eq. (9.7-9) with (9.7-6) shows that

$$\tilde{\phi}_{nn}(t-s) = \dfrac{\overline{\tilde{n}(t)\tilde{n}*(s)}}{2} \tag{9.7-10}$$

The complex envelope of noise process $\{n(t)\}$ can be expanded as

$$\tilde{n}(t) = n_I(t) + jn_Q(t) \tag{9.7-11}$$

where $n_I(t)$ and $n_Q(t)$ are the quadrature components of narrowband waveform $n(t)$. To demonstrate this result, note that

$$n(t) = \text{Re}\left[\tilde{n}(t)e^{j\omega_0 t}\right] = n_I(t)\cos\omega_0 t - n_Q(t)\sin\omega_0 t \tag{9.7-12}$$

Among the approximate relations developed in Section 4.11 for a narrow-band Gaussian noise process, it was shown that $n_I(t)$ and $n_Q(t)$ are each Gaussian, have zero means, and are uncorrelated. For a complex noise process that is analytic (see Section 2.5), the previous statements can be shown to be exact [6]—that is,

$$\overline{n_I(t)} = \overline{n_Q(t)} = 0 \tag{9.7-13a}$$

$$\overline{n_I(t)n_Q(t)} = 0 \tag{9.7-13b}$$

From Eqs. (9.7-11) and (9.7-13) it can be concluded that $\tilde{n}(t)$ is Gaussian and that

$$\overline{\tilde{n}(t)} = 0 \tag{9.7-14}$$

Consider next integral equation (4.10-1). With $\phi_{nn}(\tau)$ replaced by $\tilde{\phi}_{nn}(\tau)$, the equation becomes

$$\int_0^T \tilde{\phi}_{nn}(t - s)\psi_k(s)\, ds = \lambda_k \psi_k(t), \qquad 0 \leqslant t, s \leqslant T \qquad (9.7\text{-}15)$$

where the $\psi_k(t)$ are in general complex functions. Note that $\tilde{\phi}_{nn}(t - s)$ is Hermitian symmetric from Eq. (9.7-10) [see Eq. (2.7-6)]. This is sufficient to insure that the eigenfunction solutions $\psi_k(t)$ of Eq. (9.7-15) are orthogonal —that is,

$$\int_0^T \psi_k^*(t)\psi_j(t)\, dt = \delta_{kj} = \begin{cases} 1, & k = j \\ 0, & k \neq j \end{cases} \qquad (9.7\text{-}16)$$

It can also be shown that $\tilde{\phi}_{nn}(\tau)$ is positive definite; this condition and Hermitian symmetry are sufficient to assure the completeness of the set of eigenfunctions $\psi_k(t)$ and that eigenvalues λ_k are both real and positive.

By virtue of the *completeness* of the eigenfunction set, every member function of the complex process $\{\tilde{v}(t)\}$ can be expanded over the interval $(0, T)$ in a Karhunen-Loéve representation, as follows:

$$\tilde{v}(t) = \sum_{k=1}^{\infty} v_k \psi_k(t), \qquad 0 \leqslant t \leqslant T \qquad (9.7\text{-}17)$$

where v_k are complex coefficients given by

$$v_k = \int_0^T \tilde{v}(t)\psi_k^*(t)\, dt \qquad (9.7\text{-}18)$$

In Eq. (9.7-17) convergence of the series to the function $\tilde{v}(t)$ has been assumed in the sense of limit-in-the-mean—that is,

$$\tilde{v}(t) = \underset{N \to \infty}{\text{l.i.m.}} \sum_{k=1}^{N} v_k \psi_k(t) \qquad (9.7\text{-}19)$$

Consider first hypothesis H_0 (signal absent) where $\tilde{v}(t) = \tilde{n}(t)$. It follows from Eq. (9.7-14) that

$$\overline{v_k} = \int_0^T \overline{\tilde{v}(t)}\psi_k^*(t)\, dt = 0 \qquad (9.7\text{-}20)$$

With (9.7-18), (9.7-10), (9.7-15), and (9.7-16) we find that

$$\overline{v_k v_j^*} = \int_0^T \int_0^T \overline{\tilde{v}(t)\tilde{v}^*(s)}\psi_k^*(t)\psi_j(s)\, dt\, ds \qquad (9.7\text{-}21\text{a})$$

$$= 2\int_0^T \int_0^T \tilde{\phi}_{nn}(t - s)\psi_j(s)\psi_k^*(t)\, ds\, dt \qquad (9.7\text{-}21\text{b})$$

$$= 2\lambda_j \int_0^T \psi_j(t)\psi_k^*(t)\, dt \qquad (9.7\text{-}21\text{c})$$

$$= 2\lambda_j \delta_{kj} \qquad (9.7\text{-}21\text{d})$$

Also, from Eq. (9.7-8) it can be shown that

$$\overline{v_k v_j} = 0, \qquad \text{all } k, j \qquad (9.7\text{-}22)$$

If v_k is expanded as

$$v_k = x_k + jy_k \qquad (9.7\text{-}23)$$

where x_k and y_k are respectively the real and imaginary parts of v_k, it follows from Eq. (9.7-18) that x_k and y_k are each Gaussian, since $\tilde{v}(t)$ is a sample function of a complex Gaussian process, and from Eq. (9.7-20) that

$$\overline{x_k} = \overline{y_k} = 0 \tag{9.7-24}$$

and from (9.7-23) and (9.7-21d) that

$$\overline{|v_k|^2} = \overline{x_k^2} + \overline{y_k^2} = 2\lambda_k \tag{9.7-25}$$

From Eqs. (9.7-22) and (9.7-21d)

$$\overline{x_k y_k} = 0 \tag{9.7-26}$$

and

$$\overline{x_k^2} = \overline{y_k^2} = \lambda_k \tag{9.7-27}$$

Thus, the real and imaginary parts of coefficients v_k are uncorrelated Gaussian random variables with variances λ_k and zero means. It can also be shown, from Eqs. (9.7-22) and (9.7-21d), that $\overline{x_k x_j} = 0$ and $\overline{x_k y_j} = 0$ for all $k \neq j$.

Consider next the case when signal is present. For hypothesis H_1, complex envelope $\tilde{v}(t) = \tilde{s}(t)e^{j\phi} + \tilde{n}(t)$, where ϕ is the initial phase of narrowband signal $s(t)$. In this case, a simple computation shows that

$$\overline{v_k} = s_k e^{j\phi} \tag{9.7-28}$$

where

$$s_k = \int_0^T \tilde{s}(t)\psi_k^*(t)\,dt \tag{9.7-29}$$

and

$$\tilde{s}(t) = \operatorname*{l.i.m.}_{N\to\infty} \sum_{k=1}^{N} s_k\psi_k(t) \tag{9.7-30}$$

is the Karhunen-Loéve expansion of complex envelope $\tilde{s}(t)$. For an exactly known signal, initial phase ϕ is set equal to zero for convenience. However, for a signal of unknown phase, ϕ is a random variable.

It can be demonstrated that

$$\overline{(v_k - \bar{v}_k)(v_j - \bar{v}_j)} = 0, \qquad \text{all } k, j \tag{9.7-31}$$

$$\overline{(v_k - \bar{v}_k)(v_j^* - \bar{v}_j^*)} = 2\lambda_j\delta_{kj} \tag{9.7-32}$$

Further, when Eq. (9.7-23) is substituted into (9.7-28), we find that

$$\overline{x_k} = \operatorname{Re}(s_k e^{j\phi}) \tag{9.7-33a}$$

$$\overline{y_k} = \operatorname{Im}(s_k e^{j\phi}) \tag{9.7-33b}$$

It follows from Eqs. (9.7-31) and (9.7-32) that x_k and y_k are uncorrelated Gaussian random variables with variances λ_k and means given by Eqs. (9.7-33). In addition x_k and x_j are uncorrelated for all $k \neq j$ as are also x_k and y_j.

With the above results for hypotheses H_0 and H_1, the likelihood ratio can be formulated. Since the variances of coefficients x_k and y_k are real, positive, and equal [see Eq. (9.7-27)], it is convenient to order them with the largest first—that is,

$$\lambda_1 > \lambda_2 > \lambda_3 > \ldots \lambda_N > \ldots \qquad (9.7\text{-}34)$$

To find the joint density function for the first N complex coefficients v_k under hypothesis H_0, note that

$$p(\mathbf{x}_N, \mathbf{y}_N \mid 0) = \frac{1}{(2\pi)^N \prod_{k=1}^{N} \lambda_k} \exp\left(-\sum_{k=1}^{N} \frac{x_k^2 + y_k^2}{2\lambda_k}\right) \qquad (9.7\text{-}35)$$

Thus,

$$p(\mathbf{v}_N \mid 0) = \frac{1}{(2\pi)^N \prod_{k=1}^{N} \lambda_k} \exp\left(-\sum_{k=1}^{N} \frac{|v_k|^2}{2\lambda_k}\right) \qquad (9.7\text{-}36)$$

where $\mathbf{v}_N$ is an N-dimensional approximation to the infinite-dimensioned vector $\mathbf{v}$. For hypothesis H_1 the corresponding result is

$$p(\mathbf{v}_N \mid \mathbf{s}_N, \phi) = \frac{1}{(2\pi)^N \prod_{k=1}^{N} \lambda_k} \exp\left(-\sum_{k=1}^{N} \frac{|v_k - s_k e^{j\phi}|^2}{2\lambda_k}\right) \qquad (9.7\text{-}37)$$

The likelihood ratio can now be computed as

$$\ell(\mathbf{v}_N \mid \mathbf{s}_N, \phi) = \frac{p}{q} \exp\left[-\sum_{k=1}^{N} \frac{|s_k|^2 - 2\,\mathrm{Re}\,(v_k s_k^* e^{-j\phi})}{2\lambda_k}\right] \qquad (9.7\text{-}38)$$

Next, let N approach infinity. This is permissible when

$$\sum_{k=1}^{\infty} \frac{|s_k|^2}{\lambda_k} < \infty$$

which is the regular case discussed by Kelly *et al.* [6]; otherwise the singular case, as discussed by Grenander [7], results. It is shown in [6] that if the signal and noise can be considered the result of passing white noise and a signal of finite energy through a filter, the detection problem is regular. This will usually be true in practical problems.

As in the earlier discussion of Section 9.2 the modified likelihood ratio is actually used; this function is given by (for $N \to \infty$)

$$\ell'(\mathbf{v} \mid \mathbf{s}, \phi) = \exp\left[-\sum_{k=1}^{\infty} \frac{|s_k|^2 - 2\,\mathrm{Re}\,(v_k s_k^* e^{-j\phi})}{2\lambda_k}\right] \qquad (9.7\text{-}39)$$

In following sections, Eq. (9.7-39) is applied to: an exactly known signal, a signal of unknown phase, and a signal of unknown amplitude and phase. For unknown phase, the modified likelihood ratio of Eq. (9.7-39) is averaged as before, using a least favorable distribution (the *intuitive substitute*); with respect to amplitude, the known a priori amplitude statistics are employed (the *Bayes* criterion). The test strategy is given by

$$\overline{\ell'(\mathbf{v} \mid \mathbf{s}, \phi)_{\mathbf{s}, \phi}} \begin{cases} \geqslant \mathcal{H}' & (signal\ present) \\ < \mathcal{H}' & (signal\ absent) \end{cases} \tag{9.7-40}$$

The modified likelihood ratio will be used to derive a test statistic in each case, which permits the structure and performance of the optimum receiver to be evaluated.

9.8. Optimum Detection of an Exactly Known Signal in Colored Noise

A. Structure of optimum receiver

For an exactly known signal, substituting (9.7-39) into threshold test (9.7-40) yields (for $\phi = 0$):

$$\exp\left[\sum_{k=1}^{\infty} \frac{\mathrm{Re}\,(v_k s_k^*)}{\lambda_k} - \frac{|s_k|^2}{2\lambda_k} \right] \begin{cases} \geqslant \mathcal{H}' & (signal\ present) \\ < \mathcal{H}' & (signal\ absent) \end{cases} \tag{9.8-1a}$$

Since signal amplitude is known, the quantity $\exp\left[-\sum_{k=1}^{\infty} |s_k|^2/\lambda_k\right]$ can be incorporated into the threshold. Hence an equivalent test is

$$\sum_{k=1}^{\infty} \frac{\mathrm{Re}\,(v_k s_k^*)}{\lambda_k} \begin{cases} \geqslant \mathcal{H}_1 & (signal\ present) \\ < \mathcal{H}_1 & (signal\ absent) \end{cases} \tag{9.8-1b}$$

The quantity on the left in Eq. (9.8-1b) is the test statistic q.

To determine the structure of a receiver that performs threshold test (9.8-1b), define weighting coefficients

$$w_k = \frac{s_k}{\lambda_k} \tag{9.8-2}$$

Then, test statistic q in (9.8-1b) can be rewritten as

$$q = \sum_{k=1}^{\infty} \mathrm{Re}\,(v_k w_k^*) \tag{9.8-3}$$

Further, let $w(t)$ and $\tilde{w}(t)$ be defined by

$$w(t) = \mathrm{Re}\,[\tilde{w}(t)e^{j\omega_0 t}] \tag{9.8-4}$$

and

$$\tilde{w}(t) = \sum_{k=1}^{\infty} w_k \psi_k(t) \tag{9.8-5}$$

where the w_k are established by Eq. (9.8-2). It will now be shown that, when $v(t)$ is the input to a filter whose impulse response $h(t)$ is

$$h(t) = 2w(T - t), \qquad 0 \leqslant t \leqslant T \tag{9.8-6}$$

the filter output at time $t = T$ is the test statistic q.

A variation of the superposition integral [see Eq. (2.2-16)] describes the output $y(t)$ of filter $h(t)$ by

$$y(t) = \int_0^T h(\tau)v(t-\tau)\,d\tau \qquad (9.8\text{-}7)$$

for a finite observation interval T. Substituting (9.8-6) into (9.8-7) yields

$$y(t) = 2\int_0^T w(T-\tau)v(t-\tau)\,d\tau \qquad (9.8\text{-}8a)$$

or

$$y(t) = 2\int_0^T w(x)v(t-T+x)\,dx \qquad (9.8\text{-}8b)$$

At time $t = T$,

$$y(T) = 2\int_0^T w(x)v(x)\,dx \qquad (9.8\text{-}9)$$

Substituting Eqs. (9.7-3) and (9.8-4) into (9.8-9) results in

$$y(T) = 2\int_0^T \operatorname{Re}\left[\tilde{w}(x)e^{j\omega_0 x}\right]\operatorname{Re}\left[\tilde{v}(x)e^{j\omega_0 x}\right]dx \qquad (9.8\text{-}10a)$$

$$= 2\int_0^T \left[\frac{\tilde{w}(x)e^{j\omega_0 x} + \tilde{w}^*(x)e^{-j\omega_0 x}}{2}\right]\operatorname{Re}\left[\tilde{v}(x)e^{j\omega_0 x}\right]dx \qquad (9.8\text{-}10b)$$

$$= \operatorname{Re}\int_0^T \tilde{w}^*(x)\tilde{v}(x)\,dx \qquad (9.8\text{-}10c)$$

where the integrals of terms containing the double frequency component $e^{j2\omega_0 x}$ are neglected in passing from (9.8-10b) to (9.8-10c). (These integrals are vanishingly small, relative to the video component of Eq. (9.8-10c), for narrowband waveforms and filters.) When both $\tilde{w}(x)$ and $\tilde{v}(x)$ in (9.8-10c) are replaced by their respective Karhunen-Loéve expansions, the filter output becomes

$$y(T) = \operatorname{Re}\left[\sum_{k=1}^{\infty}\sum_{j=1}^{\infty} w_k^* v_j \int_0^T \psi_k^*(x)\psi_j(x)\,dx\right] \qquad (9.8\text{-}11)$$

With Eq. (9.7-16), (9.8-11) simplifies to

$$y(T) = \sum_{k=1}^{\infty} \operatorname{Re}\left(w_k^* v_k\right) \qquad (9.8\text{-}12)$$

and $y(T) = q$ by Eq. (9.8-3). Thus, the structure of the optimum receiver can be implemented as shown in Fig. 9.8-1.

Filter $h(t)$ is also the optimum filter that maximizes peak signal-to-noise ratio at its output (see Sections 5.2 and 6.1). To demonstrate this result, integral equation (6.1-1), whose solution yields optimum filter impulse response $h_{\text{opt}}(t)$, is repeated here with $t_m = T$ and $k = 1$:

$$\int_0^T h_{\text{opt}}(s)\phi_{nn}(t-s)\,ds = s(T-t), \qquad 0 \leqslant t \leqslant T \qquad (9.8\text{-}13)$$

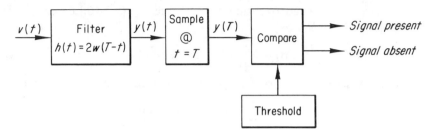

Fig. 9.8-1. Structure of optimum detector for an exactly known signal in colored noise.

It will be shown that replacing $h_{\text{opt}}(s)$ by $2w(T-s)$ in Eq. (9.8-13) results in an identity.

Since $\phi_{nn}(\tau)$ is even [see Eq. (4.4-9)], Eq. (9.8-13) can be rewritten as

$$2\int_0^T w(T-s)\phi_{nn}(s-t)\,ds = s(T-t), \qquad 0 \leqslant t \leqslant T \qquad (9.8\text{-}14a)$$

or

$$2\int_0^T w(x)\phi_{nn}(T-t-x)\,dx = s(T-t), \qquad 0 \leqslant t \leqslant T \qquad (9.8\text{-}14b)$$

When (9.7-6) and (9.8-4) are substituted into (9.8-14b), the integral on the left becomes, successively,

$$2\int_0^T \operatorname{Re}\left[\tilde{w}(x)e^{j\omega_0 x}\right]\operatorname{Re}\left[\tilde{\phi}_{nn}(T-t-x)e^{j\omega_0(T-t-x)}\right]dx \qquad (9.8\text{-}15a)$$

$$2\int_0^T \left[\frac{\tilde{w}(x)e^{j\omega_0 x} + \tilde{w}^*(x)e^{-j\omega_0 x}}{2}\right]\operatorname{Re}[\tilde{\phi}_{nn}(T-t-x)e^{j\omega_0(T-t-x)}]\,dx \qquad (9.8\text{-}15b)$$

$$\operatorname{Re}\int_0^T \tilde{w}(x)\tilde{\phi}_{nn}(T-t-x)e^{j\omega_0(T-t)}\,dx, \qquad 0 \leqslant t \leqslant T \qquad (9.8\text{-}15c)$$

where the integrals of terms containing the double frequency component $\exp(-j2\omega_0 x)$ have been neglected in (9.8-15c). If $\tilde{w}(x)$ is now replaced by its Karhunen-Loéve expansion (9.8-5), (9.8-15c) becomes

$$\operatorname{Re}\left[\sum_{k=1}^{\infty} w_k e^{j\omega_0(T-t)}\int_0^T \tilde{\phi}_{nn}(T-t-x)\psi_k(x)\,dx\right], \qquad 0 \leqslant t \leqslant T \qquad (9.8\text{-}16)$$

From Eq. (9.7-15), the integral in (9.8-16) can be replaced by $\lambda_k\psi_k(T-t)$ so that (9.8-16) can be rewritten as

$$\operatorname{Re}\left[\sum_{k=1}^{\infty} w_k\lambda_k\psi_k(T-t)e^{j\omega_0(T-t)}\right] \qquad (9.8\text{-}17)$$

Substituting (9.8-2) into (9.8-17) yields

$$\operatorname{Re}\left[\sum_{k=1}^{\infty} s_k\psi_k(T-t)e^{j\omega_0(T-t)}\right] \qquad (9.8\text{-}18)$$

which, with Eqs. (9.7-30) and (9.7-1), completes the proof:

$$\text{Re}\,[\tilde{s}(T-t)e^{j\omega_0(T-t)}] = s(T-t) \tag{9.8-19}$$

Hence, the filter defined by (9.8-6) is also the optimum filter that maximizes peak signal-to-noise ratio for an exactly known signal in stationary Gaussian noise.

B. Performance of the optimum receiver

The test strategy corresponds to a comparison of test statistic $q = y(T)$ with a specified threshold. Quantity $y(T)$ is the (real) sampled output of the optimum filter whose impulse response (9.8-6) satisfies Eq. (9.8-13). The test in this case is similar to that of Section 9.2 except for the replacement of the matched filter by the optimum colored noise filter.

From (9.8-9) $y(T)$ is a Gaussian random variable, since $v(t)$ is a sample waveform of a Gaussian process (see Section 9.7). For hypothesis H_0 (signal absent), it also follows from (9.8-9) that

$$\overline{y(T)} = 2\int_0^T w(t)\overline{v(t)}\,dt = 0 \tag{9.8-20}$$

and

$$\overline{y^2(T)} = 4\int_0^T\int_0^T w(t)w(s)\overline{v(t)v(s)}\,dt\,ds \tag{9.8-21a}$$

$$= 2\int_0^T w(t)\left[2\int_0^T \phi_{nn}(t-s)w(s)\,ds\right]dt \tag{9.8-21b}$$

The inner integral in (9.8-21b) can be replaced with Eq. (9.8-14b); the result is

$$\overline{y^2(T)} = 2\int_0^T s(t)w(t)\,dt \tag{9.8-22}$$

The integral in (9.8-22) has special significance. Consider the signal-to-noise ratio out of a filter $h(t)$ at time T with an input $v(t) = s(t) + n(t)$ defined over the observation interval T. If the output signal-to-noise ratio at time T is denoted by χ_c, it follows that (see Section 5.2)

$$\chi_c = \frac{\left[\int_0^T h(\tau)s(T-\tau)\,d\tau\right]^2}{\int_0^T\int_0^T h(\tau)h(\mu)\overline{n(T-\tau)n(T-\mu)}\,d\tau\,d\mu} \tag{9.8-23a}$$

$$\chi_c = \frac{\left[\int_0^T h(\tau)s(T-\tau)\,d\tau\right]^2}{\int_0^T\int_0^T h(\tau)h(\mu)\phi_{nn}(\tau-\mu)\,d\tau\,d\mu} \tag{9.8-23b}$$

When $h(t)$ is the optimum filter solution of Eq. (9.8-13), it follows from the definition of the optimum filter in Chapters 5 and 6 that χ_c is the peak output signal-to-noise ratio. In this case, substituting (9.8-13) into Eq. (9.8-23b) yields

$$\chi_c = \frac{\left[\int_0^T h_{\text{opt}}(\tau)s(T-\tau)\,d\tau\right]^2}{\int_0^T h_{\text{opt}}(\tau)s(T-\tau)\,d\tau} \tag{9.8-24a}$$

$$\chi_c = \int_0^T h_{\text{opt}}(\tau)s(T-\tau)\,d\tau \tag{9.8-24b}$$

Replacing $h_{\text{opt}}(\tau)$ by $2w(T-\tau)$ [see Eq. (9.8-6)] finally yields

$$\chi_c = 2\int_0^T s(t)w(t)\,dt \tag{9.8-25}$$

which is identical to the integral in Eq. (9.8-22). Hence, Eq. (9.8-22) can be rewritten as

$$\overline{y^2(T)} = \chi_c \tag{9.8-26}$$

When the noise is white, from (9.8-14a) $w(t)$ becomes $s(t)/N_0$, since $\phi_{nn}(\tau) = N_0\delta(\tau)/2$; for this case Eq. (9.8-25) becomes $\chi_c = \mathcal{R} = 2E/N_0$, which is the expected result for a matched filter.

For hypothesis H_1 (signal present), it follows from (9.8-9) that

$$\overline{y(T)} = 2\int_0^T w(t)\overline{v(t)}\,dt \tag{9.8-27a}$$

$$\overline{y(T)} = 2\int_0^T w(t)s(t)\,dt = \chi_c \tag{9.8-27b}$$

The variance of test statistic $y(T)$ is given by

$$\overline{[y(T)-\chi_c]^2} = \chi_c \tag{9.8-28}$$

which is the same result obtained for hypothesis H_0 (no signal) in Eq. (9.8-26).

With χ_c replaced by $\mathcal{R}$, the means and variances of test statistic $y(T)$, for hypotheses H_0 and H_1, are the same as those obtained in Section 9.2 for an exactly known signal immersed in white Gaussian noise; hence, the performance of the optimum colored-noise receiver is given by the graphs of Fig. 9.2-3 with $\mathcal{R}$ replaced by χ_c. The quantity χ_c can be calculated from (9.8-25) where $2w(t)$ is the solution of integral equation (9.8-14b).

9.9. Optimum Detection of a Signal of Unknown Phase in Colored Noise

A. Structure of optimum receiver

For a signal of unknown phase, likelihood ratio $\ell'(\mathbf{v}\,|\,\phi)$, given by Eq. (9.7-39), is averaged with respect to initial phase ϕ assuming a uniform probability density function. The averaged result is then compared to a threshold and a test statistic is again derived. The test is given by

$$\overline{\exp\left\{\sum_{k=1}^{\infty}\frac{\mathrm{Re}\,[v_k s_k^* e^{-j\phi}]}{\lambda_k}-\frac{|s_k|^2}{2\lambda_k}\right\}}_{\phi}\begin{cases}\geqslant \mathcal{K}' & (\textit{signal present})\\< \mathcal{K}' & (\textit{signal absent})\end{cases} \tag{9.9-1}$$

Let

$$\sum_k \frac{v_k s_k^*}{\lambda_k} = re^{j\beta} \tag{9.9-2}$$

where

$$r = \left|\sum_k \frac{v_k s_k^*}{\lambda_k}\right| \tag{9.9-3}$$

Then the likelihood ratio in Eq. (9.9-1) can be rewritten as

$$\overline{\ell'(\mathbf{v}\,|\,\phi)_\phi} = \overline{\exp\left[r\cos(\beta-\phi)-\sum_k\frac{|s_k|^2}{2\lambda_k}\right]}_\phi \tag{9.9-4a}$$

$$\overline{\ell'(\mathbf{v}\,|\,\phi)_\phi} = \exp\left(-\sum_k\frac{|s_k|^2}{2\lambda_k}\right)\frac{1}{2\pi}\int_0^{2\pi}e^{r\cos(\beta-\phi)}\,d\phi \tag{9.9-4b}$$

From Eq. (9.3-13),† (9.9-4b) becomes

$$\overline{\ell'(\mathbf{v}\,|\,\phi)_\phi} = \exp\left(-\sum_k\frac{|s_k|^2}{2\lambda_k}\right)I_0(r) \tag{9.9-5}$$

Since amplitude is assumed known, and since $I_0(r)$ is a monotonic function of its argument, the quantity r is a sufficient statistic. Thus, the test strategy corresponds to

$$r\begin{cases}\geqslant \mathcal{K}' & (\textit{signal present})\\< \mathcal{K}' & (\textit{signal absent})\end{cases} \tag{9.9-6}$$

In Section 9.8, the output $y(T)$ of the optimum receiver at time T was found to be [see Eqs. (9.8-1b) to (9.8-12)]

$$y(T) = \mathrm{Re}\left(\sum_{k=1}^{\infty}\frac{v_k s_k^*}{\lambda_k}\right) \tag{9.9-7}$$

If the complex envelope $\tilde{y}(t)$ of the optimum filter output is defined by

$$y(t) = \mathrm{Re}\,[\tilde{y}(t)e^{j\omega_0 t}] \tag{9.9-8}$$

then from Eqs. (9.9-7), and (9.9-8),

$$|\tilde{y}(T)| = \left|\sum_{k=1}^{\infty}\frac{v_k s_k^*}{\lambda_k}\right| \tag{9.9-9}$$

Combining (9.9-3) and (9.9-9) yields

$$r = |\tilde{y}(T)| \tag{9.9-10}$$

---

†The integral in (9.9-4b) is slightly different from (9.3-13); however, they are equal since $\cos x$ is an even function.

Eq. (9.9-10) states that test statistic r is the magnitude of the complex envelope of the signal out of the optimum filter for colored noise at time T. It is shown in Section 2.5 that, when $y(t)$ is band-limited about ω_0, $|\tilde{y}(t)|$ is equal to the envelope of $y(t)$; further, when $y(t)$ is narrowband, but not band-limited, the difference between $|\tilde{y}(t)|$ and env $\{y(t)\}$ is usually small. Thus, r is approximately given by

$$r \cong \text{env}\,\{y(T)\} \tag{9.9-11}$$

This result corresponds to that obtained for white noise except that the optimum filter for colored noise replaces the matched filter (see Fig. 9.3-4).

B. Performance of optimum receiver

To evaluate the performance of the optimum receiver for signals of unknown phase, the probability density function of test statistic r is required. For this purpose, note that Eqs. (9.9-8) and (9.8-10c) permit Eq. (9.9-10) to be written as

$$r = |\tilde{y}(T)| = \left| \int_0^T \tilde{w}^*(t)\tilde{v}(t)\, dt \right| \tag{9.9-12}$$

For hypothesis H_0 (signal absent)

$$\overline{\tilde{y}(T)} = \int_0^T \tilde{w}^*(t)\overline{\tilde{v}(t)}\, dt = 0 \tag{9.9-13}$$

From Eq. (2.5-17)

$$\tilde{y}(T) = y_I(T) + jy_Q(T) \tag{9.9-14}$$

where $y_I(T)$ and $y_Q(T)$ are the quadrature components of complex envelope $\tilde{y}(T)$. Hence, from (9.9-13) and (9.9-14),

$$\overline{y_I(T)} = \overline{y_Q(T)} = 0 \tag{9.9-15}$$

The variances of the real and imaginary parts of $\tilde{y}(T)$ are related to $\overline{y^2(t)}$ by Eq. (4.11-13) (see also reference [6] for an exact derivation):

$$\overline{y_I^2(T)} = \overline{y_Q^2(T)} = \overline{y^2(T)} \tag{9.9-16}$$

It follows from (9.8-22) and (9.8-25) that

$$\overline{y_I^2(T)} = \overline{y_Q^2(T)} = \chi_c \tag{9.9-17}$$

In addition, as discussed in Section 9.7, it can be shown that $y_I(T)$ and $y_Q(T)$ are each Gaussian and statistically independent. Thus, statistic r (for hypothesis H_0) is the envelope of a complex two-dimensional Gaussian random process with component means of zero and variances of χ_c. The joint density function is given by

$$p[y_I(T), y_Q(T)] = \frac{1}{2\pi\chi_c} \exp\left[-\frac{y_I^2(T) + y_Q^2(T)}{2\chi_c} \right] \tag{9.9-18}$$

It follows from (9.9-12) and (9.9-14) that

$$r = \sqrt{y_I^2(T) + y_Q^2(T)} \tag{9.9-19}$$

By a transformation of variables, the probability density function of r, for hypothesis H_0, is found from (9.9-18) to be

$$p(r) = \frac{r}{\chi_c} \exp\left(-\frac{r^2}{2\chi_c}\right) \tag{9.9-20}$$

Equation (9.9-20) can be simplified by defining a normalized envelope,

$$\mathfrak{r} = \frac{r}{\sqrt{\chi_c}} \tag{9.9-21}$$

Then, (9.9-20) reduces to

$$p(\mathfrak{r}) = \mathfrak{r} \exp\left(-\frac{\mathfrak{r}^2}{2}\right) \tag{9.9-22}$$

For hypothesis H_1 (signal present), the mean values of the quadrature components of $\tilde{y}(T)$ are nonzero. In this case

$$\overline{\tilde{y}(T)} = \int_0^T \tilde{w}^*(t)\tilde{s}(t)e^{j\phi}\,dt \tag{9.9-23a}$$

$$= e^{j\phi} \int_0^T \left[\sum_{k=1}^\infty w_k^* \psi_k^*(t)\right]\left[\sum_{j=1}^\infty s_j \psi_j(t)\right] dt \tag{9.9-23b}$$

$$= e^{j\phi} \sum_{k=1}^\infty w_k^* s_k \tag{9.9-23c}$$

$$= e^{j\phi} \sum_{k=1}^\infty \frac{|s_k|^2}{\lambda_k} \tag{9.9-23d}$$

But, from (9.7-1), (9.8-4), (9.8-25), and the Karhunen-Loéve expansions of $\tilde{w}(t)$ and signal $\tilde{s}(t)$,

$$\chi_c = \sum_{k=1}^\infty \frac{|s_k|^2}{\lambda_k} \tag{9.9-24}$$

Then Eq. (9.9-23d) can be rewritten as

$$\overline{\tilde{y}(T)} = \chi_c \cos\phi + j\chi_c \sin\phi \tag{9.9-25}$$

Hence,

$$\overline{y_I(T)} = \chi_c \cos\phi \tag{9.9-26a}$$

$$\overline{y_Q(T)} = \chi_c \sin\phi \tag{9.9-26b}$$

The variances of both $y_I(T)$ and $y_Q(T)$ can be shown to be χ_c. Thus, for this case

$$p[y_I(T), y_Q(T)] = \frac{1}{2\pi\chi_c} \exp\left\{-\frac{[y_I(T) - \chi_c \cos\phi]^2 + [y_Q(T) - \chi_c \sin\phi]^2}{2\chi_c}\right\} \tag{9.9-27}$$

By a suitable transformation of variables [see Eqs. (9.9-19) and (9.9-21)], the probability density of $\mathfrak{r}$ for hypothesis H_1 can be found from (9.9-27) to be

$$p(\mathfrak{r}) = \mathfrak{r}e^{-(\mathfrak{r}^2+\chi_c)/2}I_0(\mathfrak{r}\sqrt{\chi_c}) \qquad (9.9\text{-}28)$$

A comparison of Eqs. (9.9-22) and (9.9-28) with (9.3-25b) and (9.3-38) indicates that the probability density functions of the test statistic in the colored-noise case for hypotheses H_0 and H_1 are the same as those in the white noise case with $\mathscr{R}$ replaced by χ_c. Thus the performance graphs of Fig. 9.3-7 can be used to compute and predict the performance of the optimum detector for signals of unknown phase imbedded in colored Gaussian noise; the only change is the replacement of $\mathscr{R}$ by peak signal-to-noise ratio χ_c defined in Eq. (9.8-25).

9.10. Optimum Detection of a Signal of Unknown Phase and Amplitude in Colored Noise

In the discussion of Section 9.4 for signals of unknown phase and amplitude, it was shown that envelope r is a test statistic. The plots in Figs. 9.4-1 for the Rayleigh fluctuating signal and 9.5-1 for the one-dominant-plus-Rayleigh signal are applicable to detection in colored noise provided $\mathscr{R}$ is replaced by $\overline{\chi_c}$. The average value of χ_c is obtained as follows. Let

$$s(t) = A\Sigma(t) \qquad (9.10\text{-}1)$$
$$w(t) = A\mathfrak{w}(t) \qquad (9.10\text{-}2)$$

Then Eq. (9.8-25) can be rewritten as

$$\chi_c = 2A^2 \int_0^T \Sigma(t)\mathfrak{w}(t)\, dt \qquad (9.10\text{-}3)$$

By analogy with (9.4-4), definitions (9.10-1) and (9.10-2) become unique by the normalization

$$2 \int_0^T \Sigma(t)\mathfrak{w}(t)\, dt = 1 \qquad (9.10\text{-}4)$$

It follows from (9.10-3) and (9.10-4) that $\overline{\chi_c}$ is given by

$$\overline{\chi_c} = \int_A A^2 p(A)\, dA \qquad (9.10\text{-}5)$$

where $p(A)$ is the probability density function of signal amplitude A.

9.11. Summary

In this chapter we considered the detection of signals in noise based on a single observation. It was shown that the structure of the optimum detector for the exactly known signal in white Gaussian noise is a matched filter followed by a sampling and decision circuit.

Next the optimum processor for a signal of unknown phase was considered. In this case the optimum processor consists of a matched filter followed by an envelope detector and sampling and decision circuits. The performance of this receiver differs from that of the exactly known signal by only a small amount at high signal-to-noise ratios. A typical difference is 0.8 db in the region of 50 per cent probability of detection for a false-alarm probability of 10^{-6}.

Signals of unknown phase and amplitude were considered next. It was shown that the optimum receiver is the same as that for signals with only phase unknown.

In all of the cases above it was shown that detectability depends on peak signal-to-noise ratio $\mathscr{R} = 2E/N_0$. The computed performance curves apply to all coherently processed waveforms including long pulse trains.

For the more general case of Gaussian colored noise, all of the previous results apply provided the matched filter is replaced by the optimum colored noise filter. To predict performance it is only necessary to replace $\mathscr{R}$ by signal-to-noise ratio χ_c, or $\overline{\chi_c}$, whichever is appropriate.

PROBLEMS

9.1. The radar equation can be written as

$$\mathscr{R} \text{ (or } \chi_c) = \frac{E_T G_T G_R \lambda^2 \sigma}{(4\pi)^3 R^4 N_0 L}$$

where E_T = transmit pulse energy
         G_T = transmit antenna gain (35 db)
         G_R = receive antenna gain (35 db)
         λ = wavelength (use $f = 1000$ MHz)
         L = system loss factor = 3 db
         σ = target cross section (1 square meter)
         R = target range ($R_{\max} = 500$ nautical miles)
         $N_0 = KT_{eq}$
         K = Boltzmann's constant = 1.38×10^{-23} joule/°K
         T_{eq} = equivalent input noise temperature (1000°K)

Design a search radar for detection on a single observation of a 1-square-meter target at 500 nautical miles with $P_{fa} = 10^{-6}$ and $P_d = 1$ per cent, 50 per cent, and 90 per cent. Compute required pulse energy for: (a) an exactly known signal, (b) a signal of unknown phase, (c) a Rayleigh target, (d) a one-dominant-plus-Rayleigh target.

9.2. Using the results of problem 9.1, plot P_d versus required pulse energy (in db). Explain the cross-over and superiority of fluctuating target models over a signal of unknown phase at low probabilities of detection.

9.3. Using the results of problem 9.1, determine peak power required for a rectangular pulse with lengths of 10 μsec, 100 μsec, and 1000 μsec, for $P_d = 50$ per cent and a signal of unknown phase. Repeat for a Rayleigh target and a one-dominant-plus-Rayleigh target model.

9.4. Consider a coherent pulse-train processor (delay-line integrator) that processes 10 pulses. What is the required single-pulse energy for the various cases treated in problem 9.1 for 50 per cent detection probability?

9.5. For a signal whose amplitude A has a Rayleigh density function, derive the density function for peak signal-to-noise ratio $\mathscr{R}$.

9.6. For a signal whose amplitude A has a one-dominant-plus-Rayleigh density function, derive the density function for peak signal-to-noise ratio $\mathscr{R}$.

9.7. For P_{fa} in the range of 10^{-2} to 10^{-14}, plot peak signal-to-noise ratio $\mathscr{R}$ required for 50 per cent probability of detection for both an exactly known signal and a signal of unknown phase. Repeat for $P_d = 20$ per cent and 90 per cent. Plot the differences in $\mathscr{R}$ for the three detection levels. What conclusions can be drawn from the shape and magnitude of the differences?

9.8. For P_{fa} in the range of 10^{-2} to 10^{-14}, plot the signal-to-noise ratio required for 50 per cent probability of detection for a signal of unknown phase. Interpolate for increments of P_{fa} of 10^{-1}. Repeat for a Rayleigh signal and a one-dominant-plus-Rayleigh signal. Repeat for $P_d = 20$ per cent and 90 per cent. What can be said about the sensitivity of peak signal-to-noise ratio $\mathscr{R}$ to a change in false-alarm probability of an order of magnitude?

9.9. An alternate method for obtaining the optimum receiver structure for small signal-to-noise ratios ($Ar < 1$) is to expand the likelihood ratio given in Eq. (9.4-5) in a series. What is the test statistic that implements the test strategy for known amplitude? For unknown amplitude average the expansion with respect to an arbitrary probability density function. What is the test statistic?

9.10. Repeat problem 9.9 for large signal-to-noise ratios ($Ar \gg 1$) using an asymptotic expansion for Eq. (9.4-5) assuming known amplitude. Find the logarithm of the likelihood ratio and simplify. Plot the (nonlinear) law that operates on envelope r. Interpret the results over the regions of significant changes in the shape of the curve.

9.11. In the series approximation of Eq. (9.6-7) evaluate the error term when only the first two terms are used. When is this error a maximum? Over what regions of β and α is the approximation best?

9.12. As an exercise, derive an expression for the likelihood ratio similar to that given by Eq. (9.2-8) when samples are taken at a rate exceeding $1/2\omega_c$, where the signal spectrum is contained in the interval $(-\omega_c, \omega_c)$. Discuss the significance of the correlation terms that appear. Can a linear transformation eliminate these terms?

9.13. For small signal-to-noise ratios ($E \ll N_0$) expand the exponential of Eq. (9.2-10) in a Taylor series about zero.
(a) For the exactly known signal which term is of greatest significance? What is the structure of the optimum receiver?

(b) Which term is of greatest significance for a signal of unknown phase, assuming a narrowband signal as given in Eq. (9.3-2)? How does the structure indicated compare with that shown in Figs. 9.3-2 and 9.3-4?

9.14. As an exercise, derive the likelihood ratio for a signal with unknown phase and Rayleigh amplitude and confirm the result that the test strategy corresponds to a comparison of envelope r, defined in Eq. (9.3-10), with a threshold.

9.15. For a narrowband Gaussian process with complex envelope $\tilde{n}(t) = n_I(t) + jn_Q(t)$, prove relation (9.7-8) using results in Chapter 4.

9.16. Prove Eq. (9.7-22) using (9.7-8).

9.17. Verify Eqs. (9.7-31) and (9.7-32).

9.18. Verify Eq. (9.8-28).

REFERENCES

1. Middleton, D.: "An Introduction to Statistical Communication Theory," McGraw-Hill, New York, 1960.

2. "Handbook of Mathematical Functions," U.S. Department of Commerce, National Bureau of Standards, AMS 55 (June, 1964).

3. Marcum, J. I.: A Statistical Theory of Target Detection by Pulsed Radar, Rand Research Memo RM-754 (December, 1947). [Reissued since in a special issue, *Trans. IRE Prof. Group on Information Theory*, **IT-6**: (*2*), 59–144 (April, 1960).]

4. Rubin, W. L. and DiFranco, J. V.: Radar Detection, *Electro-Technology* (*Sci. and Eng. Ser.*), **64**: 61–92 (April, 1964).

5. Watson, G. N.: "Theory of Bessel Functions," Cambridge University Press, London and New York, 1944.

6. Kelly, E. J., Reed, I. S., and Root W. L.: The Detection of Radar Echoes in Noise, *J. Soc. Indust. Appl. Math.*, **8**: (*2*), 309–341 (June, 1960); **8**: (*3*), 481–507 (September, 1960).

7. Grenander, U.: Stochastic Processes and Statistical Inference, *Ark. Mat.*, **1** (1950).

8. Rubin, W. L., and DiFranco, J. V.: Analytic Representation of Wide-Band Radio Frequency Signals, *J. Franklin Inst.*, **275**: (*3*), 197–204 (March, 1963).

9. Woodward, P. M.: "Probability and Information Theory with Application to Radar," McGraw-Hill, New York, 1955.

10. Rice, S. O.: Mathematical Analysis of Random Noise, *Bell System Tech. J.*, **23**: 282–332 (July, 1944); **24**: 46–156 (January, 1945).

10

DETECTION BASED
ON MULTIPLE OBSERVATIONS:
NONFLUCTUATING MODEL

10.1. Introduction

In Chapter 9 we investigated the structure and performance of an opti mum receiver on the basis of a single observation. In this chapter we examine the optimum Bayes† receiver structure and performance when many observations are available for detection. The radar waveform transmitted is assumed to be identical within each observation interval; a sequence of waveform echoes from a target is called a pulse train.

A number of cases are considered in this and the next chapter. For completeness, we consider first a pulse train in which all of the pulses are coherently related. However, pulse trains whose members are incoherently related are the main subject of this study; in an incoherent pulse train each member waveform is assumed to have an initial phase that is a random variable and statistically independent of the initial phases of other pulse train members. The study also presupposes that the number of observation

†Except with respect to initial phase.

intervals (pulse-train length) is **known.**† This important assumption distinguishes the subject matter of this chapter from sequential detection, where the number of observation intervals is random; sequential detection is discussed in Chapter 16.

The target echo is assumed to be immersed in background noise which is characterized by a white Gaussian stochastic process. Results for white noise can be extended simply to colored noise by substituting an appropriately defined colored signal-to-noise ratio, as discussed in Sections 9.8, 9.9, and 9.10, for peak signal-to-noise ratio $\mathscr{R}$.

The Bayes strategy with respect to target range, doppler, and angle parameters which is described in Chapters 8, 9 and 17, requires a generalized likelihood-ratio test for target existence in each radar resolution cell in the search volume. When target doppler is known exactly, the number of resolution cells is equal to the number of range bins as defined in Chapter 17. If target doppler is not known, the strategy requires a simultaneous test in a multiplicity of parallel doppler channels which span the interval of unknown target dopplers. This strategy modifies the overall probability of false alarm, since multiple tests are performed in each range bin. It is difficult to compute the exact false-alarm probability for this case due to the correlation that exists between the noise outputs of the various doppler channels. It can be shown, however, that the probability of false alarm increases, at most, by the number of parallel doppler channels in the system.

With respect to initial phase, the detection criterion adopted is the intuitive substitute, which is discussed in Chapters 8 and 9. In this strategy, the likelihood ratio is averaged with respect to a least favorable density function for initial phase—namely, one that is uniform over the interval $(0, 2\pi)$. This applies to both coherent and incoherent pulse trains. In the latter case, however, uniform density functions are also assumed for the initial phase of every pulse in the train echo. This assumption is realistic since oscillator drifts, system instabilities, and changes in target orientation and reflectivity in the interpulse interval usually introduce phase shifts that are significant compared to 2π radians.

The results we seek depend on the selected model of target radar cross section. Fluctuating radar target models are investigated in the next chapter. In this chapter we consider a target with a constant or unchanging cross section; this characterizes stationary and moving targets that display an unchanging aspect on successive pulse transmissions. The structure of the optimum detection receiver is derived and its performance evaluated for this case. Results are presented in the form of detailed graphs that are useful in radar system computations. In addition, analytic approximations are derived that are convenient for making rapid estimates of system performance.

†The pulse-train duration for a search radar with a rotating antenna is usually established by scan rate.

10.2. Nonfluctuating Coherent Pulse Train

Consider the exactly known pulse train illustrated in Fig. 10.2-1. For simplicity each member of the train is shown as an uncoded rectangular pulse —that is, with no phase modulation. An exactly known coherent pulse

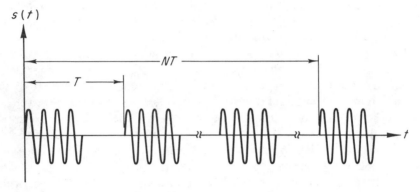

Fig. 10.2-1. Exactly known pulse train.

train is a special case of an exactly known signal; hence the results of Chapter 9 for optimum detection based on a single observation apply directly, since the discussion was completely general with respect to the radar waveform employed. The optimum receiver structure includes a filter that is matched to the received signal waveform when the background noise process is white. Matched filter processing for a pulse train is sometimes called *coherent processing* or *predetection integration.* Since detectability is a function only of signal energy and noise-power spectral density, the plots in Chapter 9 can be used to compute the performance of a coherent pulse

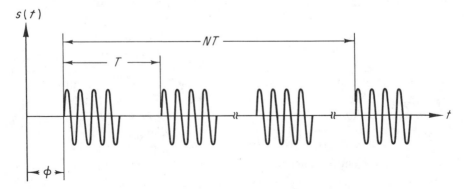

Fig. 10.2-2. Pulse train with unknown starting phase.

train of N pulses, with signal energy given by NE_p, where the energy of each pulse is E_p.

Consider next a coherent pulse train that is exactly known except for initial phase, as shown in Fig. 10.2-2 (however, phase relations between pulses in the train are assumed to be known exactly). This waveform is a special case of a coherent signal with unknown initial phase. From the results of Chapter 9, the optimum receiver is a filter matched to the pulse train, followed by an envelope detector. To calculate performance, the graphs of Chapter 9 can be used if the signal energy is once again computed by summing the energy of each pulse in the train.

10.3. Optimum Receiver Structure for a Nonfluctuating Incoherent Pulse Train

The structure of an optimum receiver is considered next for a non-fluctuating pulse-train echo composed of N identical signals, each with statistically independent random initial phases. Performance is computed in the following section. These results also apply to a coherent pulse train processed suboptimally—that is, not utilizing phase coherence between pulses.

The optimum detection receiver computes the generalized likelihood ratio, which has been averaged with respect to the initial phase of each train pulse, and compares this quantity to a threshold. For narrowband waveforms, the generalized likelihood ratio in vector notation is given by

$$\overline{\ell(\mathbf{v}\,|\,\boldsymbol{\theta})_\theta} = \frac{\mathfrak{p}}{\mathfrak{q}}\,\frac{\overline{p[\mathbf{v}\,|\,\mathbf{s}(\boldsymbol{\theta})]_\theta}}{p(\mathbf{v}\,|\,0)} \tag{10.3-1}$$

where $\mathfrak{p}$ and $\mathfrak{q}$ are the a priori probabilities of signal and no signal, respectively, vector $\mathbf{v}$ describes the sampled received waveform in the observation interval, and $\mathbf{s} = \mathbf{s}(\boldsymbol{\theta})$ describes the sampled radar signal, which is a function of the set of random variables $\boldsymbol{\theta}$. It is convenient to absorb the ratio $\mathfrak{p}/\mathfrak{q}$ into the threshold $\mathscr{K}$ and define the modified generalized likelihood ratio $\overline{\ell'(\mathbf{v}\,|\,\boldsymbol{\theta})_\theta}$ by

$$\overline{\ell'(\mathbf{v}\,|\,\boldsymbol{\theta})_\theta} = \frac{\overline{p[\mathbf{v}\,|\,\mathbf{s}(\boldsymbol{\theta})]_\theta}}{p(\mathbf{v}\,|\,0)} \tag{10.3-2}$$

The optimum receiver computes and compares $\overline{\ell'(\mathbf{v}\,|\,\boldsymbol{\theta})_\theta}$ with a threshold $\mathscr{K}' = (\mathfrak{q}/\mathfrak{p})\mathscr{K}$.

Since the observation period consists of N observation intervals, it is convenient to express waveform vectors $\mathbf{v}$ and $\mathbf{s}$ as

$$\mathbf{v} = (\mathbf{v}_1, \mathbf{v}_2, \ldots, \mathbf{v}_N) \tag{10.3-3}$$

and

$$\mathbf{s} = [\mathbf{s}_1(\boldsymbol{\theta}_1), \mathbf{s}_2(\boldsymbol{\theta}_2), \ldots, \mathbf{s}_N(\boldsymbol{\theta}_N)] \tag{10.3-4}$$

where $\mathbf{v}_i$ and $\mathbf{s}_i(\boldsymbol{\theta}_i)$ are vectors that describe the received waveform and signal, respectively, in the ith observation interval; and $\boldsymbol{\theta}_i$ denotes the signal random variables in the ith observation interval. With Eqs. (10.3-3) and (10.3-4), expression (10.3-2) can be rewritten as follows:

$$\overline{\ell'(\mathbf{v}\,|\,\boldsymbol{\theta})}_{\theta} = \frac{\overline{p[\mathbf{v}_1, \mathbf{v}_2 \ldots, \mathbf{v}_N \,|\, \mathbf{s}_1(\boldsymbol{\theta}_1), \mathbf{s}_2(\boldsymbol{\theta}_2), \ldots, \mathbf{s}_N(\boldsymbol{\theta}_N)]}_{\theta}}{p(\mathbf{v}_1, \mathbf{v}_2, \ldots, \mathbf{v}_N \,|\, 0)} \qquad (10.3\text{-}5)$$

where $\boldsymbol{\theta}$ is the set of random variables

$$\boldsymbol{\theta} = (\boldsymbol{\theta}_1, \boldsymbol{\theta}_2, \ldots, \boldsymbol{\theta}_N) \qquad (10.3\text{-}6)$$

If the noise received in the ith observation interval is assumed to be statistically independent of the noise in every other observation interval,† Eq. (10.3-5) simplifies to

$$\overline{\ell'(\mathbf{v}\,|\,\boldsymbol{\theta})}_{\theta} = \frac{\overline{\{p[\mathbf{v}_1\,|\,\mathbf{s}_1(\boldsymbol{\theta}_1)] \cdot p[\mathbf{v}_2\,|\,\mathbf{s}_2(\boldsymbol{\theta}_2)] \cdot \,\ldots\, \cdot p[\mathbf{v}_N\,|\,\mathbf{s}_N(\boldsymbol{\theta}_N)]\}}_{\theta}}{p(\mathbf{v}_1\,|\,0) \cdot p(\mathbf{v}_2\,|\,0) \cdot \,\ldots\, \cdot p(\mathbf{v}_N\,|\,0)} \qquad (10.3\text{-}7)$$

Define the (unaveraged) likelihood ratio in the ith observation interval by

$$\ell'(\mathbf{v}_i\,|\,\boldsymbol{\theta}_i) = \frac{p[\mathbf{v}_i\,|\,\mathbf{s}_i(\boldsymbol{\theta})]}{p(\mathbf{v}_i\,|\,0)} \qquad (10.3\text{-}8)$$

Then Eq. (10.3-7) can be rewritten as

$$\overline{\ell'(\mathbf{v}\,|\,\boldsymbol{\theta})}_{\theta} = \overline{\left[\prod_{i=1}^{N} \ell'(\mathbf{v}_i\,|\,\boldsymbol{\theta}_i)\right]}_{\theta} \qquad (10.3\text{-}9\text{a})$$

or

$$\overline{\ell'(\mathbf{v}\,|\,\boldsymbol{\theta})}_{\theta} = \overline{\left[\prod_{i=1}^{N} \ell'(\mathbf{v}_i\,|\,\boldsymbol{\theta}_i)\right]}_{\theta_1, \theta_2, \ldots, \theta_N} \qquad (10.3\text{-}9\text{b})$$

Further simplification of (10.3-9b) requires knowledge about the set of signals $(\mathbf{s}_1, \mathbf{s}_2, \ldots, \mathbf{s}_N)$ and random variables $(\boldsymbol{\theta}_1, \boldsymbol{\theta}_2, \ldots, \boldsymbol{\theta}_N)$. In the present case the signal is identical in each observation interval; further, the signal random variables are the set of initial phases $\phi_i, i = 1, 2, \ldots, N$, corresponding to an initial phase in each observation interval. Thus,

$$\boldsymbol{\theta} = (\phi_1, \phi_2, \ldots, \phi_N) \qquad (10.3\text{-}10)$$

Because of the assumed incoherence between signals in different observation intervals, the initial phase ϕ_i in the ith interval is statistically independent of ϕ_j for all $i \neq j$. In this case, Eq. (10.3-9b) simplifies to

$$\overline{\ell'(\mathbf{v}\,|\,\boldsymbol{\theta})}_{\theta} = \overline{\ell'(\mathbf{v}_1\,|\,\phi_1)}_{\phi_1} \cdot \overline{\ell'(\mathbf{v}_2\,|\,\phi_2)}_{\phi_2} \cdot \,\ldots\, \cdot \overline{\ell'(\mathbf{v}_N\,|\,\phi_N)}_{\phi_N} \qquad (10.3\text{-}11\text{a})$$

or

$$\overline{\ell'(\mathbf{v}\,|\,\boldsymbol{\theta})}_{\theta} = \prod_{i=1}^{N} \overline{\ell'(\mathbf{v}_i\,|\,\phi_i)}_{\phi_i} \qquad (10.3\text{-}11\text{b})$$

In order to evaluate Eq. (10.3-11b), results obtained in Chapter 9 are

---

†This condition is usually satisfied in practice; stationary clutter is an exception.

useful; equation (9.3-14) for $\overline{\ell'(\mathbf{v}\,|\,\phi)}_\phi$ is the appropriate expression to begin with and is repeated here in suitably modified notation:

$$\overline{\ell'(\mathbf{v}\,|\,\phi_i)}_{\phi_i} = e^{-\mathscr{R}_p/2}I_0[r_i(\mathbf{v}_i)\sqrt{\mathscr{R}_p}] \tag{10.3-12}$$

where $\mathscr{R}_p = 2E_p/N_0$ is the peak signal-to-noise ratio within a single observation interval and $I_0(x)$ is the modified Bessel function of the first kind and order zero. Note that the quantity $r_i(\mathbf{v}_i)\sqrt{\mathscr{R}_p}$ is the ith sampled value of the envelope of signal plus noise out of a filter matched to a (train) pulse; the sampling instant corresponds to the time at which the single-pulse matched filter output peaks due to the ith signal pulse. The sequence $r_i\sqrt{\mathscr{R}_p}$, $i = 1, 2, \ldots, N$, appears at the output of the sampler shown in Fig. 10.3-1. The first three blocks of the receiver in Fig. 10.3-1 were obtained from Fig. 9.3-4.

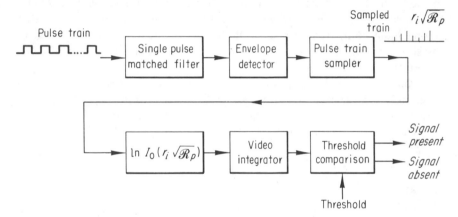

Fig. 10.3-1. Optimum detection receiver for a constant-amplitude incoherent pulse train.

Substituting Eq. (10.3-12) into (10.3-11b) results in

$$\overline{\ell'(\mathbf{v}\,|\,\boldsymbol{\theta})}_\theta = \prod_{i=1}^{N} e^{-\mathscr{R}_p/2}I_0(r_i\sqrt{\mathscr{R}_p}) \tag{10.3-13a}$$

$$= e^{-N\mathscr{R}_p/2}\prod_{i=1}^{N} I_0(r_i\sqrt{\mathscr{R}_p}) \tag{10.3-13b}$$

The test strategy is to compare the modified generalized likelihood ratio in (10.3-13b) with a threshold $\mathscr{K}'$ in each radar resolution cell. Since N and $\mathscr{R}_p$ are known, an equivalent test is

$$\prod_{i=1}^{N} I_0(r_i\sqrt{\mathscr{R}_p}) \begin{cases} \geqslant e^{N\mathscr{R}_p/2}\mathscr{K}' = \mathscr{K}_{-1} & \text{(\textit{signal present})} \\ < \mathscr{K}_{-1} & \text{(\textit{signal absent})} \end{cases} \tag{10.3-14}$$

The operating threshold $\mathscr{K}_{-1}$ is determined by specifying the false alarm probability.

An alternative test is obtained by taking the logarithm of Eq. (10.3-14); this results in

$$\sum_{i=1}^{N} \ln I_0(r_i\sqrt{\mathscr{R}_p}) \begin{cases} \geqslant \mathscr{K}_{-2} & \textit{(signal present)} \\ < \mathscr{K}_{-2} & \textit{(signal absent)} \end{cases} \tag{10.3-15}$$

where $\mathscr{K}_{-2}$ is a modified threshold equal to the logarithm of $\mathscr{K}_{-1}$. A block diagram of an optimum detection receiver that implements the operations indicated in Eq. (10.3-15) is shown in Fig. 10.3-1.

The optimum detection receiver operates in the following manner. Each pulse (waveform) in the train is processed by a matched filter. The matched filter output is envelope detected. The envelope of each pulse out of the envelope detector is sampled at a time when it is expected to peak in the absence of noise. The sampling is repeated N times for each of the N pulses in the train. This results in a string of N sampled values. Next, $\ln I_0(x)$ is computed for each of these numbers; the computed quantities are then summed and compared with a threshold. When the sum exceeds the given threshold, a signal is assumed to be present; otherwise, noise alone is presumed present.

From a practical viewpoint all of the operations described in block diagram 10.3-1 can be easily implemented except for the $\ln I_0(x)$ computation. Depending on the magnitude of x, $\ln I_0(x)$ can be approximated. For $x < 1$, $I_0(x)$ can be expanded in a power series:

$$I_0(x) = 1 + \frac{x^2}{4} + \frac{x^4}{64} + \cdots \tag{10.3-16}$$

so that

$$\ln I_0(x) = \ln\left(1 + \frac{x^2}{4} + \frac{x^4}{64} + \cdots\right) \cong \frac{x^2}{4} + O(x^4) \tag{10.3-17}$$

For small x, only the first term on the right of (10.3-17) is significant. With this approximation, Eq. (10.3-15) can be rewritten as

$$\frac{1}{4}\sum_{i=1}^{N} r_i^2 \mathscr{R}_p \begin{cases} \geqslant \mathscr{K}_{-2} \\ < \mathscr{K}_{-2} \end{cases} \tag{10.3-18a}$$

or more simply as

$$\sum_{i=1}^{N} r_i^2 \begin{cases} \geqslant \mathscr{K}_{-3} \\ < \mathscr{K}_{-3} \end{cases} \tag{10.3-18b}$$

An (approximate) optimum detection receiver for small signal-to-noise ratios based on (10.3-18b) is shown in Fig. 10.3-2.

For large signal-to-noise ratios, we have $r_i\sqrt{\mathscr{R}_p} \gg 1$; in this case, $I_0(x)$ can be approximated by an asymptotic expansion:

$$I_0(x) = \frac{e^x}{\sqrt{2\pi x}}\left(1 + \frac{1}{8x} + \cdots\right), \qquad x > 1 \tag{10.3-19}$$

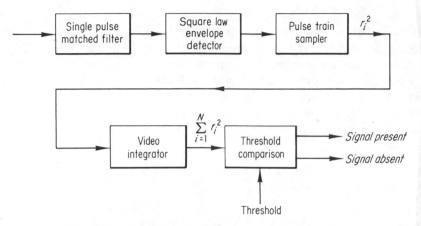

Fig. 10.3-2. Small-signal optimum-detection receiver for a constant-amplitude incoherent pulse train.

and

$$\ln I_0(x) = x - \frac{1}{2}\ln 2\pi x + \ln\left(1 + \frac{1}{8x} + \dots\right), \qquad x > 1 \qquad (10.3\text{-}20)$$

For $x \gg 1$ the first term in the expansion of Eq. (10.3-20) is a satisfactory approximation to $\ln I_0(x)$. In this case Eq. (10.3-15) becomes:

$$\sum_{i=1}^{N} r_i \sqrt{\mathscr{R}_p}\begin{cases} \geqslant \mathscr{K}_{-2} \\ < \mathscr{K}_{-2} \end{cases} \qquad (13.3\text{-}21a)$$

or

$$\sum_{i=1}^{N} r_i \begin{cases} \geqslant \mathscr{K}_{-4} & (signal\ present) \\ < \mathscr{K}_{-4} & (signal\ absent) \end{cases} \qquad (10.3\text{-}21b)$$

An approximate optimum detection receiver for large signal-to-noise ratios based on (10.3-21b) is shown in Fig. 10.3-3.

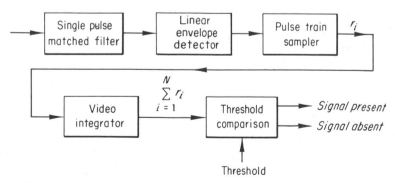

Fig. 10.3-3. Large-signal optimum-detection receiver for a constant-amplitude incoherent pulse train.

The principal difference between Figs. 10.3-2 and 10.3-3 is the type of envelope detector employed; a square-law detector is used in the small-signal optimum receiver and a linear detector in the large-signal receiver. In the next section the expected performance of an optimum receiver employing a square-law envelope detector is calculated. This choice is governed by the mathematical difficulties of evaluating performance of linear detectors, as well as greater interest in small-signal detection. It will be shown later that the performance of both detectors is practically the same for all signal-to-noise ratios.

10.4. Small-Signal Optimum Receiver Performance

In this section, the probability of detection and the probability of false alarm are calculated for the small-signal optimum detection receiver shown in Fig. 10.3-2. The analysis follows the classic work of Marcum [1, 2]. The test statistic is given in (10.3-18b)—namely, $\sum_{i=1}^{N} r_i^2$. Let

$$y = \sum_{i=1}^{N} r_i^2 \tag{10.4-1}$$

The probability of detection is the probability that y exceeds the threshold when signal is present, and the probability of false alarm is the probability that y exceeds the threshold when only noise is present. To perform the desired calculations the probability density of sum random variable y must be found.

The probability density of y can be calculated from the density function of r_i. The probability density of r_i was derived in Chapter 9 for a single observation of signal plus noise, and the result is given by Eq. (9.3-38):

$$p(r_i; \mathscr{R}_p) = \begin{cases} r_i e^{-(r_i^2 + \mathscr{R}_p)/2} I_0(r_i\sqrt{\mathscr{R}_p}), & r_i \geqslant 0 \\ 0 & r_i < 0 \end{cases} \tag{10.4-2}$$

where the parametric dependence on $\mathscr{R}_p$ is made evident by the notation $p(r_i; \mathscr{R}_p)$.

To simplify calculations, it is convenient to replace test statistic y given by (10.4-1) by a related test statistic $Y = y/2$ which is compared to a suitably modified threshold Y_b. If

$$z_i = \frac{1}{2} r_i^2 \tag{10.4-3}$$

then

$$Y = \sum_{i=1}^{N} \frac{1}{2} r_i^2 = \sum_{i=1}^{N} z_i \tag{10.4-4}$$

A scalar change of the test statistic does not affect the basic structure or performance of the optimum receiver.

To find the probability density of Y, first the probability density $p(z_i; \mathscr{R}_p)$ is found directly from (10.4-2) by a transformation of variables:

$$p(z_i; \mathscr{R}_p) = \begin{cases} \dfrac{p(r_i; \mathscr{R}_p)}{|J|}\bigg|_{r_i = \sqrt{2z_i}}, & z_i \geqslant 0 \\ 0, & z_i < 0 \end{cases} \qquad (10.4\text{-}5)$$

The magnitude of the Jacobian is

$$|J| = \left|\frac{dz_i}{dr_i}\right| = r_i, \qquad r_i \geqslant 0 \qquad (10.4\text{-}6)$$

From Eqs. (10.4-2), (10.4-5), and (10.4-6), the probability density function of z_i is

$$p(z_i; \mathscr{R}_p) = \begin{cases} e^{-(z_i + \mathscr{R}_p/2)} I_0(\sqrt{2z_i \mathscr{R}_p}), & z_i \geqslant 0 \\ 0, & z_i < 0 \end{cases} \qquad (10.4\text{-}7)$$

Random variable Y is the sum of N statistically independent random variables z_i. Equation (3.12-3) relates the characteristic function of a sum random variable to the characteristic function of the individual random variables; this relation is

$$C_Y(\xi; \mathscr{R}_p) = \prod_{i=1}^{N} C_{z_i}(\xi; \mathscr{R}_p) \qquad (10.4\text{-}8)$$

where $C_{z_i}(\xi; \mathscr{R}_p)$ is the characteristic function of random variable z_i and $C_Y(\xi; \mathscr{R}_p)$ is the characteristic function of sum variable Y. The probability density of Y is given by the anticharacteristic function of $C_Y(\xi; \mathscr{R}_p)$.

The characteristic function corresponding to $p(z_i; \mathscr{R}_p)$ is computed from the expression [see Eq. (3.12-1b)]:

$$C_{z_i}(\xi; \mathscr{R}_p) = \int_{-\infty}^{\infty} e^{j\xi z_i} p(z_i; \mathscr{R}_p)\, dz_i \qquad (10.4\text{-}9)$$

As noted in Chapter 3, the right side of Eq. (10.4-9) is the Fourier transform of $p(z_i; \mathscr{R}_p)$ with transform variable ξ replaced by $-\xi$. Hence tables of Fourier transform pairs can be used to evaluate characteristic functions and their inverses and are frequently employed in this and the following chapter.

Substituting Eq. (10.4-7) into (10.4-9) yields

$$C_{z_i}(\xi; \mathscr{R}_p) = \int_{0}^{\infty} e^{j\xi z_i} e^{-(z_i + \mathscr{R}_p/2)} I_0(\sqrt{2z_i \mathscr{R}_p})\, dz_i \qquad (10.4\text{-}10)$$

This integral can be evaluated using pair 655.1 in the Campbell and Foster tables [3] of Fourier transforms; the result is

$$C_{z_i}(\xi; \mathscr{R}_p) = \frac{e^{j\mathscr{R}_p \, \xi/2(1-j\xi)}}{1 - j\xi} \qquad (10.4\text{-}11)$$

Substituting Eq. (10.4-11) into (10.4-8) yields for the characteristic function of Y

$$C_Y(\xi; \mathscr{R}_p) = \prod_{i=1}^{N} \frac{e^{j\mathscr{R}_p \, \xi/2(1-j\xi)}}{1 - j\xi} \tag{10.4-12a}$$

$$= \frac{e^{jN\mathscr{R}_p\xi/2(1-j\xi)}}{(1 - j\xi)^N} \tag{10.4-12b}$$

The probability density $p(Y; \mathscr{R}_p)$ is the anticharacteristic function of Eq. (10.4-12b). The desired result is obtained from pair 650.0 of the Campbell and Foster tables [3] and is

$$p(Y; \mathscr{R}_p) = \begin{cases} \left(\dfrac{2Y}{N\mathscr{R}_p}\right)^{(N-1)/2} e^{-Y-N\mathscr{R}_p/2} I_{N-1}(\sqrt{2N\mathscr{R}_p Y}) & Y \geqslant 0 \\ 0, & Y < 0 \end{cases} \tag{10.4-13}$$

where $I_{N-1}(x)$ is the modified Bessel function of the first kind of order $N-1$.

To compute the probability of false alarm, it is necessary to know $p(Y; 0)$—that is, the distribution of Y when signal is absent. An obvious method for obtaining this distribution is to let $\mathscr{R}_p$ approach zero in expression (10.4-13) for $p(Y; \mathscr{R}_p)$. For this purpose $I_{N-1}(\sqrt{2N\mathscr{R}_p Y})$ is replaced by the first term of its power series expansion. The power series expansion of $I_m(x)$ for small values of x is given by

$$I_m(x) = \frac{x^m}{2^m m!} \left[1 + \frac{x^2}{2^2(m+1)} + \frac{x^4}{2 \cdot 2^4(m+1)(m+2)} + \cdots \right] \tag{10.4-14}$$

Then, for $\mathscr{R}_p$ approaching zero, $I_{N-1}(\sqrt{2N\mathscr{R}_p Y})$ is approximately given by

$$I_{N-1}(\sqrt{2N\mathscr{R}_p Y})\bigg|_{\mathscr{R}_p \to 0} \cong \frac{(2N\mathscr{R}_p Y)^{(N-1)/2}}{2^{N-1}(N-1)!} \tag{10.4-15}$$

Substituting Eq. (10.4-15) into (10.4-13) yields the following approximate expression for $p(Y; \mathscr{R}_p)$:

$$p(Y; \mathscr{R}_p)\bigg|_{\mathscr{R}_p \to 0} \cong \frac{Y^{N-1}e^{-Y-N\mathscr{R}_p/2}}{(N-1)!}, \qquad Y \geqslant 0 \tag{10.4-16}$$

In the limit, $\mathscr{R}_p = 0$ and probability density $p(Y; 0)$ is given by

$$p(Y; 0) = \begin{cases} \dfrac{Y^{N-1}e^{-Y}}{(N-1)!}, & Y \geqslant 0 \\ 0, & Y < 0 \end{cases} \tag{10.4-17}$$

The distribution of Y when noise alone is present could have been obtained by another procedure. Random variable $(2Y)$ has a χ^2-distribution with $2N$ degrees of freedom, since it is the sum of $2N$ statistically independent squared Gaussian variables—that is, N pairs of in-phase and quadrature noise components—each of zero mean and unit variance. A simple transformation of variables would then have resulted in expression (10.4-17).

The false-alarm probability is the probability that Y exceeds threshold Y_b when noise alone is present. Hence, from Eq. (10.4-17) false-alarm probability P_{fa} is given by

$$P_{fa} = \int_{Y_b}^{\infty} p(Y; 0)\, dY = \int_{Y_b}^{\infty} \frac{Y^{N-1} e^{-Y}}{(N-1)!}\, dY \qquad (10.4\text{-}18a)$$

or equivalently by

$$P_{fa} = 1 - \int_{0}^{Y_b} \frac{Y^{N-1} e^{-Y}}{(N-1)!}\, dY \qquad (10.4\text{-}18b)$$

The integral in (10.4-18b) is similar in form to the incomplete gamma function $I(u, s)$ which has been tabulated by Pearson [4]; this function is defined by

$$I(u, s) = \int_{0}^{u\sqrt{1+s}} \frac{e^{-v} v^s}{s!}\, dv \qquad (10.4\text{-}19)$$

In terms of the incomplete gamma function Eq. (10.4-18b) for P_{fa} can be written as

$$P_{fa} = 1 - I\left(\frac{Y_b}{\sqrt{N}}, N-1\right) \qquad (10.4\text{-}20)$$

To obtain Y_b for specified values of P_{fa} and N, the Pearson tables [4] are useful for $N < 50$ and $P_{fa} > 10^{-6}$. For $N > 50$ or $P_{fa} < 10^{-6}$, Marcum [2] has shown that Eq. (10.4-18a) can be successively integrated by parts to give

$$P_{fa} = \frac{Y_b^{N-1} e^{-Y_b}}{(N-1)!}\left[1 + \frac{N-1}{Y_b} + \frac{(N-1)(N-2)}{Y_b^2} + \cdots\right] \qquad (10.4\text{-}21)$$

To simplify expression (10.4-21), it can be shown for $N \gg 1$ that $Y_b > N$ and the bracketed quantity in Eq. (10.4-21) is approximately equal to

$$\left[1 + \frac{N-1}{Y_b} + \frac{(N-1)(N-2)}{Y_b^2} + \cdots\right] \cong \frac{1}{1 - \dfrac{N-1}{Y_b}} \qquad (10.4\text{-}22)$$

With (10.4-22) substituted into (10.4-21) and rewriting $(N-1)!$ as $N!/N$, P_{fa} becomes approximately

$$P_{fa} \cong \frac{N Y_b^N e^{-Y_b}}{N!(Y_b - N + 1)} \qquad (10.4\text{-}23)$$

When $N!$ is replaced by Stirling's approximation:

$$N! = \sqrt{2\pi N}\left(\frac{N}{e}\right)^N \qquad (10.4\text{-}24)$$

Eq. (10.4-23) can be rewritten as

$$P_{fa} \cong \sqrt{\frac{N}{2\pi}}\left(\frac{Y_b}{N}\right)^N \frac{e^{-Y_b + N}}{Y_b - N + 1} \qquad (10.4\text{-}25a)$$

or as

$$P_{fa} = \sqrt{\frac{N}{2\pi}} \frac{1}{Y_b - N + 1} \exp\left[-Y_b + N\left(1 + \ln\frac{Y_b}{N}\right)\right] \quad (10.4\text{-}25b)$$

Y_b can be calculated from (10.4-25b) for selected values of P_{fa} and N with good accuracy for $N \gg 1$. For $1 \leqslant N \leqslant 150$ and false-alarm probabilities equal to 10^{-m}, where m is an integer ranging between 1 and 12, tables for Y_b have been prepared by Pachares [5] that are accurate to four decimal places.

Next, probability of detection is calculated as a function of peak signal-to-noise ratio $\mathscr{R}_p$ by substituting Eq. (10.4-13) into

$$P_d = \int_{Y_b}^{\infty} p(Y; \mathscr{R}_p)\, dY \quad (10.4\text{-}26)$$

This results in

$$P_d = \int_{Y_b}^{\infty} \left(\frac{2Y}{N\mathscr{R}_p}\right)^{(N-1)/2} e^{-Y - N\mathscr{R}_p/2} I_{N-1}(\sqrt{2N\mathscr{R}_p Y})\, dY \quad (10.4\text{-}27)$$

which can be rewritten as

$$P_d = 1 - \int_0^{Y_b} \left(\frac{2Y}{N\mathscr{R}_p}\right)^{(N-1)/2} e^{-Y - N\mathscr{R}_p/2} I_{N-1}(\sqrt{2N\mathscr{R}_p Y})\, dY \quad (10.4\text{-}28)$$

This integral is similar in form to the incomplete Toronto function:

$$T_B(m, n, r) = 2r^{n-m+1} e^{-r^2} \int_0^B t^{m-n} e^{-t^2} I_n(2rt)\, dt \quad (10.4\text{-}29)$$

After some simple manipulations, expression (10.4-28) for P_d can be written as

$$P_d = 1 - T_{\sqrt{Y_b}}\left(2N - 1, N - 1, \sqrt{\frac{N\mathscr{R}_p}{2}}\right) \quad (10.4\text{-}30)$$

The incomplete Toronto function is tabulated by Heatley [6]. It is also plotted in Marcum's report [2]. In Figs. 10.4-1 through 10.4-10 probability of detection P_d is plotted as a function of peak (single-pulse) signal-to-noise ratio $\mathscr{R}_p$ for selected values of probability of false alarm P_{fa} (see discussion at end of Section 9.2 concerning units of $\mathscr{R}_p$). These plots are for $N = 1, 2, 3, 6, 10, 30, 100, 300, 1000,$ and 3000 respectively. The data for these curves were obtained from computer results of Fehlner [7].

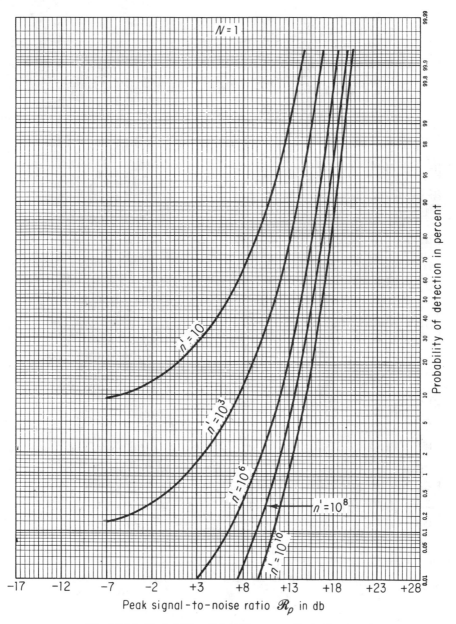

Fig. 10.4-1. Probability of detecting a nonfluctuating target (square-law detector), $N = 1$. [N = number of pulses incoherently integrated; $P_{fa} = 0.693/n'$.]

349

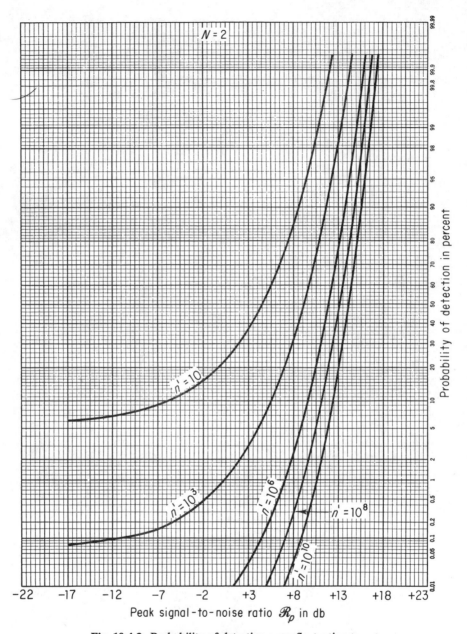

Fig. 10.4-2. Probability of detecting a nonfluctuating target (square-law detector), $N = 2$. [N = number of pulses incoherently integrated; $P_{fa} = 0.693/n'$.]

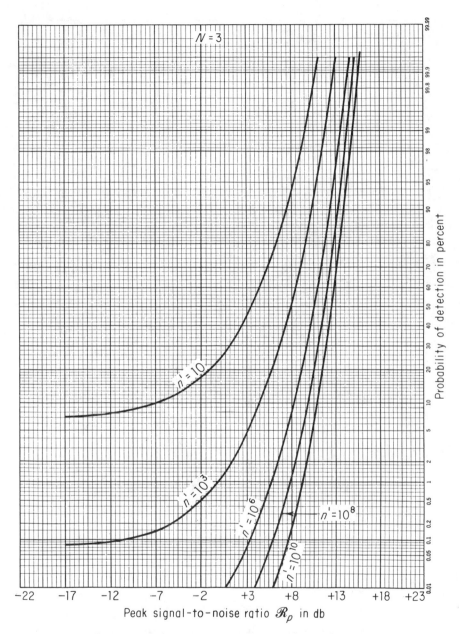

Fig. 10.4-3. Probability of detecting a nonfluctuating target (square-law detector), $N = 3$. [N = number of pulses incoherently integrated; $P_{fa} = 0.693/n'$.]

351

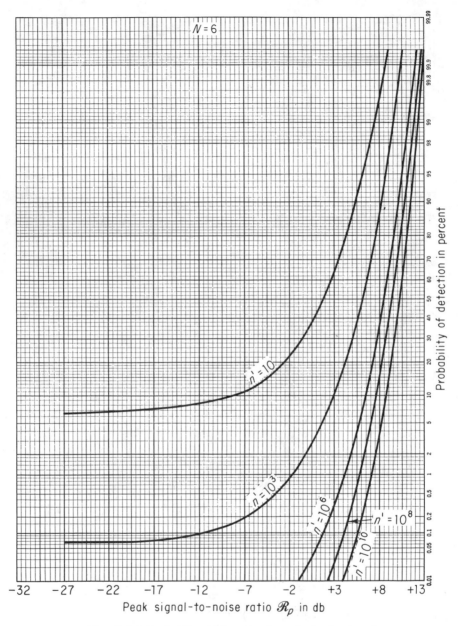

Fig. 10.4-4. Probability of detecting a nonfluctuating target (square-law detector), $N = 6$. [N = number of pulses incoherently integrated; $P_{fa} = 0.693/n'$.]

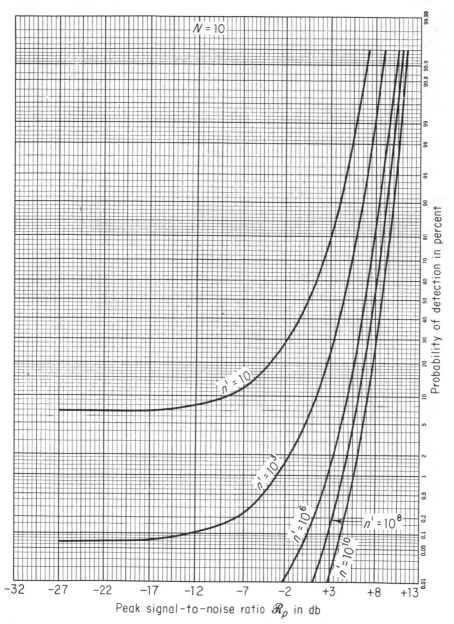

Fig. 10.4-5. Probability of detecting a nonfluctuating target (square-law detector), $N = 10$. [N = number of pulses incoherently integrated; $P_{fa} = 0.693/n'$.]

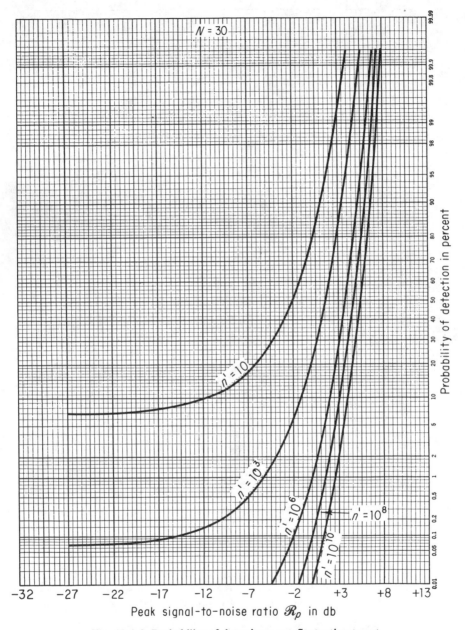

Fig. 10.4-6. Probability of detecting a nonfluctuating target (square-law detector), $N = 30$. [N = number of pulses incoherently integrated; $P_{fa} = 0.693/n'$.]

354

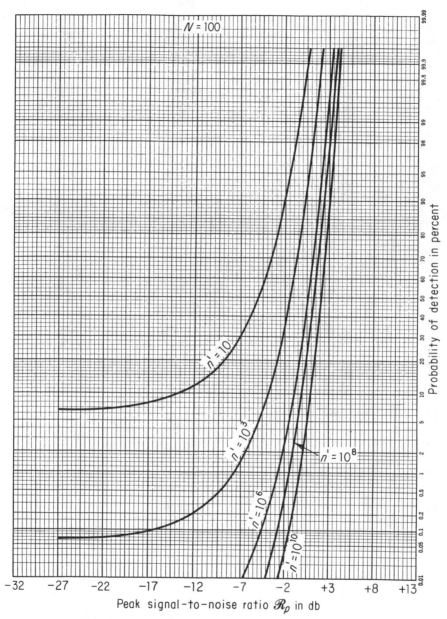

Fig. 10.4-7. Probability of detecting a nonfluctuating target (square-law detector), $N = 100$. [$N =$ number of pulses incoherently integrated; $P_{fa} = 0.693/n'$.]

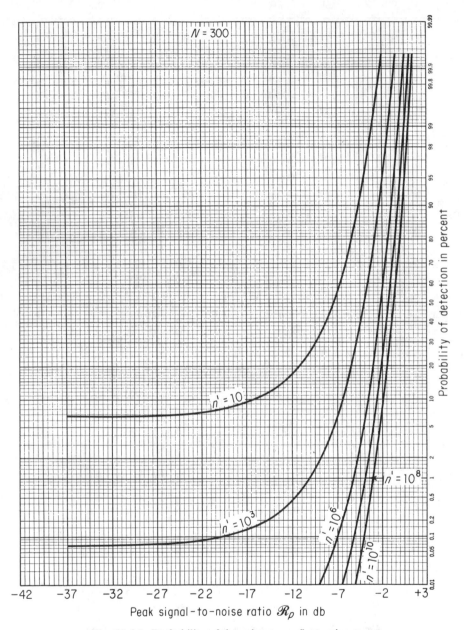

Fig. 10.4-8. Probability of detecting a nonfluctuating target (square-law detector), $N = 300$. [N = number of pulses incoherently integrated; $P_{fa} = 0.693/n'$.]

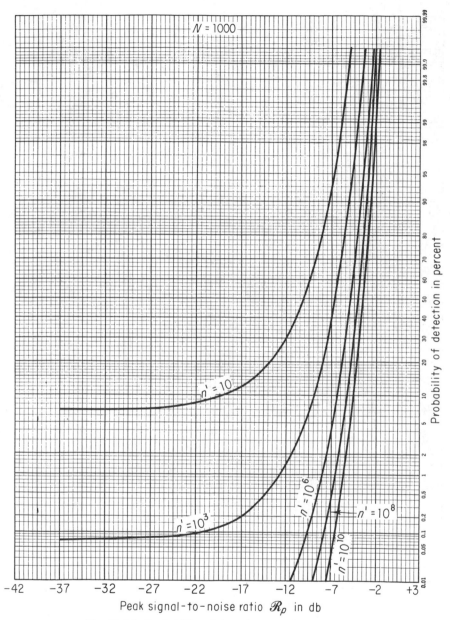

Fig. 10.4-9. Probability of detecting a nonfluctuating target (square-law detector), $N = 1000$. [$N =$ number of pulses incoherently integrated; $P_{fa} = 0.693/n'$.]

357

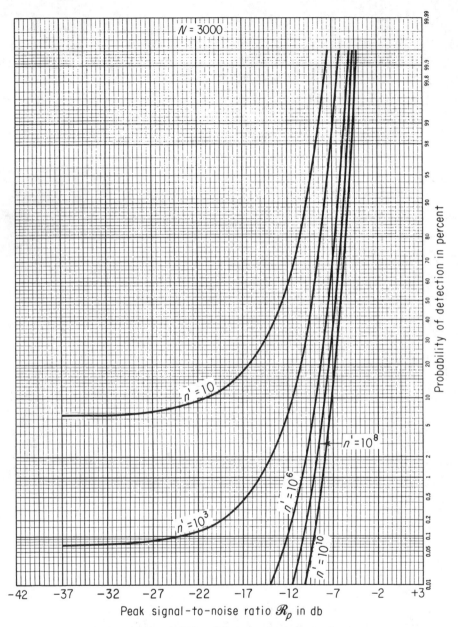

Fig. 10.4-10. Probability of detecting a nonfluctuating target (square-law detector), $N = 3000$. [N = number of pulses incoherently integrated; $P_{fa} = 0.693/n'$.]

10.5. Integration Loss for Nonfluctuating Incoherent Detector

From a practical viewpoint it is easier to transmit and process an incoherent pulse train than a coherent pulse train. However, the detectability of an incoherent pulse train is poorer than that of a coherent pulse train. It is of interest to determine the extent of this so-called *integration loss*.

Let $\mathscr{R}$ denote the peak signal-to-noise ratio of a coherently processed waveform of unknown initial phase to achieve a specified P_d and P_{fa}. When the coherent waveform is an N-pulse train, we infer that the peak signal-to-noise ratio of each pulse in the train is $\mathscr{R}/N$. For an incoherent pulse train the peak signal-to-noise ratio $\mathscr{R}_p$ of each pulse is greater than $\mathscr{R}/N$ to achieve the same P_d and P_{fa}. Hence, a measure of integration loss is given by

$$L = 10 \log \frac{\mathscr{R}_p}{\mathscr{R}/N} \qquad (10.5\text{-}1)$$

In general L is a function of P_d, P_{fa}, and N. However, it is fairly insensitive to changes in P_d and P_{fa}. This is illustrated in Figs. 10.5-1 and 10.5-2, which are plots of integration loss L versus N for selected values of P_d and P_{fa}.

The loss is small for small N. For large N, the slope of the integration loss curve approaches a value of $10 \log \sqrt{N}$; this is shown by a dotted line in Figs. 10.5-1 and 10.5-2. In this region, single-pulse peak signal-to-noise ratio $\mathscr{R}_p$ is inversely proportional to $\sqrt{N}$. For $N \gg 1$, the loss in db is approximately equal to $(10 \log \sqrt{N} - 5.5)$. This result is useful for obtaining rapid estimates of performance.

10.6. Gram-Charlier Series

When many random variables are summed, the central limit theorem of statistics states that the probability density of the sum random variable, under very general conditions, approaches the Gaussian distribution. Hence, it is reasonable to approximate the probability density of sum random variable Y for large N by a series expansion in which the Gaussian density function is the principal term. An expansion of this type may be useful for any random variable that is approximately Gaussian.

The Gram-Charlier series is an orthogonal polynomial expansion whose first term is the Gaussian density function. If $p(x)$ is a probability density function that is nearly Gaussian and x has zero mean and unit variance, then the Gram-Charlier series expansion of $p(x)$ is given by

$$p(x) = \sum_{i=0}^{\infty} a_i \overset{(i)}{\phi}(x) \qquad (10.6\text{-}1)$$

where

$$\overset{(0)}{\phi}(x) = \phi(x) = \frac{1}{\sqrt{2\pi}} e^{-x^2/2} \qquad (10.6\text{-}2)$$

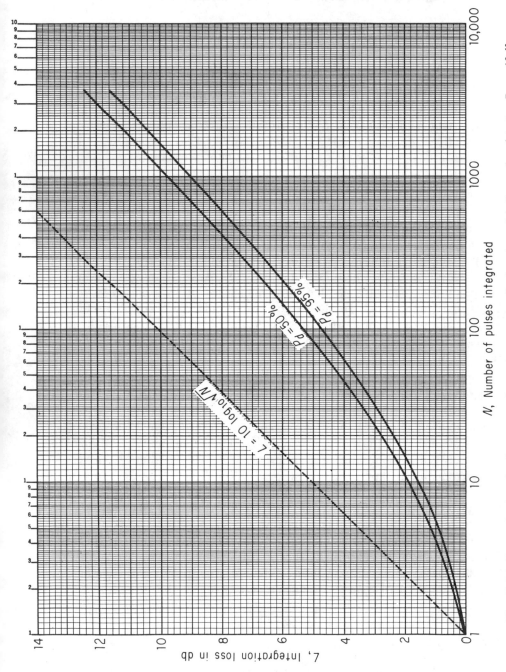

Fig. 10.5-1. Integration loss, incoherent vs. coherent (constant-amplitude pulse train). Square-law detector, $P_{fa} = 10^{-10}$.

360

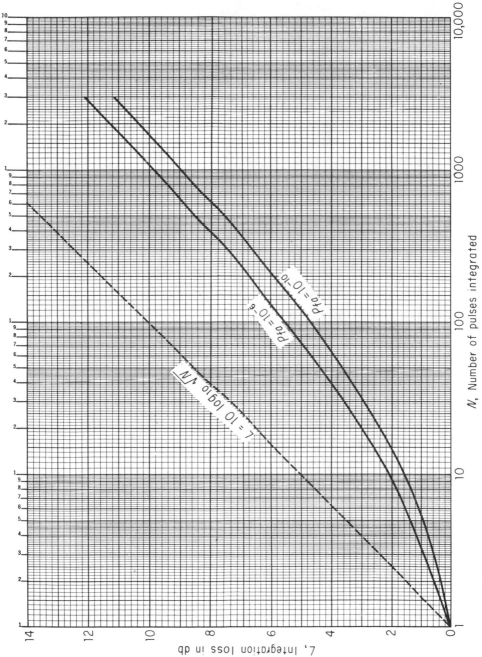

Fig. 10.5-2. Integration loss, incoherent vs. coherent (constant-amplitude pulse train). Square-law detector, $P_d = 95\%$.

Along axes of figure:

N, Number of pulses integrated

L, Integration loss in db

Curve labels:

Pfa = 10^{-10}

Pfa = 10^{-6}

$\sqrt{N}$: 10 log₁₀ $\sqrt{N}$ = 7

and

$$\overset{(i)}{\phi}(x) = \frac{d^i \phi(x)}{dx^i} \tag{10.6-3}$$

Coefficients a_i in (10.6-1) can be evaluated in terms of the Hermite polynomials:

$$\overset{(i)}{\phi}(x) = \frac{(-1)^i}{\sqrt{2\pi}} e^{-x^2/2} H_i(x) \tag{10.6-4}$$

From Eq. (10.6-4) the first few Hermite polynomials are given by

$$H_0(x) = 1 \tag{10.6-5a}$$

$$H_1(x) = x \tag{10.6-5b}$$

$$H_2(x) = x^2 - 1 \tag{10.6-5c}$$

$$H_3(x) = x^3 - 3x \tag{10.6-5d}$$

$$H_4(x) = x^4 - 6x^2 + 3 \tag{10.6-5e}$$

It can be shown that $\overset{(i)}{\phi}(x)$ and $H_j(x)$ are biorthogonal functions; that is,

$$\int_{-\infty}^{\infty} \overset{(i)}{\phi}(x) H_j(x)\, dx = (-1)^i i! \,\delta_{ij} = \begin{cases} (-1)^i i!, & i = j \\ 0, & i \neq j \end{cases} \tag{10.6-6}$$

To evaluate coefficient a_i, both sides of Eq. (10.6-1) are multiplied by $H_i(x)$ and the result is integrated from $-\infty$ to $+\infty$. With Eq. (10.6-6), the result simplifies to

$$a_i = \frac{(-1)^i}{i!} \int_{-\infty}^{\infty} p(x) H_i(x)\, dx \tag{10.6-7}$$

For example, the first three coefficients a_0, a_1, and a_2 can be found by substituting (10.6-5a), (10.6-5b) and (10.6-5c) into (10.6-7):

$$a_0 = \int_{-\infty}^{\infty} p(x)\, dx = 1 \tag{10.6-8a}$$

$$a_1 = -\int_{-\infty}^{\infty} x\, p(x)\, dx = 0 \tag{10.6-8b}$$

$$a_2 = \frac{1}{2!} \int_{-\infty}^{\infty} (x^2 - 1) p(x)\, dx = \frac{1}{2}(1 - 1) = 0 \tag{10.6-8c}$$

With these results, the Gram-Charlier series expansion of $p(x)$ in (10.6-1) can be simplified to

$$p(x) = \phi(x) + \sum_{i=3}^{\infty} a_i \overset{(i)}{\phi}(x) \tag{10.6-9}$$

Equation (10.6-9) is of the desired form; that is, the first term in the series expansion is a Gaussian density function of zero mean and unit variance. Frequently, the first few terms of the expansion provide a good approximation of $p(x)$.

When the mean of random variable x is not zero and/or its variance σ^2 is not unity, the Gram-Charlier expansion of $p(x)$ is obtained by expanding a related density function $g(t)$ in a Gram-Charlier series, where

$$t = \frac{x - \bar{x}}{\sigma} \tag{10.6-10}$$

is a standardized variable. By a simple transformation of variables it follows that

$$p(x) = \left. \frac{g(t)}{\left| \dfrac{dx}{dt} \right|} \right|_{t=(x-\bar{x})/\sigma} = \frac{1}{\sigma} g \left(\frac{x - \bar{x}}{\sigma} \right) \tag{10.6-11}$$

If the Gram-Charlier series expansion for $g(t)$ is

$$g(t) = \sum_{i=0}^{\infty} c_i \overset{(i)}{\phi}(t) \tag{10.6-12}$$

then, from (10.6-11) and (10.6-12), the expansion for $p(x)$ becomes

$$p(x) = \frac{1}{\sigma} \sum_{i=0}^{\infty} c_i \overset{(i)}{\phi} \left(\frac{x - \bar{x}}{\sigma} \right) \tag{10.6-13}$$

If both sides of (10.6-13) are multiplied by $H_j[(x - \bar{x})/\sigma]$ and the result integrated from $-\infty$ to $+\infty$, coefficient c_i can be evaluated with relation (10.6-6) as

$$c_i = \frac{(-1)^i}{i!} \int_{-\infty}^{\infty} p(x) H_i \left(\frac{x - \bar{x}}{\sigma} \right) dx \tag{10.6-14}$$

Evaluating (10.6-14) for the first few coefficients c_i yields

$$c_0 = 1 \tag{10.6-15a}$$

$$c_1 = c_2 = 0 \tag{10.6-15b}$$

$$c_3 = -\frac{1}{3!} \alpha_3 \tag{10.6-15c}$$

$$c_4 = \frac{1}{4!} (\alpha_4 - 3) \tag{10.6-15d}$$

$$c_5 = -\frac{1}{5!} (\alpha_5 - 10\alpha_3) \tag{10.6-15e}$$

$$c_6 = \frac{1}{6!} (\alpha_6 - 15\alpha_4 + 30) \tag{10.6-15f}$$

where α_i is the ith central moment of $p(x)$, normalized by σ^i:

$$\alpha_i = \frac{1}{\sigma^i} \int_{-\infty}^{\infty} (x - \bar{x})^i p(x) \, dx \tag{10.6-16}$$

The α_i can be expressed in terms of the conventional moments of the distribution $p(x)$; thus if m_n denotes the nth moment of $p(x)$—that is,

$$m_n = \int_{-\infty}^{\infty} x^n p(x) \, dx \tag{10.6-17}$$

—then α_3 through α_6 are related to noncentral moments m_n by

$$\alpha_3 = \frac{m_3 - 3m_2 m_1 + 2m_1^3}{\sigma^3} \tag{10.6-18a}$$

$$\alpha_4 = \frac{m_4 - 4m_3 m_1 + 6m_2 m_1^2 - 3m_1^4}{\sigma^4} \tag{10.6-18b}$$

$$\alpha_5 = \frac{m_5 - 5m_4 m_1 + 10m_3 m_1^2 - 10m_2 m_1^3 + 4m_1^5}{\sigma^5} \tag{10.6-18c}$$

$$\alpha_6 = \frac{m_6 - 6m_5 m_1 + 15m_4 m_1^2 - 20m_3 m_1^3 + 15m_2 m_1^4 - 5m_1^6}{\sigma^6} \tag{10.6-18d}$$

In summary, to obtain a Gram-Charlier expansion, moments m_n are found from Eq. (10.6-17), or they can be conveniently computed from the characteristic function of the distribution. Next the α_i are evaluated using Eqs. (10.6-18). The α_i are then substituted into Eq. (10.6-15) to obtain coefficients c_i of Gram-Charlier expansion (10.6-13).

Because the order of magnitude of coefficient c_i does not decrease uniformly as i increases, the Gram-Charlier series can be regrouped into terms with magnitudes of the same order. In the regrouped series, called an Edgeworth series, the terms grouped together, in descending order of magnitude, are

$$i = 0 \tag{10.6-19a}$$

$$i = 3 \tag{10.6-19b}$$

$$i = 4, 6 \tag{10.6-19c}$$

$$i = 5, 7, 9 \tag{10.6-19d}$$

Thus, a first-order approximation of $p(x)$ is given by the first term of the Gram-Charlier series. A better approximation is provided by adding the $i = 3$ term. The next order approximation is provided by the addition of terms $i = 4$ and $i = 6$, and so on.

10.7. Analytic Approximations (Nonfluctuating Incoherent Pulse Train)

In this section analytic approximations relating probability of false alarm P_{fa}, probability of detection P_d, peak (single-pulse) signal-to-noise ratio $\mathcal{R}_p$, and the number of train pulses N are derived for a constant-amplitude incoherent pulse train. The approximations are convenient for rapid estimation of desired parameters.

A first-order approximation to the density function of sum random variable Y, $p(Y; \mathcal{R}_p)$, is obtained by retaining only the $i = 0$ term of Gram-

Charlier series (10.6-13):

$$p(Y; \mathcal{R}_p)_{\text{approx}} \cong c_0 \phi(t)|_{t=(Y-\bar{Y})/\sigma} \tag{10.7-1a}$$

$$= \frac{1}{\sigma} \frac{1}{\sqrt{2\pi}} e^{-t^2/2}\bigg|_{t=(Y-\bar{Y})/\sigma} \tag{10.7-1b}$$

$$= \frac{1}{\sqrt{2\pi}\,\sigma} e^{-(Y-\bar{Y})^2/2\sigma^2} \tag{10.7-1c}$$

It follows from Eq. (10.7-1c) that a first order Gaussian approximation of probability density function $p(Y; \mathcal{R}_p)$ is characterized by the true mean and variance of Y. These parameters can be evaluated from the exact distribution $p(Y; \mathcal{R}_p)$ given by Eq. (10.4-13).

Alternatively, a method described in Section 3.12 can be used to obtain the moments of this distribution from its characteristic function. The required relation, given by Eq. (3.12-6), is

$$m_n = \frac{1}{j^n} \overset{(n)}{C}_Y(0) \tag{10.7-2}$$

where m_n is the nth moment of the distribution of Y and $\overset{(n)}{C}_Y(0)$ is the nth derivative of characteristic function $C_Y(\xi)$ evaluated at $\xi = 0$.

The characteristic function $C_Y(\xi; \mathcal{R}_p)$ of Y is given in Eq. (10.4-12b). The first derivative of $C_Y(\xi; \mathcal{R}_p)$ is

$$\frac{dC_Y(\xi; \mathcal{R}_p)}{d\xi} = \frac{jN\mathcal{R}_p}{2(1-j\xi)^{N+2}} e^{j\xi N \mathcal{R}_p/2(1-j\xi)} + \frac{jN}{(1-j\xi)^{N+1}} e^{j\xi N \mathcal{R}_p/2(1-j\xi)} \tag{10.7-3}$$

Substituting (10.7-3) into (10.7-2) yields

$$m_1 = \bar{Y} = N\left(1 + \frac{\mathcal{R}_p}{2}\right) \tag{10.7-4}$$

In similar manner the second moment is

$$m_2 = N^2\left(1 + \frac{\mathcal{R}_p}{2}\right)^2 + N(1 + \mathcal{R}_p) \tag{10.7-5}$$

Variance σ^2 is obtained from m_1 and m_2 as follows:

$$\sigma^2 = m_2 - m_1^2 = N(1 + \mathcal{R}_p) \tag{10.7-6}$$

Substituting Eqs. (10.7-4) and (10.7-6) into (10.7-1c) yields

$$p(Y; \mathcal{R}_p)_{\text{approx}} = \frac{1}{\sqrt{2\pi N(1 + \mathcal{R}_p)}} e^{-[Y-N(1+\mathcal{R}_p/2)]^2/2N(1+\mathcal{R}_p)} \tag{10.7-7}$$

For noise alone, the approximate Gaussian distribution of Y is obtained with $\mathcal{R}_p = 0$ in Eq. (10.7-7):

$$p(Y; 0)_{\text{approx}} = \frac{1}{\sqrt{2\pi N}} e^{-(Y-N)^2/2N} \tag{10.7-8}$$

From (10.7-8) the (approximate) false-alarm probability is given by

$$P_{fa} \cong \int_{Y_b}^{\infty} \frac{1}{\sqrt{2\pi N}} e^{-(Y-N)^2/2N} \, dY \qquad (10.7\text{-}9)$$

If we let

$$x = \frac{Y - N}{\sqrt{N}} \qquad (10.7\text{-}10)$$

Eq. (10.7-9) can be rewritten as

$$P_{fa} \cong \int_{(Y_b-N)/\sqrt{N}}^{\infty} \frac{1}{\sqrt{2\pi}} e^{-x^2/2} \, dx \qquad (10.7\text{-}11)$$

The following definitions were introduced in Section 9.6:

$$\Phi(T) = \int_{T}^{\infty} \frac{1}{\sqrt{2\pi}} e^{-x^2/2} \, dx = b \qquad (10.7\text{-}12)$$

and the inverse relationship

$$\Phi^{-1}(b) = T \qquad (10.7\text{-}13)$$

With definition (10.7-12), Eq. (10.7-11) for P_{fa} can be concisely stated as

$$P_{fa} \cong \Phi\left(\frac{Y_b - N}{\sqrt{N}}\right) \qquad (10.7\text{-}14)$$

With definition (10.7-13), Eq. (10.7-14) can be rewritten as

$$\Phi^{-1}(P_{fa}) \cong \frac{Y_b - N}{\sqrt{N}} \qquad (10.7\text{-}15)$$

Solving Eq. (10.7-15) for Y_b yields

$$Y_b \cong \sqrt{N} \, \Phi^{-1}(P_{fa}) + N \qquad (10.7\text{-}16)$$

The approximate value of threshold Y_b can be evaluated from Eq. (10.7-16) using Fig. 10.7-1, which is a plot of P_{fa} versus $\Phi^{-1}(P_{fa})$ for values of P_{fa} between 10^{-2} and 10^{-15}.

The (approximate) probability of detection is computed from Eq. (10.7-7) as follows:

$$P_d \cong \int_{Y_b}^{\infty} \frac{1}{\sqrt{2\pi N(1 + \mathscr{R}_p)}} e^{-[Y-N(1+\mathscr{R}_p/2)]^2/2N(1+\mathscr{R}_p)} \, dY \qquad (10.7\text{-}17)$$

If

$$z = \frac{Y - N(1 + \mathscr{R}_p/2)}{\sqrt{N(1 + \mathscr{R}_p)}} \qquad (10.7\text{-}18)$$

then Eq. (10.7-18) can be rewritten as

$$P_d \cong \int_{Y_b-N(1+\mathscr{R}_p/2)/\sqrt{N(1+\mathscr{R}_p)}}^{\infty} \frac{1}{\sqrt{2\pi}} e^{-z^2/2} \, dz \qquad (10.7\text{-}19)$$

or more concisely as

$$P_d \cong \Phi \left[\frac{Y_b - N(1 + \mathscr{R}_p/2)}{\sqrt{N(1 + \mathscr{R}_p)}} \right] \tag{10.7-20}$$

Inverting Eq. (10.7-20) using the definition of $\Phi^{-1}(\quad)$ yields

$$\Phi^{-1}(P_d) \cong \frac{Y_b - N(1 + \mathscr{R}_p/2)}{\sqrt{N(1 + \mathscr{R}_p)}} \tag{10.7-21}$$

Replacing Y_b in Eq. (10.7-21) by (10.7-16) results in the following approximation:

$$\sqrt{N(1 + \mathscr{R}_p)}\, \Phi^{-1}(P_d) \cong \sqrt{N}\, \Phi^{-1}(P_{fa}) - \frac{N\mathscr{R}_p}{2} \tag{10.7-22}$$

For $N \gg 1$, Eq. (10.7-22) is reasonably accurate. When $\mathscr{R}_p < 1$, Eq. (10.7-22) can be simplified with the approximation

$$\sqrt{1 + \mathscr{R}_p} \cong 1 + \frac{\mathscr{R}_p}{2} \tag{10.7-23}$$

Substituting Eq. (10.7-23) into (10.7-22) and solving for $\mathscr{R}_p$ gives

$$\mathscr{R}_p \cong \frac{2[\Phi^{-1}(P_{fa}) - \Phi^{-1}(P_d)]}{\Phi^{-1}(P_d) + \sqrt{N}} \tag{10.7-24}$$

Relation (10.7-24) can be further simplified. If Eq. (10.7-21) is approximated by

$$\Phi^{-1}(P_d) \cong \frac{Y_b - N}{\sqrt{N}} \tag{10.7-25}$$

then, since Y_b is of the same order of magnitude as N [see Eq. (10.7-16) and Fig. 10.7-1], it follows from (10.7-25) that

$$\Phi^{-1}(P_d) \sim \frac{1}{\sqrt{N}} \ll \sqrt{N}, \qquad N \gg 1 \tag{10.7-26}$$

Hence the denominator of Eq. (10.7-24) can be simplified to yield

$$\mathscr{R}_p = \frac{2[\Phi^{-1}(P_{fa}) - \Phi^{-1}(P_d)]}{\sqrt{N}} \tag{10.7-27}$$

Equation (10.7-27) states that for $N \gg 1$, the single-pulse peak signal-to-noise ratio decreases inversely as $\sqrt{N}$. A similar result was obtained in Section 10.5 on integration loss. Equation (10.7-27) yields values of $\mathscr{R}_p$ accurate within a db for $N > 100$, $\mathscr{R}_p < -2$ db, and over the region of P_{fa} normally of interest in radar.

When either Eq. (10.7-27) or the more accurate relation (10.7-24) is used, Fig. 10.7-2 is required. This figure is a plot of P_d versus $\Phi^{-1}(P_d)$. Figs. 10.7-1 and 10.7-2 facilitate rapid estimates of required performance and parameter values.

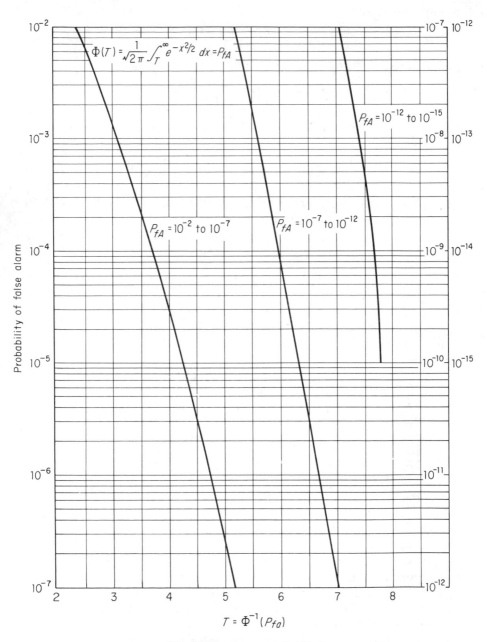

Probability of false alarm

$$\Phi(T) = \frac{1}{\sqrt{2\pi}} \int_T^\infty e^{-x^2/2}\, dx = P_{fA}$$

$P_{fA} = 10^{-12}$ to 10^{-15}

$P_{fA} = 10^{-2}$ to 10^{-7}

$P_{fA} = 10^{-7}$ to 10^{-12}

$T = \Phi^{-1}(P_{fa})$

Fig. 10.7-1. P_{fa} versus $\Phi^{-1}(P_{fa})$.

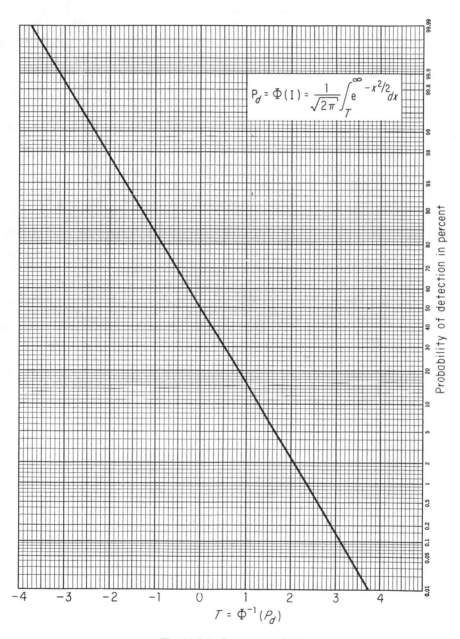

$$P_d = \Phi(I) = \frac{1}{\sqrt{2\pi}} \int_T^\infty e^{-x^2/2} dx$$

Fig. 10.7-2. P_d versus $\Phi^{-1}(P_d)$.

369

10.8. Linear versus Square-Law Detection

In Section 10.3 it was shown that the optimum $\ln I_0$ detector law for a nonfluctuating incoherent pulse train can be approximated by a square-law envelope detector at small signal-to-noise ratios $(r_i\sqrt{\mathscr{R}_p} \ll 1)$ and by a linear detector when the single-pulse signal-to-noise ratio is large $(r_i\sqrt{\mathscr{R}_p} \gg 1)$. In Section 10.4 the performance of a receiver with a square-law envelope detector was evaluated. It is of interest to compare the performance of linear and square-law detectors.

    To evaluate performance with a linear detector the probability density of sum variable Y must be obtained. If the method of characteristic functions is employed, difficulty is encountered immediately, since the characteristic function for one variate of signal-plus-noise has not been found in closed form for a linear detector. Alternatively, the desired distribution can be approximated with a Gram-Charlier expansion. Marcum [2] used this method to calculate performance of a linear detector. He then compared the results with those of a square-law detector; Marcum's calculations are plotted in Fig. 10.8-1 for $P_d = 0.5$ and $P_{fa}/N = 10^{-6}$. This graph shows that there is negligible difference between the expected performance of both types of envelope detectors. Note that for $N = 1$ and $N = 70$, both are identical in performance. For small N, the linear detector is superior by an amount not exceeding 0.11 db in signal-to-noise ratio; for large N,

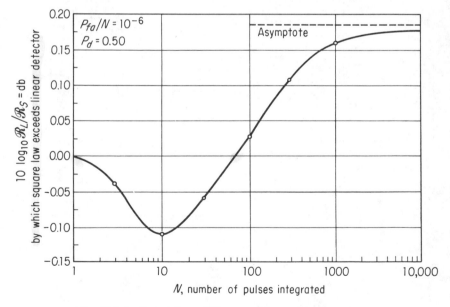

Fig. 10.8-1. Comparison of linear and square-law detectors.

the square-law detector performance exceeds the linear detector by an amount that asymptotically approaches 0.19 db as $N \rightarrow \infty$.

In practice, envelope-detector implementations often exhibit a square-law characteristic for small signals and a linear characteristic for large signals, thereby approximating the ideal $\ln I_0$ detector. The performance difference between the two detector characteristics is usually of little concern.

PROBLEMS

10.1. The radar equation may be written concisely as

$$\mathscr{R}_p = K \frac{1}{R^4}$$

where K is a constant of proportionality and R is radar range to a target. Suppose a radar is designed for 10 hits (observations) per scan on a non-fluctuating target. At maximum range, $R_{\max}$, the peak (single-pulse) signal-to-noise ratio, $\mathscr{R}_p$, is specified as 5 db and the P_{fa} as 0.7×10^{-8}. Plot the probability of detection as a function of range from below $R_{\max}$ to $R_{\max}$.

10.2. Narrowband Gaussian noise with variance σ^2 is detected by a square-law envelope detector whose output y is related to the noise waveform envelope v by

$$y = av^2$$

Find the probability density function of the noise at the detector output. Calculate the voltage at the detector output that would be indicated by a long time-constant dc meter; then a long time-constant rms meter; and last, a long time-constant rms meter preceded by a blocking capacitor.

10.3. With Eqs. (10.6-2) to (10.6-4) verify that the first five Hermite polynomials are correctly given by Eqs. (10.6-5). Verify Eq. (10.6-6), which states that $\overset{(i)}{\phi}(x)$ and $H_i(x)$ are biorthogonal functions.

10.4. With Eq. (10.4-25b) calculate and plot the threshold values of Y_b for $P_{fa} = 10^{-4}, 10^{-6}$, and 10^{-8} and $N = 100$ and 1000. From tables of the incomplete Toronto function (see references [2, 6]) plot the probability of detection as a function of $\mathscr{R}_p$, for the values just determined, using Eq. (10.4-30). Compare these results with the curves in the text.

10.5. For $P_d = 30$ per cent and $P_{fa} = 10^{-8}$, compute the integration loss as a function of N. Plot the result. Repeat for $P_d = 75$ per cent. Compare these curves to those in the text.

10.6. From Eq. (10.6-14) determine the coefficients c_i given in Eqs. (10.6-15a) to (10.6-15f).

10.7. Verify Eqs. (10.6-18) for $\alpha_3, \alpha_4, \alpha_5$, and α_6 using (10.6-16) and (10.6-17).

10.8. Integrate Eq. (10.4-18a) successively by parts to verify (10.4-21) and find the order of magnitude of the error in approximation (10.4-22).

10.9. Using both (10.7-24) and (10.7-27), plot the approximate probability of detection as a function of (single-pulse) peak signal-to-noise ratio $\mathscr{R}_p$ for the following sets of values: (a) $N = 10$, $P_{fa} = 0.7 \times 10^{-6}$; (b) $N = 100$, $P_{fa} = 0.7 \times 10^{-6}$; (c) $N = 1000$, $P_{fa} = 0.7 \times 10^{-6}$. Using these results, plot the difference in signal-to-noise ratio for the same P_d between the approximate curves and more accurate values obtained from the text for each set of values of N and P_{fa}. What conclusions can be drawn about the accuracy of the two approximations?

REFERENCES

1. Marcum, J. I.: A Statistical Theory of Target Detection by Pulsed Radar, Rand Report RM-754 (December, 1947); reissued since in a special issue, *Trans. IRE Prof. Group on Information Theory*, **IT-6:** (*2*), 59–144 (April, 1960).

2. Marcum, J. I.: A Statistical Theory of Target Detection by Pulsed Radar, Mathematical Appendix, Rand Report RM-753 (July, 1948); reissued since in a special issue, *Trans. IRE Prof. Group on Information Theory*, **IT-6:** (*2*), 145–267 (April, 1960).

3. Campbell, G. A., and Foster, R. M.: "Fourier Integrals for Practical Applications," Van Nostrand, Princeton, N.J., 1948.

4. Pearson, K.: "Tables of the Incomplete Γ-Function," Cambridge University Press, London and New York, 1934.

5. Pachares, J.: A Table of Bias Levels Useful in Radar Detection Problems, *Trans. IRE*, **IT-4** (March, 1958).

6. Heatley, A. H.: A Short Table of the Toronto Functions, *Trans. Roy. Soc. Canada*, **37** (Sec. III): 13–29 (1943).

7. Fehlner, L. F.: Target Detection by a Pulsed Radar, Report TB-451, Applied Physics Laboratory, Johns Hopkins University, Silver Spring, Maryland (July, 1962).

11

DETECTION BASED ON
MULTIPLE OBSERVATIONS:
SWERLING FLUCTUATING MODELS

11.1. Introduction

This chapter continues the study of detection based on multiple observations begun in Chapter 10. The same basic approach is now applied to the detection of fluctuating incoherent pulse trains. Fluctuating pulse trains occur often in practice. When a radar target consists of several relatively strong reflecting surfaces displaced from one another by the order of a wavelength, the amplitude and phase of the composite radar echo are sensitive to the spatial orientation of the target. (See discussion in Section 1.6.) If the target has relative motion with respect to the radar, such as translation, pitch, yaw, tumble, it presents a time varying radar cross section. Two cases can be distinguished:

(a) The target orientation changes slowly compared to the pulse-train duration; this results in a nearly undistorted pulse train except for an overall random amplitude and initial phase established by momentary target orientation. The target orientation, however, is assumed to change sufficiently from scan to scan so that echo amplitude and initial phase in each scan are statistically independent. (A scan is a complete search of the surveillance volume.) In addition, the pulse train is assumed to be incoherent; that is, owing to system instability, each train pulse has a statistically independent initial phase.

(b) Changes in target orientation are rapid compared to an interpulse period of the pulse train but are slow when compared to the duration of a single pulse; this results in a statistically independent amplitude and initial phase for each pulse in the train but practically no distortion of the waveform structure within a pulse.

Scan-to-scan and pulse-to-pulse fluctuating pulse trains are representative of many practical radar situations and comprise the main subject matter of this chapter. Pulse-train fluctuations that lie between these extremes are difficult to analyze, and limited results are available for these cases.

To apply the strategy described in Chapter 10, a priori distributions for the signal random variables must be known or assumed. It has been found experimentally that echo amplitude from a target composed of many small scattering surfaces, such as an airplane, fluctuates in accordance with a Rayleigh probability density function; in addition, the return from a target composed of one large scattering surface plus several smaller scattering surfaces, such as a spherical satellite with extensions, is approximately described by a one-dominant-plus-Rayleigh probability density function. Both distributions are applied to scan-to-scan and pulse-to-pulse fluctuating pulse trains. Further, as previously noted, whenever initial phase is a random variable, it is assumed to have a (least favorable) uniform distribution on the interval $(0, 2\pi)$.

11.2. Scan-to-Scan Rayleigh Fluctuating Incoherent Pulse Train (Swerling I)

The radar model called Swerling I (see [1]) assumes that the amplitude of an entire pulse train is a single random variable with a Rayleigh probability density function. In addition, the initial phase of each pulse is assumed to be a statistically independent random variable with a uniform probability density.

A. Optimum receiver structure

The optimum receiver strategy requires the computation of the generalized likelihood ratio, which is then compared with a threshold. The formulation in Chapter 10 is the same for fluctuating and nonfluctuating pulse trains up to Eq. (10.3-9a); this expression for the modified generalized likelihood ratio is

$$\overline{\ell'(\mathbf{v}\,|\,\boldsymbol{\theta})_\theta} = \overline{\left[\prod_{i=1}^{N}\ell'(\mathbf{v}_i\,|\,\boldsymbol{\theta}_i)\right]}_\theta \qquad (11.2\text{-}1)$$

where $\boldsymbol{\theta}$ is the set of random variables associated with the signal waveform.

To show explicitly its dependence on pulse-train amplitude, signal $s(t)$ can be related to a reference signal $\Sigma(t)$ by

$$s(t) = A\Sigma(t) \qquad (11.2\text{-}2)$$

where A denotes the random pulse-train amplitude. Let $\mathscr{E}_p$ represent the energy in one pulse of $\Sigma(t)$; then the actual pulse energy E_p in each pulse is given by

$$E_p = A^2 \mathscr{E}_p \qquad (11.2\text{-}3)$$

The definition of A becomes unique if reference pulse energy $\mathscr{E}_p$ is normalized such that

$$\frac{2\mathscr{E}_p}{N_0} = 1 \qquad (11.2\text{-}4)$$

With these definitions, Eq. (11.2-1) can be rewritten as

$$\overline{\ell'(\mathbf{v}\,|\,\boldsymbol{\theta})_\theta} = \overline{\left[\prod_{i=1}^{N}\ell'(\mathbf{v}_i\,|\,\boldsymbol{\theta}_i\right]}_{\phi_1,\,\phi_2,\,\cdots,\,\phi_N,\,A} \qquad (11.2\text{-}5)$$

where $\phi_1, \phi_2, \ldots, \phi_N$ are the random phases associated with each pulse in the train and A is a Rayleigh-distributed random variable. Equation (11.2-5) can also be written as

$$\overline{\ell'(\mathbf{v}\,|\,\boldsymbol{\theta})_\theta} = \overline{\left[\prod_{i=1}^{N}\overline{\ell'(\mathbf{v}_i\,|\,\phi_i, A)_{\phi_i}}\right]}_A \qquad (11.2\text{-}6)$$

where the statistical independence of each pulse initial phase is explicitly stated in (11.2-6). [See discussion in Chapter 10 leading to Eq. (10.3-11b).]

The expression for $\overline{\ell'(\mathbf{v}_i\,|\,\phi_i, A)_{\phi_i}}$ in (11.2-6) is given by Eq. (9.4-5) and is restated here in suitably modified notation:

$$\overline{\ell'(\mathbf{v}_i\,|\,\phi_i, A)_{\phi_i}} = \begin{cases} e^{-A^2/2} I_0[Ar_i(\mathbf{v}_i)], & r_i \geqslant 0 \\ 0, & r_i < 0 \end{cases} \qquad (11.2\text{-}7)$$

where $r_i(\mathbf{v}_i)$ is the sampled value of the envelope of the ith signal pulse plus noise out of a (single pulse) matched filter with normalized gain [see Eq. (9.3-10)]; the sampling instant is the time at which the matched filter output peaks due to the ith signal pulse alone. Substituting Eq. (11.2-7) into (11.2-6) results in

$$\overline{\ell'(\mathbf{v}\,|\,\boldsymbol{\theta})_\theta} = \overline{\left[\prod_{i=1}^{N} e^{-A^2/2} I_0(Ar_i)\right]}_A, \qquad r_i \geqslant 0 \qquad (11.2\text{-}8a)$$

$$= \overline{\left[e^{-NA^2/2}\prod_{i=1}^{N} I_0(Ar_i)\right]}_A, \qquad r_i \geqslant 0 \qquad (11.2\text{-}8b)$$

The Rayleigh probability density of random variable A is given by

$$p(A) = \begin{cases} \dfrac{A}{A_0^2} e^{-A^2/2A_0^2}, & A \geqslant 0 \\ 0, & A < 0 \end{cases} \tag{11.2-9}$$

where A_0 is the most probable value of A. When equation (11.2-8b) is averaged with respect to A using Eq. (11.2-9), the modified generalized likelihood ratio becomes

$$\overline{\ell'(\mathbf{v}\,|\,\boldsymbol{\theta})_\theta} = \int_0^\infty \frac{A}{A_0^2} e^{-(A^2/2A_0^2)-(NA^2/2)} \prod_{i=1}^N I_0(Ar_i)\,dA \tag{11.2-10a}$$

$$= \int_0^\infty \frac{A}{A_0^2} e^{-h^2A^2} \prod_{i=1}^N I_0(Ar_i)\,dA \tag{11.2-10b}$$

where

$$h^2 = \frac{1}{2A_0^2} + \frac{N}{2} \tag{11.2-11}$$

The integral in (11.2-10b) cannot be evaluated exactly; however, it can be approximated for small signal-to-noise ratios. If the upper limit of the integral in (11.2-10b) is modified to $3.05A_0$, the integration includes 99 per cent of the possible values of A. Further, for $Ar_i < 1$, it follows from Eq. (10.3-16) that

$$\prod_{i=1}^N I_0(Ar_i) \cong \prod_{i=1}^N \left(1 + \frac{A^2 r_i^2}{4}\right) \cong 1 + \sum_{i=1}^N \frac{A^2 r_i^2}{4} \tag{11.2-12}$$

Substituting Eq. (11.2-12) into (11.2-10b) with revised limits of integration gives, after simplification,

$$\overline{\ell'(\mathbf{v}\,|\,\boldsymbol{\theta})_\theta} \cong \int_0^{3.05A_0} \frac{A}{A_0^2} e^{-h^2A^2}\,dA + \sum_{i=1}^N r_i^2 \int_0^{3.05A_0} \frac{A^3}{4A_0^2} e^{-h^2A^2}\,dA \tag{11.2-13}$$

Equation (11.2-13) can be written in the following functional form:

$$\overline{\ell'(\mathbf{v}\,|\,\boldsymbol{\theta})_\theta} \cong F_1(A_0, h) + F_2(A_0, h) \sum_{i=1}^N r_i^2 \tag{11.2-14}$$

where functions F_1 and F_2 are independent of r_i.

Comparing the (modified) likelihood ratio given by (11.2-14) with a threshold corresponds to a comparison of $\sum_{i=1}^N r_i^2$ with a modified threshold —that is,

$$\sum_{i=1}^N r_i^2 \begin{cases} \geqslant \mathscr{K} & (\textit{signal present}) \\ < \mathscr{K} & (\textit{signal absent}) \end{cases} \tag{11.2-15}$$

The threshold is usually established by the choice of false-alarm probability. Since each r_i is a sample of signal plus noise out of a normalized matched filter, the optimum receiver for a Swerling I fluctuating target is essentially the same as that in Fig. 10.3-2, for small signal-to-noise ratios.

B. Optimum performance

Probability of false alarm P_{fa} for Swerling I is identical to that for a nonfluctuating target, since the receiver structures are the same and the statistics of sum random variable Y, where Y is defined in Eq. (10.4-4), are unaffected by assumptions about the signal. Hence only the probability of detection must be considered. A direct method for calculating the probability of detection in the present case is to average Eq. (10.4-30) with respect to unknown signal amplitude A. Unfortunately, this is very difficult to carry out mathematically.

Swerling [1] has been shown an alternative method for obtaining the desired result. Equation (10.4-12b), the characteristic function of sum random variable Y, is the point of departure. Since $\mathscr{R}_p$ is a random variable and not a parameter in the present context, $C_Y(\xi; \mathscr{R}_p)$ is notationally replaced by $C_Y(\xi \mid \mathscr{R}_p)$. Thus,

$$C_Y(\xi \mid \mathscr{R}_p) = \frac{1}{(1 - j\xi)^N} e^{jN\mathscr{R}_p\xi/2(1-j\xi)} \qquad (11.2\text{-}16)$$

The quantity $\mathscr{R}_p$ is related to signal amplitude A by Eqs. (11.2-3) and (11.2-4) as follows:

$$\mathscr{R}_p = \frac{2E_p}{N_0} = A^2 \frac{2\mathscr{E}_p}{N_0} = A^2 \qquad (11.2\text{-}17)$$

Averaging Eq. (11.2-17) with respect to A using (11.2-9) shows that

$$\overline{\mathscr{R}_p} = \overline{A^2} = 2A_0^2 \qquad (11.2\text{-}18)$$

Substituting (11.2-17) into Eq. (11.2-16) yields

$$C_Y(\xi \mid A) = \frac{1}{(1 - j\xi)^N} e^{jNA^2\xi/2(1-j\xi)} \qquad (11.2\text{-}19)$$

Equation (11.2-19) is averaged next with respect to A using (11.2-9) as follows:

$$\overline{C_Y(\xi)}_A = \frac{1}{(1 - j\xi)^N} \int_0^\infty \frac{A}{A_0^2} e^{-A^2 g^2} \, dA \qquad (11.2\text{-}20)$$

where

$$g^2 = \frac{1}{2A_0^2} - \frac{jN\xi}{2(1 - j\xi)} \qquad (11.2\text{-}21)$$

Evaluating the integral in (11.2-20) yields

$$\overline{C_Y(\xi)}_A = \frac{1}{2g^2 A_0^2 (1 - j\xi)^N} \qquad (11.2\text{-}22)$$

When (11.2-21) is substituted into (11.2-22), $\overline{C_Y(\xi)}_A$ can be modified successively as follows:

$$\overline{C_Y(\xi)}_A = \frac{1}{A_0^2 \left(\frac{1}{A_0^2} - \frac{j\xi N}{1 - j\xi} \right)(1 - j\xi)^N} \tag{11.2-23a}$$

$$= \frac{1}{\left(1 - \frac{jNA_0^2 \xi}{1 - j\xi} \right)(1 - j\xi)^N}$$

$$= \frac{1}{(1 - j\xi - jNA_0^2 \xi)(1 - j\xi)^{N-1}}$$

$$= \frac{1}{[1 - j\xi(1 + NA_0^2)](1 - j\xi)^{N-1}} \tag{11.2-23b}$$

$$= \frac{(-1)^N}{(1 + NA_0^2) \left(-\frac{1}{1 + NA_0^2} + j\xi \right)(-1 + j\xi)^{N-1}} \tag{11.2-23c}$$

Probability density $p(Y)$ is obtained by evaluating the anticharacteristic function of (11.2-23c) using Fourier transform pair 581.7 in the Campbell and Foster tables [3]; it is repeated here suitably modified for the present application:

$$\frac{1}{(\beta + j\xi)(\rho + j\xi)^{\alpha-1}} \longleftrightarrow \frac{e^{\beta Y}}{\Gamma(\alpha - 1)(\rho - \beta)^{\alpha-1}} \gamma[\alpha - 1, (\beta - \rho)Y] \tag{11.2-24}$$

where

$$\gamma(v, z) = \Gamma(v) - \Gamma(v, z) \tag{11.2-25}$$

$$\Gamma(v) = \int_0^\infty z^{v-1} e^{-z} \, dz = (v - 1)! \tag{11.2-26}$$

and

$$\Gamma(v, z) = \int_z^\infty z^{v-1} e^{-z} \, dz \tag{11.2-27}$$

The left side of (11.2-24) assumes a form similar to the characteristic function in (11.2-23c) for $\alpha = N$, $\rho = -1$, and $\beta = -1/(1 + NA_0^2)$; density function $p(Y)$ is obtained from the right side of Eq. (11.2-24):

$$p(Y) = \frac{\beta e^{\beta Y}}{(N - 2)!(1 + \beta)^{N-1}} \gamma[N - 1, (\beta + 1)Y] \tag{11.2-28}$$

where $\Gamma(N - 1)$ has been replaced by $(N - 2)!$ using Eq. (11.2-26). The function $\gamma[N - 1, z]$ in (11.2-28), for $z = (\beta + 1)Y$, is evaluated using (11.2-25) and (11.2-26) as follows:

$$\gamma[N - 1, z] = \Gamma(N - 1) - \int_z^\infty z^{N-2} e^{-z} \, dz \tag{11.2-29a}$$

$$= \Gamma(N - 1) - \int_0^\infty z^{N-2} e^{-z} \, dz + \int_0^z z^{N-2} e^{-z} \, dz$$

$$= \Gamma(N - 1) - \Gamma(N - 1) + (N - 2)! \int_0^z \frac{z^{N-2} e^{-z}}{(N - 2)!} \, dz$$

$$= (N - 2)! I\left(\frac{z}{\sqrt{N + 1}}, N - 2 \right) \tag{11.2-29b}$$

where $I(u, s)$ is the incomplete gamma function defined in Eq. (10.4-19). Substituting (11.2-29b) into (11.2-28) and replacing z by $(\beta + 1)Y$ yields

$$p(Y) = \frac{-\beta e^{\beta Y}}{(1 + \beta)^{N-1}} I\left[\frac{(\beta + 1)Y}{\sqrt{N-1}}, N-2\right] \tag{11.2-30}$$

Since $\beta = -1(1 + NA_0^2)$, note that

$$\beta + 1 = \frac{NA_0^2}{1 + NA_0^2} = \frac{1}{1 + \dfrac{1}{NA_0^2}} \tag{11.2-31}$$

Substituting (11.2-31) and the definition of β into (11.2-30) yields

$$p(Y) = \frac{\left(1 + \dfrac{1}{NA_0^2}\right)^{N-2} e^{-Y/(1+NA_0^2)}}{NA_0^2} I\left[\frac{Y}{\sqrt{N-1}\left(1 + \dfrac{1}{NA_0^2}\right)}, N-2\right] \tag{11.2-32}$$

The probability of detection is in general given by

$$P_d = \int_{Y_b}^{\infty} p(Y)\, dY \tag{11.2-33a}$$

$$= 1 - \int_0^{Y_b} p(Y)\, dY \tag{11.2-33b}$$

When (11.2-32) is substituted into (11.2-33b), the integration is very difficult to perform. An alternative procedure is to return to Eq. (11.2-23b) which, after crossmultiplication, can be rewritten as

$$[1 - j\xi(1 + NA_0^2)]\,\overline{C_Y(\xi)}_A = \frac{1}{(1 - j\xi)^{N-1}} \tag{11.2-34}$$

Expanding (11.2-34) and rearranging terms yields

$$\overline{C_Y(\xi)}_A = \frac{1}{(1 - j\xi)^{N-1}} + j\xi(1 + NA_0^2)\overline{C_Y(\xi)}_A \tag{11.2-35}$$

The anticharacteristic function of (11.2-35) taken term by term results in

$$p(Y) = \frac{Y^{N-2}e^{-Y}}{(N-2)!} - (1 + NA_0^2)\frac{dp(Y)}{dY} \tag{11.2-36}$$

where we have used the fact that the anticharacteristic function of $-j\xi\overline{C_Y(\xi)}_A$ is $dp(Y)/dY$ [see Table 2.2-1 and Eq. (3.12-2)]. Substituting (11.2-36) into (11.2-33b) permits the probability of detection to be written as

$$P_d = 1 - \int_0^{Y_b} \frac{Y^{N-2}e^{-Y}}{(N-2)!}\, dY + (1 + NA_0^2)[p(Y)]_0^{Y_b} \tag{11.2-37}$$

The integral in (11.2-37) is the incomplete gamma function, $I(Y_b/\sqrt{N-1}, N-2)$. When (11.2-32) is substituted into (11.2-37) and A_0^2 is replaced by $\mathscr{R}_p/2$, the result is

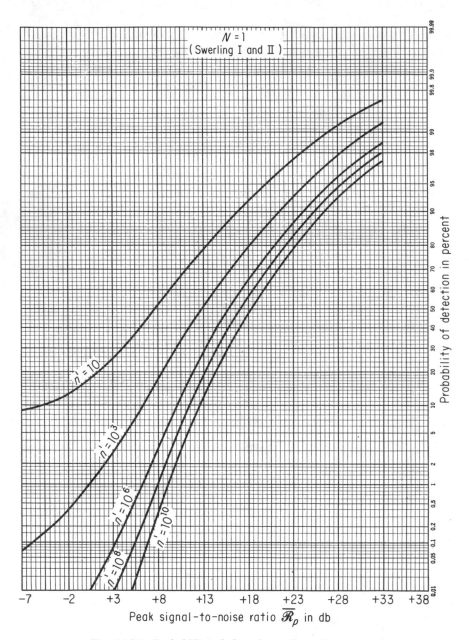

Fig. 11.2-1. Probability of detecting a fluctuating target, $N = 1$ (Swerling I and II). [N = number of pulses incoherently integrated; $P_{fa} = 0.693/n'$.]

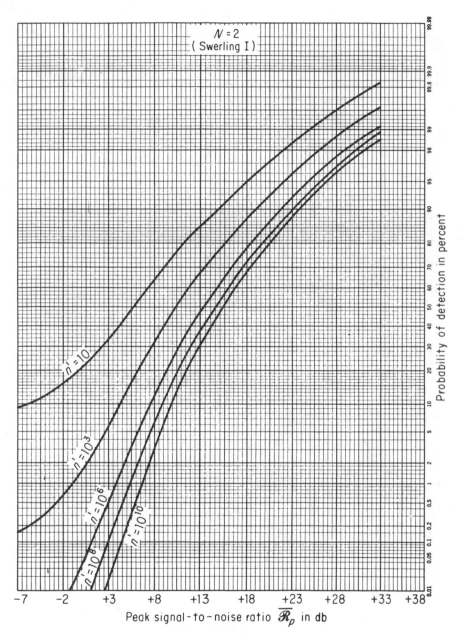

Fig. 11.2-2. Probability of detecting a fluctuating target, $N = 2$ (Swerling I). [N = number of pulses incoherently integrated; $P_{fa} = 0.693/n'$.]

381

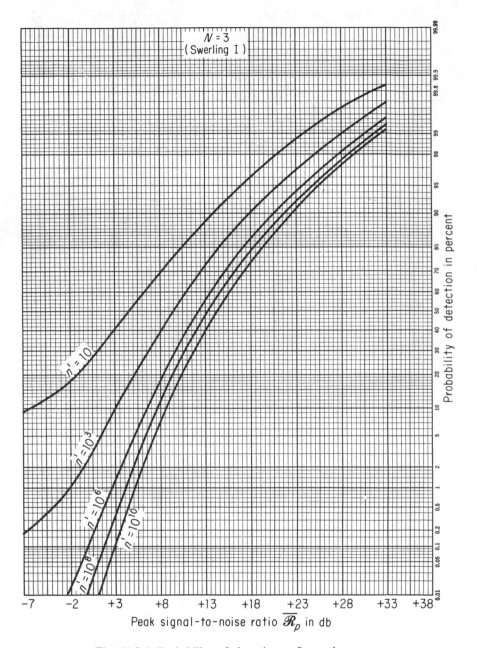

$N = 3$
(Swerling I)

$n' = 10$

$n' = 10^3$

$n' = 10^6$

$n' = 10^{10}$

$n' = 10^8$

Probability of detection in percent

Peak signal-to-noise ratio $\overline{\mathscr{R}}_p$ in db

Fig. 11.2-3. Probability of detecting a fluctuating target, $N = 3$ (Swerling I). [N = number of pulses incoherently integrated; $P_{fa} = 0.693/n'$.]

382

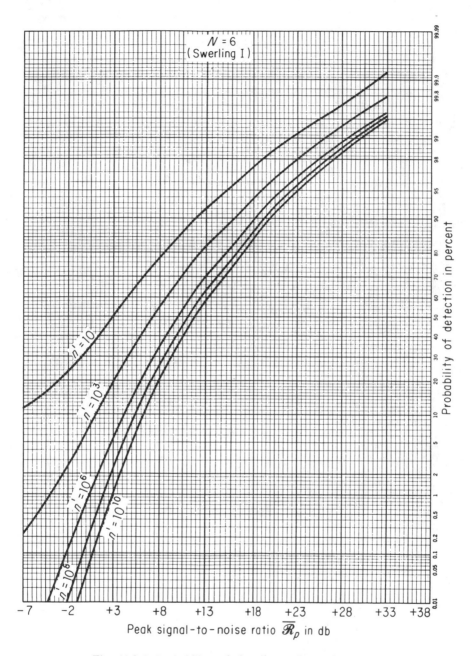

Fig. 11.2-4. Probability of detecting a fluctuating target, $N = 6$ (Swerling I). [N = number of pulses incoherently integrated; $P_{fa} = 0.693/n'$.]

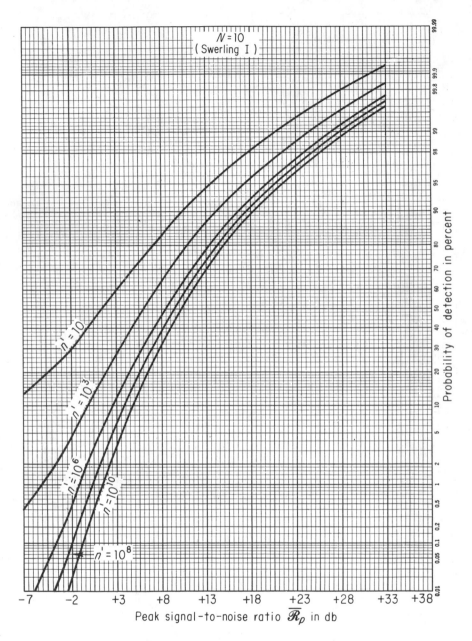

Fig. 11.2-5. Probability of detecting a fluctuating target, $N = 10$ (Swerling I). [N = number of pulses incoherently integrated; $P_{fa} = 0.693/n'$.]

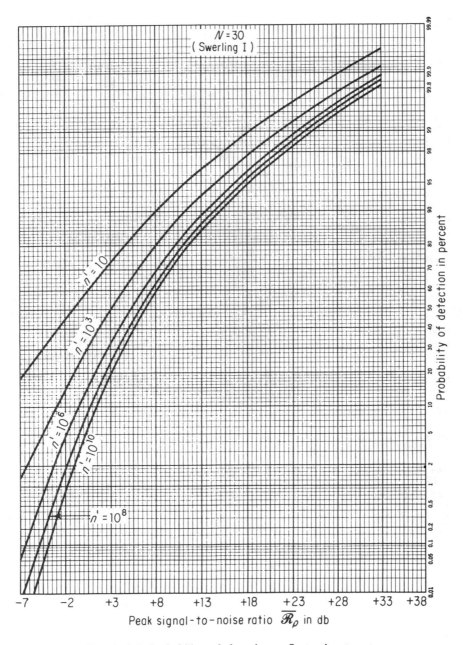

Fig. 11.2-6. Probability of detecting a fluctuating target, $N = 30$ (Swerling I). [N = number of pulses incoherently integrated; $P_{fa} = 0.693/n'$.]

385

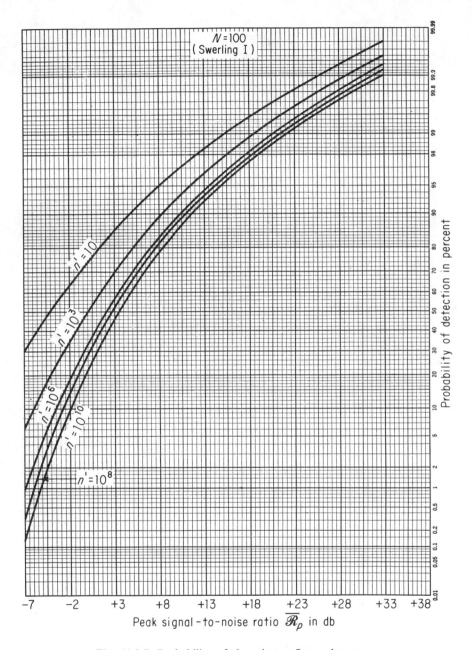

Fig. 11.2-7. Probability of detecting a fluctuating target, $N = 100$ (Swerling I). [N = number of pulses incoherently integrated; $P_{fa} = 0.693/n'$.]

386

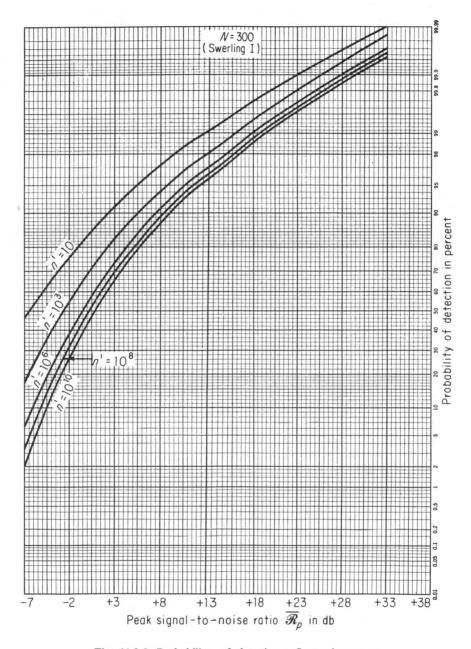

Fig. 11.2-8. Probability of detecting a fluctuating target, $N = 300$ (Swerling I). [N = number of pulses incoherently integrated; $P_{fa} = 0.693/n'$.]

387

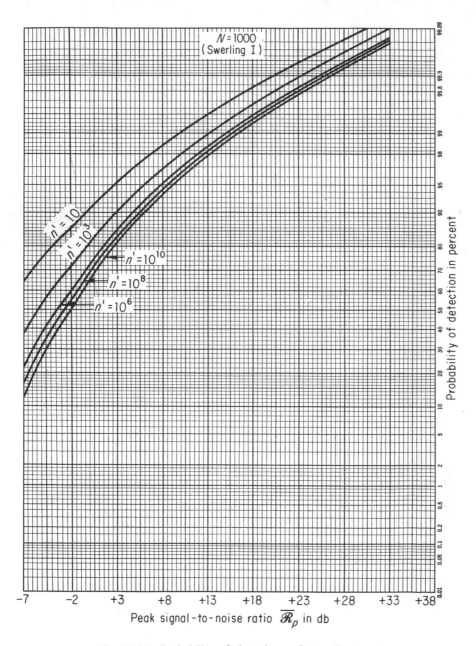

Fig. 11.2-9. Probability of detecting a fluctuating target, $N = 1000$ (Swerling I). [N = number of pulses incoherently integrated; $P_{fa} = 0.693/n'$.]

388

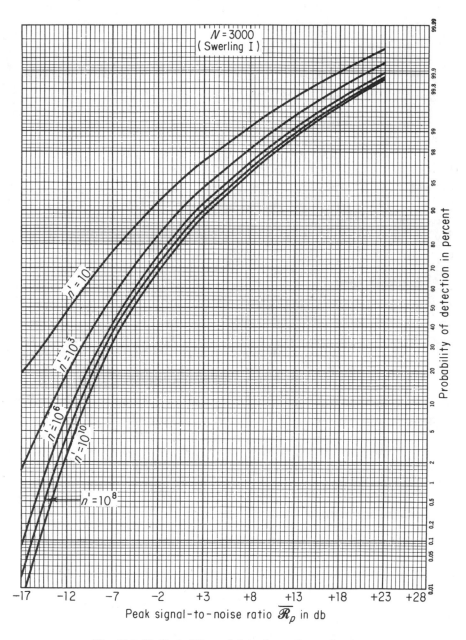

Fig. 11.2-10. Probability of detecting a fluctuating target,
$N = 3000$ (Swerling I). [N = number of pulses incoherently
integrated; $P_{fa} = 0.693/n'$.]

$$P_d = 1 - I\left(\frac{Y_b}{\sqrt{N-1}}, N-2\right) + \left(1 + \frac{1}{N\overline{\mathscr{R}}_p/2}\right)^{N-1} e^{-Y_b/(1+N\overline{\mathscr{R}}_p/2)}$$

$$\cdot I\left[\frac{Y_b}{\sqrt{N-1}\left(1 + \frac{1}{N\overline{\mathscr{R}}_p/2}\right)}, N-2\right] \qquad (11.2\text{-}38)$$

To obtain the performance characteristic for Swerling I, Y_b is evaluated using Eq. (10.4-20) or (10.4-25b) for false-alarm probability. With this value substituted into (11.2-38) P_d can be computed as a function of N and $\overline{\mathscr{R}}_p$. In Figs. 11.2-1 through 11.2-10 P_d is plotted versus average (single-pulse) signal-to-noise ratio $\overline{\mathscr{R}}_p$ for selected values of false-alarm probability P_{fa} (see discussion at end of Section 9.2 concerning units of $\overline{\mathscr{R}}_p$). The figures correspond to pulse trains with $N = 1, 2, 3, 6, 10, 30, 100, 300, 1000$, and 3000, respectively, and are based on computer results obtained by Fehlner [7].

C. Integration loss

In Fig. 11.2-11 integration loss L is plotted as a function of N for a Swerling I fluctuating target referenced to a coherent pulse train with scan-to-scan Rayleigh fluctuating amplitude and unknown initial phase. [See Eq. (10.5-1) for definition of L.] To calculate L, we require the single-pulse signal-to-noise ratio for a coherently processed Rayleigh fluctuating pulse train as a function of N. For this purpose the peak signal-to-noise ratio obtained from Fig. 11.2-1 for $N = 1$ is divided by the selected value of N. Integration loss was evaluated for $P_d = 50$ per cent and 95 per cent with $P_{fa} = 10^{-10}$. The loss is nearly the same for both values of P_d; hence only a single curve is plotted. It may be noted that for large values of N, the slope of the loss curve approaches a slope $10 \log \sqrt{N}$. This result is confirmed in the next section.

D. Analytic approximations

Equation (11.2-38) for P_d can be approximated as follows. The two incomplete gamma functions on the right side of (11.2-38) have almost identical arguments and differ only in the denominator of the first argument by the factor $[1 + 1/(N\overline{\mathscr{R}}_p/2)]$. This factor is usually close to unity for $N \gg 1$, since $N\overline{\mathscr{R}}_p/2$ must be considerably greater than one in order to obtain reliable detection with small false-alarm probabilities.† For $N\overline{\mathscr{R}}_p/2 \gg 1$ and $P_{fa} \ll 1$, the incomplete gamma functions are also close to unity; then (11.2-38) can be rewritten approximately as

$$P_d \cong \left(1 + \frac{1}{N\overline{\mathscr{R}}_p/2}\right)^{N-1} e^{-Y_b/(1+N\overline{\mathscr{R}}_p/2)} \qquad (11.2\text{-}39)$$

Note that this expression is exact for $N = 1$.

†See plots in Figs. 11.2-1 through 11.2-10.

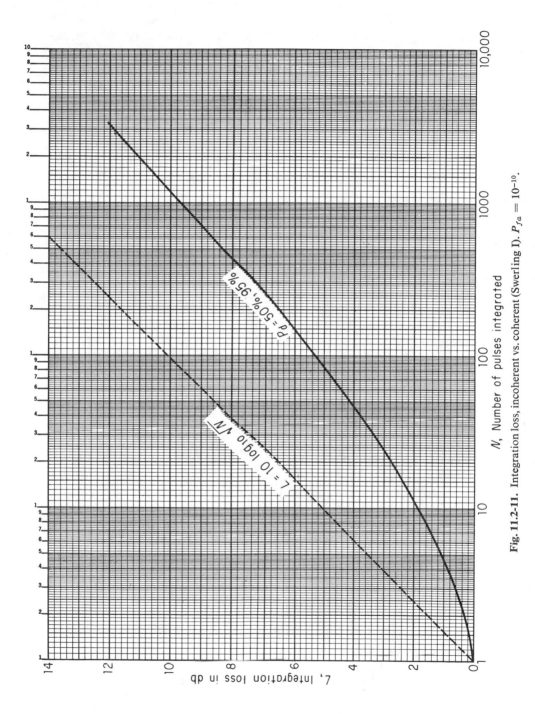

Fig. 11.2-11. Integration loss, incoherent vs. coherent (Swerling I). $P_{fa} = 10^{-10}$.

N, Number of pulses integrated

L, Integration loss in db

$P_d = 50\%, 95\%$

$L = 10 \log_{10} \sqrt{N}$

Forming the logarithm of (11.2-39) and rearranging the last term results in

$$\ln P_d \cong (N-1)\ln\left(1 + \frac{1}{N\overline{\mathscr{R}}_p/2}\right) - \frac{Y_b}{(N\overline{\mathscr{R}}_p/2)\left(1 + \frac{1}{N\overline{\mathscr{R}}_p/2}\right)} \qquad (11.2\text{-}40)$$

A series expansion of each term on the right of (11.2-40) yields (for $N\overline{\mathscr{R}}_p/2 > 1$)

$$\ln P_d \cong (N-1)\left[\frac{1}{N\overline{\mathscr{R}}_p/2} - \frac{1}{2}\frac{1}{(N\overline{\mathscr{R}}_p/2)^2} + \frac{1}{3}\frac{1}{(N\overline{\mathscr{R}}_p/2)^3} - \cdots\right]$$

$$- Y_b\left[\frac{1}{N\overline{\mathscr{R}}_p/2} - \frac{1}{(N\overline{\mathscr{R}}_p/2)^2} + \frac{1}{(N\overline{\mathscr{R}}_p/2)^3} - \cdots\right] \qquad (11.2\text{-}41)$$

or

$$\ln P_d \cong -\frac{1}{N\overline{\mathscr{R}}_p/2}(Y_b - N + 1) + \frac{1}{(N\overline{\mathscr{R}}_p/2)^2}\left(Y_b - \frac{N-1}{2}\right)$$

$$- \frac{1}{(N\overline{\mathscr{R}}_p/2)^3}\left(Y_b - \frac{N-1}{3}\right) + \cdots \qquad (11.2\text{-}42)$$

The right side of Eq. (11.2-42) can be approximated by retaining only the first term; thus

$$\ln P_d \cong -\frac{2}{N\overline{\mathscr{R}}_p}(Y_b - N + 1) \qquad (11.2\text{-}43)$$

Replacing Y_b in (11.2-43) by Eq. (10.7-16) results in

$$\ln P_d \cong -\frac{2}{N\overline{\mathscr{R}}_p}[\sqrt{N}\,\Phi^{-1}(P_{fa}) + 1] \qquad (11.2\text{-}44)$$

where $\Phi^{-1}(A)$ is defined by Eqs. (10.7-12) and (10.7-13). For $N \gg 1$, the second term in the bracket in (11.2-44) is negligible. Solving Eq. (11.2-44) for $\overline{\mathscr{R}}_p$ yields

$$\overline{\mathscr{R}}_p \cong \frac{2\Phi^{-1}(P_{fa})}{\sqrt{N}\ln(1/P_d)} \qquad (11.2\text{-}45)$$

Equation (11.2-45) with Figs. 10.7-1 and 10.7-2 permits rapid estimates of system performance parameters. The accuracy in estimating $\overline{\mathscr{R}}_p$ is of the order of a decibel for $N \geqslant 10$.

11.3. Pulse-to-Pulse Rayleigh Fluctuating Incoherent Pulse Train (Swerling II)

Swerling II differs from Swerling I in that the amplitude of *each* pulse in the train is a statistically independent random variable, with the same Rayleigh probability density function. The initial phases of each pulse in the train are assumed again to be independent with uniform probability densities.

A. Optimum receiver structure

The modified generalized likelihood ratio of Eq. (10.3-9a) can be expressed in the following form:

$$\overline{\ell'(\mathbf{v}\,|\,\boldsymbol{\theta})_\theta} = \left[\overline{\prod_{i=1}^{N} \ell'(\mathbf{v}_i\,|\,\phi_i, A_i)}\right]_{\phi_1, \phi_2, \dots, \phi_N, A_1, A_2, \dots, A_N} \tag{11.3-1}$$

where ϕ_i and A_i are the initial phase and amplitude of the ith pulse, respectively. Because all ϕ_i and A_i are statistically independent, Eq. (11.3-1) can be rewritten as

$$\overline{\ell'(\mathbf{v}\,|\,\boldsymbol{\theta})_\theta} = \prod_{i=1}^{N} \overline{\ell'(\mathbf{v}_i\,|\,\phi_i, A_i)}_{\phi_i, A_i} \tag{11.3-2}$$

To evaluate (11.3-2), Eq. (11.2-7) is a convenient starting point (with A replaced by A_i):

$$\overline{\ell'(\mathbf{v}_i\,|\,\phi_i, A_i)}_{\phi_i} = e^{-A_i^2/2} I_0[A_i r_i(\mathbf{v}_i)] \tag{11.3-3}$$

The quantity $r_i(\mathbf{v}_i)$ is the waveform envelope out of a matched filter with normalized gain sampled at the time the ith received pulse would peak when noise is absent. To obtain $\overline{\ell'(\mathbf{v}_i\,|\,\phi_i, A_i)}_{\phi_i, A_i}$, (11.3-3) is averaged with respect to random variable A_i. All of the A_i have the identical density function given by Eq. (11.2-9); thus,

$$\overline{\ell'(\mathbf{v}_i\,|\,\phi_i, A_i)}_{\phi_i, A_i} = \int_0^\infty \frac{A_i}{A_0^2} e^{-A_i^2(1/2A_0^2 + 1/2)} I_0(A_i r_i)\, dA_i \tag{11.3-4}$$

The solution of this integral is known (see Watson [4]) and is of the form:

$$\int_0^\infty t^{\mu-1} I_\nu(\alpha t) e^{-p^2 t^2}\, dt$$

$$= \frac{\Gamma\!\left(\dfrac{\mu+\nu}{2}\right)\!\left(\dfrac{\alpha}{2p}\right)^\nu}{2p^\mu \Gamma(\nu+1)}\, e^{\alpha^2/4p^2}\, {}_1F_1\!\left(\frac{\nu-\mu}{2}+1, \nu+1; -\frac{\alpha^2}{4p^2}\right) \tag{11.3-5}$$

where ${}_1F_1(a, c; z)$ is the confluent hypergeometric function. To apply (11.3-5) to (11.3-4), let $\mu = 2$, $\nu = 0$, $\alpha = r_i$, and $p^2 = (1/2A_0^2 + 1/2)$. For these values, the confluent hypergeometric function in (11.3-5) becomes ${}_1F_1(0, 1; -r_i^2/4p^2)$. The confluent hypergeometric function ${}_1F_1(a, c; z)$ has a series expansion [6] given by

$$_1F_1(a, c; z) = 1 + \frac{a}{c}\frac{z}{1!} + \frac{a(a+1)}{c(c+1)}\frac{z^2}{2!} + \cdots \tag{11.3-6}$$

From Eq. (11.3-6) the confluent hypergeometric function ${}_1F_1(0, c, z)$ is unity for $c \neq 0$. Also $\Gamma[(\mu+\nu)/2]$ reduces to $\Gamma(1) = 1$. Then the integral in (11.3-4) becomes

$$\overline{\ell'(\mathbf{v}_i\,|\,\phi_i, A_i)}_{\phi_i, A_i} = \frac{1}{2p^2} e^{r_i^2/4p^2} \tag{11.3-7}$$

Substituting Eq. (11.3-7) into (11.3-2) results in

$$\overline{\ell'(\mathbf{v}\,|\,\boldsymbol{\theta})_\theta} = \left(\frac{1}{2p^2}\right)^N \prod_{i=1}^N e^{r_i^2/4p^2} \tag{11.3-8a}$$

$$= \left(\frac{1}{2p^2}\right)^N \exp\left(\sum_{i=1}^N \frac{r_i^2}{4p^2}\right) \tag{11.3-8b}$$

Comparing the modified likelihood ratio in (11.3-8b) with a threshold is equivalent to the following tests:

$$\ln\left(\frac{1}{2p^2}\right)^N + \sum_{i=1}^N \frac{r_i^2}{4p^2} \begin{cases} \geqslant \mathscr{K} \\ < \mathscr{K} \end{cases} \tag{11.3-9}$$

$$\sum_{i=1}^N \frac{r_i^2}{4p^2} \begin{cases} \geqslant \mathscr{K}_{-1} \\ < \mathscr{K}_{-1} \end{cases} \tag{11.3-10}$$

$$\sum_{i=1}^N r_i^2 \begin{cases} \geqslant \mathscr{K}_{-2} & \textit{(signal present)} \\ < \mathscr{K}_{-2} & \textit{(signal absent)} \end{cases} \tag{11.3-11}$$

It can be concluded from (11.3-11) that the optimum receiver for Swerling II is the same as that shown in Fig. 10.3-2 without any approximation; that is, a square-law envelope detector is best for all signal-to-noise ratios.

B. Optimum performance

Since the Swerling II receiver is the same as that for a nonfluctuating pulse train, its false-alarm probability is given by Eq. (10.4-20). Probability of detection is obtained by evaluating

$$P_d = 1 - \int_0^{Y_b} p(Y)\,dY \tag{11.3-12}$$

Probability density $p(Y)$ is obtained by the method of characteristic functions previously employed. For this purpose $p(Y)$ can be written as

$$p(Y) = \overline{p(Y\,|\,\mathbf{A})}_\mathbf{A} \tag{11.3-13}$$

where $\mathbf{A}$ represents the random amplitudes $(A_1, A_2, \ldots, A_N)$ of each train pulse. Characteristic function $\overline{C_Y(\xi)}_\mathbf{A}$ of sum random variable $Y = \sum_{i=1}^N z_i$ is related by Eq. (3.12-3) to the characteristic functions of random variables $z_i,\ i = 1, 2, \ldots, N$, by

$$\overline{C_Y(\xi)}_\mathbf{A} = \prod_{i=1}^N \overline{C_{z_i}(\xi)}_{A_i} \tag{11.3-14}$$

Since $\overline{C_{z_i}(\xi)}_{A_i}$ is the same for all i,

$$\overline{C_Y(\xi)}_\mathbf{A} = \left[\overline{C_{z_i}(\xi)}_{A_i}\right]^N \tag{11.3-15}$$

An expression for $\overline{C_{z_i}(\xi)}_{A_i}$ is obtained from Eq. (11.2-23b) with $N = 1$ and with A replaced by A_i; thus,

$$\overline{C_{z_i}(\xi)}_{A_i} = \frac{1}{1 - j\xi(1 + A_0^2)} \tag{11.3-16}$$

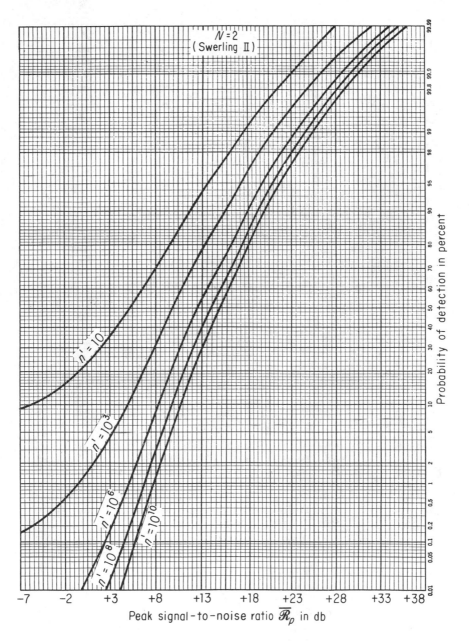

Fig. 11.3-1. Probability of detecting a fluctuating target, $N = 2$ (Swerling II). [N = number of pulses incoherently integrated; $P_{fa} = 0.693/n'$.]

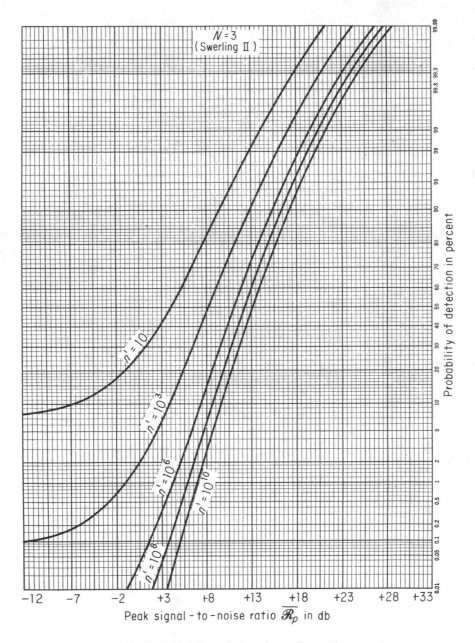

Fig. 11.3-2. Probability of detecting a fluctuating target, $N = 3$ (Swerling II). [N = number of pulses incoherently integrated; $P_{fa} = 0.693/n'$.]

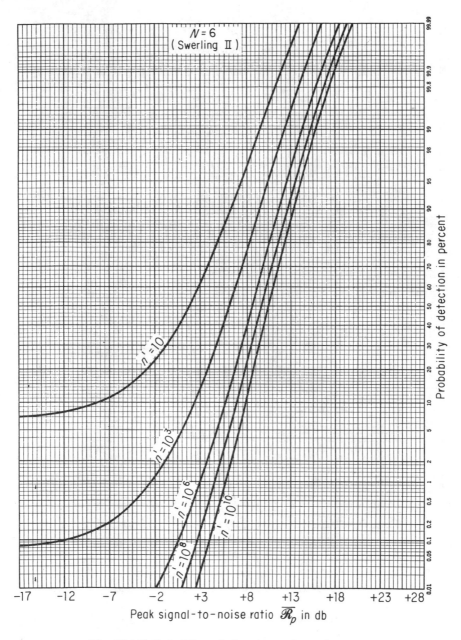

Fig. 11.3-3. Probability of detecting a fluctuating target, $N = 6$ (Swerling II). [N = number of pulses incoherently integrated; $P_{fa} = 0.693/n'$.]

397

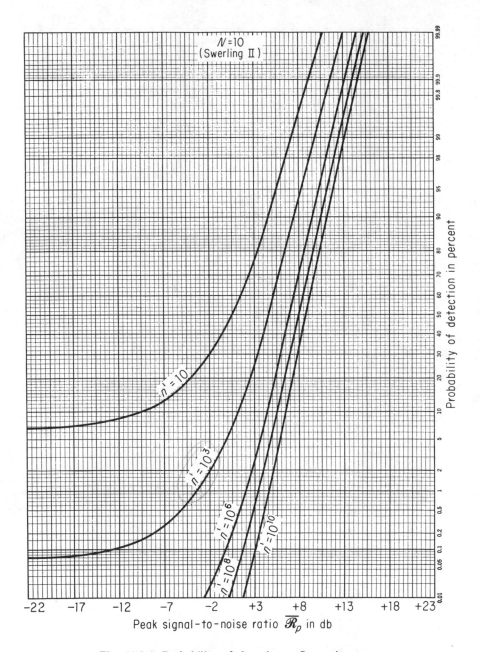

Fig. 11.3-4. Probability of detecting a fluctuating target, $N = 10$ (Swerling II). [N = number of pulses incoherently integrated; $P_{fa} = 0.693/n'$.]

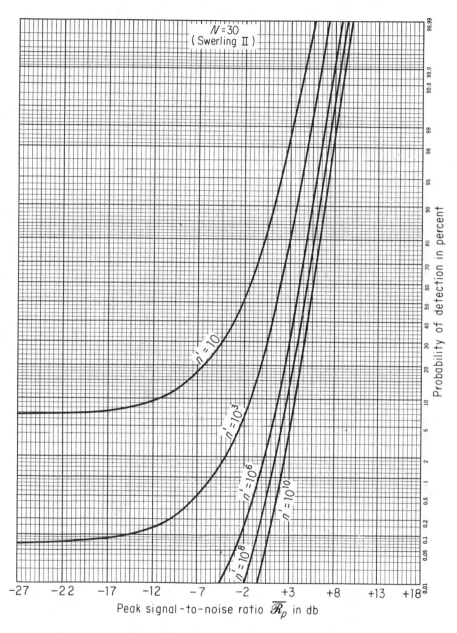

Fig. 11.3-5. Probability of detecting a fluctuating target, $N = 30$ (Swerling II). [N = number of pulses incoherently integrated; $P_{fa} = 0.693/n'$.]

399

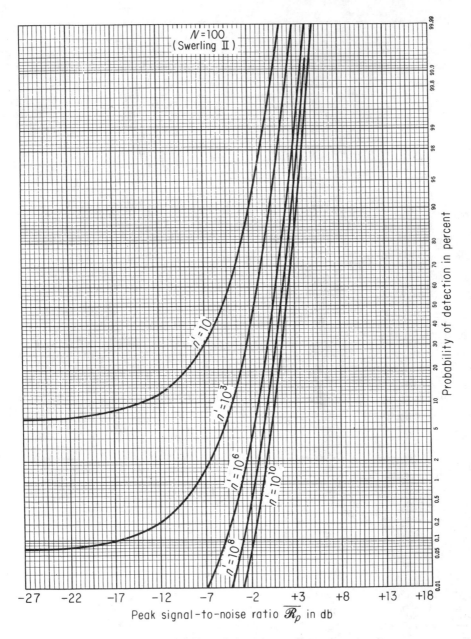

Fig. 11.3-6. Probability of detecting a fluctuating target, $N = 100$ (Swerling II). [N = number of pulses incoherently integrated; $P_{fa} = 0.693/n'$.]

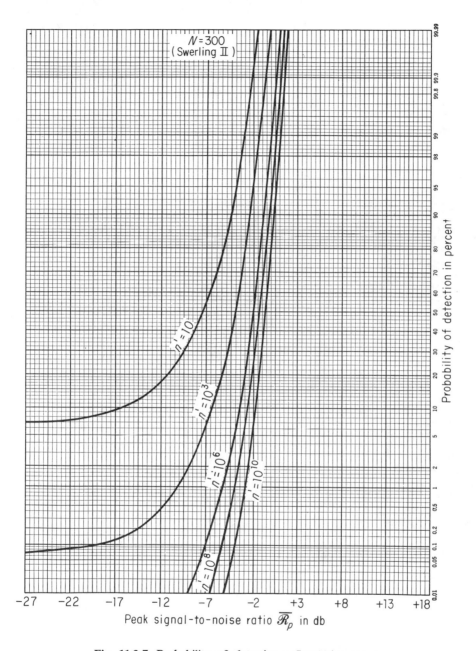

Fig. 11.3-7. Probability of detecting a fluctuating target, $N = 300$ (Swerling II). [N = number of pulses incoherently integrated; $P_{fa} = 0.693/n'$.]

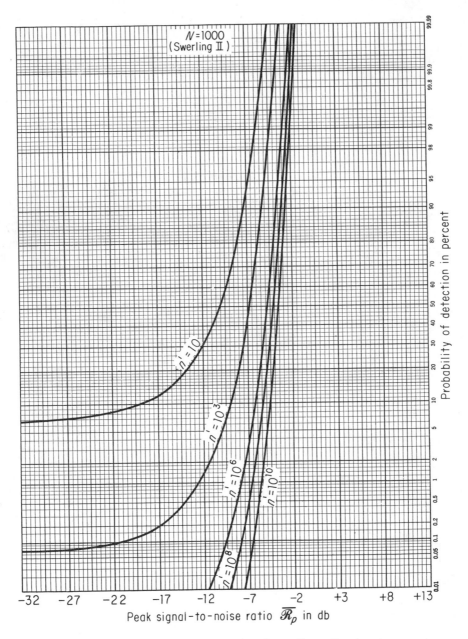

Fig. 11.3-8. Probability of detecting a fluctuating target, $N = 1000$ (Swerling II). [N = number of pulses incoherently integrated; $P_{fa} = 0.693/n'$.]

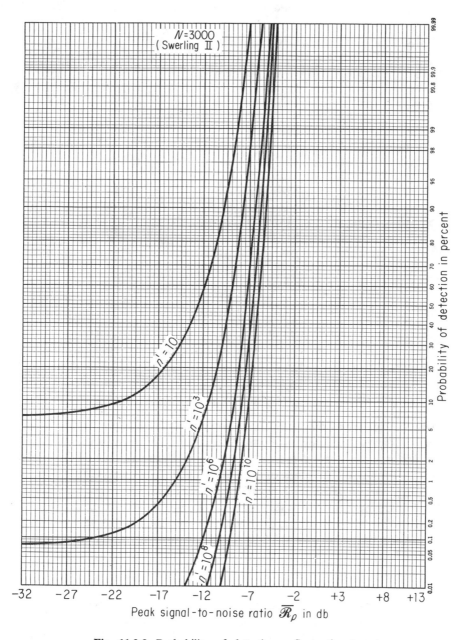

Fig. 11.3-9. Probability of detecting a fluctuating target, $N = 3000$ (Swerling II). [$N =$ number of pulses incoherently integrated; $P_{fa} = 0.693/n'$.]

Substituting (11.3-16) into (11.3-15) results in

$$\overline{C_Y(\xi)}_A = \frac{1}{[1 - j\xi(1 + A_0^2)]^N} \tag{11.3-17a}$$

$$= \frac{1}{(1 + A_0^2)^N \left[\dfrac{1}{1 + A_0^2} - j\xi\right]^N} \tag{11.3-17b}$$

The anticharacteristic function of (11.3-17b) is found using Fourier trans-form pair 431 of the Campbell and Foster tables [3]; the result is

$$p(Y) = \frac{1}{(1 + A_0^2)^N} \frac{1}{(N - 1)!} Y^{N-1} e^{-Y/(1+A_0^2)} \tag{11.3-18}$$

Substituting equation (11.3-18) into (11.3-12) yields

$$P_d = 1 - \frac{1}{(N - 1)!} \int_0^{Y_b} \left(\frac{Y}{1 + A_0^2}\right)^{N-1} e^{-Y/(1+A_0^2)} \frac{dY}{1 + A_0^2} \tag{11.3-19}$$

For

$$x = \frac{Y}{1 + A_0^2} \tag{11.3-20}$$

Eq. (11.3-19) can be rewritten as

$$P_d = 1 - \int_0^{Y_b/(1+A_0^2)} \frac{x^{N-1} e^{-x}}{(N - 1)!} dx \tag{11.3-21a}$$

$$= 1 - I\left[\frac{Y_b}{\sqrt{N}(1 + \overline{\mathscr{R}}_p/2)}, N - 1\right] \tag{11.3-21b}$$

where $I(u, s)$ is the incomplete gamma function defined by Eq. (10.4-19) and constant A_0^2 has been replaced by the quantity $\overline{\mathscr{R}}_p/2$ [see relation (11.2-18)]. The performance characteristics for Swerling II are calculated from Eq. (11.3-21b) for selected values of P_{fa} (and hence Y_b). Results for $N = 2, 3,$ 6, 10, 30, 100, 300, 1000, and 3000 are plotted in Figs. 11.3-1 through 11.3-9 based on computer evaluation of (11.3-21b) (see Fehlner [7]). Per-formance for $N = 1$ is identical to that for Swerling I and is shown in Fig. 11.2-1. (See discussion at end of Section 9.2 concerning units of $\overline{\mathscr{R}}_p$.)

C. Integration loss

Integration loss L is plotted in Fig. 11.3-10 for Swerling II referenced to a scan-to-scan Rayleigh fluctuating coherent pulse train with unknown initial phase, which is the same reference as that used for Swerling I.† The loss was

---

†In each of the Swerling cases, integration loss is referenced to a *coherent* pulse train having a similar amplitude fluctuation. Since random pulse-to-pulse amplitude fluctuation is incompatible for physical reasons with pulse-to-pulse coherence, in the case of Swerling II and IV the reference waveform is assumed to fluctuate scan-to-scan, and not pulse-to-pulse.

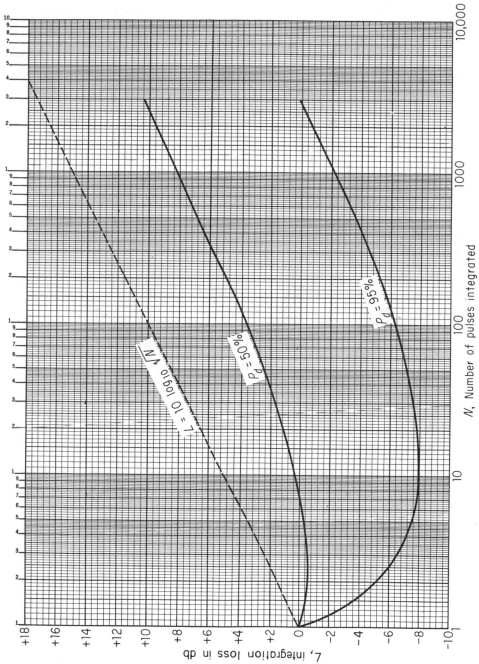

Fig. 11.3-10. Integration loss, incoherent vs. coherent, Swerling II. $P_{fa} = 10^{-10}$.

N, Number of pulses integrated

L, integration loss in db

$N \sqrt{} = 10 \log_{10} N$

$P_d = 50\%$

$P_d = 95\%$

405

evaluated as a function of N for $P_d = 50$ per cent and 95 per cent with $P_{fa} = 10^{-10}$. Fig. 11.3-10 shows that the loss is sensitive to P_d as well as N.

Note that the loss is negative for a range of values of N; that is, an improvement in performance can be expected with respect to a scan-to-scan Rayleigh fluctuating coherent pulse train signal. This is most noticeable at high probabilities of detection corresponding to high signal-to-noise ratios. Since most radars are designed for a probability of detection of 50 per cent or more, this phenomenon results in enhanced detectability of fluctuating targets. Figure 11.3-10 also shows that for large N, the slope of the loss curves approaches $10 \log \sqrt{N}$.

D. Analytic approximations

Approximate performance of the Swerling II model can be obtained by approximating $p(Y)$, the probability density of Y, by the first term of a Gram-Charlier expansion. In Section 10.6 it was shown that this corresponds to approximating $p(Y)$ by a Gaussian density function with mean and variance equal to the true mean and variance of Y. The first and second moments of Y can be found by substituting Eq. (11.3-17b) into (10.7-2); this yields

$$m_1 = \bar{Y} = \frac{1}{j}\left[\frac{jN}{(1+A_0^2)^N\left(\frac{1}{1+A_0^2} - j\xi\right)^{N+1}}\right]_{\xi=0} \tag{11.3-22a}$$

$$= N(1 + A_0^2) \tag{11.3-22b}$$

and

$$m_2 = \overline{Y^2} = \frac{1}{j^2}\left[\frac{j^2 N(N+1)}{(1+A_0^2)^N\left(\frac{1}{1+A_0^2} - j\xi\right)^{N+2}}\right]_{\xi=0} \tag{11.3-23a}$$

$$= N(N+1)(1 + A_0^2)^2 \tag{11.3-23b}$$

With m_1 and m_2 above, the variance of Y, σ_Y^2, is computed as

$$\sigma_Y^2 = m_2 - m_1^2 \tag{11.3-24a}$$

$$= (N^2 + N)(1 + A_0^2)^2 - N^2(1 + A_0^2)$$

$$= N(1 + A_0^2)^2 \tag{11.3-24b}$$

With the mean and variance of Y given by Eqs. (11.3-22b) and (11.3-24b), respectively, the Gaussian approximation of $p(Y)$ is given by

$$p(Y) \cong \frac{1}{\sqrt{2\pi N}\,(1+A_0^2)} \exp\left\{-\frac{[Y - N(1+A_0^2)]^2}{2N(1+A_0^2)^2}\right\} \tag{11.3-25}$$

Then

$$P_d \cong \int_{Y_b}^{\infty} p(Y)\,dY \tag{11.3-26a}$$

$$\cong \int_{Y_b - N(1+A_0^2)/\sqrt{N}\,(1+A_0^2)}^{\infty} e^{-x^2/2}\,dx \tag{11.3-26b}$$

where (11.3-26b) results when (11.3-25) is substituted into (11.3-26a) with

$$x = \frac{Y - N(1 + A_0^2)}{\sqrt{N}\,(1 + A_0^2)} \tag{11.3-27}$$

Using definitions (10.7-12) and (10.7-13), we can rewrite Eq. (11.3-26b) as

$$P_d \cong \Phi\left[\frac{Y_b - N(1 + A_0^2)}{\sqrt{N}\,(1 + A_0^2)}\right] \tag{11.3-28}$$

or

$$\Phi^{-1}(P_d) \cong \frac{Y_b - N(1 + A_0^2)}{\sqrt{N}\,(1 + A_0^2)} \tag{11.3-29}$$

When Y_b is replaced by Eq. (10.7-16) and A_0^2 by $\overline{\mathscr{R}}_p/2$ [see Eq. (11.2-18)], Eq. (11.3-29) becomes

$$\Phi^{-1}(P_d) \cong \frac{\sqrt{N}\,\Phi^{-1}(P_{fa}) - N\overline{\mathscr{R}}_p/2}{\sqrt{N}(1 + \overline{\mathscr{R}}_p/2)} \tag{11.3-30}$$

Solving Eq. (11.3-30) for signal-to-noise ratio $\overline{\mathscr{R}}_p$ yields

$$\overline{\mathscr{R}}_p \cong \frac{2[\Phi^{-1}(P_{fa}) - \Phi^{-1}(P_d)]}{\sqrt{N} + \Phi^{-1}(P_d)} \tag{11.3-31}$$

For $N \gg 1$, $\Phi^{-1}(P_d)$ is much less than $\sqrt{N}$ over the range of values of P_d of interest; then (11.3-31) can be further approximated by

$$\overline{\mathscr{R}}_p \cong \frac{2[\Phi^{-1}(P_{fa}) - \Phi^{-1}(P_d)]}{\sqrt{N}} \tag{11.3-32}$$

Note that (11.3-32) predicts that performance will vary inversely as $\sqrt{N}$ for large N; this agrees with the plots in Fig. 11.3-10. Approximations (11.3-31) and (11.3-32) are the same as approximations (10.7-24) and (10.7-26), respectively, obtained for the nonfluctuating target model using a similar approximation technique. The more accurate approximation, Eq. (11.3-31), yields an estimate of $\overline{\mathscr{R}}_p$ accurate to the order of a decibel for $N > 100$.

11.4. Scan-to-Scan One-Dominant-Plus-Rayleigh Fluctuating Incoherent Pulse Train (Swerling III)

Swerling III is similar to Swerling I in that each pulse in the train has the same amplitude. In the Swerling III model, however, pulse-train amplitude A is assumed to be a random variable with a one-dominant-plus-Rayleigh probability density function given by

$$p(A) = \begin{cases} \dfrac{9A^3}{2A_0^4}\, e^{-3A^2/2A_0^2}, & A \geqslant 0 \\ 0, & A < 0 \end{cases} \tag{11.4-1}$$

By differentiating (11.4-1) and equating it to zero it can be shown that A_0 is the most probable value of A.

A. Optimum receiver structure

Because of the similarity of Swerling III to Swerling I the development can be started with expression (11.2-8b) for $\overline{\ell'(\mathbf{v}\,|\,\boldsymbol{\theta})}_\theta$; this equation is then averaged with respect to amplitude A using density (11.4-1). Thus,

$$\overline{\ell'(\mathbf{v}\,|\,\boldsymbol{\theta})}_\theta = \overline{\ell'(\mathbf{v})}_{\phi,A} \tag{11.4-2a}$$

$$= \int_0^\infty \frac{9A^3}{2A_0^4} e^{-A^2(3/2A_0^2 + N/2)} \prod_{i=1}^{N} I_0(Ar_i)\, dA, \qquad r_i \geqslant 0 \tag{11.4-2b}$$

For Ar_i less than unity (small signal-to-noise ratio), the product of the I_0 functions in (11.4-2b) can be approximated by the sum given in Eq. (11.2-12). Following the approximation procedure of Section 11.2, we obtain the result

$$\overline{\ell'(\mathbf{v}\,|\,\boldsymbol{\theta})}_\theta \cong F_1(A_0, h') + F_2(A_0, h') \sum_{i=1}^{N} r_i^2 \tag{11.4-3}$$

where functions F_1 and F_2 are independent of r_i, and h' is the coefficient of A^2 in the exponential term of Eq. (11.4-2b). Comparing the likelihood ratio in (11.4-3) with a threshold is equivalent to a comparison of $\sum_{i=1}^{N} r_i^2$ with a suitably altered threshold—that is,

$$\sum_{i=1}^{N} r_i^2 \begin{cases} \geqslant \mathcal{K} & \textit{(signal present)} \\ < \mathcal{K} & \textit{(signal absent)} \end{cases} \tag{11.4-4}$$

It follows from (11.4-4) that the optimum receiver structure for small signal-to-noise ratio is again identical to that of Fig. 10.3-2.

B. Optimum performance

The false-alarm probability of the optimum receiver is the same as in previous cases since the processor is unchanged. To compute the probability of detection, density $\overline{p(Y\,|\,A)}_A$ must be found. It is obtained by averaging characteristic function $C_Y(\xi\,|\,A)$, given by Eq. (11.2-19), with respect to A, and then evaluating the anticharacteristic function.

With the probability density of A given by Eq. (11.4-1), constant A_0 can be related to (average) peak signal-to-noise ratio $\overline{\mathcal{R}}_p$ as follows. From (11.4-1),

$$\overline{A^2} = \int_0^\infty A^2 p(A)\, dA \tag{11.4-5a}$$

$$= \int_0^\infty \frac{9A^5}{2A_0^4} e^{-3A^2/2A_0^2}\, dA \tag{11.4-5b}$$

If we let

$$x = \frac{3A^2}{2A_0^2} \tag{11.4-6}$$

Eq. (11.4-5b) can be rewritten as

$$\overline{A^2} = \frac{2}{3} A_0^2 \int_0^\infty x^2 e^{-x} \, dx \qquad (11.4\text{-}7)$$

The integral in (11.4-7) is the gamma function which, from Eq. (11.2-26), is equal to $\Gamma(3) = 2! = 2$. Thus Eq. (11.4-7) reduces to

$$\overline{A^2} = \frac{4}{3} A_0^2 \qquad (11.4\text{-}8)$$

Combining Eqs. (11.2-17) and (11.4-8) yields the desired relationship:

$$\mathscr{R}_p = \frac{4A_0^2}{3} \qquad (11.4\text{-}9)$$

Characteristic function $C_Y(\xi \,|\, A)$ given by Eq. (11.2-19) is averaged with respect to A using (11.4-1) as follows:

$$\overline{C_Y(\xi)_A} = \frac{1}{(1 - j\xi)^N} \int_0^\infty \frac{9A^3}{2A_0^4} e^{-h^2 A^2} \, dA \qquad (11.4\text{-}10)$$

where

$$h^2 = \frac{3}{2A_0^2} - \frac{jN\xi}{2(1 - j\xi)} \qquad (11.4\text{-}11)$$

With

$$z = h^2 A^2 \qquad (11.4\text{-}12)$$

Eq. (11.4-10) can be rewritten as

$$\overline{C_Y(\xi)_A} = \frac{9}{4A_0^4 h^4 (1 - j\xi)^N} \int_0^\infty z e^{-z} \, dz \qquad (11.4\text{-}13)$$

The integral in (11.4-13) is a gamma function which, from (11.2-26), has a value of $\Gamma(1) = 1$. Substituting (11.4-11) for h^2 into (11.4-13) yields

$$\overline{C_Y(\xi)_A} = \frac{9}{4A_0^4 (1 - j\xi)^N \left(\dfrac{3}{2A_0^2} - \dfrac{jN\xi}{2(1 - j\xi)} \right)^2} \qquad (11.4\text{-}14a)$$

$$= \frac{1}{(1 - j\xi)^{N-2} \left(1 - j\xi - \dfrac{jNA_0^2 \xi}{3} \right)^2}$$

$$= \frac{1}{\left(1 + \dfrac{NA_0^2}{3} \right)^2 (1 - j\xi)^{N-2} \left(\dfrac{1}{1 + NA_0^2/3} - j\xi \right)^2} \qquad (11.4\text{-}14b)$$

To obtain the anticharacteristic function of (11.4-14b) from Campbell and Foster tables [3], we replace ξ in (11.4-14b) by $(-\xi)$ and identify $(-j\xi)$ with the Campbell and Foster transform variable s; then Eq. (11.4-14b) can be rewritten as

$$\overline{C_Y(s)_A} = \frac{1}{\left(1 + \dfrac{NA_0^2}{3} \right)^2 (1 + s)^{N-2} \left(\dfrac{1}{1 + NA_0^2/3} + s \right)^2} \qquad (11.4\text{-}15)$$

Fourier transform pair 581.1 in the Campbell and Foster tables is given by

$$\frac{1}{(s+\rho)^{\alpha+\nu}(s+\sigma)^{\alpha-\nu}} \longleftrightarrow \frac{1}{\Gamma(2\alpha)(\rho-\sigma)^{\alpha}} Y^{\alpha-1}e^{-(\rho+\sigma)Y/2}M_{\nu,\alpha-1/2}[(\rho-\sigma)Y],$$

$$Y>0 \qquad (11.4\text{-}16)$$

where

$$M_{\nu,\mu}(z) = z^{\mu+1/2}e^{-z/2}\,{}_1F_1(\mu-\nu+\tfrac{1}{2}, 2\mu+1; z) \qquad (11.4\text{-}17)$$

With relations (11.4-16) and (11.4-17), density $p(Y)$ is found from Eq. (11.4-15) to be

$$p(Y) = \frac{Y^{N-1}e^{-Y}}{(1+NA_0^2/3)^2(N-1)!}\,{}_1F_1\!\left[2, N; \frac{1}{\left(1+\frac{1}{NA_0^2/3}\right)}\right] \qquad (11.4\text{-}18)$$

Expression (11.4-18) can be simplified using the following relationships for the confluent hypergeometric function [5]:

$$_1F_1(2, N; z) = (z+2-N)\,{}_1F_1(1, N; z) + N - 1 \qquad (11.4\text{-}19)$$

and

$$_1F_1(1, N; z) = e^z z^{-N+1}(N-1)!\,I\!\left[\frac{z}{\sqrt{N-1}}, N-2\right] \qquad (11.4\text{-}20)$$

where $I(u, s)$ is the incomplete gamma function defined in Eq. (10.4-19). Substituting (11.4-20) into (11.4-19), and the result into (11.4-18), yields (with $A_0^2/3$ replaced by $\bar{\mathcal{R}}_p/4$)

$$p(Y) = \frac{\left(1+\dfrac{1}{N\bar{\mathcal{R}}_p/4}\right)^{N-2}}{(1+N\bar{\mathcal{R}}_p/4)^2}\,Ye^{-Y/(1+N\bar{\mathcal{R}}_p/4)}I\!\left[\frac{Y}{\left(1+\dfrac{1}{N\bar{\mathcal{R}}_p/4}\right)\sqrt{N-1}}, N-2\right]$$

$$-\frac{(N-2)\left(1+\dfrac{1}{N\bar{\mathcal{R}}_p/4}\right)^{N-1}}{(1+N\bar{\mathcal{R}}_p/4)^2}\,e^{-Y/(1+N\bar{\mathcal{R}}_p/4)}$$

$$\cdot I\!\left[\frac{Y}{\left(1+\dfrac{1}{N\bar{\mathcal{R}}_p/4}\right)\sqrt{N-1}}, N-2\right] + \frac{Y^{N-1}e^{-Y}}{(N-2)!(1+N\bar{\mathcal{R}}_p/4)^2}$$

$$(11.4\text{-}21)$$

The probability of detection is obtained by substituting (11.4-21) into

$$P_d = \int_{Y_b}^{\infty} p(Y)\,dY \qquad (11.4\text{-}22)$$

and performing the indicated integration. A computer evaluation of (11.4-22) has been carried out by Fehlner [7]; the results of this computation have been used to plot the probability of detection for a Swerling III target in Figs. 11.4-1 through 11.4-10 for selected values of P_{fa} and N. (See discussion at end of Section 9.2 concerning units of $\bar{\mathcal{R}}_p$.)

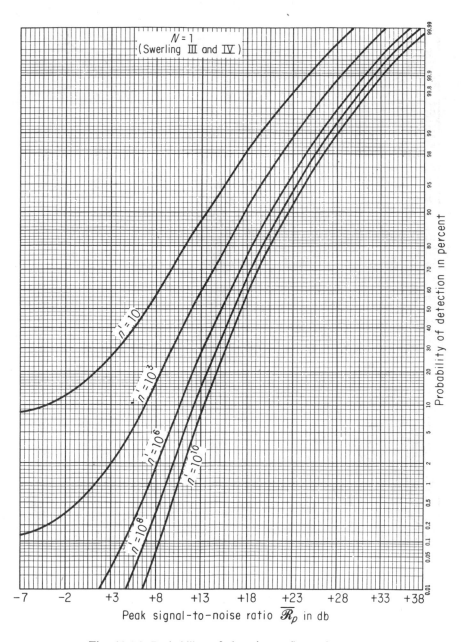

Fig. 11.4-1. Probability of detecting a fluctuating target, $N = 1$. (Swerling III and IV). [N = number of pulses incoherently integrated; $P_{fa} = 0.693/n'$.]

411

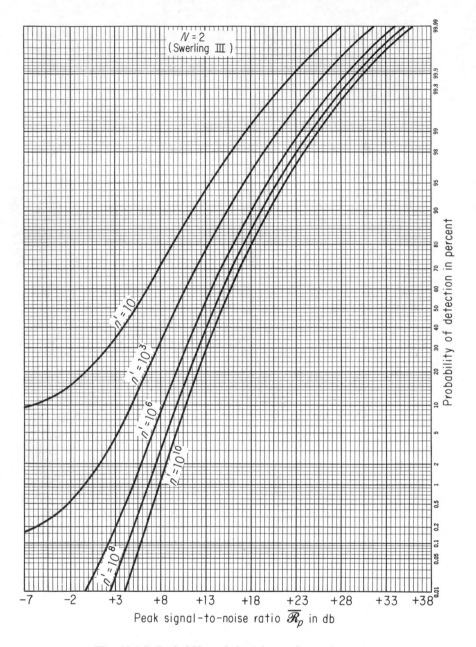

Fig. 11.4-2. Probability of detecting a fluctuating target, $N = 2$ (Swerling III). [N = number of pulses incoherently integrated; $P_{fa} = 0.693/n'$.]

412

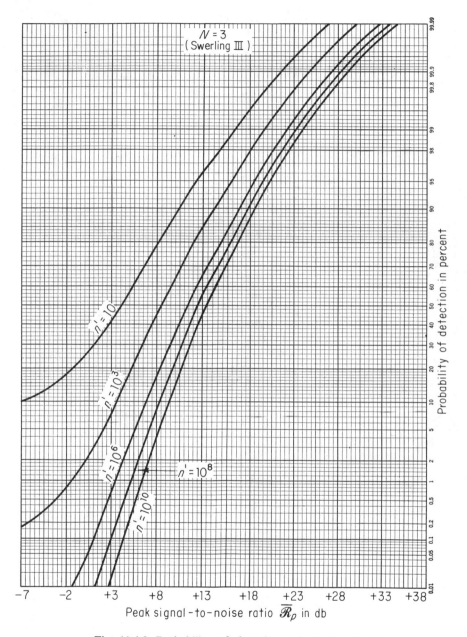

Fig. 11.4-3. Probability of detecting a fluctuating target, $N = 3$ (Swerling III). [N = number of pulses incoherently integrated; $P_{fa} = 0.693/n'$.]

413

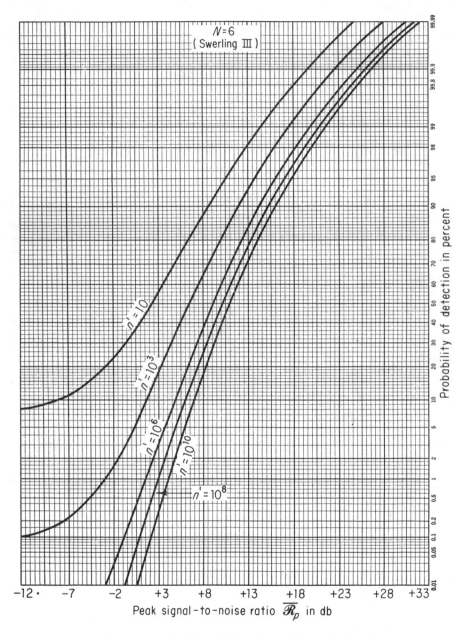

Fig. 11.4-4. Probability of detecting a fluctuating target, $N = 6$ (Swerling III). [N = number of pulses incoherently integrated; $P_{fa} = 0.693/n'$.]

414

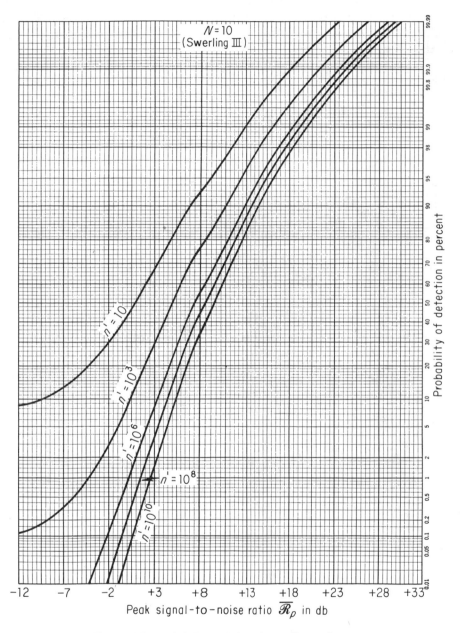

Fig. 11.4-5. Probability of detecting a fluctuating target, $N = 10$ (Swerling III). [N = number of pulses incoherently integrated; $P_{fa} = 0.693/n'$.]

415

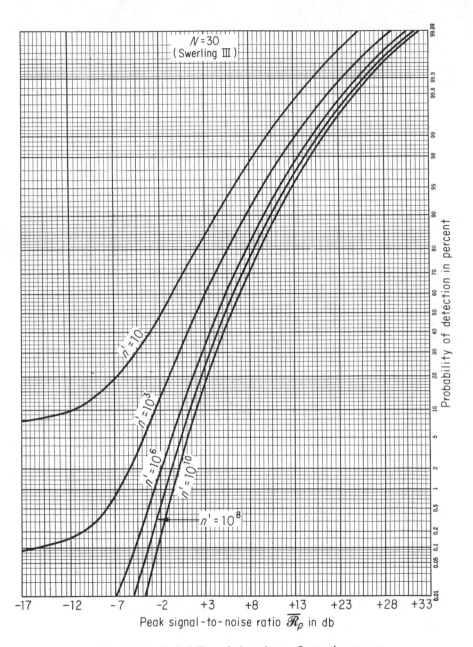

Fig. 11.4-6. Probability of detecting a fluctuating target, $N = 30$ (Swerling III). [N = number of pulses incoherently integrated; $P_{fa} = 0.693/n'$.]

416

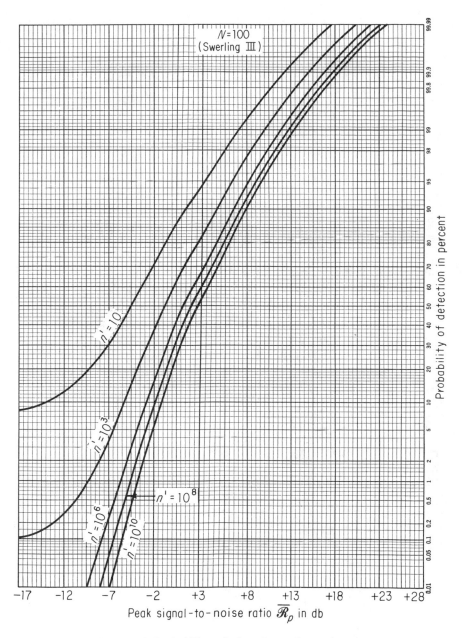

Fig. 11.4-7. Probability of detecting a fluctuating target,
$N = 100$ (Swerling III). [N = number of pulses incoherently
integrated; $P_{fa} = 0.693/n'$.]

417

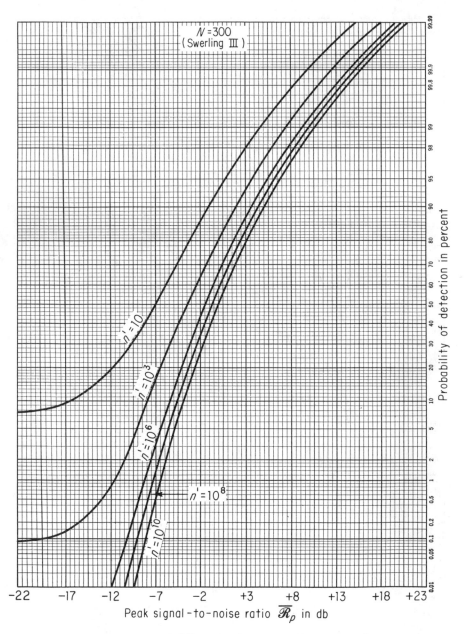

Fig. 11.4-8. Probability of detecting a fluctuating target, $N = 300$ (Swerling III). [$N = $ number of pulses incoherently integrated; $P_{fa} = 0.693/n'$.]

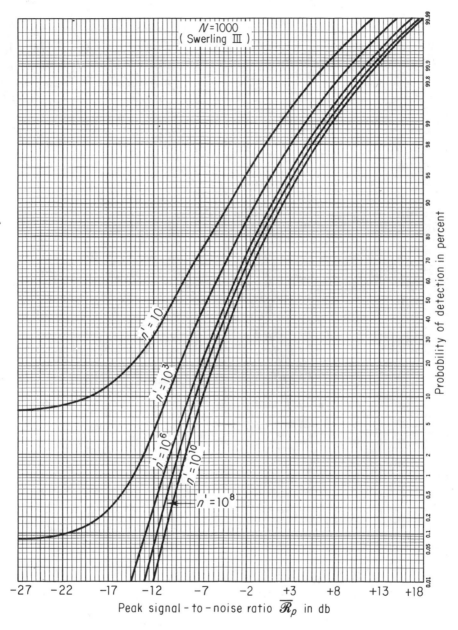

Fig. 11.4-9. Probability of detecting a fluctuating target, $N = 1000$ (Swerling III). [N = number of pulses incoherently integrated; $P_{fa} = 0.693/n'$.]

419

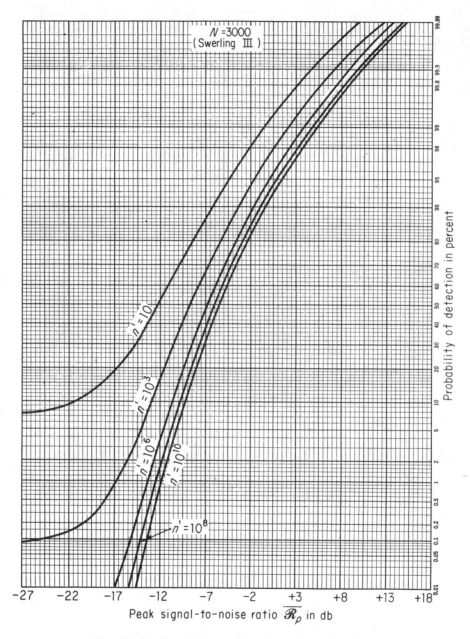

Probability of detection in percent

Peak signal-to-noise ratio $\overline{\mathscr{R}_p}$ in db

$N = 3000$
(Swerling III)

$n'=10$

$n'=10^3$

$n'=10^6$

$n'=10^{10}$

$n'=10^8$

Fig. 11.4-10. Probability of detecting a fluctuating target, $N = 3000$ (Swerling III). [N = number of pulses incoher ently integrated; $P_{fa} = 0.693/n'$.]

420

Swerling [1] obtained an approximate expression for P_d that is fairly accurate in radar applications. For false-alarm probability much less than one (the usual radar case), and for $N\overline{\mathscr{R}}_p/4$ greater than one (which corresponds to reasonable detection probabilities), the values of the incomplete gamma functions in (11.4-21) are practically unity for $Y > Y_b$. Substituting Eq. (11.4-21) into (11.4-22) yields with these approximations

$$P_d \cong \frac{\left(1 + \dfrac{1}{N\overline{\mathscr{R}}_p/4}\right)^{N-2}}{(1 + N\overline{\mathscr{R}}_p/4)^2}$$

$$\cdot \left[\int_{Y_b}^{\infty} Y e^{-Y/(1+N\overline{\mathscr{R}}_p/4)}\, dY - (N-2)\int_{Y_b}^{\infty}\left(1 + \frac{1}{N\overline{\mathscr{R}}_p/4}\right) e^{-Y/(1+N\overline{\mathscr{R}}_p/4)}\, dY\right]$$

$$+ \frac{1}{(N-2)!\,(1+N\overline{\mathscr{R}}_p/4)}\int_{Y_b}^{\infty} Y^{N-1} e^{-Y}\, dY \qquad (11.4\text{-}23)$$

For $N > 2$, the last integral in (11.4-23) can be neglected. Performing the integrations in (11.4-23) results in

$$P_d \cong \left(1 + \frac{1}{N\overline{\mathscr{R}}_p/4}\right)^{N-2}\left(1 + \frac{Y_b}{1 + N\overline{\mathscr{R}}_p/4} - \frac{N-2}{N\overline{\mathscr{R}}_p/4}\right) e^{-Y_b/(1+N\overline{\mathscr{R}}_p/4)} \qquad (11.4\text{-}24)$$

Swerling [1] shows that (11.4-24), which is approximate for $N > 2$, is exact for $N = 1$ and $N = 2$.

C. Integration loss

In Fig. 11.4-11 integration loss L is plotted as a function of N for the Swerling III model referenced to a coherent pulse train with unknown initial phase and a scan-to-scan one-dominant-plus-Rayleigh fluctuating amplitude. The loss curves for both 50 per cent and 95 per cent probability of detection are very close and a single curve is sufficient to describe both. Note that the slope of the loss curve approaches $10 \log \sqrt{N}$ for large N.

D. Analytic approximations

Approximation (11.4-24) can be further simplified for $N \gg 1$. The logarithm of (11.4-24) is given by

$$\ln P_d \cong (N-2)\ln\left(1 + \frac{1}{N\overline{\mathscr{R}}_p/4}\right) + \ln\left(1 + \frac{Y_b}{1 + N\overline{\mathscr{R}}_p/4} - \frac{N-2}{N\overline{\mathscr{R}}_p/4}\right)$$

$$- \frac{Y_b}{1 + N\overline{\mathscr{R}}_p/4} \qquad (11.4\text{-}25)$$

In the usual radar case for large N, $N\overline{\mathscr{R}}_p/4 \gg 1$; with this assumption (11.4-25) can be approximated by

$$\ln P_d \cong (N-2)\left(\frac{1}{N\overline{\mathscr{R}}_p/4}\right) + \ln\left(1 + \frac{Y_b - N + 2}{N\overline{\mathscr{R}}_p/4}\right) - \frac{Y_b}{N\overline{\mathscr{R}}_p/4} \qquad (11.4\text{-}26)$$

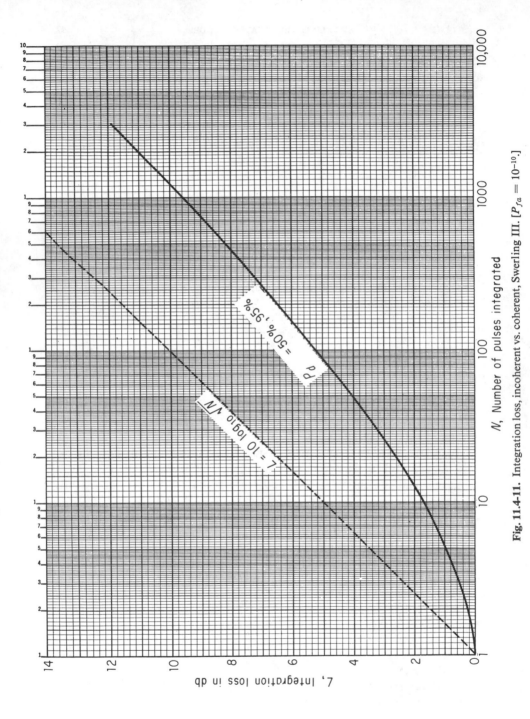

Fig. 11.4-11. Integration loss, incoherent vs. coherent, Swerling III. $[P_{fa} = 10^{-10}.]$

N, Number of pulses integrated

L, Integration loss in db

$P_D = 50\%, 95\%$

$10 \log_{10} N$

The first term in (11.4-26) was obtained from (11.4-25) using the approximation,

$$\ln (1 + x) \cong x, \qquad x \ll 1 \tag{11.4-27}$$

The second and third terms were obtained using the approximation,

$$1 + \frac{N\bar{\mathscr{R}}_p}{4} \cong \frac{N\bar{\mathscr{R}}_p}{4} \tag{11.4-28}$$

Combining terms in (11.4-26), we have

$$\ln P_d \cong - \frac{Y_b - N + 2}{N\bar{\mathscr{R}}_p/4} + \ln \left(1 + \frac{Y_b - N + 2}{N\bar{\mathscr{R}}_p/4}\right) \tag{11.4-29}$$

Next, Y_b in (11.4-29) is replaced by Eq. (10.7-16); then (11.4-29) can be rewritten as

$$-\ln P_d = \ln \left(\frac{1}{P_d}\right) \cong \frac{\sqrt{N}\,\Phi^{-1}(P_{fa}) + 2}{N\bar{\mathscr{R}}_p/4} - \ln \left(1 + \frac{\sqrt{N}\Phi^{-1}(P_{fa}) + 2}{N\bar{\mathscr{R}}_p/4}\right) \tag{11.4-30}$$

where $\Phi^{-1}(A)$ is defined by Eqs. (10.7-12) and (10.7-13). For $N \gg 1$, $\sqrt{N}\,\Phi^{-1}(P_{fa})$ is much greater than 2 over the range of P_{fa} normally encountered in radar systems. Hence (11.4-30) can be further simplified as follows:

$$\ln \left(\frac{1}{P_d}\right) \cong M - \ln (1 + M) \tag{11.4-31}$$

where

$$M = \frac{4\Phi^{-1}(P_{fa})}{\sqrt{N}\,\bar{\mathscr{R}}_p} \tag{11.4-32}$$

A graphical solution of transcendental equation (11.4-32) is plotted in Fig. 11.4-12. First M is determined from Fig. 11.4-12 for a selected probability of detection. Then $\bar{\mathscr{R}}_p$ is obtained from (11.4-32), which can be rewritten as

$$\bar{\mathscr{R}}_p = \frac{4\Phi^{-1}(P_{fa})}{\sqrt{N}\,M} \tag{11.4-33}$$

The quantity $\Phi^{-1}(P_{fa})$ in (11.4-33) is obtained from Fig. 10.7-1. This approximation yields values of $\bar{\mathscr{R}}_p$ accurate within a decibel for $N > 100$.

11.5. Pulse-to-Pulse One-Dominant-Plus-Rayleigh Fluctuating Incoherent Pulse Train (Swerling IV)

The Swerling IV pulse-train model is similar to Swerling II. In this case the amplitude of each pulse in the train is assumed to be a statistically

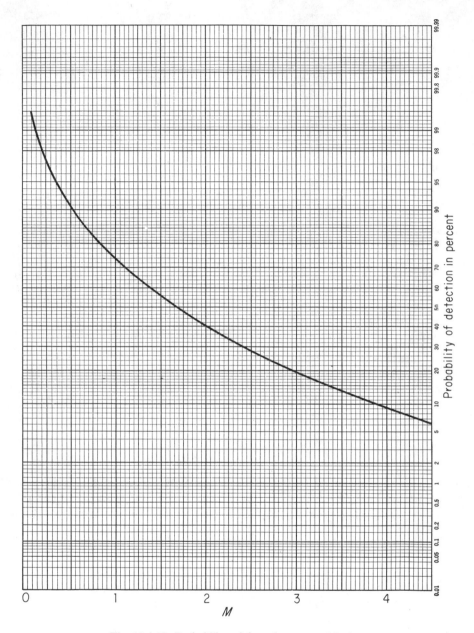

Fig. 11.4-12. Probability of detection versus M where

$$M = \frac{4\Phi^{-1}(P_{fa})}{\sqrt{N}\,\mathscr{R}_p}$$

independent random variable characterized by a one-dominant-plus-Rayleigh probability density. This density is described by Eq. (11.4-1).

A. Optimum receiver structure

To compute the modified generalized likelihood ratio for Swerling IV, $\overline{\ell'(v_i \mid \phi_i, A_i)}_{\phi_i}$, given by Eq. (11.3-3), is averaged with respect to A_i using (11.4-1). Thus,

$$\overline{\ell'(v_i \mid \phi_i, A_i)}_{\phi_i, A_i} = \int_0^\infty \frac{9A_i^3}{2A_0^4} \exp\left[-A_i^2\left(\frac{1}{2} + \frac{3}{2A_0^2}\right)\right] I_0(A_i r_i)\, dA_i \qquad (11.5\text{-}1)$$

Expression (11.5-1) can be evaluated with tabulated integral (11.3-5) to yield

$$\overline{\ell'(v_i \mid \phi_i, A_i)}_{\phi_i, A_i} = \frac{9}{4A_0^4 p^4}\, e^{r_i^2/4p^2}\, {}_1F_1\left(-1, 1; -\frac{r_i^2}{4p^2}\right) \qquad (11.5\text{-}2)$$

where

$$p^2 = \frac{1}{2} + \frac{3}{2A_0^2} \qquad (11.5\text{-}3)$$

From the series expansion of the confluent hypergeometric function in Eq. (11.3-6), we have

$$_1F_1(-1, 1; z) = 1 + \frac{(-1)}{1}\frac{z}{1!} + 0 \qquad (11.5\text{-}4a)$$

$$= 1 - z \qquad (11.5\text{-}4b)$$

Substituting (11.5-4b) into Eq. (11.5-2) yields

$$\overline{\ell'(v_i \mid \phi_i, A_i)}_{\phi_i, A_i} = c_1 e^{r_i^2/4p^2}[1 + r_i^2 \mid 4p^2] \qquad (11.5\text{-}5)$$

where c_1 is a constant.

The modified generalized likelihood ratio is obtained by substituting (11.5-5) into Eq. (11.3-2) as follows:

$$\overline{\ell'(\mathbf{v} \mid \boldsymbol{\phi}, \mathbf{A})}_{\phi, A} = \prod_{i=1}^N \overline{\ell'(v_i \mid \phi_i, A_i)}_{\phi_i, A_i} \qquad (11.5\text{-}6a)$$

$$= c_2 \exp\left(\sum_{i=1}^N \frac{r_i^2}{4p^2}\right)\prod_{i=1}^N \left(1 + \frac{r_i^2}{4p^2}\right) \qquad (11.5\text{-}6b)$$

where c_2 is a constant. The natural logarithm of Eq. (11.5-6b) is given by

$$\ln \overline{[\ell'(\mathbf{v} \mid \boldsymbol{\phi}, \mathbf{A})}_{\phi, A}] = c_3 + \sum_{i=1}^N \frac{r_i^2}{4p^2} + \sum_{i=1}^N \ln\left(1 + \frac{r_i^2}{4p^2}\right) \qquad (11.5\text{-}7)$$

where $c_3 = \ln c_2$. For small signal-to-noise ratios, $r_i^2/4p^2 < 1$; in this case the last term in (11.5-7) can be approximated by $\sum_{i=1}^N r_i^2/4p^2$; then Eq. (11.5-7) becomes approximately

$$\ln \overline{[\ell'(\mathbf{v} \mid \boldsymbol{\phi}, \mathbf{A})}_{\phi, A}] \cong c_3 + \sum_{i=1}^N \frac{r_i^2}{4p^2} + \sum_{i=1}^N \frac{r_i^2}{4p^2} \qquad (11.5\text{-}8a)$$

$$\cong c_3 + 2\sum_{i=1}^N \frac{r_i^2}{4p^2} \qquad (11.5\text{-}8b)$$

Comparing the modified generalized likelihood ratio in (11.5-8b) to a threshold corresponds to a comparison of $\sum_{i=1}^{N} r_i^2$ with a suitably altered threshold $\mathcal{K}$:

$$\sum_{i=1}^{N} r_i^2 \begin{cases} \geqslant \mathcal{K} & \text{(signal present)} \\ < \mathcal{K} & \text{(signal absent)} \end{cases} \tag{11.5-9}$$

It follows from (11.5-9) that, for small signal-to-noise ratios, the optimum receiver is again described by Fig. 10.3-2. The invariance of the optimum receiver structure in Fig. 10.3-2 is of considerable practical importance; this receiver is optimum for small signal-to-noise ratios for a nonfluctuating target and for the Swerling I through IV fluctuating radar models. Hence only one radar receiver is required to optimally detect pulse-train echoes from a target with a cross section characterized by the five models treated.

B. Optimum performance

As in previous cases, the false-alarm probability is unchanged. Calculation of the probability of detection requires knowledge of the density $p(Y)$, which can also be expressed as

$$p(Y) = \overline{p(Y|\mathbf{A})}_{\mathbf{A}} \tag{11.5-10}$$

Probability density $\overline{p(Y|\mathbf{A})}_{\mathbf{A}}$ is computed most simply from characteristic function $\overline{C_Y(\xi)}_{\mathbf{A}}$. This function is given by Eq. (11.3-15), since Swerling II and IV differ only in the probability density function of A_i. To evaluate $\overline{C_Y(\xi)}_{\mathbf{A}}$, an expression for characteristic function $\overline{C_{z_i}(\xi)}_{A_i}$, that has been averaged with respect to a one-dominant-plus-Rayleigh density function, is obtained from Eq. (11.4-14b) with $N = 1$ and $A = A_i$:

$$\overline{C_{z_i}(\xi)}_{A_i} = \frac{(1 - j\xi)}{(1 + A_0^2/3)^2 \left(\dfrac{1}{1 + A_0^2/3} - j\xi \right)^2} \tag{11.5-11}$$

Substituting (11.5-11) into (11.3-15) results in

$$\overline{C_Y(\xi)}_{\mathbf{A}} = [\overline{C_{z_i}(\xi)}_{A_i}]^N \tag{11.5-12a}$$

$$= \frac{(1 - j\xi)^N}{(1 + A_0^2/3)^{2N} \left(\dfrac{1}{1 + A_0^2/3} - j\xi \right)^{2N}} \tag{11.5-12b}$$

When $(-j\xi)$ is replaced by s in (11.5-12b), we have

$$\overline{C_Y(s)}_{\mathbf{A}} = \frac{1}{(1 + A_0^2/3)^{2N} \left(s + \dfrac{1}{1 + A_0^2/3} \right)^{2N} (s + 1)^{-N}} \tag{11.5-13}$$

The inverse Fourier transform of (11.5-13) is obtained from transform pair 581.1 of Campbell and Foster tables [3] [see Eqs. (11.4-16) and (11.4-17)]; this yields the following expression for $p(Y)$:

$$p(Y) = \frac{Y^{N-1} e^{-Y/(1 + A_0^2/3)}}{(1 + A_0^2/3)^{2N} (N - 1)!} \, {}_1F_1\left(-N, N; \frac{-A_0^2/3}{1 + A_0^2/3} Y \right) \tag{11.5-14}$$

From the expansion of the confluent hypergeometric function in (11.3-6), $_1F_1(-N, N; z)$ is given by

$$_1F_1(-N, N; z) = 1 + \frac{(-N)}{N} \frac{z}{1!} + \frac{(-N)(-N+1)}{N(N+1)} \frac{z^2}{2!}$$
$$+ \frac{(-N)(-N+1)(-N+2)}{N(N+1)(N+2)} \frac{z^3}{3!} + \cdots \qquad (11.5\text{-}15)$$

Note that beginning with the $(N + 2)$ term all higher-order terms in (11.5-15) are identically zero since $(N - N)$ appears in the numerator of each such term. Equation (11.5-15) can therefore be rewritten as

$$_1F_1(-N, N; z) = \sum_{k=0}^{N} \frac{(-1)^k[N!/(N-k)!]}{[(N+k-1)!/(N-1)!]} \frac{z^k}{k!} \qquad (11.5\text{-}16)$$

Substituting (11.5-16) into (11.5-14) yields

$$p(Y) = \frac{Y^{N-1}e^{-Y/(1+A_0^2/3)}N!}{(1 + A_0^2/3)^{2N}} \sum_{k=0}^{N} \left(\frac{A_0^2/3}{1+A_0^2/3}\right)^k \frac{Y^k}{k!(N+k-1)!(N-k)!}$$
$$(11.5\text{-}17)$$

The probability of detection is obtained by substituting (11.5-17) into

$$P_d = 1 - \int_0^{Y_b} p(Y)\, dY \qquad (11.5\text{-}18)$$

From (11.4-9) and the definition of the incomplete gamma function in (10.4-19), the probability of detection can be written as

$$P_d = 1 - \frac{N!}{(1+\bar{\mathscr{R}}_p/4)^N} \sum_{k=0}^{N} \left(\frac{\bar{\mathscr{R}}_p}{4}\right)^k \frac{I\left[\dfrac{Y_b}{(1+\bar{\mathscr{R}}_p/4)\sqrt{N+k}}, N+k-1\right]}{k!(N-k)!}$$
$$(11.5\text{-}19)$$

With Y_b obtained from Eq. (10.4-20), performance graphs of P_d versus $\bar{\mathscr{R}}_p$ can be calculated from (11.5-19) for the Swerling IV model with N and P_{fa} as parameters. Figures 11.5-1 through 11.5-9 contain plots of P_d versus $\bar{\mathscr{R}}_p$ for $N = 2, 3, 6, 10, 30, 100, 300, 1000,$ and 3000 for selected values of P_{fa} based on computer results obtained by Fehlner [7]. The plot for $N = 1$ is identical to that for Swerling III shown in Fig. 11.4-1. (See discussion at end of Section 9.2 concerning units of $\bar{\mathscr{R}}_p$.)

C. Integration loss

Figure 11.5-10 is a plot of integration loss L as a function of N for the Swerling IV model. The loss is computed with respect to a coherent scan-to-scan one-dominant-plus-Rayleigh fluctuating signal with unknown initial phase.† The loss curve for Swerling IV is similar to that of Swerling II—

---

†See footnote on page 404.

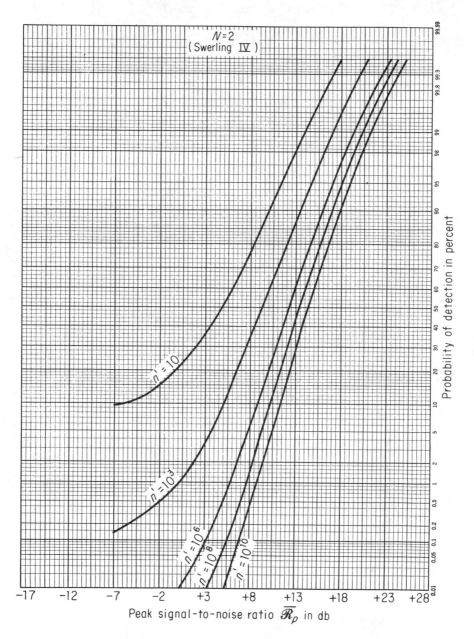

Fig. 11.5-1. Probability of detecting a fluctuating target, $N = 2$ (Swerling IV). [N = number of pulses incoherently integrated; $P_{fa} = 0.693/n'$.]

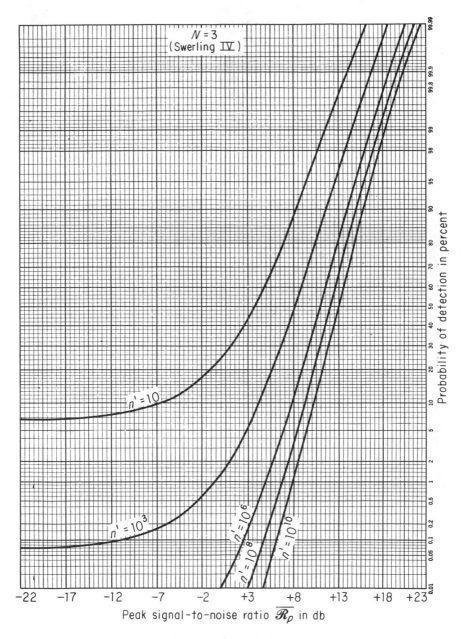

Fig. 11.5-2. Probability of detecting a fluctuating target, $N = 3$ (Swerling IV). [N = number of pulses incoherently integrated; $P_{fa} = 0.693/n'$.]

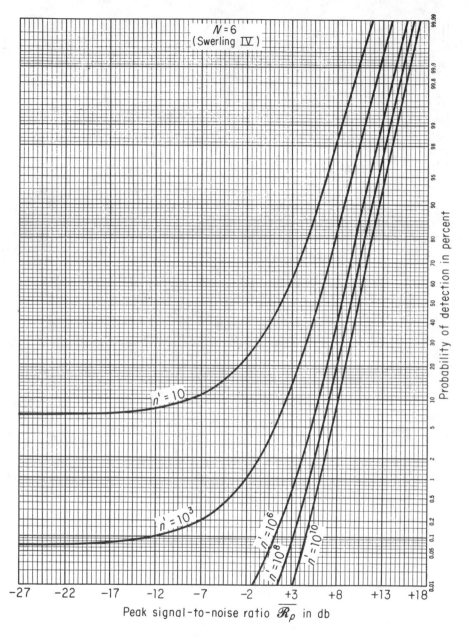

Fig. 11.5-3. Probability of detecting a fluctuating target, $N = 6$ (Swerling IV). [N = number of pulses incoherently integrated; $P_{fa} = 0.693/n'$.]

430

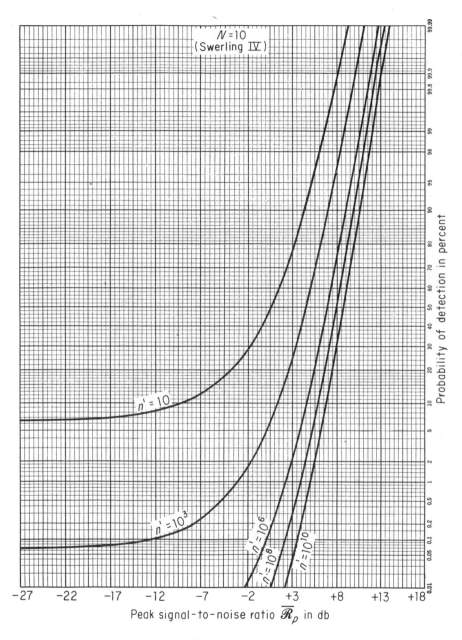

Fig. 11.5-4. Probability of detecting a fluctuating target, $N = 10$ (Swerling IV). [N = number of pulses incoherently integrated; $P_{fa} = 0.693/n'$.]

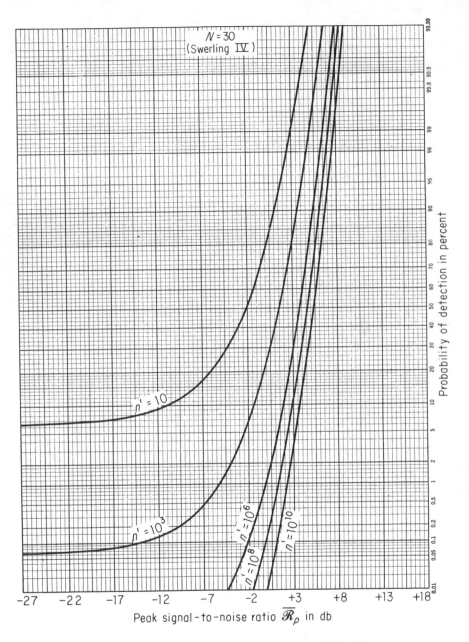

Fig. 11.5-5. Probability of detecting a fluctuating target, $N = 30$ (Swerling IV). [N = number of pulses incoherently integrated; $P_{fa} = 0.693/n'$.]

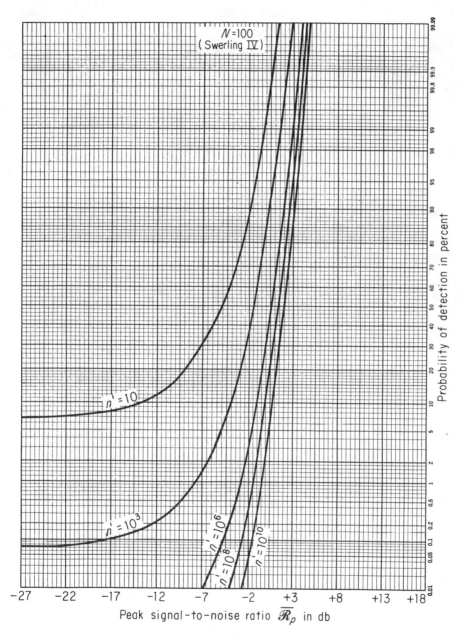

Fig. 11.5-6. Probability of detecting a fluctuating target, $N = 100$ (Swerling IV). [N = number of pulses incoherently integrated; $P_{fa} = 0.693/n'$.]

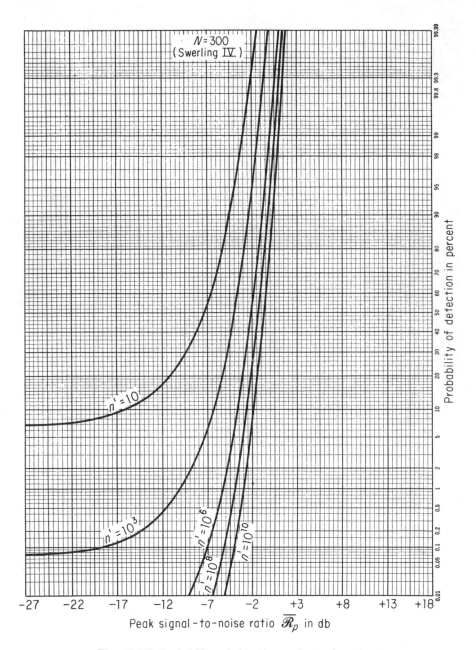

Fig. 11.5-7. Probability of detecting a fluctuating target, $N = 300$ (Swerling IV). [$N =$ number of pulses incoherently integrated; $P_{fa} = 0.693/n'$.]

434

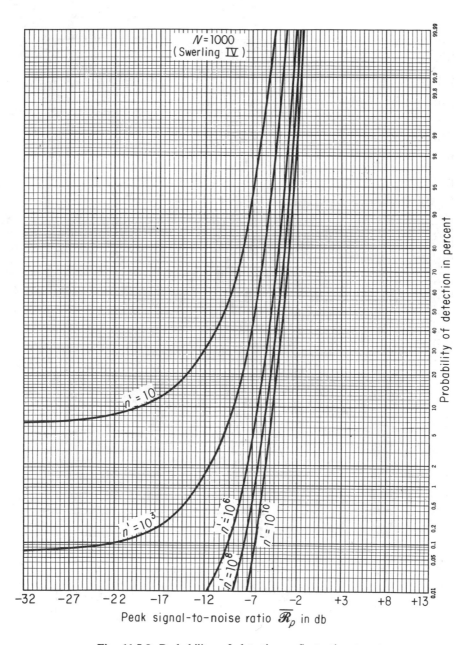

Fig. 11.5-8. Probability of detecting a fluctuating target, $N = 1000$ (Swerling IV). [N = number of pulses incoherently integrated; $P_{fa} = 0.693/n'$.]

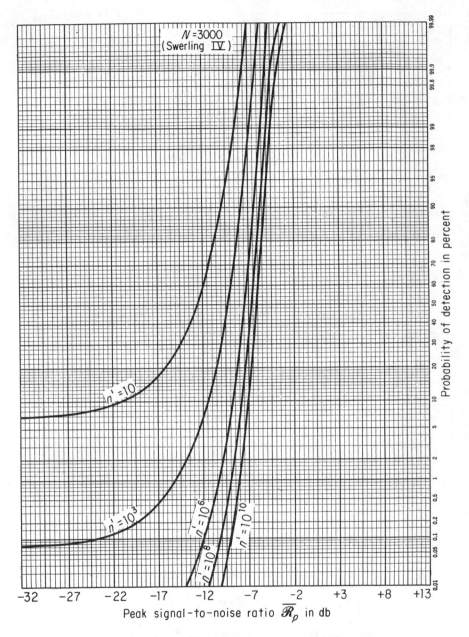

Fig. 11.5-9. Probability of detecting a fluctuating target, $N = 3000$ (Swerling IV). [N = number of pulses incoherently integrated; $P_{fa} = 0.693/n'$.]

436

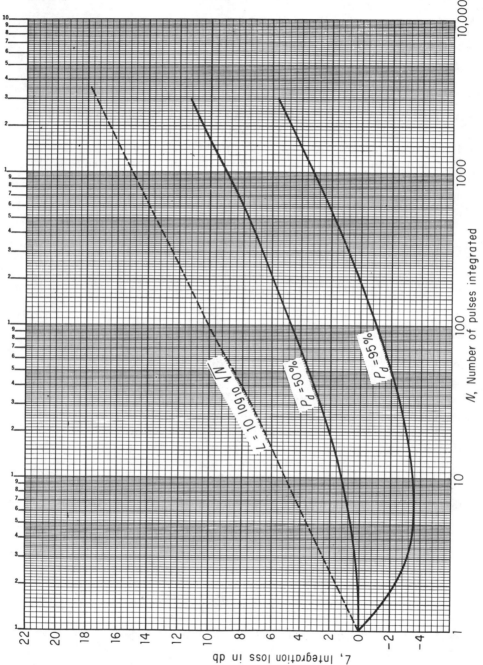

Fig. 11.5-10. Integration loss, incoherent vs. coherent (Swerling IV). [$P_{fa} = 10^{-10}$.]

N, Number of pulses integrated

ℓ, Integration loss in db

$N\sqrt{}$ = 10 log$_{10}$ N

P_d = 50%

P_d = 95%

437

namely, a gain in detectability occurs at high probabilities of detection over a wide range of N. Note that the slope of both loss curves approaches a slope of $10 \log \sqrt{N}$ for $N \gg 1$.

D. Analytic approximations

An analytic approximation for Swerling IV can be obtained by once again approximating $p(Y)$ by the first term of its Gram-Charlier expansion. This corresponds to a Gaussian approximation of the probability density of Y with mean and variance equal to the true mean and variance of Y. The first and second moments of Y are found by substituting Eq. (11.5-12b) into (10.7-2):

$$m_1 = \bar{Y} = N\left(1 + \frac{2A_0^2}{3}\right) \tag{11.5-20}$$

$$m_2 = \overline{Y^2} = N^2\left(1 + \frac{2A_0^2}{3}\right)^2 + N\left[1 + \frac{2A_0^2}{3}\left(2 + \frac{A_0^2}{3}\right)\right] \tag{11.5-21}$$

From (11.5-20) and (11.5-21) the variance is computed as

$$\sigma_Y^2 = m_2 - m_1^2 \tag{11.5-22a}$$

$$= N\left[2\left(1 + \frac{A_0^2}{3}\right)^2 - 1\right] \tag{11.5-22b}$$

With mean and variance given by (11.5-20) and (11.5-22b), the Gaussian approximation of $p(Y)$ is

$$p(Y) \cong \frac{\exp\left\{-\dfrac{[Y - N(1 + 2A_0^2/3)]^2}{2N[2(1 + A_0^2/3)^2 - 1]}\right\}}{\sqrt{2\pi N[2(1 + A_0^2/3)^2 - 1]}} \tag{11.5-23}$$

If we let

$$z = \frac{Y - N(1 + 2A_0^2/3)}{\sqrt{N[2(1 + A_0^2/3)^2 - 1]}} \tag{11.5-24}$$

the probability of detection can be written as

$$P_d = \int_{Y_b}^{\infty} p(Y)\, dY \tag{11.5-25a}$$

$$\cong \int_{Y_b - N(1 + 2A_0^2/3)/\sqrt{N[2(1 + A_0^2/3)^2 - 1]}}^{\infty} \frac{1}{\sqrt{2\pi}} e^{-z^2/2}\, dz \tag{11.5-25b}$$

$$\cong \Phi\left[\frac{Y_b - N(1 + 2A_0^2/3)}{\sqrt{N[2(1 + A_0^2/3)^2 - 1]}}\right] \tag{11.5-25c}$$

where $\Phi(T)$ is defined by Eq. (10.7-12). Inverting (11.5-25c) using (10.7-13) and replacing Y_b by Eq. (10.7-16), we have

$$\Phi^{-1}(P_d) \cong \frac{\sqrt{N}\,\Phi^{-1}(P_{fa}) - 2NA_0^2/3}{\sqrt{N[2(1 + A_0^2/3)^2 - 1]}} \tag{11.5-26}$$

Cross-multiplying, squaring both sides, and rearranging (11.5-26) yields

$$\{4N - 2[\Phi^{-1}(P_d)]^2\}\left(\frac{A_0^2}{3}\right)^2 - \{\sqrt{N}\ \Phi^{-1}(P_{fa}) - [\Phi^{-1}(P_d)]^2\}\left(\frac{4A_0^2}{3}\right)$$

$$+ [\Phi^{-1}(P_{fa})]^2 - [\Phi^{-1}(P_d)]^2 \cong 0 \qquad (11.5\text{-}27)$$

The bracketed coefficient of $(A_0^2/3)^2$ is approximately equal to $4N$ for $N \gg 1$. In addition, the first term in the bracketed coefficient of $(4A_0^2/3)$ is substantially greater than the second term for $N \gg 1$. Hence Eq. (11.5-27) can be further approximated by

$$\left(\frac{A_0^2}{3}\right)^2 - \frac{\Phi^{-1}(P_{fa})}{\sqrt{N}}\left(\frac{A_0^2}{3}\right) + \frac{[\Phi^{-1}(P_{fa})]^2 - [\Phi^{-1}(P_d)]^2}{4N} \cong 0 \qquad (11.5\text{-}28)$$

Solving the quadratic in (11.5-28) for $\bar{\mathscr{R}}_p = 4A_0^2/3$ leads to two solutions:

$$\bar{\mathscr{R}}_p \cong \frac{2[\Phi^{-1}(P_{fa}) \pm \Phi^{-1}(P_d)]}{\sqrt{N}} \qquad (11.5\text{-}29)$$

Physical reasoning leads to the choice of the minus sign in (11.5-29), since $\bar{\mathscr{R}}_p$ should increase with increasing P_d when P_{fa} and N are constant. Thus,

$$\bar{\mathscr{R}}_p \cong \frac{2[\Phi^{-1}(P_{fa}) - \Phi^{-1}(P_d)]}{\sqrt{N}} \qquad (11.5\text{-}30)$$

Equation (11.5-30) is identical to approximation (11.3-32) previously obtained for a pulse-to-pulse Rayleigh fluctuating pulse train with $N \gg 1$. It is also identical to approximation (10.7-27) for a constant-amplitude train with $\bar{\mathscr{R}}_p < 1$ and $N \gg 1$. Expression (11.5-30) yields values of $\bar{\mathscr{R}}_p$ that are accurate within a decibel for $N > 100$.

11.6. Performance of Partially Correlated Fluctuating Pulse Trains

Up to this point we have dealt with two types of fluctuating pulse trains: those with uncorrelated pulse-to-pulse amplitude fluctuations and those with completely correlated pulse-to-pulse fluctuations. In practice, situations arise that lie between these two extremes.

Swerling [8] has developed a method for evaluating the performance of fluctuating pulse trains with partially correlated pulse-to-pulse amplitude fluctuations. The method is applicable to a large family of probability density functions and for fairly general correlation properties of the signal fluctuations. The analysis assumes a receiver structure like that shown in Fig. 10.3-2. Swerling's method yields the characteristic function of the probability density of sum random variable Y. In general, to obtain $p(Y)$ and P_d with Swerling's technique requires computer evaluation.

Using a different approach, Schwartz [9] considered the effect of partial correlation on detectability for $N = 2$ and a Rayleigh amplitude fluctuation. Schwartz's curves [9] show that the performance plots for varying degrees of correlation fall between the extremes of zero correlation pulse-to-pulse and complete correlation pulse-to-pulse. Similar results can be anticipated for N greater than two.

Hence, to estimate performance for partially correlated pulse trains, interpolation between the curves in this chapter for the two bounding cases can be used. More accurate results can be obtained from a computer evaluation of performance using Swerling's method [8].

11.7. Summary

In this and the preceding chapter we have studied the structure and performance of an optimum detection receiver based on multiple observations. For simplicity the complete radar signal over a multiple observation interval is called a pulse train. Detection performance was evaluated for six pulse-train models, which differed in assumptions about the initial phase and amplitude of the train pulses.

First, a coherent pulse train was considered. Coherent pulse trains, with known or unknown initial phases, are particular examples of coherent waveforms. In this case, the relative phase of each pulse is known with respect to every other pulse. The optimum detection receiver employs predetection integration, which coherently sums the train pulses. Performance for such waveforms depends on total signal energy.

The five other pulse-train models are incoherent pulse-to-pulse. As a result, the optimum detection receiver for these cases employs postdetection or video integration, which does not depend on relative phase information between pulses. In practice postdetection integration is usually simpler to implement then predetection integration.

For a nonfluctuating pulse train, it was shown in Chapter 10 that the optimum detection characteristic is $\ln I_0(x)$. This detection characteristic is approximated by a square-law detector when the single-pulse signal-to-noise ratio is small and by a linear detector for large signal-to-noise ratios. Receiver performance was computed for a square-law characteristic. The difference in performance between both types of detectors is small for the usual values of P_d and P_{fa} employed in radar.

Four different fluctuating pulse-train models were treated in this chapter. In Swerling I and Swerling III the amplitude of the entire pulse train is assumed to fluctuate randomly from scan to scan; however, all pulses within a train have the same amplitude. In Swerling I the amplitude fluctuation is Rayleigh and in Swerling III the fluctuation is one-dominant-plus-Rayleigh. In Swerling II and Swerling IV the amplitude of each pulse in the train is a statistically independent random variable with a Rayleigh or a one-dominant-plus-Rayleigh density, respectively.

It was shown that the optimum detection receiver structure for small signal-to-noise ratios is the same for all the Swerling models considered. As a result the calculation of false-alarm probability is the same for each case. However, the probability of detection differs considerably. To show relative performance, the probability of detection as a function of (average) peak signal-to-noise ratio $\overline{\mathscr{R}}_p$ is plotted in Figs. 11.7-1 and 11.7-2 for $N = 10$ and $N = 100$, respectively, for Swerling I through IV and for both coherent and incoherent nonfluctuating pulse trains. In both plots, false-alarm probability is 10^{-6}.

A study of Figs. 11.7-1 and 11.7-2 leads to a number of pertinent observations. At high probabilities of detection, a nonfluctuating pulse train has the least loss with respect to a coherent pulse train; the loss is greater for $N = 100$ than for $N = 10$. The Swerling II and IV plots do not differ greatly from the performance curve of a nonfluctuating incoherent train over a wide range of values of P_d. The figures also show that the performance curves for Swerling I and III deviate considerably from that for a nonfluctuating incoherent pulse train. The difference between Swerling I and III and the curve for a coherently processed waveform is large at high P_d and diminishes at low P_d.

Integration loss provides a measure of the decrease in detectability incurred by incoherent processing. It was shown in Chapter 10 that, for equal P_d and P_{fa}, the required signal-to-noise ratio for a nonfluctuating incoherent train exceeds by less than 3 db the signal-to-noise ratio for a nonfluctuating coherent pulse train, for $N < 10$. For larger values of N, the integration loss is greater. A comparison of Figs. 11.2-11 and 11.4-11 with Fig. 10.5-1 shows that integration loss for the Swerling I and III models is nearly identical to that for a nonfluctuating target for all N. This similarity between the three curves results from the fact that integration loss in each case is computed for an incoherent pulse train relative to a coherent pulse train having the same amplitude characteristic. This is not true of the integration loss curves for Swerling II and IV; in these cases the reference waveforms are assumed to fluctuate in amplitude from scan-to-scan, while Swerling II and IV have a pulse-to-pulse fluctuation characteristic.

Despite the losses incurred by incoherent processing with respect to coherent processing, incoherent processing is often preferable since it is generally simpler to implement.

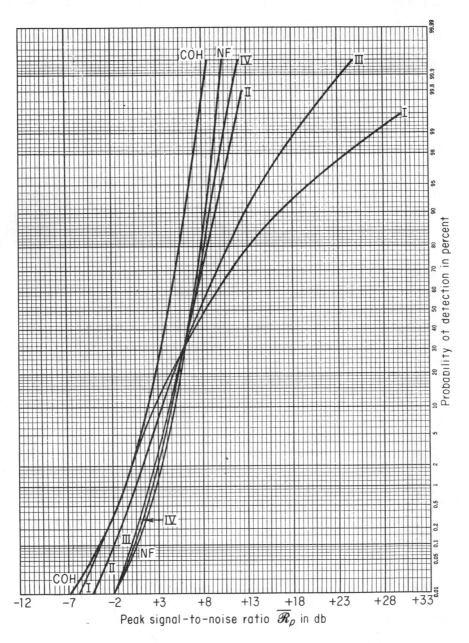

Fig. 11.7-1. Comparison of coherent, nonfluctuating, and Swerling I, II, III, IV cases: $N = 10$, $P_{fa} \cong 10^{-6}$.

442

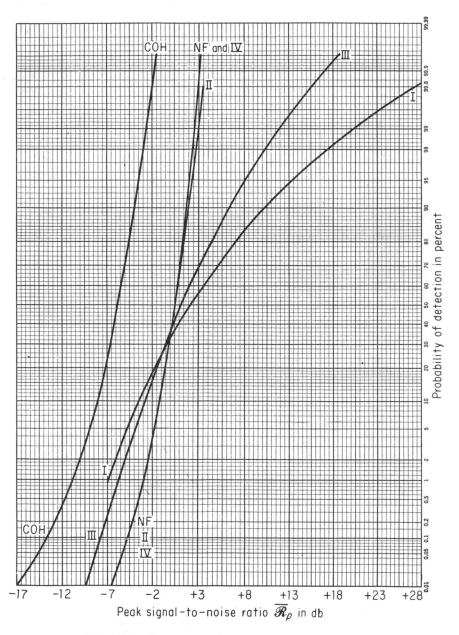

Fig. 11.7-2. Comparison of coherent, nonfluctuating, and Swerling I, II, III, IV cases: $N = 100$, $P_{fa} \cong 10^{-6}$.

443

PROBLEMS

11.1. Plot approximate P_d as a function of $\bar{\mathcal{R}}_p$ using Swerling I approximation (11.2-45) for $P_{fa} = 0.7 \times 10^{-6}$ and $N = 10, 100$, and 1000. With these results, plot the difference in $\bar{\mathcal{R}}_p$, for the same P_d, between the approximate curves and the more accurate plots in Section 11.2. What conclusions can be drawn about the accuracy of the approximation?

11.2. Repeat problem 11.1 using the Swerling II approximations given by (11.3-31) and (11.3-32).

11.3. Repeat problem 11.1 for Swerling III using approximation (11.4-33) and Fig. 11.4-12.

11.4. Repeat problem 11.1 for Swerling IV using approximation (11.5-30).

11.5. Verify Eq. (11.4-21) using the power series expansion for the confluent hypergeometric function given in Eq. (11.3-6).

11.6. Starting with (11.4-15), fill in the steps required to obtain (11.4-21).

11.7. Fill in the steps required to obtain (11.5-19) from (11.5-17).

11.8. Verify that the values of the incomplete gamma functions in (11.4-21) are less than $(1 - P_{fa})$ for $Y > Y_b$ when $N\bar{\mathcal{R}}_p/4 \gg 1$. [Hint: See Eqs. (10.4-19) and (10.4-20).]

11.9. Compute the Swerling II integration loss for $P_d = 10$ per cent and 75 per cent with $P_{fa} = 10^{-10}$. Repeat for $P_{fa} = 10^{-4}$. How does the loss change in the region where the loss is negative?

11.10. Confirm Eqs. (11.3-22b) and (11.3-23b) by computing moments m_1 and m_2 of Y directly from the probability density function given in (11.3-18).

11.11. For $\mathcal{R}_p = A^2$ [see Eq. (11.2-17)], find the probability density of $\mathcal{R}_p$ when A is a Rayleigh random variable with a distribution given by (11.2-9). What is the form of the density of $\mathcal{R}_p$? Since $\mathcal{R}_p$ is proportional to radar cross section, this probability density describes a Swerling I target cross section.

11.12. Repeat problem 11.11 when A has a one-dominant-plus-Rayleigh probability density function. Compare the result with that obtained in problem 11.11.

11.13. Fill in the steps required to obtain (11.4-24) from (11.4-23).

11.14. Compute m_1 and m_2 of Y by substituting (11.5-12b) into Eq. (10.7-2) to verify Eqs. (11.5-20) and (11.5-21).

11.15. Discuss the similarity in performance between the Swerling II and IV pulse-train models and the nonfluctuating model, as shown in Figs. 11.7-1 and 11.7-2.

11.16. Discuss the performance deviations between Swerling I and III and the other cases at high detection probabilities.

REFERENCES

1. Swerling, P.: Probability of Detection for Fluctuating Targets, Rand Report RM-1217 (March, 1954); reissued since in a special issue, *Trans. IRE Prof. Group on Information Theory*, **IT-6**: (*2*), 269–308 (April, 1960).

2. Edrington, T. S.: The Amplitude Statistic of Aircraft Radar Echoes, *IEEE Trans. on Military Electronics*, **MIL-9**: (*1*), 10–16 (January, 1965).

3. Campbell, G. A., and Foster, R. M.: "Fourier Integrals for Practical Applications," Van Nostrand, Princeton, N.J., 1948.

4. Watson, G. N.: "Theory of Bessel Functions," Cambridge University Press, London and New York, 1944, p. 394.

5. Marcum, J. I.: A Statistical Theory of Target Detection by Pulsed Radar: Mathematical Appendix, Rand Report RM-753 (July, 1948); reissued since in a special issue, *Trans. IRE Prof. Group on Information Theory*, **IT-6**: (*2*), 145–267 (April, 1960).

6. Magnus, W., and Oberhettinger, F.: "Special Functions of Mathematical Physics," Chelsea, New York, 1949.

7. Fehlner, L. F.: Target Detection by a Pulsed Radar, Report TB-451, Applied Physics Laboratory, Johns Hopkins University, Silver Spring, Maryland (July, 1962).

8. Swerling, P.: Detection of Fluctuating Pulsed Signals in the Presence of Noise, *IRE Trans. on Information Theory*, **IT-3**: (*3*), 175–178 (September, 1957).

9. Schwartz, M.: Effects of Signal Fluctuation in the Detection of Pulsed Signals in Noise, *IRE Trans. on Information Theory*, **IT-2**: 66–71 (June, 1956).

12

THE RADAR EQUATION

12.1. Introduction

The basic quantities that determine radar range performance are related by the radar equation. With this equation, received radar echo power can be calculated from known characteristics of the transmitter, receiver, and antenna and with knowledge of target range, target cross section, and the propagation characteristics of the transmission medium. Other forms of the radar equation are useful for computing signal-to-noise ratio, which permits detection performance to be predicted from the P_d-P_{fa} curves of Chapters 9 through 11. (Signal-to-noise ratio also enters into the prediction of parameter estimation accuracy when the fundamental radar system parameters have been specified.)

12.2. Classical Radar Equation

The signal power returned from a target of radar cross section σ at a distance R from the radar antenna can be derived as follows. Assume initially that transmitted power† P_T is radiated isotropically from an omnidirectional antenna in free space. Then the power density on the surface of an imaginary sphere in space, at a distance R from the antenna, is equal to the radiated power divided by the surface area of the sphere:

$$\left(\begin{array}{c}\text{power density at range } R \\ \text{from an isotropic radiator}\end{array}\right) = \frac{P_T}{4\pi R^2} \qquad (12.2\text{-}1)$$

To increase the detection range, directive antennas are employed that concentrate the radiated power in a smaller region of space. (See Fig. 12.2-1

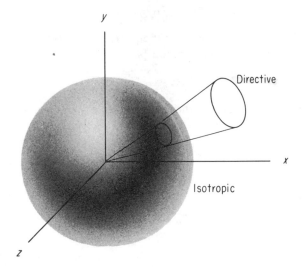

Fig. 12.2-1. Isotropic and directive antenna patterns.

and Chapter 1.) A measure of the increased power concentration of a directive antenna is its gain. Gain G_T is defined as the ratio of the maximum power density provided by a directive antenna on the surface of a sphere (of arbitrary radius) to the power density available from a lossless isotropic antenna with the same total radiated power; thus,

$$\left(\begin{array}{c}\text{maximum power density at range } R \\ \text{from a directive antenna}\end{array}\right) = \left(\frac{P_T}{4\pi R^2}\right)\cdot G_T \qquad (12.2\text{-}2)$$

A portion of the radiated power is intercepted by the target and reradiated back toward the radar. The target radar cross section σ, also called the backscattering coefficient, has the dimensions of area and is a measure of the ability of the radar target to reflect energy to the radar receiving antenna.

---

†Unless otherwise indicated, effective power units are employed in this chapter.

For a directive antenna pointing directly at the target, we have

$$\begin{pmatrix} \text{reradiated power density at} \\ \text{radar from a target at range } R \end{pmatrix} = \left(\frac{P_T G_T}{4\pi R^2} \right) \cdot \left(\frac{\sigma}{4\pi R^2} \right) \quad (12.2\text{-}3)$$

Usually, σ is a function of target aspect angle with the effect that relative motion between target and radar results in a fluctuating radar cross section, as discussed in Chapters 1, 9, and 11.

A portion of the reflected power is intercepted by the receiving antenna. An effective aperture area A_e of the receiving antenna can be defined such that the echo power P_R intercepted by this antenna is given by

$$P_R = \left[\frac{P_T G_T \sigma}{(4\pi R^2)^2} \right] \cdot A_e \quad (12.2\text{-}4)$$

Equation (12.2-4) is the classical radar equation. A variation of this equation is obtained with the antenna relation [see Eq. (12.4-20)]

$$A_e = \frac{G_R \lambda^2}{4\pi} \quad (12.2\text{-}5)$$

where G_R is the power gain of the receiving antenna and λ is the radar wavelength. Combining Eqs. (12.2-4) and (12.2-5) yields

$$P_R = \frac{P_T G_T G_R \lambda^2 \sigma}{(4\pi)^3 R^4} \quad (12.2\text{-}6)$$

When the same antenna is used for both transmission and reception, (12.2-6) simplifies to

$$P_R = \frac{P_T G^2 \lambda^2 \sigma}{(4\pi)^3 R^4} \quad (12.2\text{-}7)$$

12.3. Radar Equation and Signal-to-Noise Ratio

The classical radar equation can be modified to relate radar range performance to signal-to-noise ratio. Peak signal-to-noise ratio $\mathscr{R}$ at the output of a matched filter is given by

$$\mathscr{R} = \frac{2E}{N_0} \quad (12.3\text{-}1)$$

where E is the energy of the received echo and $N_0/2$ is the (two-sided) spectral density of the background white noise (see Eq. 9.2-20). For a transmitted signal of duration T_S, transmitted signal energy E_T is related to transmitted power $P_T(t)$ by

$$E_T = \int_0^{T_s} P_T(t) \, dt \quad (12.3\text{-}2)$$

Similarly, received signal energy E_R is related to received power $P_R(t)$ by

$$E = E_R = \int_{\tau_0}^{\tau_0 + T_s} P_R(t) \, dt \quad (12.3\text{-}3)$$

where τ_0 is the round-trip delay of the echo relative to the transmitted signal. With (12.3-2) and (12.3-3), integration of Eq. (12.2-7) within the interval zero to infinity results in

$$E = \frac{E_T G^2 \lambda^2 \sigma}{(4\pi)^3 R^4} \tag{12.3-4}$$

Substituting Eq. (12.3-4) into (12.3-1) yields

$$\mathscr{R} = \frac{E_T G^2 \lambda^2 \sigma}{(4\pi)^3 R^4 (N_0/2)} \tag{12.3-5}$$

When target cross section σ is a random variable, Eq. (12.3-5) is written in terms of expected values as follows:

$$\bar{\mathscr{R}} = \frac{E_T G^2 \lambda^2 \bar{\sigma}}{(4\pi)^3 R^4 (N_0/2)} \tag{12.3-6}$$

In practice, the value of power spectral density $N_0/2$ depends on the noise characteristics of the radar receiver and antenna. Johnson noise is generated by the fluctuation of charges or polarized molecules in lossy materials and characterizes the thermal noise available from resistors, lossy networks, and lossy dielectrics; its spectral density is nearly white. The (two-sided) power spectral density of Johnson noise is $k_B \mathscr{T}/2$,† where k_B is Boltzmann's constant and $\mathscr{T}$ is the (absolute) temperature of the lossy material [1, 2]. A lossless receiving antenna is also a source of Johnson noise; $\mathscr{T}$ in this case is the average temperature of the environment that the antenna "sees." At standard temperature, $\mathscr{T} = 290°K$ for which spectral density $k_B \mathscr{T}/2$ is equal to 2×10^{-21} watt/Hz.

There is more noise in a practical receiver than can be ascribed to Johnson noise alone. Two methods have been developed to account for the added noise. In one method‡ a *noise factor* F_n is defined to account for the increase in output noise power per unit bandwidth relative to the noise out of an ideal (noise-free) receiver with an input termination at an assumed standard temperature of $\mathscr{T}_0 = 290°K$. In this case the equivalent background power spectral density at the input to the receiver can be expressed as

$$\frac{N_0}{2} = \frac{k_B \mathscr{T}_0 F_n}{2} \tag{12.3-7}$$

Since environmental factors such as sky temperature and antenna elevation angle affect the actual noise level, a more useful method is to define an *effective input temperature* $\mathscr{T}_i$. This temperature is defined for a receiving

---

†The power spectral density of Johnson noise is less than $k_B \mathscr{T}/2$ at very high frequencies. Nyquist and others [1] have developed more exact quantum-mechanical formulas in this region. However, the expression above is accurate to tens of thousands of MHz, which is well above the radar frequency band.

‡Institute of Radio Engineers, *IRE Dictionary of Electronic Terms and Symbols* (1961), p. 96.

system by equating actual output noise to that encountered in an ideal (noise-free) receiver with its input termination at $\mathcal{T}_i$; this results in

$$\frac{N_0}{2} = \frac{k_B \mathcal{T}_i}{2} \tag{12.3-8}$$

The effective input temperature is the sum of the (effective) temperatures of the various sources contributing to receiver noise. For example, $\mathcal{T}_i$ can be separated into components $\mathcal{T}_a$ and $\mathcal{T}_r$, where $\mathcal{T}_a$ is the effective noise temperature of the antenna and $\mathcal{T}_r$ is the effective noise temperature of the receiver. Antenna noise temperature $\mathcal{T}_a$ is a function of sky temperature and varies with antenna pointing angle and beam pattern.

In actual radar systems there are many losses present that degrade the system performance from that predicted by the radar equation. Some of these are signal-processing loss, antenna losses, atmospheric propagation losses, operator losses, and so on. They are included in the radar equation by introducing an overall system loss factor L (greater than unity). With Eqs. (12.3-7) and (12.3-8), and loss factor L, radar equation (12.3-5) can be modified to

$$\mathcal{R} = \frac{2E_T G^2 \lambda^2 \sigma}{(4\pi)^3 R^4 k_B \mathcal{T}_0 F_n L} \tag{12.3-9}$$

when the receiver input termination is at standard temperature, or to the more useful form,

$$\mathcal{R} = \frac{2E_T G^2 \lambda^2 \sigma}{(4\pi)^3 R^4 k_B \mathcal{T}_i L} \tag{12.3-10}$$

Equations (12.3-9) and (12.3-10) relate radar performance to signal-to-noise ratio $\mathcal{R}$. Some of the terms appearing in the equations above warrant further discussion and are considered briefly in the following sections.

12.4. Antenna Gain and Effective Aperture

The electromagnetic field of a radiating body has two principal components: an *induction field* and a *radiation field*. The first is significant only in the immediate vicinity of the radiating body; the latter describes the electromagnetic far field of the radiating body and the outward flow of energy from the radiator. The performance of an antenna is usually characterized by the directive properties of its radiation field.

If $P(\theta, \phi)$ is the power radiated per unit solid angle in direction (θ, ϕ) and P_T is the total power radiated, the directive properties of an antenna are described by its *gain function* $G(\theta, \phi)$, defined by

$$G(\theta, \phi) = \frac{P(\theta, \phi)}{(P_T/4\pi)} \tag{12.4-1}$$

Gain function $G(\theta, \phi)$ is a measure of the concentration of radiated power in direction (θ, ϕ) relative to an isotropic antenna that radiates the same total power. The maximum value of this function is called *antenna gain* G_m; from (12.4-1)

$$G_m = G(\theta, \phi)_{\text{max}} = \frac{P(\theta, \phi)_{\text{max}}}{(P_T/4\pi)} \tag{12.4-2}$$

Antenna *solid beamwidth* β is defined as that solid angle through which all of the radiated power would pass if the antenna gain within this angle were constant and equal to its maximum value. It follows that

$$\beta \cdot P(\theta, \phi)_{\text{max}} = P_T \tag{12.4-3a}$$

or

$$\beta = \frac{1}{P(\theta, \phi)_{\text{max}}/P_T} \tag{12.4-3b}$$

From Eqs. (12.4-2) and (12.4-3b),

$$\beta = \frac{4\pi}{G_m} \tag{12.4-4}$$

The antenna *radiation pattern* is defined as the gain function normalized with respect to G_m:

$$(\text{antenna radiation pattern}) = \frac{G(\theta, \phi)}{G_m} \tag{12.4-5}$$

The antenna radiation pattern has a maximum value of unity and provides a relative measure of the directive properties of an antenna. Some typical radiation patterns for microwave antennas are shown in Fig. 12.4-1.

Both the antenna gain function and radiation pattern are measures of the directional characteristics of an antenna. They are unaffected by antenna losses such as impedance mismatch losses, ohmic losses, and rf losses, since their definitions depend on total radiated power rather than antenna input power. In practical applications, however, it is frequently desirable to relate maximum radiated power density to the power into the antenna input terminals. Such a quantity (which appears in the radar equation) is the *power gain G* of an antenna and is defined by

$$G = \frac{P(\theta, \phi)_{\text{max}}}{(P_{\text{input}}/4\pi)} \tag{12.4-6}$$

Power gain G and antenna gain G_m are related by *radiation efficiency factor* ρ_e† as follows:

$$G = \rho_e G_m \tag{12.4-7}$$

---

†ρ_e is a measure of ohmic and feed losses. Spillover losses which occur in lens or reflector type antennas can be accounted for by defining a spillover efficiency.

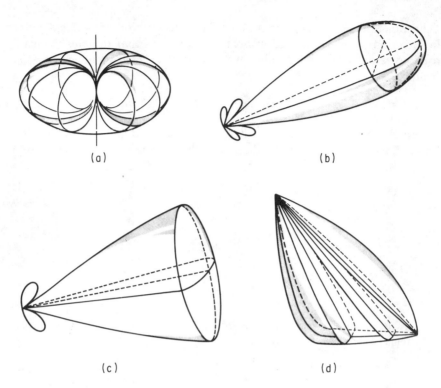

(a)   (b)

(c)   (d)

Fig. 12.4-1. Typical gain functions for microwave antennas
(*after Silver* [5], *courtesy of McGraw-Hill Book Company*).

In many applications the radiation efficiency factor is near unity, and the difference between G and G_m is small. Antenna beamwidth β can be related to power gain G by substituting (12.4-7) into (12.4-4), as follows:

$$\beta = \frac{4\pi \rho_e}{G} \qquad (12.4\text{-}8)$$

The performance of a receiving antenna is described by its *receiving cross section* and *receiving pattern*. If P_r is the total power absorbed by a lossless antenna from an incident radiation field in direction (θ, ϕ), then receiving cross section $A(\theta, \phi)$ is defined by

$$A(\theta, \phi) = \frac{P_r}{\left[\begin{array}{c}\text{power density of incident field} \\ \text{in direction } (\theta, \phi)\end{array}\right]} \qquad (12.4\text{-}9)$$

The antenna receiving pattern, analogous to the antenna radiation pattern, is defined by

$$(\text{antenna receiving pattern}) = \frac{A(\theta, \phi)}{A_m} \qquad (12.4\text{-}10)$$

where

$$A_m = A(\theta, \phi)_{\text{max}} \qquad (12.4\text{-}11)$$

The antenna receiving pattern and radiation pattern are identical by virtue of the classical reciprocity theorem [5]; hence, from (12.4-5) and (12.4-10),

$$\frac{G(\theta, \phi)}{G_m} = \frac{A(\theta, \phi)}{A_m} \qquad (12.4\text{-}12)$$

It can be shown† that, for matched antennas, A_m/G_m is a constant given by

$$\frac{A_m}{G_m} = \frac{\lambda^2}{4\pi} \qquad (12.4\text{-}13)$$

Then, from (12.4-12) and (12.4-13), the antenna receiving pattern is related to its gain function by

$$A(\theta, \phi) = \frac{\lambda^2}{4\pi} G(\theta, \phi) \qquad (12.4\text{-}14)$$

Since the antenna receiving cross section and receiving pattern do not depend on the power actually delivered to the receiver termination, these definitions do not include antenna losses. Hence, it is convenient to define *effective antenna cross section A_e* by

$$A_e = \frac{(\text{actual power received at antenna termination})}{(\text{power density of incident field})} \qquad (12.4\text{-}15)$$

where the direction of the incident radiation field corresponds to that for maximum antenna cross section A_m. Effective antenna cross section A_e appears in the radar equation; it is related to maximum cross section A_m by the antenna efficiency factor ρ_e, which was introduced in Eq. (12.4-7), as follows:

$$A_e = \rho_e A_m \qquad (12.4\text{-}16)$$

Another efficiency factor that applies to both transmitting and receiving antennas is the *aperture efficiency*, designated by ρ_a. It can be shown [5] that antenna power gain is a maximum when the physical antenna aperture is uniformly illuminated. Aperture efficiency ρ_a is defined as the ratio of the power gain of an antenna with a nonuniform aperture illumination to the (maximum) power gain obtained with uniform aperture illumination. For a receiving antenna, maximum antenna cross section A_m is related to the physical antenna cross section A by aperture efficiency ρ_A, as follows:‡

$$A_m = \rho_a A \qquad (12.4\text{-}17)$$

The effect of the illumination function on the radiation pattern of an antenna is illustrated in Fig. 12.4-2, which shows the pattern produced by

---

†See reference [5], p. 51.

‡This relation does not apply to elementary radiators such as dipoles, loops, and so on.

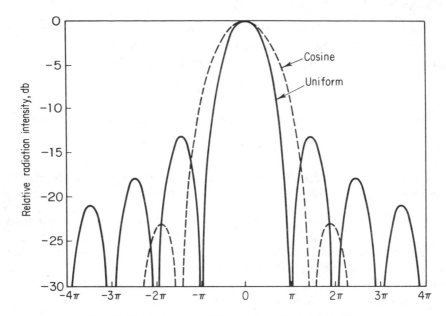

Fig. 12.4-2. Antenna radiation pattern for uniform illumination and cosine weighted aperture illumination.

uniform illumination and the pattern resulting from a cosine-weighted aperture illumination. Note that the main lobe is wider with cosine weighting than with uniform weighting; in addition, the adjacent sidelobes are reduced in amplitude. This figure illustrates a general result, namely: tapering the aperture illumination function to a small value at the edges results in lower sidelobes, a broadened mainlobe, and reduced aperture efficiency.

Overall antenna efficiency ρ is sometimes called the antenna *radiation efficiency;* it is the product of ρ_e, which is a measure of antenna losses, and ρ_a, which depends only on the aperture illumination. Thus,

$$\rho = \rho_e \rho_a \tag{12.4-18}$$

The following relations between the physical aperture area A and effective aperture A_e result from (12.4-16) through (12.4-18):

$$A_e = \rho_e \rho_a A = \rho A \tag{12.4-19}$$

With Eqs. (12.4-7), (12.4-13), (12.4-16), and (12.4-19), the following relations between power gain G and antenna cross section can be derived:

$$G = \frac{4\pi A_e}{\lambda^2} = \frac{4\pi \rho A}{\lambda^2} \tag{12.4-20}$$

12.5. Radar Cross Section

Radar cross section is a measure of the energy transfer from a target back to the receiving antenna. It is sometimes called target *effective echoing area* or *backscattering coefficient*. The radar cross section of a target is the area assumed to intercept the incident radiation which, when isotropically reradiated, yields the actual power density at the receiving aperture.

The radar cross section of irregularly shaped bodies depends on target aspect relative to the direction of the incident radiation. The radar cross section of a symmetric body such as a sphere is independent of aspect angle; in Fig. 12.5-1 the radar cross section of a conductive sphere of radius a is shown as a function of its circumference (measured in wavelengths $2\pi a/\lambda$). The region where $2\pi a/\lambda$ is much less than one is called the *Rayleigh region*.

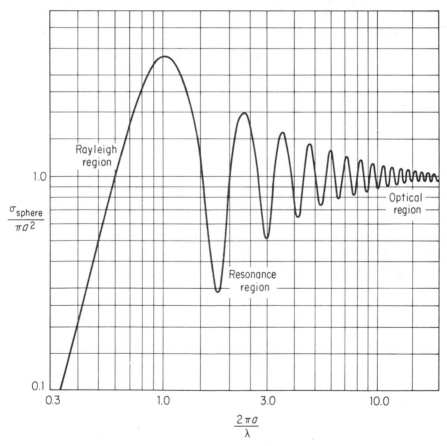

Fig. 12.5-1. Normalized cross section for conducting sphere (*after Barton* [10]).

Radar cross section in the Rayleigh region is proportional to λ^{-4}; because of this dependence, rain and cloud drops are essentially invisible to a low-frequency microwave radar.

The region where $2\pi a/\lambda$ is much greater than one is called the *optical region*. In this region the radar cross section of a sphere is equal to its optical cross section πa^2. The region separating the Rayleigh from the optical region is called the *resonance region* and is characterized by an oscillatory cross section as a function of wavelength.

Theoretically, the radar cross section of an object can be obtained by solving Maxwell's equations with appropriate boundary conditions. The computation, however, is usually difficult and has been carried out only for simple shapes and with body dimensions large compared to wavelength (optical region). Some results of these computations [4] for several simple bodies are shown for selected aspect angles in Table 12.5-1. Except for the sphere, the radar cross sections of these objects are aspect-sensitive. The aspect sensitivity of real targets is shown in Figs. 1.6-1 and 1.6-2.

TABLE 12.5-1

Radar Cross Sections of Simple Bodies with Large Dimensions [4]

Object	Aspect	Radar cross section σ
Sphere of radius a	–	πa^2
Large flat plate with area A and arbitrary shape	normal	$\dfrac{4\pi A^2}{\lambda^2}$
Infinite cone with half angle θ_0	axial	$\dfrac{\lambda^2}{16\pi} \tan^4 \theta_0$
Prolate spheroid with semimajor axis a_0 and semiminor axis b_0	axial	$\dfrac{\pi b_0^4}{a_0^2}$

For large objects with smooth curved surfaces, average radar cross section is of the order of one-fourth of actual surface area, when resonant edge irregularities do not exist.† A sphere is representative of this type of object.

12.6. Noise Factor and Noise Temperature

Noise factor and noise temperature are two methods for computing noise spectral density $N_0/2$ [see Eqs. (12.3-7) and (12.3-8)].

A. Noise factor

Receiver noise factor F_n is introduced to account for the additional noise introduced by a receiver. If the overall gain of a receiver is denoted

†See reference [3], p. 467.

by G and its input termination is at 290°K, then the receiver noise factor is defined by

$$F_n = \frac{\text{(output noise power)}}{\text{(input noise power)} \cdot G} \qquad (12.6\text{-}1)$$

The output noise power and gain are measured quantities; the input noise power can be calculated from the expression

$$\text{(input noise power)} = \left(\frac{k_B \mathcal{T}_0}{2}\right) \cdot (2\Delta_f) \qquad (12.6\text{-}2)$$

where $\mathcal{T}_0$ is the temperature of the receiver input termination, and Δ_f is the (one-sided) noise bandwidth of the receiver.

B. Noise temperature

Noise factor is not a convenient measure of noise performance for low-noise receivers such as those employing masers or parametric amplifiers. A more suitable approach is to assign an effective noise temperature $\mathcal{T}_i$ to the input termination of an ideal receiver that yields the actual measured noise output power; thus,

$$\mathcal{T}_i = \frac{\text{(output noise power)}}{(k_B \Delta_f) \cdot G} \qquad (12.6\text{-}3)$$

The output noise power consists of two components: the first due to noise from the antenna environment, and the second from internally generated receiver noise. Equation (12.6-3) can be rewritten as

$$\mathcal{T}_i = \mathcal{T}_a + \mathcal{T}_r \qquad (12.6\text{-}4)$$

where $\mathcal{T}_a$, the noise temperature of the antenna, is given by

$$\mathcal{T}_a = \frac{\text{(output noise power due to antenna noise)}}{(k_B \Delta_f) \cdot G} \qquad (12.6\text{-}5)$$

and $\mathcal{T}_r$, the effective noise temperature of the receiver, is given by

$$\mathcal{T}_r = \frac{\text{(output noise power due to internal receiver noise)}}{(k_B \Delta_f) \cdot G} \qquad (12.6\text{-}6)$$

Receiver noise factor F_n can be related to receiver noise temperature $\mathcal{T}_r$ by rewriting Eq. (12.6-1) as follows:

$$F_n = \frac{\left(\begin{array}{c}\text{output noise power due}\\ \text{to external input noise}\end{array}\right)}{\text{(input noise power)} \cdot G} + \frac{\left(\begin{array}{c}\text{output noise power due to}\\ \text{internal receiver noise}\end{array}\right)}{\text{(input noise power)} \cdot G} \qquad (12.6\text{-}7)$$

The first term in (12.6-7) is unity; substituting Eqs. (12.6-6) and (12.6-2) into (12.6-7) yields

$$F_n = 1 + \frac{(k_B \Delta_f \mathcal{T}_r) \cdot G}{(k_B \Delta_f \mathcal{T}_0) \cdot G} \qquad (12.6\text{-}8a)$$

$$F_n = 1 + \frac{\mathscr{T}_r}{\mathscr{T}_0} \qquad\qquad (12.6\text{-}8b)$$

Equation (12.6-8b) can also be written as

$$\mathscr{T}_r = (F_n - 1)\mathscr{T}_0 \qquad\qquad (12.6\text{-}9)$$

Note that a perfect receiver (no internal noise) has a receiver noise temperature of 0°K and a noise factor of unity. Equation (12.6-9) is plotted in Fig. 12.6-1.

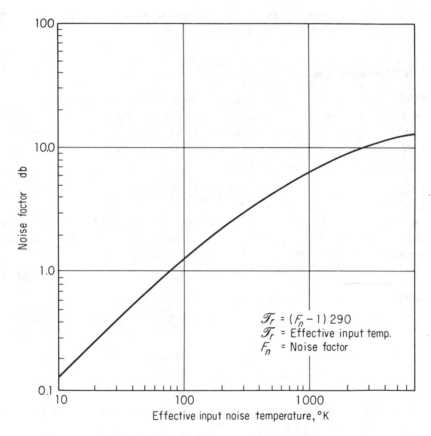

Fig. 12.6-1. Noise factor vs. noise temperature.

C. Antenna noise temperature

At microwave frequencies, external antenna noise is frequently small compared to internal receiver noise ($\mathscr{T}_a \ll \mathscr{T}_r$); for this case radar performance is limited by receiver noise. However, this is not true for low-noise receivers employing masers or parametric amplifiers. Antenna noise temperature is a measure of the environmental noise power absorbed by the antenna and represents the average temperature of all noise sources in the antenna pattern (including antenna sidelobes). Antenna noise temperature can be calculated by integrating the temperature of all objects illuminated by the antenna, each weighted by an appropriate antenna gain function. In practice this is difficult to do with great precision.

The principal source of antenna noise is sky noise. Sky noise consists of cosmic noise, which predominates at the lower frequencies, and atmospheric absorption noise, which predominates at the higher frequencies. Maximum cosmic noise occurs in the direction of the galactic center and is minimum at the galactic pole. Maximum atmospheric noise occurs with the antenna beam pointed along the horizon and minimum with the antenna

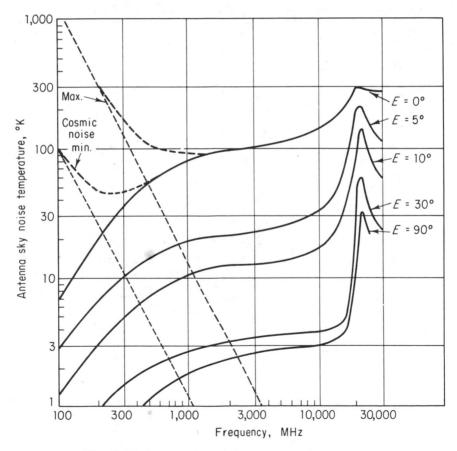

Fig. 12.6-2. Antenna sky-noise temperature due to atmospheric absorption noise and cosmic noise (*after Barton* [10]).

pointed at the zenith. Figure 12.6-2 shows approximate values of antenna sky temperature for various antenna elevation angles.

In addition to cosmic and atmospheric absorption noise, there are "hot spots" in the sky that cause antenna noise temperature to vary with beam pointing angle. These localized noise sources include the sun, certain radio stars (one of the largest is located in the Cassiopeia constellation), and the moon.

Blake [6] calculated typical values of antenna noise temperature by adding sky noise to an average contribution of 36°K from ground noise that enters the lower sidelobes or mainlobe of a search radar antenna. The results are shown in Fig. 12.6-3 for several antenna elevation angles.

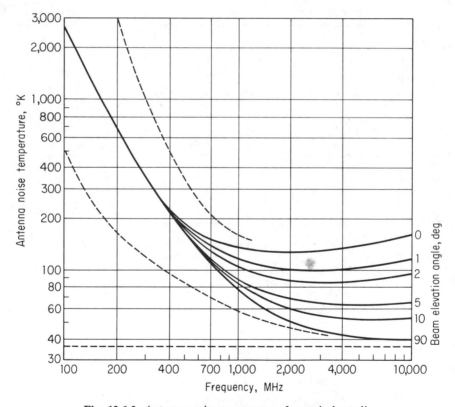

Fig. 12.6-3. Antenna noise temperature for typical conditions of sky and ground noise (the sloping dashed curves indicate the maximum and minimum levels of cosmic and atmospheric noise; the horizontal dashed line is the assumed ground-noise contribution of 36°K) (*courtesy of L. V. Blake* [6], *Naval Research Laboratory*).

12.7. System Loss Factor

The principal losses that affect radar range performance are discussed next. These losses are incorporated in system loss factor L, which appears in Eqs. (12.3-9) and (12.3-10).

A. Atmospheric attenuation

In deriving the radar equation, it was assumed that the radar and target are located in free space. For surface radars, a propagation loss is encountered due to the absorption of radar signals by the propagating medium. Figure 12.7-1 shows theoretical loss curves obtained by Van Vleck [7] due to atmospheric absorption by oxygen and water vapor. Figure 12.7-2 shows the theoretical attenuation of radar waves in rain and fog obtained by Goldstein [8].

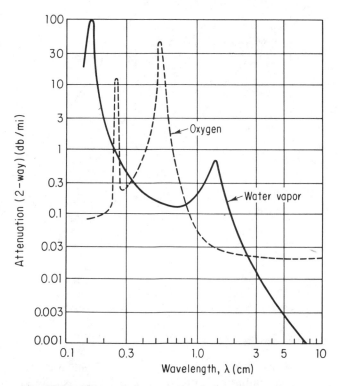

Fig. 12.7-1. Theoretical attenuation due to oxygen and water vapor (*after Van Vleck* [7], *courtesy of McGraw-Hill Book Company*).

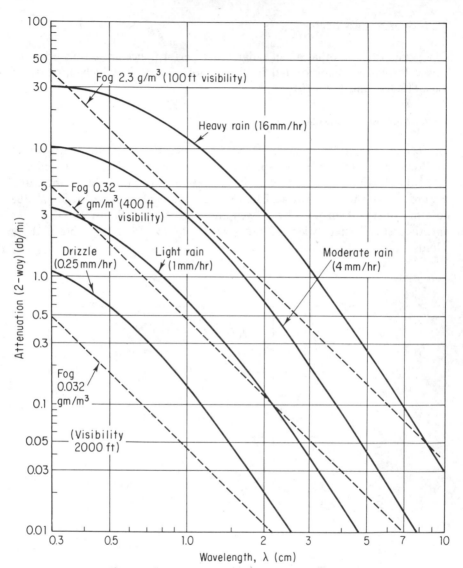

Fig. 12.7-2. Theoretical attenuation in rain and fog (*after Goldstein* [8], *courtesy of McGraw-Hill Book Company*).

B. Propagation loss

At low antenna elevation angles reflections from the earth's surface modify the antenna radiation pattern. These effects usually are accounted for by a propagation factor F. If effective radar range is defined as F times the free-space value, an equivalent propagation loss factor equal to $1/F^4$

can be included in system loss factor L. When total reflection from the earth's surface occurs, the direct and reflected rays can add constructively or destructively; this results in a maximum value of $F = 2$ and a minimum of $F = 0$. An average value of $F = 1.2$ is frequently used near the horizon; this corresponds to a propagation loss factor of 0.5, or -3 db (which represents a gain in effective radar range).

C. Beam shape loss and scanning loss

Antenna gain G, which appears in the radar equation, is the maximum value of the antenna gain function [see Eq. (12.4-1)]. In general, radar targets are not always illuminated or observed at maximum antenna gains. For example, in the case of an antenna beam of a simple search radar scanning past a target, successive radar transmissions and received echoes are modulated by the antenna beam pattern. Beam shape loss is a measure of the additional signal strength required to compensate for the reduced antenna gain when the target is off beam center.

Consider first a beam that scans in one coordinate with the beam axis passing through the target. Assuming a one-way $(\sin x/x)^2$ antenna radiation

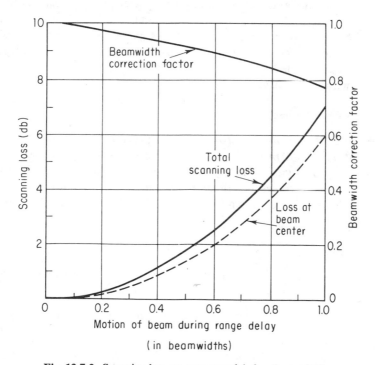

Fig. 12.7-3. Scanning loss vs. scan speed (*after Barton* [10]).

pattern, corresponding to a uniformly illuminated aperture, Blake [9] has shown that the average (two-way) beam shape loss is 1.6 db when the number of received echoes during a single scan is large.† This value was confirmed experimentally for scanning radars with postdetection integration with as few as two echoes occurring within a beamwidth.‡ When the antenna beam axis does not pass through the target during a scan, an additional loss occurs. The reduced antenna gain in this second coordinate must be included in total beam shape loss.

Scanning loss results from the change in the position of a scanning antenna between transmission and reception. The product of transmit gain G_T and receive gain G_R for a scanning antenna is always less than G_{max}^2, where G_{max} is maximum antenna gain. A measure of scanning loss is provided by the following definition:

$$\text{scanning loss} = \frac{G_{max}^2}{(G_T G_R)_{max}} \qquad (12.7\text{-}1)$$

where $(G_T G_R)_{max}$ is the maximum value of the gain product during a scan. Scanning loss is illustrated by the (dotted) curve in Fig. 12.7-3, which shows scanning loss as a function of fractional beam motion during round-trip delay time.

For rapid scanning, the effective antenna pattern $G_T G_R$ is narrower than the product pattern of a slow-scanning antenna [10]. This beam narrowing reduces the effective number of radar returns per beamwidth; a beamwidth correction factor is used to account for this effect and is shown in Fig. 12.7-3. The combined loss due to beam narrowing and scanning loss at beam center is also shown in Fig. 12.7-3.

D. Filter mismatch loss

Matched filters are usually approximated in practice (see Chapter 5). Such approximations result in a reduced peak output signal-to-noise ratio. The reduction in peak signal-to-noise ratio below optimum is called the filter mismatch loss. The losses for three typical matched filter approximations treated in Chapter 5 are shown in Table 12.7-1. (See Chapter 5 for further discussion.)

E. Collapsing loss

Collapsing loss often occurs in radar applications where detection is based on a visual radar display.§ It is the detection loss that results when extra-

---

†Blake [9] uses the one-way, 3-db beamwidth as a reference for the number of pulses received.

‡See reference [10], p. 147.

§A similar loss occurs in automatic detectors when wider than optimum range gates are employed.

TABLE 12.7-1

Signal	Matched filter approximation	Loss in peak signal-to-noise ratio
Rectangular monochromatic pulse	Ideal rectangular filter $(\Delta_\tau \Delta_f = 1.4)$	0.84 db
Linear fm rectangular pulse	Rectangular filter with quadratic phase	0.6 db
Coherent train of rectangular monochromatic pulses	North type of weighted comb filter	1.9 db

neous noise is added to the noise already present in a resolution cell. This occurs, for example, if three-dimensional radar data are compressed onto a two-dimensional PPI presentation. In this case the PPI noise traces corresponding to antenna beams at the same azimuth, but different elevation angles, are superimposed on each other. The increased noise level results in a reduction in effective signal-to-noise ratio and a detection loss in each of the elevation beams. A similar loss occurs in a radar that employs a PPI display with insufficient sweep speed; in this case the spot on the PPI cathode ray tube moves less than a spot diameter in the time interval corresponding to a range resolution cell. This results in the collapse of noise originating in different resolution cells onto the same spot. (A similar effect is obtained when the video amplifier bandwidth, following envelope detection, is less than that corresponding to a range resolution cell.)

For detection based on a single observation, collapsing loss effectively reduces signal-to-noise ratio and can be accounted for in the system loss factor. In the two examples cited above, collapsing loss also results in a reduction in the effective number of resolution cells in a range sweep. In such cases, the reduced signal-to-noise ratio is partially offset by an allowable increase in the probability of false alarm in each enlarged resolution cell for a specified false-alarm time.†

For detection based on multiple incoherent observations, the calculation of collapsing loss has been evaluated for nonfluctuating target echoes by Marcum [11]. Marcum defines collapsing ratio ρ_c by

$$\rho_c = \frac{M+N}{N} \qquad (12.7\text{-}2)$$

where N is the number of integrated signal-plus-noise variates and M is the number of additional noise variates integrated. The characteristic function $C_Y(\xi; \mathscr{R}_p)_{M+N}$ for the sum of N signal-plus-noise variates and M independent noise variates can be found by multiplying their respective characteristic functions:

$$C_Y(\xi; \mathscr{R}_p)_{M+N} = C(\xi; \mathscr{R}_p)_N \cdot C(\xi; 0)_M \qquad (12.7\text{-}3)$$

†False-alarm time is the average time between false alarms.

Characteristic functions $C(\xi; \mathcal{R}_p)_N$ and $C(\xi; 0)_M$ are obtained† from Eq. (10.4-12b); then (12.7-3) becomes

$$C_Y(\xi; \mathcal{R}_p)_{M+N} = \left[\frac{e^{-N\mathcal{R}_p/2}}{(1-j\xi)^N} e^{jN\mathcal{R}_p\xi/2(1-j\xi)}\right] \cdot \left[\frac{1}{(1-j\xi)^M}\right] \quad (12.7\text{-}4a)$$

$$= \frac{e^{-N\mathcal{R}_p/2}}{(1-j\xi)^{M+N}} e^{jN\mathcal{R}_p\xi/2(1-j\xi)} \quad (12.7\text{-}4b)$$

$$= \frac{e^{-(\rho_c N)(\mathcal{R}_p/\rho_c)/2}}{(1-j\xi)^{M+N}} \exp\left[\frac{j(\rho_c N)(\mathcal{R}_p/\rho_c)\xi}{2(1-j\xi)}\right] \quad (12.7\text{-}4c)$$

With (12.7-2), Eq. (12.7-4c) can be rewritten as

$$C_Y(\xi; \mathcal{R}_p)_{M+N} = \frac{e^{-(M+N)(\mathcal{R}_p/\rho_c)/2}}{(1-j\xi)^{M+N}} \exp\left[\frac{j(M+N)(\mathcal{R}_p/\rho_c)\xi}{2(1-j\xi)}\right] \quad (12.7\text{-}5)$$

Characteristic function (12.7-5) has the same form as that resulting from the addition of $(M + N)$ signal-plus-noise variates, each with modified peak signal-to-noise ratio $(\mathcal{R}_p/\rho_c)$.

In this case, collapsing loss can be defined by

$$(\text{collapsing loss}) = \frac{(\mathcal{R}_p)_{M+N}}{(\mathcal{R}_p)_N} \quad (12.7\text{-}6)$$

where $(\mathcal{R}_p)_{M+N}$ is the required peak signal-to-noise ratio for a specified probability of detection with M additional noise variates, and $(\mathcal{R}_p)_N$ is the

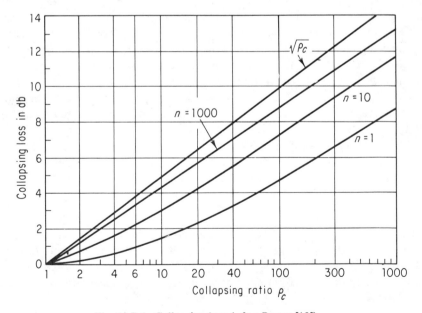

Fig. 12.7-4. Collapsing loss (*after Barton* [10]).

†Assuming a square-law detector.

required peak signal-to-noise ratio for the same probability of detection with no additional noise variates. A procedure for calculating collapsing loss follows. First $(\mathscr{R}_p)_N$ is found in the usual manner from the curves in Chapter 10 for a desired P_d and P_{fa}. Next, for the same P_d but modified P_{fa}, peak signal-to-noise ratio is obtained from these curves for $(M + N)$ signal-plus-noise variates; for the examples cited, the probability of false alarm is increased by ρ_c to maintain constant system false-alarm time.† Peak signal-to-noise ratio $(\mathscr{R}_p)_{M+N}$ is found by multiplying the signal-to-noise ratio just obtained by ρ_c [see Eq. (12.7-5)]. Figure 12.7-4 gives Marcum's results [11] for collapsing loss versus ratio ρ_c for $P_d = 50$ per cent, $P_{fa} = 0.7 \times 10^{-10}$, and for several values of N. Collapsing loss varies only slightly with P_d and P_{fa}; therefore, the losses shown in Fig. 12.7-4 are useful in evaluating radar performance.

F. Operator loss

Although automatic radars are currently of great interest, there are still many applications where the human operator is employed for target detection. The detection threshold of a human operator observing a PPI presentation varies considerably and is influenced by radar data rate, ambient light level, indicator spot size, operator fatigue, and other factors. Operator loss can be measured by an equivalent degradation in signal-to-noise ratio. A typical value for operator loss under good conditions is about 2 db [12]; under poor conditions, the loss may be as high as 10 db.

12.8. Example of Radar Equation Calculation

Calculations with the radar equation can be simplified by the use of a mixed system of units wherein the transmitted signal energy is expressed in joules, wavelength in centimeters, target cross section in square meters, and range in nautical miles. Then Eq. (12.3-9) can be rewritten as

$$\mathscr{R} = \left(\frac{2E_T G^2 \lambda^2 \sigma}{R^4 F_n L}\right) \cdot \left(\frac{m}{100 \text{ cm}}\right)^2 \cdot \left(\frac{\text{n.mi.}}{1853.2 \text{ m}}\right)^4 \cdot \left(\frac{1}{4\pi}\right)^3 \cdot \left(\frac{1}{290 \times 1.38 \times 10^{-23}}\right)$$

$$\tag{12.8-1a}$$

$$= \left(\frac{2E_T G^2 \lambda^2 \sigma}{R^4 F_n L}\right) \cdot (1.07) \tag{12.8-1b}$$

with the receiver input termination at standard temperature. Usually a nautical mile is approximated by 6000 feet; for this case the last constant in (12.8-1b)

---

†Where collapsing loss does not result in a reduction of the number of effective resolution cells, P_{fa} is unchanged.

is nearly unity. (This approximation introduces an error of 0.3 db in $\mathscr{R}$.) With a constant of unity, peak signal-to-noise $\mathscr{R}$ in decibels is approximately given by

$$10 \log_{10} \mathscr{R} \cong 3 + 10 \log_{10} E_T + 2(10 \log_{10} G) + 2(10 \log_{10} \lambda) + 10 \log_{10} \sigma$$
$$-4(10 \log_{10} R) - 10 \log_{10} F_n - 10 \log_{10} L \qquad (12.8\text{-}2)$$

For computational simplicity Eq. (12.8-2) can be rewritten as

$$(\mathscr{R})_{db} \cong 3 + (E_T)_{dbj} + 2(G)_{db} + 2(\lambda)_{dbcm} + (\sigma)_{dbm^2}$$
$$- 4(R)_{dbnmi} - (F_n)_{db} - (L)_{db} \qquad (12.8\text{-}3)$$

where *dbj* denotes decibels with reference to one joule, *dbcm* denotes decibels with reference to one centimeter, *dbm²* denotes decibels with reference to a square meter, and *dbnmi* denotes decibels with reference to a nautical mile; the remaining terms are simply $10 \log_{10}$ of the corresponding quantity.

Example: Determine peak signal-to-noise ratio $\mathscr{R}$ for a ten square meter target at a range of 100 nautical miles and for a radar with the following parameters:

$$E_T = 4 \text{ joules (6 dbj)}$$
$$G = 38 \text{ db}$$
$$\lambda = 10 \text{ cm (10 dbcm)}$$
$$F_n = 8 \text{ db}$$
$$L = 10 \text{ db}$$

Also

$$\sigma = 10 \text{ dbm}^2$$
$$R = 10 \text{ dbnmi}$$

Substituting these values into Eq. (12.8-3) yields

$$(\mathscr{R})_{db} = +17 \text{ db}$$

For a Rayleigh fluctuating target a 17 db (average) peak signal-to-noise ratio corresponds to a probability of detection of 50 per cent with a probability of false alarm of 10^{-8} (see Fig. 9.4-1).

12.9. Search Radar Equation

Average transmitter power and antenna aperture size are limited by practical radar economics. By introducing several modifications, the radar equation can be rewritten in a form that demonstrates the fundamental dependence of the detection range of a search radar on these two quantities.

Consider a search radar with solid antenna beamwidth β that scans a solid angle Ω in time T_f, called the *frame time* (see Fig. 12.9-1). Then the data rate on a target observed on consecutive scans is $1/T_f$. During each scan of the search volume, a target will be illuminated by the scanning beam for an interval T_d, called the *dwell time*. Dwell time is approximately related

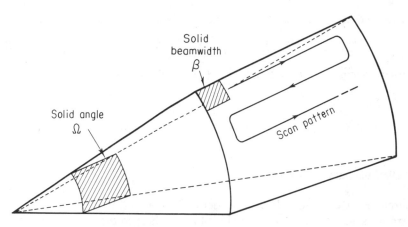

Fig. 12.9-1. Simplified scan configuration.

to the previous quantities by†

$$T_d \cong \frac{\beta}{\Omega} T_f \qquad (12.9\text{-}1)$$

The dwell time is usually much longer than the maximum range delay expected; this results in multiple observations of a target on a single scan. If average transmitted power is denoted by P_{av}, the total transmitted energy E_T in dwell time T_d is

$$E_T = P_{av}T_d \qquad (12.9\text{-}2)$$

If the signals received within the dwell interval are coherently processed, the maximum detection range‡ can be computed by solving Eq. (12.3-10) for R_{max} with E_T replaced by $P_{av}T_d$; thus,

$$R_{max}^4 = \frac{2P_{av}T_dG^2\lambda^2\sigma}{(4\pi)^3 \mathscr{R}_{min}k_B\mathscr{T}_iL} \qquad (12.9\text{-}3)$$

For an incoherently processed pulse train an integration loss (relative to coherent processing) occurs (see Chapters 10 and 11). If this integration loss is included in system loss factor L, Eq. (12.9-3) is still applicable.

To obtain the desired form of the radar equation, T_d is replaced by Eq. (12.9-1); then (12.9-3) becomes

$$R_{max}^4 = \frac{2P_{av}\beta G^2\lambda^2\sigma T_f}{(4\pi)^3 \Omega \mathscr{R}_{min}k_B\mathscr{T}_iL} \qquad (12.9\text{-}4)$$

With β replaced by $4\pi\rho_e/G$ [see Eq. (12.4-8)] and with antenna efficiency

---

†Equation (12.9-1) assumes perfect coverage of Ω by beamwidth β without overlap.

‡Maximum range is defined for minimum signal-to-noise ratio $\mathscr{R}_{min}$ (corresponding to a specified P_d and P_{fa}).

ρ_e included in system loss factor L, Eq. (12.9-4) can be rewritten as

$$R_{\max}^4 = \frac{2P_{av}G\lambda^2\sigma T_f}{(4\pi)^2\Omega\mathcal{R}_{\min}k_B\mathcal{T}_i L} \tag{12.9-5}$$

With power gain G replaced by Eq. (12.4-20), (12.9-5) becomes

$$R_{\max}^4 = \left(\frac{2\sigma T_f}{4\pi\Omega\mathcal{R}_{\min}k_B\mathcal{T}_i L}\right)\cdot(P_{av}A_e) \tag{12.9-6}$$

Equation (12.9-6) is often called the *search radar equation*.

12.10. Summary of Search Radar Status

Search radar equation (12.9-6) is useful for indicating evolutionary trends in radar system design. Barton [13] plotted the product (and the ratio) of average power and receiving aperture for a number of ground radars dating

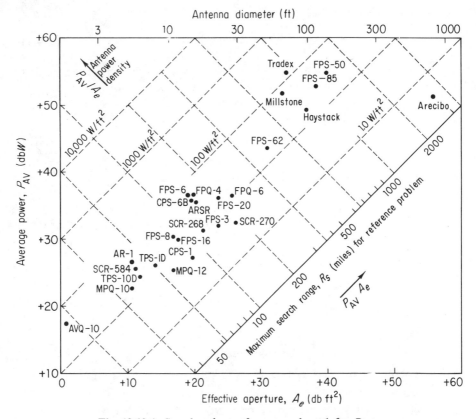

Fig. 12.10-1. Search radar performance chart (*after Barton* [13], *courtesy of IEEE*).

from 1938 to the present. This plot is reproduced in Fig. 12.10-1. The axis extending toward the upper right is $10 \log_{10}(P_{av}A_e)$. There is a spread of 90 db in power-aperture product between the smallest radar, the AVQ-10, and the largest, the ARECIBO radar.

For an assumed reference search problem, Barton [13] related $P_{av}A_e$ to a maximum detection range for each radar using search radar equation (12.9-6). Although this approach is somewhat arbitrary, it provides a reference for a comparison of all the radars. For the assumed reference search problem, the bracketed quantities in Eq. (12.9-6) are specified in Table 12.10-1. Some typical radar parameters corresponding to the reference search problem are given in Table 12.10-2.

TABLE 12.10-1

Reference Search Problem

Frame time, T_f	10 sec
Nonfluctuating target cross section, σ	1.0 m²
Search angle, Ω	$\pi/4$ steradians
Input temperature, $\mathscr{T}_i$	1150°K
Peak signal-to-noise ratio, $\mathscr{R}_{min}$	40 (16 db)
System loss factor, L	16 (12 db)

TABLE 12.10-2

*Typical Radar Parameters Corresponding
to Reference Search Problem*

Azimuth coverage	2π rad (360 deg)
Elevation coverage	1/8 rad (7 deg)
Antenna temperature	150°K
Receiver noise factor	6.5 db
Probability of detection	50%
Probability of false alarm	10^{-9}
Transmitting loss[a]	2 db
Receiving loss[a]	2 db
Mismatch loss	1 db
Integration loss	1 db
Collapsing loss	1 db
Antenna loss	2 db
Beam shape loss[b]	3 db
Scanning loss	0 db
Propagation factor, F	1.0 (0 db)

[a]Includes one-way atmospheric attenuation.
[b]Average for targets distributed between 0 and 7 degree elevation.

The maximum range shown in Fig. 12.10-1 on the reference scale does not necessarily correspond to that achieved by the actual radars; however, it is sufficient to provide a measure of relative performance. One interesting aspect of the chart is the gap in maximum range between the "conventional" radars and the "space-age" radars such as Tradex and FPS-85.

The antenna power-density axis is plotted at right angles to the power-aperture (or maximum-range) axis and extends to the upper left of the chart in Fig. 12.10-1. It is of interest to note that nearly all of the radars, independent of size, lie between 10 and 100 watts/ft² on this scale. Significant departures from this region are the SCR-270, CPS-1, and FPS-3, which are the three earliest search radars, the MPQ-12, which is the first "missile instrumentation radar," and the ARECIBO, which is a fixed reflector radar telescope. It appears that the region between 10 and 100 watts/ft², at least historically, has represented an economic balance between transmitter power and antenna size for most ground radars.

12.11. The Beacon and Jamming Radar Equations

Radar beacons are employed to extend the radar range of suitably equipped cooperative targets. Upon reception of a radar interrogation, the beacon automatically transmits a reply, which is usually coded for target-identification purposes. Only one-way propagation losses are encountered in a beacon system, since the "interrogation," or forward link, and the return, or response link, are independent.

The received signal power P_R in a one-way link is calculated by a method similar to that used to derive the radar equation in Sec. 12.2. It is equal to the radiated power intercepted by an antenna with effective aperture A_{e2} from a source with transmitter power P_{T1} and antenna gain G_{T1}; thus,

$$P_{R2} = \left(\frac{P_{T1}G_{T1}}{4\pi R^2} \right) \cdot A_{e2} \qquad (12.11\text{-}1)$$

With A_{e2} replaced by Eq. (12.2-5), (12.11-1) becomes

$$P_{R2} = \frac{P_{T1}G_{T1}G_{R2}\lambda^2}{(4\pi)^2 R^2 L_{12}} \qquad (12.11\text{-}2)$$

where L_{12} is the system loss factor for a one-way link. For matched filter signal processing, the peak output signal-to-noise ratio $\mathcal{R}_2$ at the receiver is given by

$$\mathcal{R}_2 = \left(\frac{2E}{N_0} \right)_2 = \frac{2E_{T1}G_{T1}G_{R2}\lambda^2}{(4\pi)^2 R^2 k_B \mathcal{T}_{i2} L_{12}} \qquad (12.11\text{-}3)$$

By substitution of the appropriate values, Eq. (12.11-3) can be used for both the forward and return radar-beacon links.

A radar operating in a hostile environment may be subjected to noise jamming. Most modern radars can be tuned rapidly over several hundred MHz so that a noise jammer, to be effective, must distribute its available transmitter power uniformly over the band covered by the radar. For this reason jammer strength is usually described in terms of an available (one-sided) output noise spectral density N_j (watts/Hz), which corresponds to a two-sided density of $N_j/2$. The effect of broadband noise jamming is to increase the background white-noise spectral density in the radar receiver. If a jammer radiates average power P_j (watts) within (two-sided) bandwidth $2\Delta_f$ (Hz), from an antenna with power gain G_j, then, from (12.11-1), the noise spectral density $N_j/2$ at the radar due to the jammer is given by

$$N_j/2 = \frac{P_j}{2\Delta_f} \cdot \frac{G_j A_{er}}{4\pi R^2 L_j} \qquad (12.11\text{-}4)$$

where A_{er} is the radar receiving antenna effective aperture and L_j includes jammer and propagation losses.

A jammer is often carried by a vehicle to prevent a radar from establishing its location or determining the number of hostile vehicles present. When the jammer is within radar detection range, the jammer noise is usually much greater than normal receiver noise. For this case the radar output peak signal-to-noise ratio can be found by replacing $k_B \mathcal{T}_i/2$ in radar equation (12.3-10) by $N_j/2$; the result is

$$\mathcal{R} = \left(\frac{E_{Tr}}{(P_j/2\Delta_f)}\right) \cdot \left(\frac{G_r}{G_j}\right) \cdot \left(\frac{\sigma}{4\pi R^2}\right) \cdot \left(\frac{L_j}{L_r}\right) \cdot \left(\frac{R_j}{R}\right)^2 \qquad (12.11\text{-}5)$$

where subscript r denotes radar parameters and subscript j jammer parameters. When peak signal-to-noise ratio $\mathcal{R}$ is established by P_d and P_{fa}, Eq. (12.11-5) can be solved for the maximum range R_j at which a target-jammer can be detected.

PROBLEMS

12.1. With $N = 1000$ and $\rho_c = 10$, determine collapsing loss from the curves of Chapter 10 for (a) $P_d = 90$ per cent and $P_{fa} = 0.7 \times 10^{-10}$; (b) $P_d = 50$ per cent and $P_{fa} = 0.7 \times 10^{-6}$. Assume that collapsing loss is caused by a narrow video bandwidth relative to the i-f bandwidth. Repeat for $\rho_c = 100$. Discuss results and compare with Fig. 12.7-4.

12.2. A ground-based surveillance radar with a rotating antenna is required to detect aircraft out to 250 nautical miles and up to an altitude of 100,000 feet. The smallest target to be detected with a minimum detection probability of 50 per cent has a mean radar cross section of one square meter and a Swerling I fluctuation characteristic. The frame time is 10 seconds. A false-alarm rate of one per frame time is permissible. The radar antenna has an azimuth (half-power) beamwidth of 1.6 degrees. The vertical antenna pattern is shaped to provide complete elevation coverage with maximum gain directed at a point 250 miles away and 100,000 feet in altitude (corresponding to an elevation angle of about 2 degrees). For radar performance computations, an equivalent vertical beamwidth of 4.5 degrees can be assumed. To resolve radar targets separated in range by one-half mile, an uncoded 6-microsecond pulse is transmitted. The carrier frequency is 1300 MHz.

(a) Calculate the highest pulse repetition frequency for an unambiguous search range of 250 nautical miles. Assuming a pulse repetition frequency of 225 pulses per second, calculate the dwell time and the number of echoes received within the half-power beamwidth as the antenna scans by a target.

(b) Calculate the false-alarm probability in a range cell. (The probability of false alarm is related to N_{fa}, the effective number of range cells searched in a false-alarm time; false-alarm time is the average time between false alarms. For $P_{fa} \ll 1$, $P_{fa} \cong 1/N_{fa}$.† To obtain N_{fa}, first calculate the number of (6-microsecond) range cells searched in a false-alarm time. Dividing the total number of range cells by the number of pulses in a pulse train yields N_{fa}, assuming that multiple echoes received from a target on each scan are integrated prior to thresholding.)

(c) From the curves in Chapter 11 and the results of (a) and (b), find the value of $\mathcal{R}_p$ required to achieve $P_d = 50$ per cent and the calculated P_{fa}.

(d) Compute the transmit pulse energy assuming the following radar parameters:

Antenna efficiency	50%
Overall receiver noise temperature	900°K
Atmospheric loss (round trip)	1.2 db
Mismatch loss	1.0 db
Beam shape loss	1.6 db

Scanning loss can be obtained from Fig. 12.7-3.

(e) Calculate the peak power and average power of the radar transmitter.

(f) At what range will a target with a 10-square-meter (Swerling I) cross section be detected with probability of detection equal to 95 per cent?

(g) How is this range diminished if the target carries a noise jammer with an antenna gain of 3 db, an average power of 100 watts, and a noise bandwidth of 100 MHz which is centered in the radar operating band?

(h) Assume that the radar antenna beam can be tilted up. Select representative two-way system parameters for a beacon on a cooperative target so that it is detectable 1500 miles away. Repeat for a beacon 10,000 miles away. Assume the beacon antenna aperture is limited to 4 feet in diameter.

12.3. Derive an expression for the peak output signal-to-noise ratio of a radar that receives a signal from a cooperative beacon that amplifies the radar signal prior to retransmission back to the radar. [Hint: Assume that the noise in the beacon simply adds to the noise level of the radar receiver.]

12.4. The AN/FPS-3 radar, which operates at 1300 MHz, has an average power of 1500 watts, a receiving cross section of 280 square feet, and a transmit

†See discussion in Appendix B.

gain of 38 db. Verify the maximum range of this radar on Fig. 12.10-1, using the parameters of the reference problem in Tables 12.10-1 and 12.10-2. How much is maximum range reduced when the target carries a jammer with the capability described in problem 12.2(g)?

REFERENCES

1. Van der Ziel, A.: "Noise," Prentice-Hall, Englewood Cliffs, N.J., 1954.

2. Pierce, J. R.: Physical Sources of Noise, *Proc. IRE*, **44**: 601–608 (May, 1956).

3. Kerr, D. E.: "Propagation of Short Radio Waves," MIT Radiation Laboratory Series, Vol. 13, McGraw-Hill, New York, 1951.

4. Mentzer, J. R.: "Scattering and Diffraction of Radio Waves," Pergamon, Oxford, England, 1955.

5. Silver, S.: "Microwave Antenna Theory and Design," MIT Radiation Laboratory Series, Vol. 12, McGraw-Hill, New York, 1949.

6. Blake, L. V.: Curves of Atmospheric-Absorption Loss for Use in Radar-Range Calculation, Naval Res. Lab. Report 5601 (March, 1961).

7. Van Vleck, J. H.: The Absorption of Microwaves by Oxygen, and the Absorption of Microwaves by Uncondensed Water Vapor, *Phys. Rev.*, **71**: (*7*), 413–433 (April, 1947).

8. Goldstein, A.: Attenuation by Condensed Water, Sec. 8.6 in D. E. Kerr, ed., "Propagation of Short Radio Waves," pp. 671–692, McGraw-Hill, New York, 1951.

9. Blake, L. V.: The Effective Number of Pulses per Beamwidth for a Scanning Radar, *Proc. IRE*, **41**: (*6*), 770–774 (June, 1953).

10. Barton, D. K.: "Radar System Analysis," Prentice-Hall, Englewood Cliffs, N.J., 1964.

11. Marcum, J. I.: A Statistical Theory of Target Detection by Pulsed Radar, Mathematical Appendix, *Rand Research Memo*, RM-753 (July, 1948); reissued since in a special issue, *Trans. IRE Prof. Group on Information Theory*, **IT-6**: (*2*), 145–267 (April, 1960).

12. Hall, W. M.: Prediction of Pulse Radar Performance, *Proc. IRE*, **44**: 224–231 (February, 1956).

13. Barton, D. K.: Radar System Performance Charts, *IEEE Trans. on Military Electronics*, **MIL-9**: (*3, 4*), 255–263 (July/October, 1965).

CUMULATIVE
DETECTION PROBABILITY
OF STATIONARY AND MOVING TARGETS

13.1. Introduction

Optimum detection, treated in Chapters 9, 10, and 11, dealt with the *blip-scan* ratio† of a radar—that is, detection performance during a single scan past the target. In each scan all of the received data from a target, consisting of either a single observation or multiple observations, are subjected to a single threshold test. In this chapter target detection is considered when data from several scans are available for processing. Cumulative detection probability denotes the probability that a target is detected at least once in M scans. We will compute the cumulative detection probability for targets that are essentially stationary with respect to the radar and for targets with appreciable motion relative to the radar.

Target models considered are the Rayleigh and one-dominant-plus-Rayleigh fluctuating models. In each case zero scan-to-scan correlation is

---

†The blip-scan ratio is equal to the single-scan probability of detection.

assumed. This implies that sufficient time elapses between successive scans past a target so that target returns as well as noise are decorrelated. This is usually true for a target that is scanned once per frame interval, the time required by a radar to interrogate the entire search volume. Frame times are generally of the order of seconds, while the correlation interval for piston and jet aircraft is of the order of fractions of a second [1]. When correlation is present on successive scans, the results of Chapter 9 for coherently processed trains or the results of Chapters 10 and 11 for incoherently processed trains can be applied, whichever is appropriate.†

13.2. Cumulative Detection Probability

Suppose the received waveform in each of M successive scans past a target is statistically independent of that in all other scans. The probability that a target is missed in the ith scan is $(1 - P_{di})$, where P_{di} is the probability of detection during the ith scan; the probability that a target is missed on all M scans is

$$\begin{pmatrix} \text{probability of} \\ \text{missed target} \end{pmatrix} = \prod_{i=1}^{M} (1 - P_{di}) \tag{13.2-1}$$

The probability P_{cd} that a target is detected at least once in M scans is then given by

$$P_{cd} = 1 - \prod_{i=1}^{M} (1 - P_{di}) \tag{13.2-2}$$

P_{cd} is called the cumulative probability of detection. It follows from (13.2-2) that the average cumulative probability of detection is given by

$$\bar{P}_{cd} = 1 - \prod_{i=1}^{M} (1 - \bar{P}_{di}) \tag{13.2-3}$$

where the averaging is with respect to independent random signal parameters such as unknown range, as discussed in Section 13.4.

13.3. Cumulative Detection Probability of Stationary Targets

Consider a target that is essentially stationary with respect to the radar during M successive scans. Then blip-scan ratio $P_{di} = P_d$ for all i. The cumulative probability of detection P_{cd} can be computed from Eq. (13.2-2) and is plotted in Fig. 13.3-1 for blip-scan ratios of 10 per cent, 50 per cent, and

---

†For a nonfluctuating target, optimum detection is obtained if the observations from the M scans are stored and integrated as MN observations, where N is the number of observations per scan.

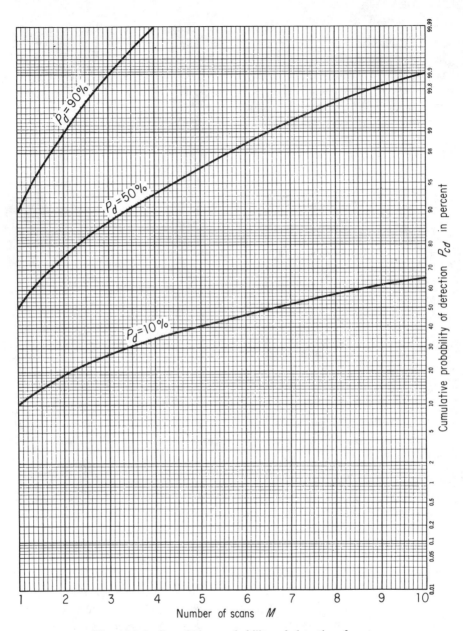

Fig. 13.3-1. Cumulative probability of detection for stationary targets.

478

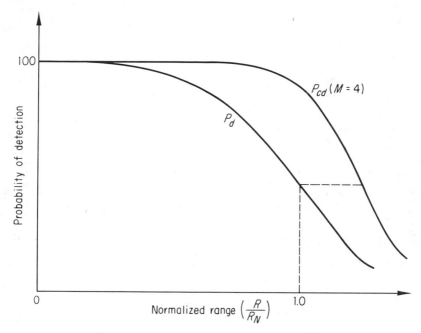

Fig. 13.3-2. Cumulative detection probability of a stationary
Rayleigh target as a function of range (coherent detector).

90 per cent. As M increases, the cumulative probability of detection increases
asymptotically toward 100 per cent. The rate of increase is more pronounced
for higher blip-scan ratios.

In Fig. 13.3-2 the blip-scan and cumulative detection probabilities for
a coherent detector† are plotted for $M = 4$ as a function of normalized‡
range. The curves show that cumulative detection extends radar range;
this result is obtained, however, for $M = 4$ which corresponds to four times
the energy in a single pulse (or pulse burst).

To evaluate the efficiency of this type of detection, assume for simplicity
that the observations within each scan are coherently processed. If the scan
rate past a target is varied, the number of observations N within a scan
and therefore the energy NE_p on target during a scan is also varied (for a
constant pulse repetition rate). An increase in scan rate results in a smaller
value of energy NE_p on target and hence in a reduced blip-scan ratio. To
obtain the same cumulative probability of detection, a larger number of

---

†These curves are also applicable to the incoherent detector if the integration loss for
N observations is nearly constant over the region of P_d plotted.

‡The normalization is arbitrary and corresponds to the choice of P_d at the reference
range.

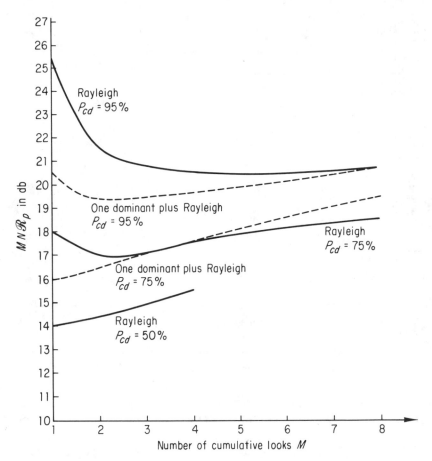

Fig. 13.3-3. Signal energy vs. cumulative observations (P_{fa} = 0.7 × 10⁻⁸).

independent scans is necessary. The quantity $MN\mathscr{R}_p = MN(2E_p/N_0)$ is a measure of total energy on target in M scans to achieve a specified P_{cd} level; this quantity is plotted as a function of M in Fig. 13.3-3 for both Rayleigh and one-dominant-plus-Rayleigh fluctuating targets and for several values of cumulative detection probability. It is apparent from Fig. 13.3-3 that the energy on target is minimized for $M = 1$ at low cumulative probabilities of detection; however, at high values of P_{cd}, it is more efficient to scan rapidly and obtain more echo samples of a fluctuating target with lower energy per scan. For a Rayleigh target, four scans is close to optimum, while two or three scans is optimum for a one-dominant-plus-Rayleigh target at a P_{cd} of 95 per cent. This behavior is used to minimize radar power-aperture product for a moving target in the next section.

While the results plotted in Fig. 13.3-3 assume coherently processed pulse trains within each scan, they can also be computed for incoherent pulse trains by evaluating the appropriate integration loss. This loss can be added to Fig. 13.3-3 and a new set of curves obtained for the Swerling I, II, III, and IV target models. This is left as an exercise (see problems 13.2, 13.3, 13.4, and 13.5).

13.4. Cumulative Detection Probability of Moving Targets with a Uniformly Scanning Radar

The cumulative detection probability of a stationary target has only academic interest since it is generally desirable in a search radar to reject stationary targets such as trees, buildings, and so forth. In this section the cumulative detection probability of moving targets is considered for a uniformly scanning search radar. Approximate expressions for two target models are derived that permit the power-aperture product to be minimized and the search frame time to be optimized.

Suppose we want to detect a target that is approaching the radar with constant velocity v at least once before it reaches range R_0 with some specified probability. As shown in Fig. 13.4-1 the distance traversed by the target in frame time T_f is vT_f. The cumulative probability of detecting the target

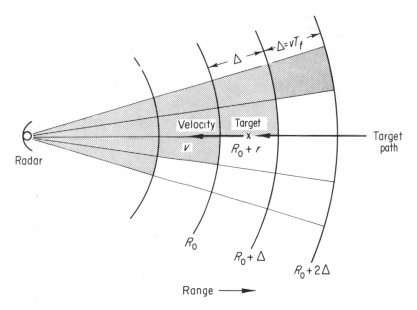

Fig. 13.4-1. Closing target geometry.

by final range $(R_0 + r)$, where $0 < r < \Delta$, is given by

$$P_{cd} = 1 - \prod_{i=0}^{M-1} [1 - P_d(R_0 + r + i\Delta)] \qquad (13.4\text{-}1)$$

The quantity r is introduced to denote the uncertainty in range relative to $(R_0 + i\Delta)$ when the target is first observed. If it is assumed that r is uniformly distributed over the interval $(R_0, R_0 + \Delta)$, the average cumulative probability of detection is given by [2]

$$\bar{P}_{cd} = \frac{1}{\Delta} \int_{R_0}^{R_0+\Delta} \left\{ 1 - \prod_{i=0}^{M-1} [1 - P_d(R_0 + r + i\Delta)] \right\} dr \qquad (13.4\text{-}2)$$

This is the average probability that the target is detected at least once by range R_0.

Equation (13.4-2) can be solved exactly only by numerical evaluation; however, an approximation can be made that permits the calculation of useful results.† Let $P_d(R_0 + r + i\Delta)$ be approximated by the first term of a Taylor series expansion about mean range $\bar{R}_i = R_0 + \Delta/2 + i\Delta$, as follows:

$$P_d(R_0 + r + i\Delta) \cong P_d(\bar{R}_i) \qquad (13.4\text{-}3)$$

Then Eq. (13.4-2) can be rewritten approximately as

$$\bar{P}_{cd} \cong 1 - \prod_{i=0}^{M-1} [1 - P_d(\bar{R}_i)] \qquad (13.4\text{-}4)$$

This approximation permits a simple evaluation of system behavior. The error introduced can be bounded by computing the cumulative detection probability at $r = 0$ and $r = \Delta$.

Equation (13.4-4) can be rewritten by noting that

$$\bar{R}_i = R_0 + \frac{\Delta}{2}(2i + 1) \qquad (13.4\text{-}5a)$$

$$= R_0 \left[1 + \frac{\Delta}{2R_0}(2i + 1) \right] \qquad (13.4\text{-}5b)$$

$$= R_0 [1 + k(2i + 1)] \qquad (13.4\text{-}5c)$$

where

$$k = \frac{\Delta}{2R_0} = \frac{vT_f}{2R_0} \qquad (13.4\text{-}6)$$

The result is

$$\bar{P}_{cd} \cong 1 - \prod_{i=0}^{M-1} (1 - P_d\{R_0[1 + k(2i + 1)]\}) \qquad (13.4\text{-}7)$$

Approximate expressions for P_d for the Swerling I and III fluctuating target models were developed in Chapter 11. For the Swerling I model

---

†The following analysis and computations are due to Mr. Aaron A. Abston of the Sperry Gyroscope division of Sperry Rand Corporation.

(scan-to-scan Rayleigh fluctuating target), the probability of detection is approximately given by [see Eq. (11.2-44)]

$$P_d \cong \exp\left(-2\,\frac{\sqrt{N}\,\Phi^{-1}(P_{fa}) + 1}{N\bar{\mathscr{R}}_p}\right) \tag{13.4-8}$$

Let R_1 denote a normalizing range for which

$$2\frac{\sqrt{N}\,\Phi^{-1}(P_{fa}) + 1}{N(\bar{\mathscr{R}}_p)_1} = 1 \tag{13.4-9}$$

From (13.4-8), this corresponds to a P_d of 37 per cent. It follows from radar equation (12.3-10) that

$$\bar{\mathscr{R}}_p = 2\frac{\sqrt{N}\,\Phi^{-1}(P_{fa}) + 1}{N}\left(\frac{R_1}{R}\right)^4 \tag{13.4-10}$$

With (13.4-10), Eq. (13.4-8) simplifies to

$$P_d \cong \exp\left[-\left(\frac{R}{R_1}\right)^4\right] \tag{13.4-11}$$

For the Swerling III model (scan-to-scan one-dominant-plus-Rayleigh fluctuating target), an approximate expression for the probability of detection [see Eq. (11.4-30)] is given by

$$P_d \cong \left(1 + 4\frac{\sqrt{N}\,\Phi^{-1}(P_{fa}) + 2}{N\bar{\mathscr{R}}_p}\right)\exp\left[-4\frac{\sqrt{N}\,\Phi^{-1}(P_{fa}) + 2}{N\bar{\mathscr{R}}_p}\right] \tag{13.4-12}$$

For $\sqrt{N}\,\Phi^{-1}(P_{fa}) \gg 1$ (small values of P_{fa}) it is convenient to modify Eq. (13.4-12) as follows:

$$P_d \cong \left(1 + 4\frac{\sqrt{N}\,\Phi^{-1}(P_{fa}) + 1}{N\bar{\mathscr{R}}_p}\right)\exp\left[-4\frac{\sqrt{N}\,\Phi^{-1}(P_{fa}) + 1}{N\bar{\mathscr{R}}_p}\right] \tag{13.4-13}$$

Substituting Eq. (13.4-10) into (13.4-13) results in

$$P_d \cong \left[1 + 2\left(\frac{R}{R_1}\right)^4\right]\exp\left[-2\left(\frac{R}{R_1}\right)^4\right] \tag{13.4-14}$$

Approximations (13.4-11) and (13.4-14) are evaluated by Swerling [3] and are shown to be reasonably accurate.

When Eqs. (13.4-11) and (13.4-14) are substituted into Eq. (13.4-7), a single expression for cumulative detection probability can be written for both the Swerling I and III fluctuating target models as follows:

$$\bar{P}_{cd} = 1 - \prod_{i=0}^{M-1}\left\{1 - \left[1 + 2\left(\frac{R_0}{R_1}\right)^4\left(1 + k[2i + 1]\right)^4\right]^{l-1}\right.$$
$$\left. \cdot \exp\left[-l\left(\frac{R_0}{R_1}\right)^4(1 + k[2i + 1])^4\right]\right\} \tag{13.4-15}$$

where $l = 1$ applies to a Rayleigh target and $l = 2$ to a one-dominant-plus-Rayleigh target. Equation (13.4-15) describes the cumulative detection probability for given M and k and for a particular minimum detection range R_0 (relative to R_1). It is in a convenient computational form. (See problems 13.7 and 13.8.) In the next section this expression is used to minimize the power-aperture product of a search radar and to optimize frame time.

13.5. Minimization of Power-Aperture Product†

Solving the search radar equation given by Eq. (12.9-6) for power-aperture product yields

$$P_{av}A_e = \frac{4\pi L k_B \mathscr{T}_i \Omega}{2\bar{\sigma} T_f} \bar{\mathscr{R}}_p R^4 \tag{13.5-1}$$

Since $\Delta = vT_f$, Eq. (13.5-1) can be rewritten as

$$P_{av}A_e = \frac{4\pi L k_B \mathscr{T}_i v \Omega}{2\bar{\sigma}\Delta} \bar{\mathscr{R}}_p R^4 \tag{13.5-2}$$

Substituting Eqs. (13.4-6) and (13.4-10) into (13.5-2), and assuming $\sqrt{N}\ \Phi^{-1}(P_{fa}) \gg 1$, results in

$$P_{av}A_e = \frac{2\pi L k_B \mathscr{T}_i v \Omega \Phi^{-1}(P_{fa}) R_0^3}{\sqrt{N}\ \bar{\sigma}} \cdot \frac{(R_1/R_0)^4}{k} \tag{13.5-3a}$$

$$= d \frac{(R_1/R_0)^4}{k} \tag{13.5-3b}$$

where all of the terms in coefficient d are constants that must be specified for a surveillance radar.

   The following procedure was used to minimize power-aperture product. For a specific cumulative probability of detection $\bar{P}_{cd}$ and M, the quantity (R_0/R_1) satisfying Eq. (13.4-15) was found by an iterative process on a digital computer for selected values of k; $(R_1/R_0)^4$ is plotted in Figs. 13.5-1 through 13.5-3 for a Rayleigh target and in 13.5-5 through 13.5-7 for a one-dominant-plus-Rayleigh target. With these results, relative power-aperture product was calculated using (13.5-3b) and is plotted as a function of k on the same curves. The calculations were repeated for various values of $\bar{P}_{cd}$ and M. Minimum power-aperture product was then established from each of these curves; the result is plotted in Figs. 13.5-4 and 13.5-8. From these two figures it can be observed that it is more efficient, for high values of $\bar{P}_{cd}$, to distribute the transmitted energy on a target over a large number of scans.

   For a specified $\bar{P}_{cd}$ and target model, the optimum M is obtained from Fig. 13.5-4 or 13.5-8; the optimum $k = k_0$ can then be read from the appropriate curve in Figs. 13.5-1 through 13.5-3 or 13.5-5 through 13.5-7. This

---

†Unless otherwise noted, effective power units are used in this section.

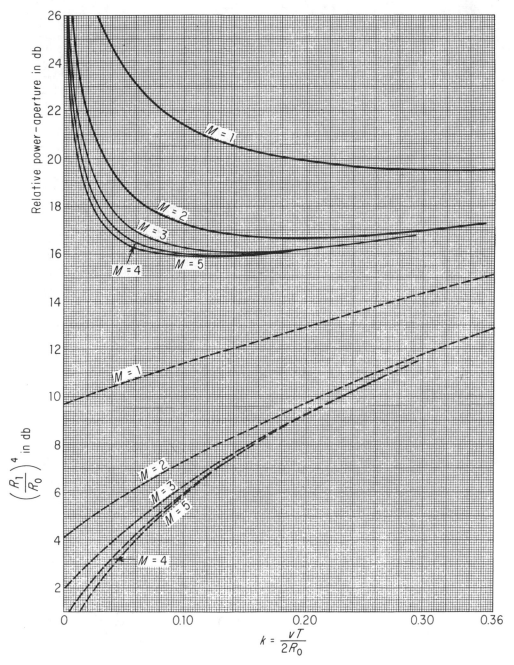

Fig. 13.5-1. Power-aperture vs. k (Rayleigh target, cumulative probability = 90 per cent, M = number of cumulative hits) (*courtesy of A. A. Abston*).

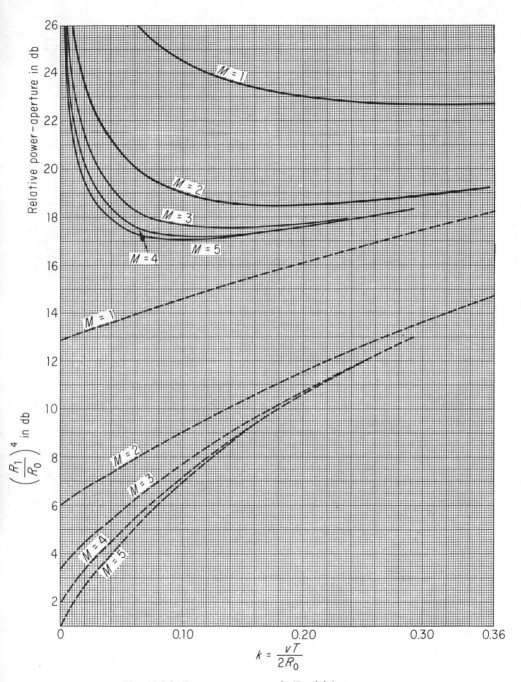

Fig. 13.5-2. Power-aperture vs. k (Rayleigh target, cumulative probability = 95 per cent, M = number of cumulative hits) (*courtesy of A. A. Abston*).

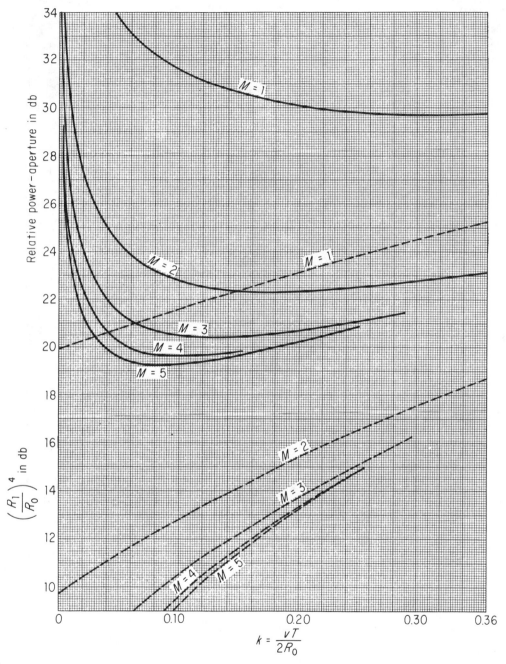

Fig. 13.5-3. Power-aperture vs. k (Rayleigh target, cumulative probability = 99 per cent, M = number of cumulative hits) (*courtesy of A. A. Abston*).

487

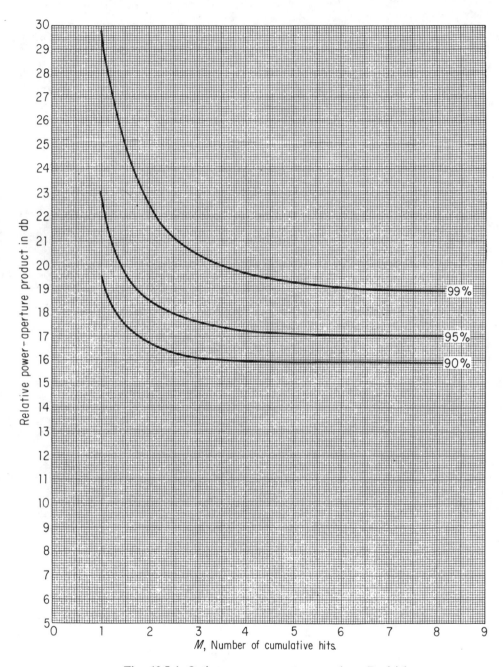

Fig. 13.5-4. Optimum power-aperture product, Rayleigh fading target (*courtesy of A. A. Abston*).

488

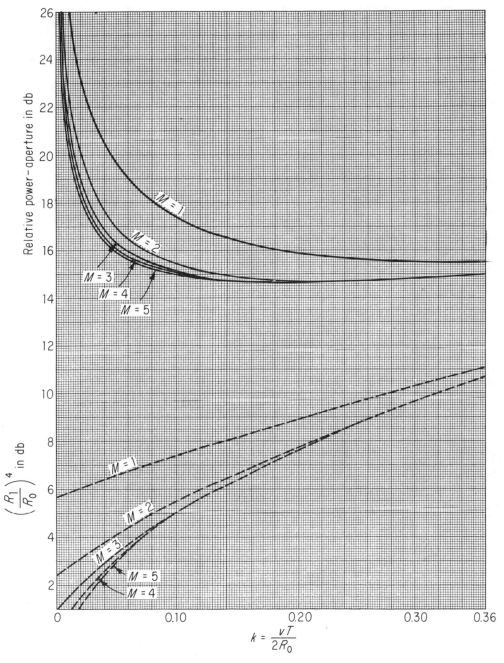

Fig. 13.5-5. Power-aperture vs. k (one-dominant-plus-Rayleigh target, cumulative probability = 90 per cent, M = number of cumulative hits) (*courtesy of A. A. Abston*).

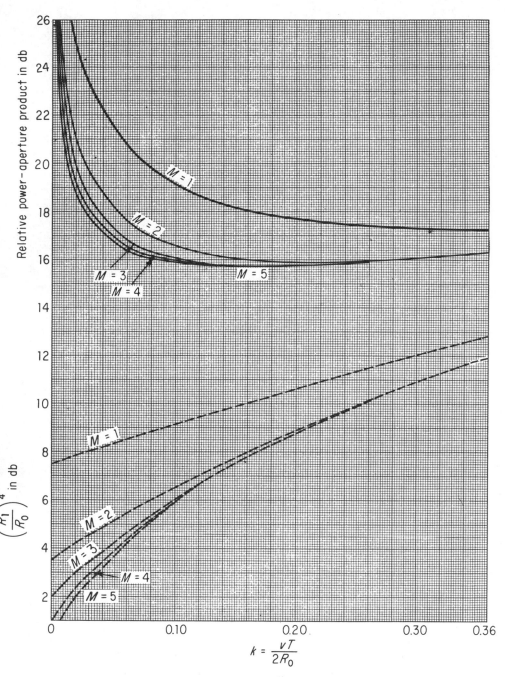

Fig. 13.5-6. Power-aperture vs. k (one-dominant-plus-Rayleigh target, cumulative probability = 95 per cent, M = number of cumulative hits) (*courtesy of A. A. Abston*).

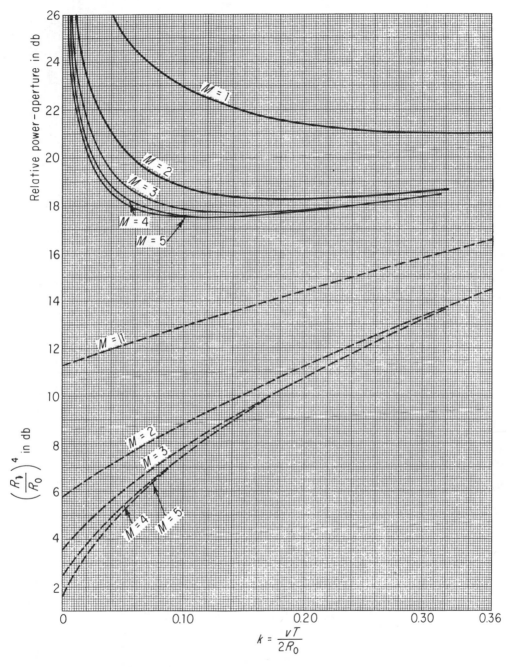

Fig. 13.5-7. Power-aperture vs. k (one-dominant-plus-Rayleigh target, cumulative probability = 99 per cent, M = number of cumulative hits) (*courtesy of A. A. Abston*).

491

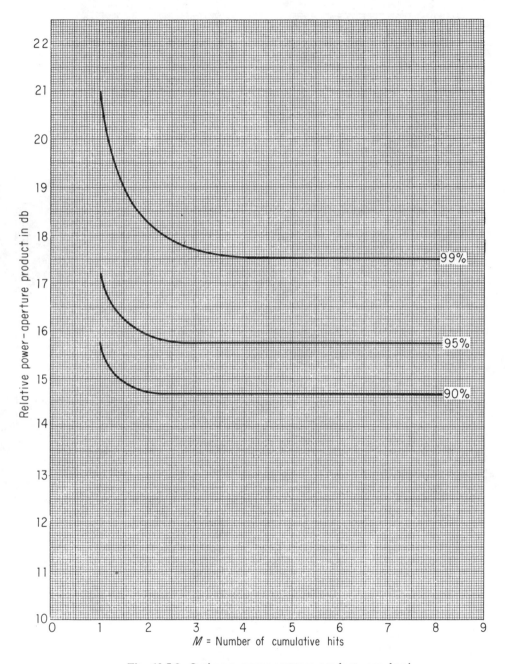

Fig. 13.5-8. Optimum power-aperture product, one-dominant-plus-Rayleigh fading target (*courtesy of A. A. Abston*).

selection permits the frame time to be calculated for a particular target velocity v and minimum detection range R_0 from the relation $k_0 = vT_f/2R_0$. From the curves of $(R_1/R_0)^4$ shown in Figs. 13.5-1 through 13.5-3 and 13.5-5 through 13.5-7, together with Eq. (13.5-3a), the actual power-aperture product can be computed for a specific case.

From examination of the data an interesting approximate relationship holds. If the region between R_0 and the maximum unambiguous range $R_m = R_0 + MvT_f$ is defined as the cumulative detection zone, the optimum combination of M and T_f occurs when

$$MvT_f = R_0 \qquad (13.5\text{-}4)$$

—that is, when the detection zone is equal to the minimum detection range R_0.

Example: Assume a Rayleigh closing target with velocity $v = 300$ miles/minute is to be detected at a minimum range of 500 miles. If a detection level $\bar{P}_{cd}$ of 95 per cent is specified, M is selected as 5 from Fig. 13.5-4; from Fig. 13.5-2, $k_0 = 0.1$ and $(R_1/R_0)^4$ is 7 db. From Eq. (13.5-4), the frame time is

$$T_f = \frac{R_0}{5v} = 20 \text{ seconds}$$

This frame time minimizes the power-aperture product. From (13.5-4) the maximum range is approximately twice R_0, or 1000 miles. If one hit per scan is assumed ($N = 1$), the maximum range delay is equal to the dwell time; in this case T_d is approximately 10 milliseconds. With these values and Eq. (12.9-1), and assuming a hemispheric surveillance volume of 2π steradians, an antenna beamwidth of approximately 0.0031 steradians is required; this can be realized with a square beam of 3.2 degrees and is equivalent to an ideal antenna gain of 36 db. Remaining system parameters are obtained from the radar equation using the optimum value of $(R_1/R_0)^4$ of 7 db.

Example: Assume a Rayleigh closing target with velocity $v = 1500$ miles/hour is to be detected at a minimum range of 50 miles with $\bar{P}_{cd}$ of 90 per cent. The optimum $R_{\max}$ is 100 miles. Then, from Fig. 13.5-4, $M = 4$ and from Fig. 13.5-1, $k_0 = 0.12$. These numbers result in a frame time of approximately 30 seconds. Assuming 4 hits per scan, dwell time T_d is approximately 5 milliseconds (four times the maximum unambiguous range delay). For a surveillance volume of π steradians, the antenna beamwidth is 0.00052 steradians which is equivalent to a square beam of 1.3 degrees. This beamwidth corresponds to an ideal antenna gain of about 43.8 db.

PROBLEMS

13.1. For single hit P_d of 50 per cent at $R = R_1$ and a P_{fa} of 10^{-8}, compute and plot a set of curves similar to those shown in Fig. 13.3-2 for $M = 4$ and 8 for a one-dominant-plus-Rayleigh fluctuating target; assume coherent processing. Discuss the threshold behavior of the radar.

13.2. For a Swerling I fluctuating target model and $P_{fa} = 0.7 \times 10^{-8}$, compute curves similar to those shown in Fig. 13.3-3 for $P_{cd} = 95$ per cent with $N = 1000$. Repeat the calculation for $N = 100$. Plot and discuss the results.

13.3. Repeat problem 13.2 for a Swerling II fluctuating target model.

13.4. Repeat problem 13.2 for a Swerling III fluctuating target model.

13.5. Repeat problem 13.2 for a Swerling IV fluctuating target model.

13.6. Compute $\bar{P}_{cd}$ approximately using Eq. (13.4-2) for $M = 2$ when two terms of the Taylor series expansion of $P_d(R_0 + r + i\Delta)$ about $\bar{R}_i$ are retained. Find the difference between a one-term and a two-term approximation.

13.7. For a Rayleigh fluctuating target ($l = 1$), calculate $\bar{P}_{cd}$ from Eq. (13.4-15) as a function of $(R_0/R_1)^4$ given $k = 0.1$ and $M = 2$. Repeat for $M = 4$ and 6. Plot the results.

13.8. Repeat problem 13.7 for a one-dominant-plus-Rayleigh fluctuating target.

13.9. From the results obtained in the first example of Section 13.5, design a radar with reasonable values for P_{fa}, $\mathscr{T}_i$, and L, assuming a 10-square-meter target. Compare the resulting power-aperture product with the operating radars shown in Fig. 12.10-1.

REFERENCES

1. Edrington, T. S.: The Amplitude Statistics of Aircraft Radar Echoes, *IEEE Trans. on Military Electronics*, **MIL-9**: (*1*), 10–16 (January, 1965).

2. Mallett, J. D., and Brennan, L. E.: Cumulative Probability of Detection for Targets Approaching a Uniformly Scanning Search Radar, *Proc. IEEE*, **51**: (*4*), 596–601 (April, 1963).

3. Swerling, P.: Probability of Detection for Fluctuating Targets, Special Monograph, *Trans. IRE on Information Theory*, **IT-6**: (*2*), 269–308 (April, 1960).

PART V

SUBOPTIMUM DETECTION
TECHNIQUES

Implementation of a postdetection integrator requires some form of video signal storage. In early radar equipments, storage was accomplished by displaying the detected video waveform on an "A" scope. In such a display, the envelope of the received waveform produces vertical electron-beam deflections on a cathode ray tube which are proportional to signal amplitude. The time base of the cathode ray tube is synchronized to the repetition rate of the pulse train so that all pulses received from a particular target are superimposed to yield a single vertical "pip" at one position of the horizontal trace. Integration is accomplished by a combination of the persistence properties of the cathode ray tube phosphor and the storage capability of the observer's eye and brain. Similar storage is employed in a PPI display. In this case, the detected signal modulates the intensity of the electron beam as it is deflected from the center of the scope to its outer edge. The polar angle of deflection of the electron trace is synchronized with the instantaneous direction of the rotating radar antenna. Integration is accomplished by superimposing echo pulses from a target at the same position on a cathode ray tube face that contains phosphors with additive afterglow properties. Neither type of cathode ray tube integration, however, can be conveniently employed in automatic detection systems.

In the next two chapters, two types of postdetection integrators are considered that do not use cathode ray tube storage. The performance of both integrators is slightly poorer than that of the ideal integrator evaluated in Chapters 10 and 11. However, the first integrator considered is superior to the ideal integrator in a pulse-jamming environment; the second is more suited to analogue processing of pulse-train echoes that are modulated by the radar antenna pattern as a result of antenna scanning.

14

THE BINARY
DETECTOR

14.1. Introduction

In previous chapters it was shown that substantial improvements in target detectability can be achieved with postdetection integration (video integration) of a pulse train when the relative phase between pulses is unknown and random. Of course, greater target detectability is obtained with predetection integration (coherent detection) of a pulse train with known relative phases between pulses. However, because of the cost and difficulties associated with the generation and coherent processing of long pulse-train waveforms, coherent pulse trains are often avoided; simplicity and low cost often dictate the use of postdetection integrators even in applications where relative phase is known a priori.

In this chapter a postdetection integrator is considered that provides slightly poorer performance than the ideal integrator discussed in Chapters 10 and 11. In this integrator the received signal is quantized so that it can be processed by simple digital circuits. The integration technique is called *binary integration* [1], or *double-threshold detection* [2]; both titles are descriptive of its operation.

14.2. Description of Binary Detector

It was shown in Chapters 10 and 11 that the optimum detection receiver for an incoherent pulse train consists of a matched filter (matched to a pulse waveform) followed by an envelope detector, a sampler, a postdetection integrator, and a threshold test. If the sum at the ideal integrator output exceeds a preassigned threshold, a target is declared present. The binary detector differs from the optimum incoherent detector in the following respects. Instead of a linear addition of N waveform envelope samples out of the matched filter corresponding to an N pulse train, the number of signals that exceed a first threshold at the sampling instants are counted in a binary integrator. If this number equals or exceeds a second threshold, a target is declared present.

The first threshold can be viewed as an amplitude quantizer, since the sampled waveforms at its output are normalized to zero or one, depending on whether or not the amplitude of the input waveform at the sampling instants exceeds the quantizing threshold. The output of the first threshold consists of N (one-bit) binary numbers corresponding to the N pulse envelope samples. The N binary numbers are counted and, if their sum exceeds a second threshold, a target is declared present. From this description, it is apparent why this technique is called both binary integration as well as double-threshold detection.

The false-alarm rate of a binary detector is less sensitive to randomly arriving impulse interference than the optimum receiver processor described in Chapters 10 and 11. Such interfering pulses may be unintentionally generated by nearby radars or intentionally by pulse jammers seeking to destroy the radar's visibility. Random interfering pulses, with an average repetition frequency comparable to that of the radar, will coincide in time with only a small fraction of the N pulses in any received pulse train. In a binary integrator one or two coincident interfering pulses, no matter how large in amplitude, will not by themselves generate a false alarm if the second threshold exceeds two. An extensive treatment of the performance of a binary integrator in the presence of both random pulse interference and Gaussian noise may be found in reference [3]. (See also the discussion of coincidence detection in Section 14.5 below.)

A simplified block diagram of a receiver employing binary integration is shown in Fig. 14.2-1. It operates as follows. The received waveform is match-filtered, envelope detected, and amplitude quantized. For a quantizer input less than a preset threshold, the quantizer output is zero; when the input is greater than the threshold, the quantizer output is unity. Each of the samplers following the quantizer sample the quantized video at times corresponding to the expected pulse-train peaks for a target at a particular range; thus the samplers separate pulse-train echoes from targets at different ranges. The output of each sampler is summed in a binary counter and the result is compared to a second threshold.

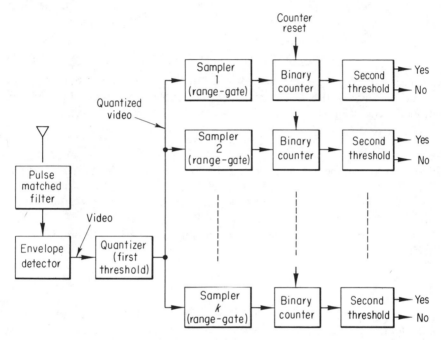

Fig. 14.2-1. Block diagram of range-gated binary integrator for pulse-train waveforms.

14.3. Performance of Binary Integrator

The performance of a binary integrator depends on both the first and second thresholds. If the first threshold is too high, weak signals are poorly detected. If the first threshold is too low, the system false-alarm rate may be excessive. Similar statements can be made about the second threshold. In the present analysis, an optimization procedure is employed that is related to the Neyman-Pearson optimization criterion discussed in Chapter 8; the two threshold values are chosen to maximize probability of detection for a selected probability of false alarm.

The additive background noise is assumed to be white and Gaussian with a two-sided spectral density $N_0/2$. Results can be extended to colored Gaussian noise as discussed in Sections 9.8, 9.9, and 9.10; for colored noise the matched filter is replaced by a colored noise optimum filter and signal-to-noise ratio is suitably redefined. A nonfluctuating target model is assumed. The results of the present analysis are compared later with those of Chapter 10 to assess the performance loss of a binary integrator relative to the optimum postdetection integrator.

The probability $p(K)$ that exactly K out of N statistically independent quantized samples (with identical probability densities) exceed a fixed threshold is given by the binomial distribution

$$p(K) = {}_NC_Kp^K[1 - p]^{N-K} \qquad (14.3\text{-}1)$$

where p is the probability that a single sample exceeds the threshold, $(1 - p)$ is the probability that a sample does not exceed the threshold, and the binomial coefficients

$$_NC_K = \frac{N!}{K!(N - K)!} \qquad (14.3\text{-}2)$$

describe the number of ways in which K successes in N trials can occur. Probability p is a function of the single-pulse rf signal-to-noise ratio $\mathscr{R}_p$ defined in Eq. (9.2-20). When $\mathscr{R}_p$ is zero, p denotes the probability that noise alone will exceed the fixed threshold.

Mean $\bar{K}$ and standard deviation σ_K of binomial distribution (14.3-1) are given by

$$\bar{K} = Np \qquad (14.3\text{-}3)$$

$$\sigma_K = \sqrt{Np(1 - p)} \qquad (14.3\text{-}4)$$

Figures 14.3-1 and 14.3-2 illustrate two binomial distributions for $N = 10$ corresponding to noise alone and signal-plus-noise, respectively.

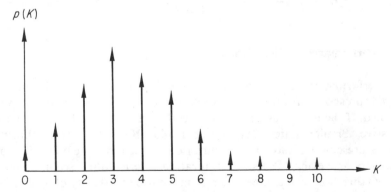

Fig. 14.3-1. Probability density (binomial distribution) for noise alone ($N = 10$).

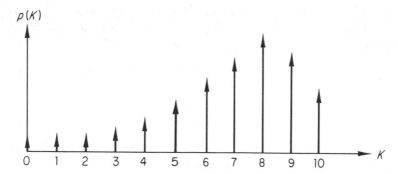

Fig. 14.3-2. Probability density (binomial distribution) for signal-plus-noise ($N = 10$).

The probability that K, the number of times the first threshold is exceeded, is equal to or greater than a second threshold M $(M \leqslant N)$ is given by

$$\Pr(K \geqslant M) = \sum_{K=M}^{N} p(K) \tag{14.3-5a}$$

$$= \sum_{K=M}^{N} {}_N C_K p^K [1 - p]^{N-K} \tag{14.3-5b}$$

Probability $\Pr(K \geqslant M)$ depends on the value of p. To obtain p, note that the probability density of the ith envelope sample r_i out of the matched filter is given by Eq. (10.4-2) as follows:

$$p(r_i; \mathcal{R}_p) = r_i \exp\left(-\frac{r_i^2 + \mathcal{R}_p}{2}\right) I_0(r_i\sqrt{\mathcal{R}_p}) \tag{14.3-6}$$

The notation $p(r_i; \mathcal{R}_p)$ indicates the parametric dependence of the density function on peak signal-to-noise ratio $\mathcal{R}_p$. Let p_s denote the probability that r_i exceeds (first) threshold r_0 when signal-plus-noise is present, and p_n the probability that r_i exceeds r_0 when noise alone is present. It follows that

$$p_s = \int_{r_0}^{\infty} p(r_i; \mathcal{R}_p)\, dr_i \tag{14.3-7a}$$

$$= \int_{r_0}^{\infty} r_i e^{-(r_i^2 + \mathcal{R}_p)/2} I_0(r_i\sqrt{\mathcal{R}_p})\, dr_i \tag{14.3-7b}$$

$$= 1 - \int_{0}^{r_0} r_i e^{-(r_i^2 + \mathcal{R}_p)/2} I_0(r_i\sqrt{\mathcal{R}_p})\, dr_i \tag{14.3-7c}$$

and

$$p_n = \int_{r_0}^{\infty} p(r_i; 0)\, dr_i \tag{14.3-8a}$$

$$= \int_{r_0}^{\infty} r_i e^{-r_i^2/2}\, dr_i \tag{14.3-8b}$$

$$= e^{-r_0^2/2} \tag{14.3-8c}$$

Probability of detection P_d is found by replacing p in Eq. (14.3-5b) by p_s and probability of false alarm P_{fa} is computed with p replaced by p_n, as follows:

$$P_d = \sum_{K=M}^{N} {}_N C_K p_s^K [1 - p_s]^{N-K} \qquad (14.3\text{-}9)$$

and

$$P_{fa} = \sum_{K=M}^{N} {}_N C_K p_n^K [1 - p_n]^{N-K} \qquad (14.3\text{-}10)$$

Equations (14.3-9) and (14.3-10) describe the performance of a binary detector with first and second thresholds r_0 and M, respectively. Unfortunately, these expressions are not convenient for optimizing performance.

Consider the problem of optimizing thresholds r_0 and M. The Neyman-Pearson criterion of optimality requires P_d to be maximized for fixed P_{fa}. Equation (14.3-10) shows that P_{fa} is a function of M and p_n. Through its dependence on p_n, P_{fa} is implicitly a function of threshold r_0; hence thresholds r_0 and M are deterministically related once P_{fa} is selected. Thus, optimizing P_d with respect to one threshold affects the other threshold. To properly take into account the relationship between the two thresholds is, in general, difficult.

Another method due to Harrington [1] is employed instead. In this approach, the integrated video sample preceding the second threshold is characterized by a "voltage signal-to-noise ratio" ρ_v, defined by

$$\rho_v = \frac{\text{(mean of signal plus noise)} - \text{(mean of noise alone)}}{\text{(standard deviation of noise alone)}} \qquad (14.3\text{-}11)$$

To optimize threshold r_0 with this method, ρ_v is maximized as a function of peak signal-to-noise ratio $\mathscr{R}_p$.

This procedure is not completely rigorous since the distributions of signal-plus-noise and noise alone preceding the second threshold are not uniquely characterized by ρ_v. However, maximizing ρ_v can be justified heuristically; we expect the probability of exceeding the second threshold to increase and the false-alarm probability to decrease when (1) the difference between the mean of integrated signal-plus-noise and the mean of integrated noise alone is maximized [see the numerator of Eq. (14.3-11)], and (2) the standard deviation of integrated noise, a measure of spread about the mean, is simultaneously minimized [see the denominator of Eq. (14.3-11)]. Further, when N becomes very large, the probability density functions of integrated signal-plus-noise and integrated noise alone approach a Gaussian distribution as a consequence of the central limit theorem. It can be shown [2] that for this case the results obtained by Harrington's method are identical with those obtained by a direct application of the Neyman-Pearson criterion, namely, maximizing P_d for fixed P_{fa}.

When Eqs. (14.3-3) and (14.3-4) are substituted into (14.3-11) with p replaced by p_s or p_n as required, ρ_v can be written as

$$\rho_v = \frac{Np_s - Np_n}{\sqrt{Np_n(1 - p_n)}} \tag{14.3-12a}$$

$$= \sqrt{N} \, \frac{p_s - p_n}{\sqrt{p_n(1 - p_n)}} \tag{14.3-12b}$$

$$= \sqrt{N} \, \rho_0 \tag{14.3-12c}$$

where

$$\rho_0 = \frac{p_s - p_n}{\sqrt{p_n(1 - p_n)}} \tag{14.3-13}$$

Equation (14.3-12c) suggests that ρ_0 can be interpreted as an *equivalent video signal-to-noise ratio* for a single sample that is increased by $\sqrt{N}$ when N independent samples are integrated.

From Eq. (14.3-12c) we see that maximizing ρ_v corresponds to maximizing ρ_0. Figure 14.3-3 is a plot of ρ_0 as a function of p_n for selected values of $\mathscr{R}_p$. To obtain this graph, tables [4] were used to evaluate Eqs. (14.3-7b) and (14.3-8b) for p_s and p_n, as functions of threshold r_0 and parameter $\mathscr{R}_p$. These values were substituted into Eq. (14.3-13) to obtain ρ_0 and the curves of Fig. 14.3-3. Note that the location of the maximum value of ρ_0 with respect to p_n varies as a function of signal-to-noise ratio $\mathscr{R}_p$ and that the maximum is very broad—that is, optimum performance is not critically dependent on p_n. The approximate values of p_n that maximize ρ_0 are plotted in Fig. 14.3-4 as a function of signal-to-noise ratio $\mathscr{R}_p$. This curve was obtained by locating the peaks in Fig. 14.3-3 for selected values of $\mathscr{R}_p$.

For small values of $\mathscr{R}_p$, the optimum value of p_n is independent of $\mathscr{R}_p$ and asymptotically approaches 0.2035. To demonstrate this result, an expression for ρ_0 in terms of p_n is derived first, using the fact that both p_s and p_n are functions of threshold r_0. With

$$z = e^{-r^2/2} \tag{14.3-14}$$

Eq. (14.3-7c) can be rewritten as

$$p_s = 1 - \int_{e^{-r_0^2/2}}^{1} e^{-\mathscr{R}_p/2} I_0(\sqrt{-2\mathscr{R}_p \ln z}) \, dz \tag{14.3-15}$$

From Eq. (14.3-8c), the lower limit in (14.3-15) is equal to p_n, so that

$$p_s = 1 - e^{-\mathscr{R}_p/2} \int_{p_n}^{1} I_0(\sqrt{-2\mathscr{R}_p \ln z}) \, dz \tag{14.3-16}$$

To integrate (14.3-16), $I_0(y)$ can be replaced by the power series expansion:

$$I_0(y) = 1 + \sum_{l=1}^{\infty} \frac{(y/2)^{2l}}{(l!)^2} \tag{14.3-17}$$

Substituting (14.3-17) into Eq. (14.3-16) and interchanging the order of integration and summation yields

$$p_s = 1 - e^{-\mathscr{R}_p/2} \int_{p_n}^{1} dz - e^{-\mathscr{R}_p/2} \sum_{l=1}^{\infty} \int_{p_n}^{1} \frac{\mathscr{R}_p^l (-\ln z)^l}{2^l (l!)^2} \, dz \tag{14.3-18}$$

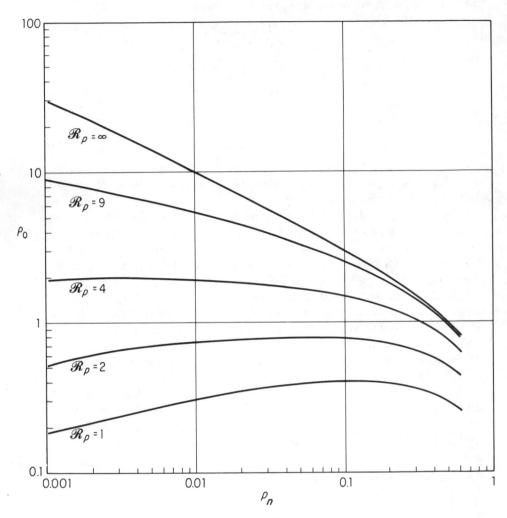

Fig. 14.3-3. ρ_0 vs. p_n for constant $\mathscr{R}_p$.

The third term in (14.3-18) can be integrated with the relationship†

$$\int (\ln z)^l \, dz = \sum_{m=0}^{l} (-1)^{l-m} \frac{l!}{m!} z(\ln z)^m \qquad (14.3\text{-}19)$$

With (14.3-19) substituted into Eq. (14.3-18), p_s is given by

†See reference [5], integral 428.

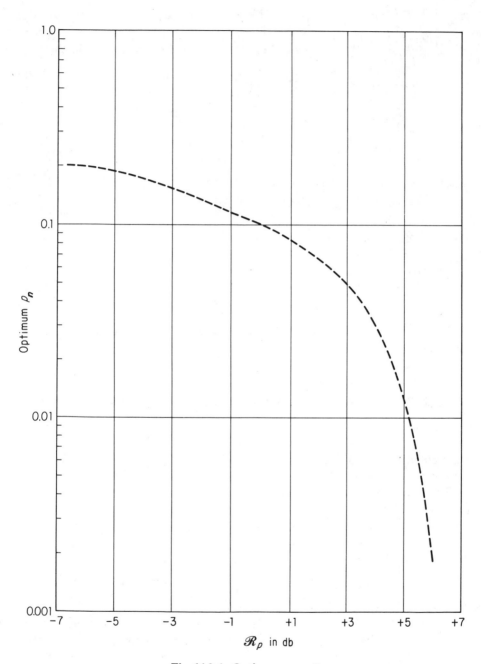

Fig. 14.3-4. Optimum p_n vs. $\mathscr{R}_p$.

$$p_s = 1 - e^{-\mathscr{R}_p/2}(1 - p_n) - \frac{\mathscr{R}_p}{2} e^{-\mathscr{R}_p/2}[1 - p_n - p_n(-\ln p_n)]$$

$$- \frac{1}{2!}\left(\frac{\mathscr{R}_p}{2}\right)^2 e^{-\mathscr{R}_p/2}\left[1 - p_n - p_n(-\ln p_n) - \frac{1}{2!}p_n(-\ln p_n)^2\right] \cdots \quad (14.3\text{-}20a)$$

$$= 1 - (1 - p_n)e^{-\mathscr{R}_p/2}\left[1 + \frac{\mathscr{R}_p}{2} + \frac{1}{2!}\left(\frac{\mathscr{R}_p}{2}\right)^2 + \cdots\right]$$

$$+ \frac{\mathscr{R}_p}{2} e^{-\mathscr{R}_p/2}[p_n(-\ln p_n)]$$

$$+ \frac{1}{2!}\left(\frac{\mathscr{R}_p}{2}\right)^2 e^{-\mathscr{R}_p/2}\left[p_n(-\ln p_n) + \frac{1}{2!}p_n(-\ln p_n)^2\right] + \cdots \quad (14.3\text{-}20b)$$

Since

$$e^{\mathscr{R}_p/2} = 1 + \frac{\mathscr{R}_p}{2} + \frac{1}{2!}\left(\frac{\mathscr{R}_p}{2}\right)^2 + \cdots \quad (14.3\text{-}21)$$

Eq. (14.3-20b) simplifies to

$$p_s = p_n + \frac{\mathscr{R}_p}{2} e^{-\mathscr{R}_p/2}[p_n(-\ln p_n)]$$

$$+ \frac{1}{2!}\left(\frac{\mathscr{R}_p}{2}\right)^2 e^{-\mathscr{R}_p/2}\left[p_n(-\ln p_n) + \frac{1}{2!}p_n(-\ln p_n)^2\right] + \cdots \quad (14.3\text{-}22)$$

For small $\mathscr{R}_p$, the first few terms of Eq. (14.3-22) yield a good approximation of p_s:

$$p_s \cong p_n + \frac{\mathscr{R}_p}{2}[p_n(-\ln p_n)] \qquad \text{for } \mathscr{R}_p \ll 1 \quad (14.3\text{-}23)$$

When (14.3-23) is substituted into Eq. (14.3-13), ρ_0 is approximately given by

$$\rho_0 \cong \frac{\mathscr{R}_p}{2}\sqrt{\frac{p_n}{1 - p_n}}(-\ln p_n) \qquad \text{for } \mathscr{R}_p \ll 1 \quad (14.3\text{-}24)$$

Differentiating Eq. (14.3-24) with respect to p_n and equating the result to zero yields $p_n = 0.2035$ as that value which maximizes ρ_0 for $\mathscr{R}_p \ll 1$.

To summarize, the value of p_n that maximizes ρ_0 can be found from Fig. 14.3-4 as a function of peak signal-to-noise ratio $\mathscr{R}_p$. Substituting this value of p_n into Eq. (14.3-8c) establishes the value of first threshold r_0. Probability p_s is calculated by substituting this value of r_0 into Eq. (14.3-7c) together with the value of $\mathscr{R}_p$ used to determine the optimum p_n. In this manner, optimum values for p_s and p_n can be obtained for selected peak signal-to-noise ratios. When an optimum pair of p_n and p_s are substituted into Eqs. (14.3-9) and (14.3-10), P_d and P_{fa} are obtained as functions of second threshold M and the number of train pulses N. Specifying P_{fa} permits second threshold M to be evaluated for selected values of N using Eq. (14.3-10); for each value of M obtained in this manner, P_d is calculated from (14.3-9). These steps must be repeated to obtain P_d and P_{fa} as a function of both $\mathscr{R}_p$ and N for a binary integrator.

The discrete sums in Eqs. (14.3-9) and (14.3-10) are not convenient for the computations we wish to perform. Instead, a Gram-Charlier expansion, grouped in the Edgeworth scheme (see discussion in Section 10.6), is used to approximate the discrete binomial distribution $p(K)$ of Eq. (14.3-1) with a continuous distribution $p_A(K)$. P_d and P_{fa} can then be approximated by integrating the continuous distribution $p_A(K)$ over the appropriate interval. The Edgeworth series for $p_A(K)$ can be written [see Eqs. (10.6-13) and (10.6-19)] in terms of the normalized central moments α_i and standard deviation σ_K of random variable K as follows:

$$p_A(K) = \frac{1}{\sigma_K} \left[\phi\left(\frac{K - \bar{K}}{\sigma_K}\right) \right] + \left[C_3 \overset{(3)}{\phi}\left(\frac{K - \bar{K}}{\sigma_K}\right) \right]$$

$$+ \left[C_4 \overset{(4)}{\phi}\left(\frac{K - \bar{K}}{\sigma_K}\right) + C_6 \overset{(6)}{\phi}\left(\frac{K - \bar{K}}{\sigma_K}\right) \right]$$

$$+ \left[C_5 \overset{(5)}{\phi}\left(\frac{K - \bar{K}}{\sigma_K}\right) + C_7 \overset{(7)}{\phi}\left(\frac{K - \bar{K}}{\sigma_K}\right) + C_9 \overset{(9)}{\phi}\left(\frac{K - \bar{K}}{\sigma_K}\right) \right] + \ldots \quad (14.3\text{-}25)$$

where

$$C_3 = \frac{-\alpha_3}{3!} \tag{14.3-26a}$$

$$C_4 = \frac{(\alpha_4 - 3)}{4!} \tag{14.3-26b}$$

$$C_5 = -\frac{(\alpha_5 - 10\alpha_3)}{5!} \tag{14.3-26c}$$

$$C_6 = \frac{(\alpha_6 - 15\alpha_4 + 30)}{6!} \tag{14.3-26d}$$

$$C_7 = -\frac{(\alpha_7 - 21\alpha_5 + 105\alpha_3)}{7!} \tag{14.3-26e}$$

$$C_9 = -\frac{(\alpha_9 - 36\alpha_7 + 378\alpha_5 - 1260\alpha_3)}{9!} \tag{14.3-26f}$$

and

$$\phi(x) = \frac{1}{\sqrt{2\pi}} e^{-x^2/2} \tag{14.3-27}$$

$$\overset{(i)}{\phi}(x) = \frac{d^i}{dx^i} \phi(x) \tag{14.3-28}$$

Each pair of brackets in (14.3-25) contains terms of the same order of magnitude in the expansion.

The probability that $K \geqslant M$ is given (approximately) by

$$\Pr\left(K \geqslant M\right) \cong \int_{M-1/2}^{\infty} p_A(K)\, dK \tag{14.3-29}$$

where a factor of $\frac{1}{2}$ is introduced in the lower limit to give better agreement between the distribution functions computed for the discrete and continuous density functions [6]. Substituting Eq. (14.3-25) into Eq. (14.3-29) and with

$$y = \frac{K - \bar{K}}{\sigma_K} \qquad (14.3\text{-}30)$$

and

$$Y = \frac{M - \frac{1}{2} - \bar{K}}{\sigma_K} \qquad (14.3\text{-}31)$$

we can write Eq. (14.3-29) as

$$\Pr(K \geqslant M) \cong \int_Y^\infty dy \, \{[\phi(y)] + [C_3 \overset{(3)}{\phi}(y)] + [C_4 \overset{(4)}{\phi}(y) + C_6 \overset{(6)}{\phi}(y)]$$
$$+ [C_5 \overset{(5)}{\phi}(y) + C_7 \overset{(7)}{\phi}(y) + C_9 \overset{(9)}{\phi}(y)] + \ldots\} \qquad (14.3\text{-}32a)$$
$$\cong [1 - \overset{(-1)}{\phi}(Y)] - [C_3 \overset{(2)}{\phi}(Y)] - [C_4 \overset{(3)}{\phi}(Y) + C_6 \overset{(5)}{\phi}(Y)]$$
$$- [C_5 \overset{(4)}{\phi}(Y) + C_7 \overset{(6)}{\phi}(Y) + C_9 \overset{(8)}{\phi}(Y)] - \ldots \qquad (14.3\text{-}32b)$$

where

$$\overset{(-1)}{\phi}(Y) = \int_{-\infty}^Y \phi(y) \, dy \qquad (14.3\text{-}33)$$

Functions $\overset{(-1)}{\phi}(Y)$ and $\overset{(i)}{\phi}(Y)$ are tabulated in [7]. The transition from (14.3-32a) to (14.3-32b) used the following relation:

$$\int_Y^\infty \overset{(i)}{\phi}(y) \, dy = - \overset{(i-1)}{\phi}(Y) \qquad (14.3\text{-}34)$$

To obtain sufficient accuracy for P_{fa} using the Edgeworth approximation, all terms shown in (14.3-32b) must be used. The coefficients in Eqs. (14.3-26) require knowledge of the normalized central moments of the binomial distribution up through the ninth as a function of the number of pulses N. Cramer [8] shows, however, that coefficients C_6, C_7, and C_9 can be evaluated in terms of the third and fourth normalized central moments, as follows:

$$C_6 = \frac{10}{6!} (\alpha_3)^2 \qquad (14.3\text{-}35a)$$

$$C_7 = -\frac{35}{7!} \alpha_3 (\alpha_4 - 3) \qquad (14.3\text{-}35b)$$

$$C_9 = -\frac{280}{9!} \alpha_3^3 \qquad (14.3\text{-}35c)$$

With these relations, coefficients C_3 through C_9 can be calculated from normalized central moments α_3 through α_5. The α_i are related to central moments μ_i and standard deviation σ by

$$\alpha_i = \frac{\mu_i}{\sigma} \qquad (14.3\text{-}36)$$

To evaluate μ_i, we will use the fact that the central moments of a random variable with zero mean are identical to its noncentral moments. The characteristic function $C_K(\xi)$ of binomial random variable K with density function (14.3-1) is [7]

$$C_K(\xi) = [1 - p + pe^{j\xi}]^N \qquad (14.3\text{-}37a)$$
$$= [q + pe^{j\xi}]^N \qquad (14.3\text{-}37b)$$

where

$$q = 1 - p \qquad (14.3\text{-}38)$$

The characteristic function $C_{K'}(\xi)$ of a related zero mean variable K'

$$K' = K - \bar{K} \qquad (14.3\text{-}39)$$

can be related to characteristic function $C_K(\xi)$ as follows. From Eq. (14.3-3)

$$K' = K - Np \qquad (14.3\text{-}40)$$

From the definition [see Eq. (3.12-1b)] of the characteristic function of K',

$$C_{K'}(\xi) = E[e^{j\xi K'}] \qquad (14.3\text{-}41)$$

Substituting Eq. (14.3-40) into (14.3-41) yields

$$C_{K'}(\xi) = E[e^{j\xi(K-Np)}] \qquad (14.3\text{-}42a)$$
$$= e^{-j\xi Np} E[e^{j\xi K}] \qquad (14.3\text{-}42b)$$
$$= e^{-j\xi Np} C_K(\xi) \qquad (14.3\text{-}42c)$$

Substituting Eq. (14.3-37b) for $C_K(\xi)$ into (14.3-42c) results in

$$C_{K'}(\xi) = e^{-j\xi Np}[q + pe^{j\xi}]^N \qquad (14.3\text{-}43a)$$
$$= [qe^{-j\xi p} + pe^{j\xi q}]^N \qquad (14.3\text{-}43b)$$

The central moments of K equal the noncentral moments of K'; hence, the first five central moments of K can be found by successive differentiation of $C_{K'}(\xi)$ at $\xi = 0$ [see Eq. (3.12-6)]; the result is

$$\mu_1 = 0 \qquad (14.3\text{-}44)$$
$$\mu_2 = \sigma^2 = Npq \qquad (14.3\text{-}45)$$
$$\mu_3 = Npq(q^2 - p^2) \qquad (14.3\text{-}46)$$
$$\mu_4 = 3N(N-1)(pq)^2 + Npq(q^3 + p^3) \qquad (14.3\text{-}47)$$
$$\mu_5 = 10N(N-1)(pq)^2(q^2 - p^2) + Npq(q^4 - p^4) \qquad (14.3\text{-}48)$$

Substituting Eqs. (14.3-44) through (14.3-48) into Eq. (14.3-36) yields the normalized central moments α_1 through α_5. These values are substituted into Eqs. (14.3-26a) through (14.3-26c) and (14.3-35a) through (14.3-35c) to obtain the coefficients of the Edgeworth series.

Performance curves for the binary integrator were obtained by the following procedure:

(a) The coefficients of the Edgeworth series were first evaluated, as a function of N, by substituting the optimum values of p_s obtained for $\mathscr{R}_p = -6, 0$, and $+3$ db into the appropriate expressions. Then, employing the first four bracketed terms of the Edgeworth series in Eq. (14.3-32b) to approximate P_d, the value of the argument $Y = Y_s$ was obtained for $P_d = 10, 30, 50, 70$, and 90 per cent with tables of the normal function and its derivatives [7]; this was repeated for $N = 10, 100$, and 1000. With knowledge of Y_s, the values of second threshold M were calculated for the selected values of P_d. The values of M were obtained by rewriting Eq. (14.3-31) as

$$M = \sigma_K Y_s + \bar{K} + \tfrac{1}{2} \tag{14.3-49}$$

With Eqs. (14.3-3) and (14.3-4), Eq. (14.3-49) becomes

$$M = \sqrt{Np_s(1 - p_s)}\, Y_s + Np_s + \tfrac{1}{2} \tag{14.3-50}$$

(b) For each calculated second threshold M, there is a corresponding P_{fa}. To calculate P_{fa} with the required accuracy, all terms of the Edgeworth series in (14.3-25) were found necessary. (Accuracy deteriorates rapidly for values of P_{fa} less than 10^{-5}; hence, P_{fa} is not plotted in this range. Accuracy also diminishes rapidly for p_n less than 0.05; hence, performance curves are plotted only for peak signal-to-noise ratios of $+3$ db or less.) To calculate P_{fa}, the value of p_n corresponding to a particular $\mathcal{R}_p$ is substituted into Eq. (14.3-31); then, with M computed as described in step (a), the parameter $Y = Y_n$ is obtained from

$$Y_n = \frac{M - Np_n - \tfrac{1}{2}}{\sqrt{Np_n(1 - p_n)}} \tag{14.3-51}$$

The values of Y_n obtained are substituted into the Edgeworth series, Eq. (14.3-25), to obtain P_{fa}.

Performance results for the optimum binary integrator are shown in Figs. 14.3-5 through 14.3-7, for $N = 10, 100$, and 1000, respectively. The ratio M/N is related to P_{fa} as a function of signal-to-noise ratio; corresponding values of P_d are the points plotted on the curves. From these curves, the following set of optimum radar system parameters can be obtained: $P_d, P_{fa}, \mathcal{R}_p, p_n$ (from which r_0 can be computed), and second threshold M.

When the performance of a binary integrator is compared to that of the optimum postdetection integrator,[†] we find that over the range of parameter values shown in Figs. 14.3-5 through 14.3-7, the binary integrator requires an rf signal-to-noise ratio 0.8 to 1.4 db larger than the optimum integrator for the same P_d and P_{fa}. This is shown in Table 14.3-1. Because of the inaccuracies involved in computing the performance of the binary integrator, the data in Table 14.3-1 do not permit conclusions to be drawn about absolute loss variation with P_d, P_{fa}, and N. What is significant is the relative insensitivity of the loss to wide variations in these parameters.

†See Chapter 10.

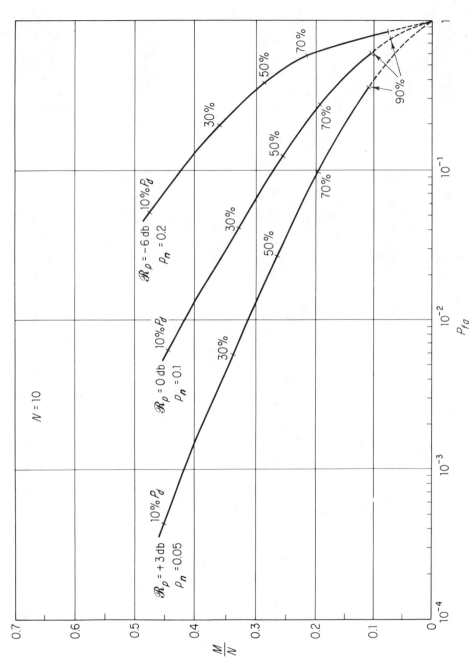

Fig. 14.3-5. Binary-detection performance (optimized), $N = 10$.

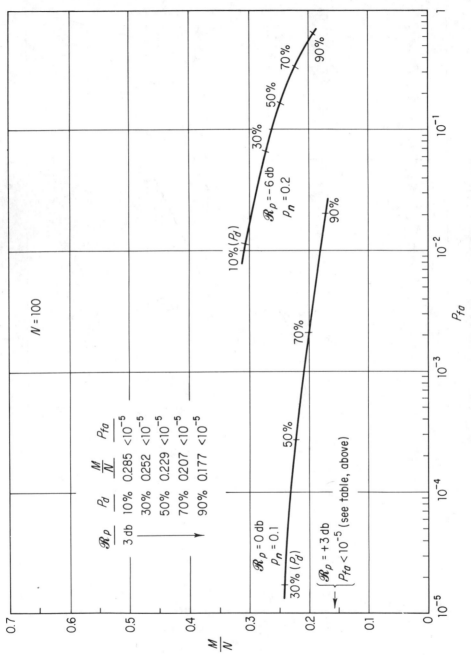

Fig. 14.3-6. Binary-detection performance (optimized), $N = 100$.

$N = 100$

$\mathscr{R}_p$	P_d	$\dfrac{M}{N}$	P_{fa}
3 db	10%	0.285	$<10^{-5}$
	30%	0252	$<10^{-5}$
	50%	0.229	$<10^{-5}$
	70%	0207	$<10^{-5}$
	90%	0.177	$<10^{-5}$

$\mathscr{R}_p = -6\,\mathrm{db}$
$p_n = 0.2$

$10\%\,(P_d)$
30%
50%
70%
90%

90%

$\mathscr{R}_p = 0\,\mathrm{db}$
$p_n = 0.1$
$30\%\,(P_d)$

50%
70%

$\mathscr{R}_p = +3\,\mathrm{db}$
$P_{fa} < 10^{-5}$ (see table, above)

P_{fa}

$\dfrac{M}{N}$

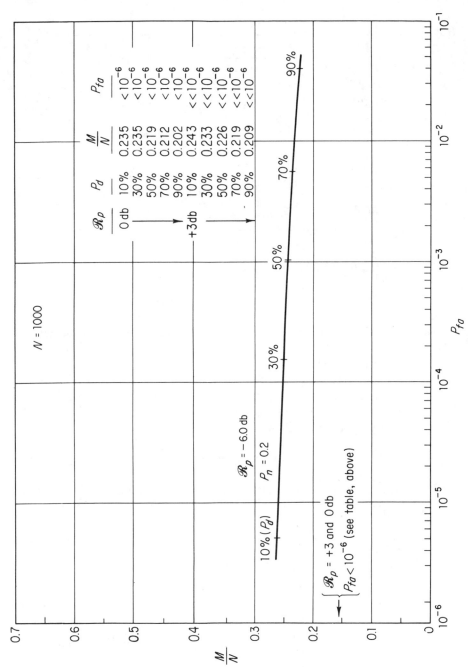

The table within the figure:

$\mathscr{R}_\rho$	P_d	$\dfrac{M}{N}$	P_{fa}
0 db	10%	0.235	$< 10^{-6}$
	30%	0.235	$< 10^{-6}$
	50%	0.219	$< 10^{-6}$
	70%	0.212	$< 10^{-6}$
	90%	0.202	$< 10^{-6}$
+3db	10%	0.243	$<< 10^{-6}$
	30%	0.233	$<< 10^{-6}$
	50%	0.226	$<< 10^{-6}$
	70%	0.219	$<< 10^{-6}$
	90%	0.209	$<< 10^{-6}$

$N = 1000$

$\mathscr{R}_\rho = -6.0$ db

$P_n = 0.2$

10% (P_d)

$\begin{cases} \mathscr{R}_\rho = +3 \text{ and } 0 \text{ db} \\ P_{fa} < 10^{-6} \text{ (see table, above)} \end{cases}$

Fig. 14.3-7. Binary-detection performance (optimized), $N = 1000$.

TABLE 14.3-1

N	P_d	P_{fa}	Loss
10	0.5	0.38	1.0 db
		0.13	0.9
		0.03	1.4
	0.9	0.83	0.8
		0.61	1.0
		0.36	1.4
100	0.5	0.17	1.2
		2×10^{-4}	1.2
	0.9	0.63	1.1
		0.02	1.3
1000	0.5	0.001	0.9
	0.9	0.039	1.0

14.4 Extension to Fluctuating Targets

In the preceding analysis, target cross section was assumed to be nonfluctuating. For a fluctuating target cross section, signal-to-noise ratio is a random variable. There are two limiting cases: scan-to-scan fluctuation and pulse-to-pulse fluctuation. (See Chapter 11 for a complete discussion of these two target models.)

The calculations performed for nonfluctuating targets can be repeated for fluctuating targets. For a pulse-to-pulse Rayleigh fluctuating target the probability that the envelope of signal-plus-noise exceeds the first threshold is given by [see Eqs. (9.4-14b) and (9.4-16b)],

$$p_s = \exp\left(-\frac{r_0^2}{2 + \bar{\mathcal{R}}_p}\right) \tag{14.4-1}$$

where $\bar{\mathcal{R}}_p$ is the mean peak signal-to-noise ratio. For optimized r_0, probability of detection P_d can be found by substituting Eq. (14.4-1) into (14.3-9). For targets with a scan-to-scan fluctuation, P_d can be found by averaging Eq. (14.3-9) with respect to the assumed target cross-section probability density function.

Linder and Swerling [3] carried out a limited number of numerical calculations assuming a Rayleigh target fluctuation for both the pulse-to-pulse and scan-to-scan models. Their results indicate that the optimum thresholds for the scan-to-scan case are the same as those obtained in the nonfluctuating case. For the pulse-to-pulse model, the optimum setting of the second threshold is slightly lower than for the nonfluctuating case. They also showed that the loss in signal-to-noise ratio of a binary integrator, relative to the optimum postdetection integrator, is relatively unaffected by target fluctuations (assuming the same target model for both integrators).

14.5. Coincidence Detection

The double-threshold method of detection, discussed in Sections 14.3 and 14.4, requires that M or more detections occur (in any order) in N consecutive trials in order to detect a signal. Coincidence detection requires M or more consecutive pairs, or triples, to occur in N consecutive trials. The application of coincidence to a double-threshold system is illustrated in Fig. 14.5-1. Figure 14.5-1(a) shows a double-threshold system without coincidence (for simplicity, range gating is not shown). The system in Fig. 14.5-1(b) requires that two consecutive pulses, having the same range delay, exceed the first threshold before an output can pass to the counter. In similar manner the system in Fig. 14.5-1(c) requires that three consecutive

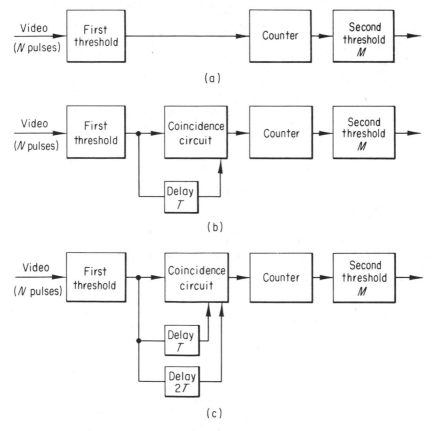

Fig. 14.5-1. Principle of coincidence detection: (a) no coincidence; (b) double coincidence; (c) triple coincidence (*after Endresen et al.* [9], *courtesy of IEEE*).

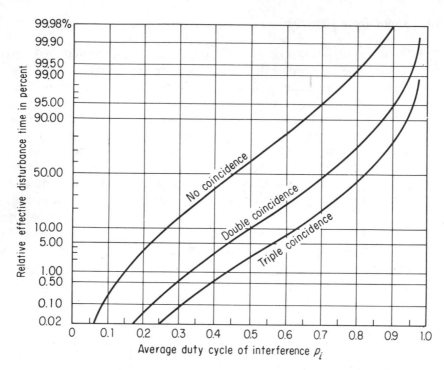

Fig. 14.5-2. Relative disturbance time; $N = 10$, $M = 5$ in all three cases (*after Endresen et al.* [9], *courtesy of IEEE*).

pulses, corresponding to the same range, exceed the first threshold before an output passes to the counter.

Coincidence detection is of interest because of its apparent effectiveness in suppressing false alarms generated by large random-pulse interference. Endresen and Hedemark [9] calculated the vulnerability of the three signal processors in Fig. 14.5-1 to strong random-pulse interference. (In the calculation an interfering pulse was always assumed to exceed the first threshold.) Performance is shown on the curves in Figs. 14.5-2 through 14.5-4 for $N = 10, 20$, and 30, respectively. These plots show the relative time that a radar is disturbed as a function of the mean duty cycle p_i of the interference—that is, the fraction of time occupied by interfering pulses. In each of the three figures, the second threshold M is (approximately) optimum under normal noise conditions for no-coincidence double-threshold detection. The curves for $N = 10$ show that, for a moderate-interference duty cycle of $p_i = 0.2$, a no-coincidence system is disturbed 3.5 per cent of the time, a double-coincidence system is disturbed 0.05 per cent of the time, and a triple-coincidence system 0.005 per cent of the time.

Improved performance in the presence of random-pulse interference is

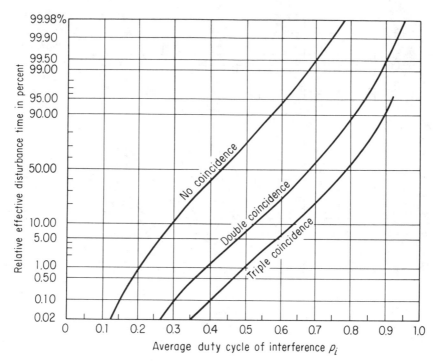

Fig. 14.5-3. Relative disturbance time; $N = 20$, $M = 9$ in all three cases (*after Endresen et al.* [9], *courtesy of IEEE*).

obtained at a penalty of poorer performance in Gaussian noise. Endresen and Hedemark [9] evaluated the loss by calculating the required signal-to-noise ratio for a constant-amplitude pulse train, and a pulse-to-pulse Rayleigh fluctuating pulse train, of length $N = 10$, for equal detection performance of each detector type. Computations were performed for $P_{fa} = 10^{-5}$, $P_d = 50$ per cent and 90 per cent, and a second threshold of $M = 5$. The results are given in Table 14.5-1.

TABLE 14.5-1

$(P_{fa} = 10^{-5}; N = 10; M = 5)$

	Signal-to-noise ratio $\mathscr{R}_p$		
Conditions	No coincidence (db)	Double coincidence (db)	Triple coincidence (db)
$P_d = 50\%$; constant signal	7.0	7.6	8.2
$P_d = 90\%$; constant signal	9.0	9.6	10.2
$P_d = 50\%$; Rayleigh signal	8.0	10.3	12.2
$P_d = 90\%$; Rayleigh signal	11.4	14.6	18.0

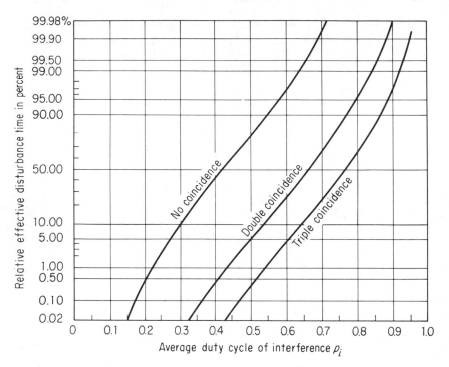

Fig. 14.5-4. Relative disturbance time; $N = 30$, $M = 13$ in all three cases (*after Endresen et al.* [9], *courtesy of IEEE*).

For a constant signal and $P_d = 50$ per cent, double coincidence requires a signal-to-noise ratio increase of about 0.6 db, and triple coincidence of about 1.2 db, with respect to a no-coincidence system; this result is the same at $P_d = 90$ per cent. For a Rayleigh fluctuating signal and $P_d = 50$ per cent, double and triple coincidence require 2.3 db and 4.2 db, respectively, greater signal-to-noise ratio than a no-coincidence system; for $P_d = 90$ per cent, the corresponding differences are 3.2 db and 6.6 db.

Commenting on the results obtained by Endresen and Hedemark [9], Hansen [10] suggests that the same reduction in sensitivity to random-pulse interference can be obtained by properly adjusting the two thresholds in a no-coincidence double-threshold detection system and with essentially the same degradation in performance in the presence of Gaussian noise. Hansen [10] cites a specific example to support his statement; however, a general proof has not yet been given.

PROBLEMS

14.1. The signal-to-noise ratio required by a binary integrator, for a given level of performance, is approximately one decibel greater than that required by an ideal integrator. However, if the video signal-to-noise ratios preceding the final threshold are compared for both integrators, the difference is considerably greater. Explain the discrepancy. Which figure of merit is more meaningful? Why?

14.2. For large N, the probability density function of noise and signal-plus-noise preceding the second threshold in a binary integrator can be approximated by a Gaussian distribution. Verify that $p_n \longrightarrow 0.2035$ for an optimum binary integrator for large N by direct application of the Neyman-Pearson criterion; that is, maximize P_d for fixed P_{fa}. (See reference [2] for solution.)

14.3. Fill in the required steps between Eqs. (14.3-16) and (14.3-22). Verify that $p_n \longrightarrow 0.2035$ for $\mathscr{R}_p \ll 1$ by differentiating (14.3-24).

14.4. (a) Derive the characteristic function for a binomial distribution given by Eq. (14.3-37a).

    (b) Show that the central moments of a random variable K, with nonzero mean, are identical to the noncentral moments of random variable $(K - \bar{K})$.

14.5. Verify expressions (14.3-3) and (14.3-4) for the mean and variance of a binomial distribution.

14.6. For small $\mathscr{R}_p$, expand the I_0 function in Eq. (14.3-16) and, by integrating only the first two terms in the expansion, show that Eq. (14.3-23) results.

14.7. For $\mathscr{R}_p = 0$ db, find the optimum p_n using Fig. 14.3-4. Compute p_s from the small signal-to-noise approximation of Eq. (14.3-23). Calculate P_d and P_{fa} from Eqs. (14.3-9) and (14.3-10) for $N = 10$ and $M = 1, 2, 3, 4$. Compare these results with the plot for $\mathscr{R}_p = 0$ db in Fig. 14.3-5.

14.8. With the value of p_s obtained in problem 14.7 for $N = 10$ and $\mathscr{R}_p = 0$ db compute the coefficients of Edgeworth expansion (14.3-25) using expressions (14.3-26a) through (14.3-26c), (14.3-35a) through (14.3-35c), (14.3-36), and (14.3-44) through (14.3-48). With these results and the tables in reference [7], find Y_s for $P_d = 50$ per cent by plotting (14.3-32b) as a function of Y_s. Compute the corresponding value of M using Eq. (14.3-50). Compute Y_n using (14.3-51) and P_{fa} using (14.3-32b). Compare the values obtained with the appropriate plot in Fig. 14.3-5.

REFERENCES

1. Harrington, J. V.: An Analysis of the Detection of Repeated Signals in Noise by Binary Integration, *IRE Trans. on Information Theory*, **1T-1**: (*1*), 1–9 (March, 1955).

2. Swerling, P.: The 'Double Threshold' Method of Detection, *Rand Corp. Memo*, RM-1008 (December, 1952).

3. Linder, I. W. Jr., and Swerling, P.: Performance of the 'Double Threshold' Radar Receiver in the Presence of Interference, *Rand Corp. Memo*, RM-1719 (May, 1956).

4. Marcum, J. I.: Table of Q Functions, *Rand Corp. Memo*, RM-339 (January, 1950).

5. Pierce, B. O.: "A Short Table of Integrals," Ginn, Boston, 1929.

6. Fry, T. C.: "Probability and its Engineering Uses," Van Nostrand, Princeton, N.J., 1928.

7. Abramowitz, M., and Stegun, I. A.: "Handbook of Mathematical Functions," National Bureau of Standards Applied Mathematics Series 55 (June, 1964).

8. Cramer, Harold: "Mathematical Methods of Statistics," Princeton University Press, Princeton, N.J., 1945.

9. Endresen, K., and Hedemark, R.: Coincidence Techniques for Radar Receivers Employing a Double-Threshold Method of Detection, *Proc. IRE*, **49**: (*10*), 1561–1567 (October, 1961).

10. Hansen, V. G.: Coincidence Techniques for Radar Receivers, *Proc. IRE*, **50**: (*4*), 480 (April, 1962).

15

WEIGHTED
INTEGRATORS

15.1. Introduction

Elementary search radars employ a continuous periodic pulse transmission with a rotating antenna. As the antenna scans across a target, the amplitudes of both transmitted and received pulses are modified by the antenna beam pattern; this converts the uniform periodic pulse train generated by the radar transmitter to an amplitude-modulated pulse train of finite duration. The pulse train duration is determined by the antenna scan rate and beamwidth; the number of pulses in the train is a function of both train duration and the radar pulse repetition frequency. Pulse-train amplitude modulation results in a variation of received pulse signal-to-noise ratio as a function of time. In previous chapters, optimum receiver processing and performance were investigated for uniform pulse trains; these results are extended to amplitude-modulated trains in this chapter.

The optimum Bayes receiver performs a generalized likelihood-ratio test. To formulate the (modified) generalized likelihood ratio for an amplitude-modulated pulse train, it is convenient to begin with expression (9.4-5) for the likelihood ratio $\overline{\ell'(\mathbf{v}\,|\,\phi, A)_\phi}$ of a single pulse with arbitrary amplitude A in white Gaussian noise that has been averaged with respect to unknown initial phase ϕ. The generalized likelihood ratio for an incoherent pulse train is a product of likelihood ratios of the type described†—that is,

$$\overline{\ell'(\mathbf{v}\,|\,\boldsymbol{\theta})_\theta} = \prod_{i=1}^{N} \overline{\ell'(\mathbf{v}_i\,|\,\phi_i, A_i)_{\phi_i}} \qquad (15.1\text{-}1)$$

where the set of random phases $\phi_1, \ldots, \phi_N$ correspond to the N pulses in the train. Substituting Eq. (9.4-5) into (15.1-1) yields

$$\overline{\ell'(\mathbf{v}\,|\,\boldsymbol{\theta})_\theta} = \prod_{i=1}^{N} \exp{(-A_i^2/2)} \cdot I_0[A_i r_i(\mathbf{v}_i)] \qquad (15.1\text{-}2)$$

where $r_i(\mathbf{v}_i)$ is the normalized envelope defined in Eq. (9.3-10). Since the A_i are known, it is convenient to replace the exponential in Eq. (15.1-2) by a constant K_i, as follows:

$$\overline{\ell'(\mathbf{v}\,|\,\boldsymbol{\theta})_\theta} = \prod_{i=1}^{N} K_i I_0(A_i r_i) \qquad (15.1\text{-}3)$$

The Bayes test strategy corresponds to a comparison of the logarithm of the generalized likelihood ratio with a threshold, as follows:

$$\sum_{i=1}^{N} \ln K_i + \sum_{i=1}^{N} \ln I_0(A_i r_i) \begin{cases} \geqslant \mathscr{H} & \text{(signal present)} \\ < \mathscr{H} & \text{(signal absent)} \end{cases} \qquad (15.1\text{-}4)$$

or

$$\sum_{i=1}^{N} \ln I_0(A_i r_i) \begin{cases} \geqslant \mathscr{H}' \\ < \mathscr{H}' \end{cases} \qquad (15.1\text{-}5)$$

where $\mathscr{H}'$ is a suitably modified threshold.

For small signal-to-noise ratio, $\ln I_0(x)$ can be approximated by $x^2/4$

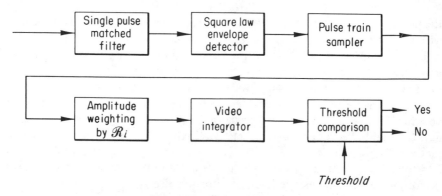

Fig. 15.1-1. Small-signal optimum detection receiver for amplitude-modulated incoherent pulse train.

†See Eq. (10.3-11b).

[see Eq. (10.3-17)]; then (15.1-5) reduces to

$$\sum_{i=1}^{N} A_i^2 r_i^2 \begin{cases} \geqslant \mathscr{K}'' \\ < \mathscr{K}'' \end{cases} \tag{15.1-6}$$

From Eq. (9.4-8), A_i^2 is equal to peak signal-to-noise ratio $\mathscr{R}_i$ for the ith pulse. Hence, expression (15.1-6) can be rewritten as

$$\sum_{i=1}^{N} \mathscr{R}_i r_i^2 \begin{cases} \geqslant \mathscr{K}'' & \textit{(signal present)} \\ < \mathscr{K}'' & \textit{(signal absent)} \end{cases} \tag{15.1-7}$$

A receiver that performs the test described in (15.1-7) for a target at known range is shown in Fig. 15.1-1. The optimum receiver consists of a filter matched to the waveform of a single pulse followed by a square-law

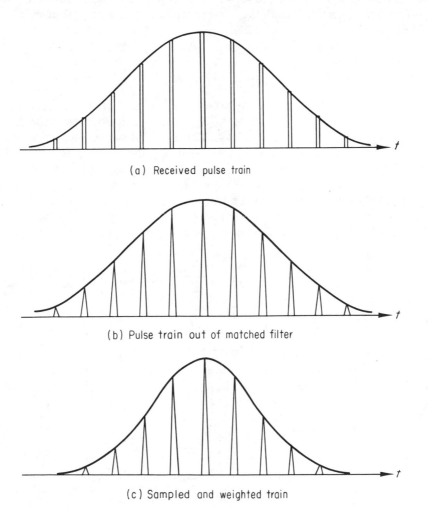

(a) Received pulse train

(b) Pulse train out of matched filter

(c) Sampled and weighted train

Fig. 15.1-2. Optimum small-signal pulse-train processing.

envelope detector. Sampled peak outputs r_i^2 of the envelope detector, $i = 1, 2, \ldots, N$, are weighted by expected peak signal-to-noise ratio $\mathcal{R}_i$ of the ith pulse and then summed by an integrator. This sum is compared to the threshold and a decision is reached. The block diagram in Fig. 15.1-1 differs from that of Fig. 10.3-2 by the addition of pulse train amplitude weighting. (At large signal-to-noise ratios, a $\ln I_0(x)$ envelope detector is approximately equivalent to a linear envelope detector; for this case the amplitude weighting of the ith pulse should be $\sqrt{\mathcal{R}_i}$ instead of $\mathcal{R}_i$.)

For a pulse train with an assumed Gaussian amplitude modulation, the pulse train at different stages of processing is shown in Fig. 15.1-2. Figure 15.1-2(a) shows the pulse train envelope at the input. Figure 15.1-2(b) shows the pulse train at the output of the matched filter. The pulse train after square-law detection, sampling, and amplitude weighting (but prior to integration) is shown in Fig. 15.1-2(c).

15.2. Implementation

In practice, both amplitude weighting and video integration are often incorporated into a single processor. One such implementation is the delay-line integrator shown in Fig. 15.2-1. The group delay of the delay line is equal to the interpulse period. As a result, a pulse entering the integrator is added to preceding train pulses which are recirculating in the delay line in a synchronous manner. This results in the superposition of all pulses just as the Nth pulse enters the integrator. The output gate is opened for one interpulse period following the Nth pulse. During this period, the input gate is closed to clear the delay line of stored information; the delay-line

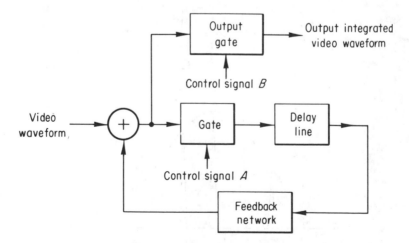

Fig. 15.2-1. Delay-line integrator.

integrator then repeats the integration process for another pulse train.

The synchronous delay-line integrator of Fig. 15.2-1 is often not convenient, since gates are used to clear the delay line after $(N - 1)$ recirculations of the first pulse. This can be done without input or output gates by employing a feedback loop around the delay line with loop gain β less than unity, as shown in Fig. 15.2-2(a). In this integrator, old information decays below the ambient noise level at a rate determined by the loop gain.

The weighting function of a single-loop integrator with loop gain β_1 can be obtained from the impulse response of the network. For an integrator with loop delay Δ_T, the impulse response consists of a train of impulses

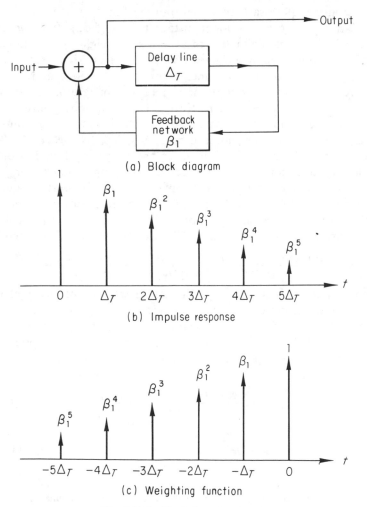

(a) Block diagram

(b) Impulse response

(c) Weighting function

Fig. 15.2-2. Single-loop integrator.

at the delay-line output at times 0, Δ_T, $2\Delta_T$, $3\Delta_T$, and so on, with respective amplitudes 1, β_1, β_1^2, β_1^3, ... This is illustrated in Fig. 15.2-2(b). Amplitude modulation results from attenuation each time an impulse completes a transit around the integrator loop. The integrator weighting function is equal to the impulse response run backward in time [see Eq. (2.2-16)]. If the integrator output is observed at time $t = 0$, Fig. 15.2-2(c) shows the relative weighting as a function of time.

The two-way pattern of a narrow beam antenna is nearly Gaussian around the peak.† Antenna scanning results in a pulse train with peak rf signal-to-noise ratios out of the matched filter modulated by an approximately Gaussian envelope. It follows from Eq. (15.1-7) that for optimum performance an ideal integrator requires a Gaussian weighting function. It is evident from Fig. 15.2-2(c) that the single-loop integrator weighting function is not Gaussian. A closer approximation to a Gaussian weighting function can be achieved by a double loop integrator, with loop gain β_2 in each loop; a block diagram of a double-loop integrator is shown in Fig. 15.2-3. The impulse response and weighting function of this integrator are shown in Figs. 15.2-3(b) and 15.2-3(c), respectively. The impulse response is found by introducing an impulse at time $t = 0$ into the integrator; the output at $t = 0$, Δ_T, $2\Delta_T$, ..., can be shown to be 1, $2\beta_2$, $3\beta_2^2$, $4\beta_2^3$, $5\beta_2^4$, ...

Loop gain β determines the shape of the weighting function of both single- and double-loop integrators. Hence, to optimize integrator performance the value of loop gain β must be optimized; for this purpose a method due to Cooper and Griffiths [1] is used. The optimization criterion is the maximization of voltage signal-to-noise ratio ρ_v out of the postdetection integrator prior to thresholding, as in Chapter 14, where ρ_v is defined by

$$\rho_v = \frac{\text{(mean of signal-plus-noise)} - \text{(mean of noise alone)}}{\text{(standard deviation of noise alone)}} \quad (15.2\text{-}1)$$

An expression for ρ_v is obtained by representing the pulse-train signal amplitudes into the integrator by the sequence $s_1, s_2, s_3, \ldots, s_N$ where each signal amplitude is defined as the mean of (video) signal-plus-noise less the mean of noise alone. Let $k_1, k_2, \ldots, k_N$ be weighting factors operating on signals $s_1, s_2, \ldots, s_N$, respectively. Then, for noise of variance σ^2, it follows from (15.2-1) that ρ_v is given by

$$\rho_v = \frac{k_1 s_1 + k_2 s_2 + \cdots + k_N s_N}{(k_1^2 \sigma^2 + k_2^2 \sigma^2 + \cdots + k_N^2 \sigma^2)^{1/2}} \quad (15.2\text{-}2\text{a})$$

$$= \frac{\sum\limits_{i=1}^{N} k_i s_i}{\sigma \left(\sum\limits_{i=1}^{N} k_i^2 \right)^{1/2}} \quad (15.2\text{-}2\text{b})$$

†A one-way Gaussian antenna pattern results in a round-trip antenna pattern which is again Gaussian with a smaller beamwidth.

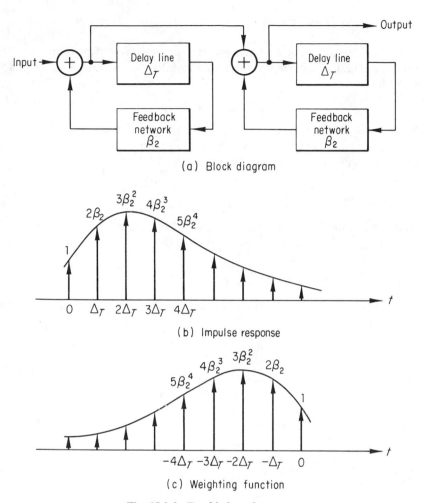

(a) Block diagram

(b) Impulse response

(c) Weighting function

Fig. 15.2-3. Double-loop integrator.

For large N, the differences between successive values of s_i and k_i are small and the sums in Eq. (15.2-2b) can be approximated by integrals. Denoting the signal amplitude function by $s(t)$ and the weighting function by $k(t)$, we can rewrite Eq. (15.2-2b) as

$$\rho_v = \frac{\dfrac{1}{\Delta_T}\displaystyle\int_{-\infty}^{\infty} k(t)s(t)dt}{\sigma\left(\dfrac{1}{\Delta_T}\displaystyle\int_{-\infty}^{\infty} k^2(t)dt\right)^{1/2}} \tag{15.2-3}$$

where Δ_T is the time interval between train pulses.

15.3. Parameter Optimization

For convenience, assume that the pulse train is symmetrical about $t = 0$; then a Gaussian signal amplitude function into the integrator can be described by

$$s(t) = s_{\max}e^{-(\alpha t)^2} \qquad (15.3\text{-}1)$$

where $s_{\max}$ denotes the maximum signal amplitude at $t = 0$ and α is a constant that depends on antenna beamwidth and scan rate. Impulse response $h_1(t)$ of a single-loop integrator is [from Fig. 15.2-2(b)]

$$h_1(t) = \begin{cases} \beta^{t/\Delta_T}, & t \geqslant 0 \\ 0, & t < 0 \end{cases} \qquad (15.3\text{-}2)$$

It follows from (15.3-2) and Fig. 15.2-2(c) that weighting function $k_1(t; t_1)$ for an integrated output at time t_1 is given by

$$k_1(t\,;\,t_1) = \begin{cases} \beta_1^{(t_1-t)/\Delta_T}, & t \leqslant t_1 \\ 0, & t > t_1 \end{cases} \qquad (15.3\text{-}3)$$

When expressions (15.3-1) and (15.3-3) are substituted into Eq. (15.2-3), ρ_v becomes

$$\rho_v(t_1) = \frac{\dfrac{s_{\max}}{\Delta_T} \displaystyle\int_{-\infty}^{t_1} \beta_1^{(t_1-t)/\Delta_T} e^{-(\alpha t)^2}\, dt}{\sigma \left(\dfrac{1}{\Delta_T} \displaystyle\int_{-\infty}^{t_1} \beta_1^{2(t_1-t)/\Delta_T}\, dt \right)^{1/2}} \qquad (15.3\text{-}4)$$

With

$$\tau = \frac{t_1 - t}{\Delta_T} \qquad (15.3\text{-}5)$$

Eq. (15.3-4) can be rewritten as

$$\rho_v(t_1) = \frac{s_{\max} \displaystyle\int_0^{\infty} \beta_1^{\tau} e^{-\alpha^2(t_1 - \Delta_T\tau)^2}\, d\tau}{\sigma \left(\displaystyle\int_0^{\infty} \beta_1^{2\tau}\, d\tau \right)^{1/2}} \qquad (15.3\text{-}6)$$

The integral in the denominator of (15.3-6), denoted by F_1, can be evaluated with the identity

$$\beta_1^{2\tau} = e^{2\tau \ln \beta_1} \qquad (15.3\text{-}7)$$

Since $(-2 \ln \beta_1)$ is positive for $\beta_1 < 1$, we have

$$F_1 = \int_0^{\infty} e^{-(-2 \ln \beta_1)\tau}\, d\tau = \frac{1}{2 \ln \beta_1} \qquad (15.3\text{-}8)$$

The integral in the numerator of (15.3-6), denoted by F_2, can be evaluated as follows:

$$F_2 = \int_0^\infty \exp\left(\ln \beta_1\right)\tau \cdot \exp\left[-(\alpha^2 t_1^2 - 2\alpha^2 \Delta_T t_1 \tau + \alpha^2 \Delta_T^2 \tau^2)\right] d\tau \qquad (15.3\text{-}9a)$$

$$= \exp\left[-(\alpha t_1)^2\right] \int_0^\infty \exp\left(-[\alpha^2 \Delta_T^2 \tau^2 - (2\alpha^2 \Delta_T t_1 + \ln \beta_1)\tau]\right) d\tau \quad (15.3\text{-}9b)$$

$$= \exp\left[-(\alpha t_1)^2\right] \int_0^\infty \exp\left[-(b_1^2 \tau^2 - b_2 \tau)\right] d\tau \qquad (15.3\text{-}9c)$$

where

$$b_1^2 = \alpha^2 \Delta_T^2 \qquad (15.3\text{-}10)$$

and

$$b_2 = 2\alpha^2 \Delta_T t_1 + \ln \beta_1 \qquad (15.3\text{-}11)$$

Let

$$y = -b_1 \tau + \frac{b_2}{2b_1} \qquad (15.3\text{-}12)$$

Then Eq. (15.3-9c) can be rewritten as

$$F_2 = \frac{e^{-(\alpha t_1)^2} e^{(b_2/2b_1)^2}}{b_1} \int_{-\infty}^{b_2/2b_1} e^{-y^2} \, dy \qquad (15.3\text{-}13)$$

With the definition of $\Phi(T)$ in Eqs. (9.6-1) and (9.6-2), F_2 can be expressed as

$$F_2 = \frac{\sqrt{\pi}}{\alpha \Delta_T} e^{-(\alpha t_1)^2} e^{[\alpha t_1 + (\ln \beta_1/2\alpha \Delta_T)]^2} \left[1 - \Phi\left(\sqrt{2}\,\alpha t_1 + \frac{\ln \beta_1}{\sqrt{2}\,\alpha \Delta_T}\right)\right] \quad (15.3\text{-}14)$$

When the integrals in Eq. (15.3-6) are replaced by (15.3-8) and (15.3-14), the expression for ρ_v becomes

$$\rho_v(t_1) = \frac{S_{\max}}{\sigma} \frac{\sqrt{-2\pi \ln \beta_1}}{\alpha \Delta_T} e^{-(\alpha t_1)^2} e^{[\alpha t_1 + (\ln \beta_1/2\alpha \Delta_T)]^2}$$

$$\cdot \left[1 - \Phi\left(\sqrt{2}\,\alpha t_1 + \frac{\ln \beta_1}{\sqrt{2}\,\alpha \Delta_T}\right)\right] \qquad (15.3\text{-}15)$$

or

$$\rho_v(\theta_1) = \frac{S_{\max}}{\sigma} \sqrt{\frac{2\pi}{\alpha \Delta_T}} \sqrt{-q_1}\, e^{-\theta_1^2}\, e^{[\theta_1 + (q_1/2)]^2} \left[1 - \Phi(\sqrt{2}\,\theta_1 + \frac{q_1}{\sqrt{2}})\right]$$

$$(15.3\text{-}16)$$

where

$$q_1 = \frac{\ln \beta_1}{\alpha \Delta_T} \qquad (15.3\text{-}17)$$

and

$$\theta_1 = \alpha t_1 \qquad (15.3\text{-}18)$$

To maximize ρ_v, Cooper and Griffiths [1] plotted ρ_v as a function of both q_1 and θ_1. They showed that ρ_v is maximized for the single-loop integrator when

$$q_1 = -0.6838 \qquad (15.3\text{-}19)$$

and

$$\theta_1 = 0.7312 \tag{15.3-20}$$

Since α and Δ_T are known for a given radar, the optimum loop gain β_1 can be calculated from (15.3-17) and (15.3-19). The optimum delay t_1, found from Eqs. (15.3-18) and (15.3-20), is the interval between the time of arrival of the peak of the pulse train amplitude function and the time at which the output of the integrator reaches its peak value.

A similar optimization can be performed for the double-loop integrator. For this case, impulse response $h_2(t)$ is [see Fig. 15.2-3(b)]

$$h_2(t) = \begin{cases} \left(\dfrac{t}{\Delta_T} + 1 \right) \beta_2^{t/\Delta_T}, & t \geqslant 0 \\ 0, & t < 0 \end{cases} \tag{15.3-21}$$

The weighting function $k_2(t; t_2)$ for an output at time t_2 can be found from (15.3-21) and Fig. 15.2-3(c) to be

$$k_2(t; t_2) = \begin{cases} \left[\dfrac{(t_2 - t)}{\Delta_T} + 1 \right] \beta_2^{(t_2-t)/\Delta_T} & t \leqslant t_2 \\ 0, & t > t_2 \end{cases} \tag{15.3-22}$$

Maximizing ρ_v is considerably simplified if the factor $[(t_2 - t)/\Delta_T] + 1$ in (15.3-22) is approximated by $(t_2 - t)/\Delta_T$; then $k_2(t; t_2)$ becomes

$$k_2(t; t_2) \cong \begin{cases} \dfrac{(t_2 - t)}{\Delta_T} \beta_2^{(t_2-t)/\Delta_T} & t \leqslant t_2 \\ 0, & t > t_2 \end{cases} \tag{15.3-23}$$

The error that results from this approximation is small when N is large. Substituting (15.3-21) and (15.3-23) into (15.2-3) yields

$$\rho_v(t_2) = \frac{\dfrac{S_{\max}}{\Delta_T} \displaystyle\int_{-\infty}^{t_2} \left(\dfrac{t_2 - t}{\Delta_T} \right) \beta_2^{(t_2-t)/\Delta_T} e^{-(\alpha t)^2} \, dt}{\sigma \left[\dfrac{1}{\Delta_T} \displaystyle\int_{-\infty}^{t_2} \left(\dfrac{t_2 - t}{\Delta_T} \right)^2 \beta_2^{2(t_2-t)/\Delta_T} \, dt \right]^{1/2}} \tag{15.3-24}$$

The integrals in (15.3-24) can be evaluated by methods analogous to those used for the single-loop integrator; the result is

$$\rho_v(t_2) = \frac{S_{\max}}{\sigma} \frac{e^{-\theta_2^2}}{\sqrt{\alpha \Delta_T}} (-q_2)^{3/2} \Bigg\{ 1 + 2\sqrt{\pi} \left(\frac{q_2}{2} + \theta_2 \right) e^{(q_2/2 + \theta_2)^2}$$
$$\cdot \left[1 - \Phi\left(\sqrt{2}\, \theta_2 + \frac{q_2}{\sqrt{2}} \right) \right] \Bigg\} \tag{15.3-25}$$

where

$$q_2 = \frac{\ln \beta_2}{\alpha \Delta_T} \tag{15.3-26a}$$

$$\theta_2 = \alpha t_2 \tag{15.3-26b}$$

By plotting Eq. (15.3-25) as a function of q_2 and θ_2, Cooper and Griffiths [1] maximized $\rho_v(t_2)$; the maximum occurs for

$$q_2 = -1.4826 \tag{15.3-27}$$

$$\theta_2 = 1.0117 \tag{15.3-28}$$

For known α and Δ_T, the optimum values of β_2 and t_2 can be found from Eqs. (15.3-26), (15.3-27), and (15.3-28).

15.4. Performance Evaluation

The performance of single- and double-loop integrators can be evaluated using an approximate procedure developed by Palmer and Cooper [2]. This procedure permits the false-alarm probability to be related to the system threshold. Once the system threshold is determined, the probability of detection is evaluated by approximating the distribution of sum signal-plus-noise at the integrator output by a Gaussian probability density. The integrator is preceded by matched filtering, square-law detection, and pulse-train sampling and weighting as shown in Fig. 15.1-1.

In this section an expression is obtained for P_{fa}. Equation (10.4-7) gives the probability density of the sampled envelope of signal-plus-noise out of the square-law detector. With $\mathscr{R}_p = 0$, the probability density $p(v_i)$ of a noise sample v_i is given by

$$p(v_i) = \begin{cases} e^{-v_i}, & v_i \geqslant 0 \\ 0, & v_i < 0 \end{cases} \tag{15.4-1}$$

Each noise sample v_i is amplitude weighted by a factor k_i, $i = 1, 2, \ldots, N$, prior to summing. If the weighted noise sample is denoted by y_i,

$$y_i = k_i v_i \tag{15.4-2}$$

the probability density of y_i can be found from (15.4-1) and (15.4-2) to be

$$p(y_i) = \frac{1}{k_i} e^{-y_i/k_i} \tag{15.4-3}$$

To calculate the probability density of sum random variable Z at the integrator output, the method of characteristic functions is used. The characteristic function $C_{y_i}(\xi)$, corresponding to density function $p(y_i)$, is [see (3.12-1b)]

$$C_{y_i}(\xi) = \int_{-\infty}^{\infty} p(y_i) e^{j\xi y_i} \, dy_i \tag{15.4-4a}$$

$$= \int_{0}^{\infty} \frac{1}{k_i} e^{-y_i/k_i} e^{j\xi y_i} \, dy_i \tag{15.4-4b}$$

$$= \frac{1}{1 - j\xi k_i} \tag{15.4-4c}$$

It follows from

$$Z = \sum_{i=1}^{N} y_i \tag{15.4-5}$$

that characteristic function $C_Z(\xi)$ of random variable Z is computed by substituting (15.4-4c) into expression (3.12-3). Since the y_i are independent, the result is

$$C_Z(\xi) = \prod_{i=1}^{N} C_{y_i}(\xi) = \prod_{i=1}^{N} \frac{1}{1 - j\xi k_i} \tag{15.4-6a}$$

$$= \exp\left[-\sum_{i=1}^{N} \ln (1 - j\xi k_i)\right] \tag{15.4-6b}$$

$C_Z(\xi)$ can be approximated by replacing the sum in (15.4-6b) by an integral. Let $k(t)$ be a continuous function which has values $k_1, k_2, \ldots, k_N$ at the sampling instants $\tau_1, \tau_2, \ldots, \tau_N$, respectively, where the interval between samples is Δ_T; that is,

$$\tau_{m+1} - \tau_m = \Delta_T, \qquad m = 1, \ldots, N-1 \tag{15.4-7}$$

The integral approximation of Eq. (15.4-6b) is

$$C_Z(\xi) \cong \exp\left\{-\frac{1}{\Delta_T} \int_0^\infty \ln\left[1 - j\xi k(\tau)\right] d\tau\right\} \tag{15.4-8}$$

Equation (15.4-8) can be written more compactly as

$$C_Z(\xi) \cong \frac{W(j\xi)}{\Delta_T} \tag{15.4-9}$$

where

$$W(j\xi) = -\int_0^\infty \ln\left[1 - j\xi k(\tau)\right] d\tau \tag{15.4-10}$$

An approximate expression for probability density function $p(Z)$ can be obtained from the anticharacteristic function of $C_Z(\xi)$, as follows:

$$p(Z) = \frac{1}{2\pi} \int_{-\infty}^{\infty} C_Z(\xi) e^{j\xi Z} d\xi \tag{15.4-11a}$$

$$\cong \frac{1}{2\pi j} \int_{-j\infty}^{j\infty} \exp\left[\frac{W(j\xi)}{\Delta_T} - j\xi Z\right] d(j\xi) \tag{15.4-11b}$$

To evaluate integral (15.4-11b), the saddle-point method of integration is employed [3]. In this approximation technique, the path of integration is modified so that only a small portion of the new path contributes significantly to the value of the integral. To evaluate an integral of the form

$$\int_C g(s) e^{vh(s)} ds$$

where $\dot{h}(s) = 0$ defines the saddle point s_0 of $h(s)$, an integration contour C_1 is sought that passes through saddle point s_0 with the properties that

(a) the imaginary part of $h(s)$ is nearly constant on C_1 in the vicinity of s_0, (b) the real part of $h(s)$ has a maximum at s_0 that decays rapidly on both sides of the saddle point, and (c) the value of the integral is unchanged by the modified contour of integration. When these conditions are satisfied, an approximate relation [3] holds for large v:

$$\int_C g(s)e^{vh(s)}\,ds = \int_{C_1} g(s)e^{vh(s)}\,ds \cong e^{vh(s_0)}\,g(s_0)\sqrt{\frac{-2\pi}{v\ddot{h}(s_0)}} \qquad (15.4\text{-}12)$$

Comparing (15.4-12) with (15.4-11b), we let $s = j\xi$, $g(j\xi) = 1$, $v = 1/\Delta_T$ and

$$h(j\xi) = [W(j\xi) - j\xi Z\Delta_T] \qquad (15.4\text{-}13)$$

The saddle point $j\xi_0$ is found by equating the derivative of $h(j\xi)$† to zero:

$$\frac{dh(j\xi)}{d(j\xi)}\bigg|_{\xi=\xi_0} = \frac{d}{d(j\xi)}[W(j\xi) - j\xi Z\Delta_T]_{\xi=\xi_0} = 0 \qquad (15.4\text{-}14)$$

which yields

$$\dot{W}(j\xi)|_{\xi=\xi_0} = \dot{W}(j\xi_0) = \Delta_T Z \qquad (15.4\text{-}15)$$

From (15.4-10), $\dot{W}(j\xi_0)$ is

$$\dot{W}(j\xi_0) = \int_0^\infty \frac{k(\tau)}{1 - j\xi_0 k(\tau)}\,d\tau \qquad (15.4\text{-}16)$$

Substituting (15.4-16) into (15.4-15) yields

$$\int_0^\infty \frac{k(\tau)}{1 - j\xi_0 k(\tau)}\,d\tau = \Delta_T Z \qquad (15.4\text{-}17)$$

By inspection the right side of Eq. (15.4-17) is real; hence the integral in (15.4-17) must be real. It follows that

$$\text{Im} \int_0^\infty \frac{k(\tau)}{1 - j\xi_0 k(\tau)}\,d\tau = 0 \qquad (15.4\text{-}18a)$$

$$\text{Im} \int_0^\infty \frac{k(\tau)[1 + j\xi_0^* k(\tau)]}{1 + |\xi_0|^2 k^2(\tau) + 2k(\tau)\,\text{Im}\,\xi_0}\,d\tau = 0 \qquad (15.4\text{-}18b)$$

Denoting saddle point ξ_0 in the ξ-plane by

$$\xi_0 = a + jb \qquad (15.4\text{-}19)$$

we can rewrite Eq. (15.4-18b) as

$$\int_0^\infty \frac{ak^2(\tau)}{1 + (a^2 + b^2)k^2(\tau) + 2bk(\tau)}\,d\tau = 0 \qquad (15.4\text{-}20)$$

For arbitrary $k(\tau)$, Eq. (15.4-20) can be satisfied only if $a = 0$. Hence, the saddle point ξ_0 is a point on the imaginary axis in the ξ-plane.

The new contour of integration C_1, which must pass through $\xi_0 = (0, b)$,

---

†The function $h(j\xi)$ can be shown to be analytic in the complex plane with a simple pole at $j\xi = 1/k(\tau)$.

is a line parallel to the real axis in the ξ-plane defined by $\xi = u + jb$. To show that contour C_1 satisfies the requirements for saddle-point integration, let any point in the ξ-plane be

$$\xi = u + jv \qquad (15.4\text{-}21)$$

Because of (15.4-14) and the analyticity of $h(j\xi)$, the derivative of the real part of $h(j\xi)$ is zero on the new contour at the saddle point—that is,

$$\left. \frac{d \operatorname{Re}[h(j\xi)]}{du} \right|_{\xi=jb} = 0 \qquad (15.4\text{-}22)$$

The real part of $h(j\xi)$ is differentiated twice with respect to u to determine whether the saddle point is a maximum or a minimum of the function $\operatorname{Re} h(j\xi)$ on the new contour; the result is

$$\left. \frac{d^2 \operatorname{Re}[h(j\xi)]}{du^2} \right|_{\xi=jb} = -\int_0^\infty \frac{k^2(\tau)}{[1 - bk(\tau)]^2} \, d\tau \qquad (15.4\text{-}23)$$

For arbitrary $k(\tau)$, (15.4-23) is always negative. Hence, the real part of $h(j\xi)$ is a maximum on C_1 at the saddle point and decreases on either side of this point on C_1; thus condition (b) for saddle-point integration (as stated above) is satisfied.

Consider the imaginary part of $h(j\xi)$ along the contour $\xi = u + jb$:

$$\operatorname{Im}[h(j\xi)]|_{\xi=u+jb} = \frac{1}{\Delta_T} \int_0^\infty \left\{ \left[\tan^{-1} \frac{uk(\tau)}{1 + bk(\tau)} \right] - \frac{uk(\tau)}{1 + bk(\tau)} \right\} d\tau \qquad (15.4\text{-}24)$$

Note that for small θ, $\tan^{-1}\theta \cong \theta$. In the vicinity of the saddle point, u is small; hence the integrand in (15.4-24) is essentially zero in this region. Since the imaginary part of $h(j\xi)$ is nearly constant on C_1 in the vicinity of ξ_0, condition (a) for saddle-point integration is satisfied. Condition (c) is also satisfied, since the change in contour from the u-axis to C_1 ($\xi = u + jb$) does not alter the value of the integral.†

From relation (15.4-12), $p(Z)$ is approximately given by

$$p(Z) \cong \frac{1}{2\pi j} \, e^{W(j\xi_0)/\Delta_T - j\xi_0 Z} \left[\frac{-2\pi\Delta_T}{\ddot{W}(j\xi_0)} \right]^{1/2} \qquad (15.4\text{-}25)$$

Replacing Z in (15.4-25) by Eq. (15.4-15) yields

$$p(Z) \cong \frac{1}{2\pi} \, e^{W(j\xi_0)/\Delta_T - j\xi_0 \dot{W}(j\xi_0)/\Delta_T} \left[\frac{2\pi\Delta_T}{\ddot{W}(j\xi_0)} \right]^{1/2} \qquad (15.4\text{-}26)$$

Note that the location of saddle point ξ_0 on the imaginary axis of the ξ-plane depends on the value of Z through Eq. (15.4-15). Equation (15.4-26) can be expressed in terms of functions of real variables by letting

$$x = j\xi_0 \qquad (15.4\text{-}27)$$

Then, Eq. (15.4-26) becomes

†Since $h(j\xi)$ contains no poles in the region between initial and final contours of integration, Cauchy's theorem on contour integration in a region of analyticity is applicable [3].

$$p(Z) \cong \frac{1}{2\pi} e^{W(x)/\Delta_T - x\dot{W}(x)/\Delta_T} \left[\frac{2\pi\Delta_T}{\ddot{W}(x)} \right]^{1/2} \tag{15.4-28}$$

Equation (15.4-28) establishes that $p(Z)$ is (approximately) determined by $W(x)$ and its first two derivatives. Variables x and Z are related by (15.4-15):

$$Z = \frac{\dot{W}(x)}{\Delta_T} \tag{15.4-29}$$

The probability of a false alarm is the probability that sum random variable Z at the integrator output exceeds a threshold Z_0. To calculate P_{fa}, Eq. (15.4-28) can be simplified by rewriting it as

$$\ln p(Z) \cong \ln \frac{1}{2\pi} + \frac{W(x)}{\Delta_T} - \frac{x\dot{W}(x)}{\Delta_T} + \frac{1}{2}\ln 2\pi\Delta_T - \frac{1}{2}\ln \ddot{W}(x) \tag{15.4-30}$$

The derivative of (15.4-30) yields

$$\frac{d\ln p(Z)}{dZ} = \frac{d\ln p(Z)}{dx} \cdot \frac{dx}{dZ} \tag{15.4-31a}$$

$$\cong -\frac{dx}{dZ}\left[\frac{x\ddot{W}(x)}{\Delta_T} + \frac{\dddot{W}(x)}{2\ddot{W}(x)} \right] \tag{15.4-31b}$$

For weighting functions $k(\tau)$ usually encountered, $\dot{W}(x)$ is strictly monotonic and differentiable, and $\ddot{W}(x) \neq 0$; then from (15.4-29),

$$\frac{dx}{dZ} = \frac{1}{dZ/dx} = \frac{\Delta_T}{\ddot{W}(x)} \tag{15.4-32}$$

With (15.4-32) substituted into (15.4-31b),

$$\frac{d\ln p(Z)}{dZ} \cong -x - \left(\frac{\Delta_T}{2} \right)\frac{\dddot{W}(x)}{[\ddot{W}(x)]^2} \tag{15.4-33}$$

The last term in (15.4-33) is small for small Δ_T. Hence, Eq. (15.4-33) can be approximated by

$$\frac{d\ln p(Z)}{dZ} \cong -x \tag{15.4-34}$$

Equation (15.4-34) states that $p(Z)$ is approximately exponential in the vicinity of Z—that is,

$$p(Z) \cong e^{-xZ} \tag{15.4-35}$$

With (15.4-35), the probability of false alarm is approximately given by

$$P_{fa} \cong \int_{Z_0}^{\infty} p(Z)dZ = \int_{Z_0}^{\infty} e^{-xZ} dZ \tag{15.4-36a}$$

$$= \frac{e^{-xZ_0}}{x} = \frac{p(Z_0)}{x} \tag{15.4-36b}$$

where x is related to Z_0 by Eq. (15.4-29); $p(Z_0)$ can be calculated from (15.4-28).

To determine the accuracy of (15.4-36b) for evaluating threshold Z_0 for a chosen value of P_{fa}, Palmer and Cooper [2] employed a digital computer to calculate Z_0 for selected values of P_{fa} for an integrator with uniform weighting. (This case is considered in Section 15.5 below.) An integrator with uniform weighting was selected, since the results could be compared with those in published tables. They found that the values of Z_0 computed by the approximate method were in error by less than one per cent for values of P_{fa} less than 10^{-2}; this is the region of greatest interest in radar applications.

15.5. Detection Probability for Uniform Integration
of Constant-Amplitude Pulses

After threshold Z_0 has been evaluated for a given P_{fa}, probability of detection P_d can be evaluated approximately by assuming that the probability density of signal plus noise at the integrator output is Gaussian. As a check on the overall accuracy of this procedure, the probability of detection is calculated for a nonfluctuating pulse train that has been processed by an integrator with uniform weighting. This case was evaluated exactly in Chapter 10, and a direct comparison of results is possible.

For this example, it is convenient to normalize pulse-train duration so that $k(\tau) = 1$ for $-\frac{1}{2} < \tau < \frac{1}{2}$ and zero elsewhere. Then for a pulse train of large N, interpulse spacing Δ_T is approximately equal to $1/N$. $W(x)$ can be found from (15.4-10) with $j\xi$ replaced by x, as follows:

$$W(x) = -\int_{-1/2}^{1/2} \ln (1 - x) \, d\tau \qquad (15.5\text{-}1a)$$

$$= -\ln (1 - x) \qquad (15.5\text{-}1b)$$

Differentiating (15.5-1b) twice with respect to x yields

$$\dot{W}(x) = \frac{1}{1 - x} \qquad (15.5\text{-}2)$$

and

$$\ddot{W}(x) = \frac{1}{(1 - x)^2} \qquad (15.5\text{-}3)$$

Substituting (15.5-2) into (15.4-29) yields

$$Z = \frac{N}{1 - x} \qquad (15.5\text{-}4)$$

Substituting (15.5-1b), (15.5-2), and (15.5-3) into (15.4-28) and the result into (15.4-36b) permits the calculation of threshold Z_0 as a function of P_{fa}.

With Z_0 computed, the probability of detection is calculated, assuming a Gaussian probability density of signal plus noise. This distribution is completely characterized by the mean and variance of sum random variable Z. The mean m_1 and variance σ_1^2 of a sample of signal plus noise at the integrator input is given by Eqs. (10.7-4) and (10.7-6) (for $N = 1$) as follows:

$$m_1 = 1 + \frac{\mathscr{R}_p}{2} \tag{15.5-5}$$

$$\sigma_1^2 = 1 + \mathscr{R}_p \tag{15.5-6}$$

where $\mathscr{R}_p$ is the rf (single pulse) peak signal-to-noise ratio. For uniform integrator weighting, the mean and variance of Z is N times the mean and variance of a single sample. Thus,

$$\bar{Z} = N\left(1 + \frac{\mathscr{R}_p}{2}\right) = \frac{1}{\Delta_T}\left(1 + \frac{\mathscr{R}_p}{2}\right) \tag{15.5-7}$$

$$\mathrm{var}\, Z = \sigma_Z^2 = N(1 + \mathscr{R}_p) = \frac{1}{\Delta_T}(1 + \mathscr{R}_p) \tag{15.5-8}$$

Then the probability of detection is given by

$$P_d = \int_{Z_0}^{\infty} \frac{1}{\sqrt{2\pi}\,\sigma_Z}\, e^{-(Z-\bar{Z})^2/2\sigma_Z^2} dZ \tag{15.5-9}$$

Computations for $P_d = 50, 90,$ and 99 per cent were performed on a digital computer with $P_{fa} = 10^{-6}$.[†] The computed results are plotted in Fig. 15.5-1 as small circles together with more exact plots obtained from

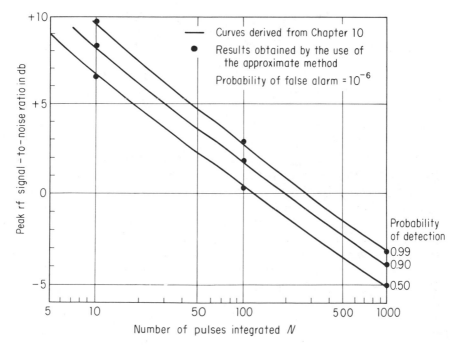

Fig. 15.5-1. Comparison of results with those of Chapter 10 (*after Palmer and Cooper* [2], *courtesy of IEEE*).

†See reference [2].

Chapter 10. Comparison of the results indicates that the overall accuracy of the approximation procedure is good. This same procedure is used next to evaluate the performance of single- and double-loop integrators.

15.6. Detection Probability for a Single-Loop Integrator Assuming a Gaussian Pulse Train Envelope

Equation (15.3-3) for the weighting function of a single-loop integrator can be simplified with

$$\tau = -\frac{t_1 - t}{\Delta_T} \ln \beta_1 \qquad (15.6\text{-}1)$$

Substituting (15.6-1) into (15.3-3) yields

$$k_1(\tau) = \begin{cases} e^{-\tau}, & \tau \geqslant 0 \\ 0, & \tau < 0 \end{cases} \qquad (15.6\text{-}2)$$

When Eq. (15.6-2) is substituted into (15.4-10), we have (for $x = j\xi$)

$$W(x) = x + \frac{x^2}{2^2} + \frac{x^3}{3^2} + \frac{x^4}{4^2} + \cdots \qquad (15.6\text{-}3)$$

which is convergent for $x < 1$. Also,

$$\dot{W}(x) = -\frac{\ln (1 - x)}{x} \qquad (15.6\text{-}4)$$

and

$$\ddot{W}(x) = \frac{\ln (1 - x)}{x^2} + \frac{1}{x(1 - x)} \qquad (15.6\text{-}5)$$

With Eqs. (15.6-3) through (15.6-5) the system threshold can be calculated as a function of false-alarm probability.

To calculate probability of detection, we approximate successive peak rf signal-to-noise ratios at the matched filter output by continuous function $\mathscr{R}_p(t)$. It follows from (15.5-5) that, after envelope detection, the mean value of signal-plus-noise samples at the integrator input is given by $1 + \mathscr{R}_p(t)/2$ as a function of time. It follows from the definition of signal $s(t)$ (see Section 15.2) as the change in mean value of the video waveform into the integrator when an rf signal is added to input noise that

$$s(t) = \left[1 + \frac{\mathscr{R}_p(t)}{2} \right] - 1 = \frac{\mathscr{R}_p(t)}{2} \qquad (15.6\text{-}6)$$

and

$$s_{\max} = \frac{\mathscr{R}_{p\text{-max}}}{2} \qquad (15.6\text{-}7)$$

Substituting (15.6-6) and (15.6-7) into Eq. (15.3-1) yields

$$\mathscr{R}_p(t) = \mathscr{R}_{p\text{-max}} \, e^{-(\alpha t)^2} \tag{15.6-8}$$

The function $\mathscr{R}_p(t)$ can be expressed in terms of τ by solving (15.6-1) for t and substituting the result into (15.6-8), which yields

$$\mathscr{R}_p(\tau) = \mathscr{R}_{p\text{-max}} \, \exp\left[-\left(\frac{\alpha\Delta_T}{\ln \beta_1}\,\tau + \alpha t_1\right)^2\right] \tag{15.6-9}$$

In terms of q_1 and θ_1, defined in Eqs. (15.3-17) and (15.3-18), (15.6-9) can be rewritten as

$$\mathscr{R}_p(\tau) = \mathscr{R}_{p\text{-max}} \, e^{-(\tau/q_1+\theta_1)^2} \tag{15.6-10}$$

The optimum values of q_1 and θ_1 for a single-loop integrator were determined in Section 15.3 as -0.6838 and 0.7312, respectively [see Eqs. (15.3-19) and (15.3-20)]. When these values are substituted into (15.6-10), Eq. (15.6-10) can be written as

$$\mathscr{R}_p(\tau) = \mathscr{R}_{p\text{-max}} \, e^{-[1.463(\tau-0.5)]^2} \tag{15.6-11}$$

The mean value of sum signal-plus-noise variate Z, at the integrator output, is found by weighting the mean of the integrator input waveform [see (15.5-5)] by function $k_1(\tau)$ [see (15.6-2)] and integrating the product. Thus,

$$\bar{Z} = \frac{1}{\Delta_T}\int_0^\infty k_1(\tau)\left[1 + \frac{\mathscr{R}_p(\tau)}{2}\right]d\tau \tag{15.6-12a}$$

$$= \frac{1}{\Delta_T}\int_0^\infty e^{-\tau}\left(1 + \frac{\mathscr{R}_{p\text{-max}}}{2}\, e^{-[1.463(\tau-0.5)]^2}\right)d\tau \tag{15.6-12b}$$

$$= \frac{1 + 0.2930\,\mathscr{R}_{p\text{-max}}}{\Delta_T} \tag{15.6-12c}$$

Similarly, the variance of Z is

$$\sigma_Z^2 = \frac{1}{\Delta_T}\int_0^\infty k^2(\tau)[1 + \mathscr{R}_p(\tau)]\,d\tau \tag{15.6-13a}$$

$$= \frac{1}{\Delta_T}\int_0^\infty e^{-2\tau}(1 + \mathscr{R}_{p\text{-max}}\, e^{-[1.463(\tau-0.5)]^2})\,d\tau \tag{15.6-13b}$$

$$= \frac{0.5 + 0.3748\,\mathscr{R}_{p\text{-max}}}{\Delta_T} \tag{15.6-13c}$$

For the assumed Gaussian distribution of Z, with mean given by (15.6-12c) and variance by (15.6-13c), the probability of detection as a function of $\mathscr{R}_{p\text{-max}}$ can be calculated for a given threshold, or for a specified P_{fa}.

Computer computations have been made [2] for a single-loop integrator, with $P_{fa} = 10^{-6}$ and $P_d = 0.5$. In Fig. 15.6-1, the value of $\mathscr{R}_{p\text{-max}}$ required to achieve this performance is plotted as a function of N, the number of pulses between the half-power points of the rf pulse train. A comparison between the curves of Figs. 15.6-1 and 15.5-1 indicates that, for $P_{fa} = 10^{-6}$, $P_d = 0.5$ and for N between 5 and 500, a single-loop integrator with a Gaussian modulated pulse train at the input requires a maximum rf signal-to-noise ratio approximately 1.1 decibels greater than the signal-to-noise ratio of an ideally integrated uniform-amplitude pulse train (of equal N).

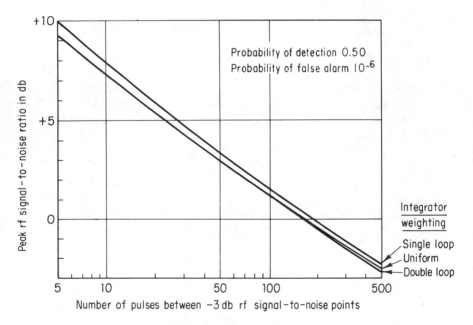

Fig. 15.6-1. Weighted integrator performance (*after Palmer and Cooper* [2], *courtesy of IEEE*).

15.7. Detection Probability for a Double-Loop Integrator Assuming a Gaussian Pulse Train Envelope

For this case, let

$$\tau = -\frac{t_2 - t}{\Delta_T} \ln \beta_2 \tag{15.7-1}$$

so that (15.3-23) becomes

$$k_2(\tau) = \begin{cases} \dfrac{1}{\ln (1/\beta_2)} \tau e^{-\tau}, & \tau \geqslant 0 \\ 0, & \tau < 0 \end{cases} \tag{15.7-2}$$

Since the constant factor $1/\ln (1/\beta_2)$ in (15.7-2) modifies both signal and noise, it can be neglected. Then double-loop weighting function $k_2(\tau)$ simplifies to

$$k_2(\tau) = \begin{cases} \tau e^{-\tau}, & \tau \geqslant 0 \\ 0, & \tau < 0 \end{cases} \tag{15.7-3}$$

Substituting Eq. (15.7-3) into (15.4-10), we find $W(x)$, $\dot{W}(x)$ and $\ddot{W}(x)$ to be

$$W(x) = \frac{1!}{1^3}x + \frac{2!}{2^4}x^2 + \frac{3!}{3^5}x^3 + \cdots \tag{15.7-4}$$

$$\dot{W}(x) = \frac{1!}{1^2} + \frac{2!}{2^3}x + \frac{3!}{3^4}x^2 + \cdots \tag{15.7-5}$$

$$\ddot{W}(x) = \frac{(1)(2!)}{2^3} + \frac{(2)(3!)}{3^4}x + \frac{(3)(4!)}{4^5}x^2 + \cdots \tag{15.7-6}$$

$W(x)$, $\dot{W}(x)$, and $\ddot{W}(x)$ are all convergent for $x < 1$. With Eqs. (15.7-4) through (15.7-6), the system threshold can be calculated for a specified false-alarm probability.

Except for the replacement of t_1 by t_2, the definitions of τ in Eqs. (15.6-1) and (15.7-1) are the same. Hence, expression (15.6-10) for $\mathscr{R}_p(\tau)$ also applies in this case. Substituting the optimum values of $q_2 = -1.4826$ and $\theta_2 = 1.0117$ for a double-loop integrator [see Eqs. (15.3-27) and (15.3-28)] into (15.6-10) yields

$$\mathscr{R}_p(\tau) = \mathscr{R}_{p\text{-max}} \, e^{-[0.674(\tau - 1.5)]^2} \tag{15.7-7}$$

The mean value of Z is given by

$$\bar{Z} = \frac{1}{\Delta_T} \int_0^\infty k_2(\tau)[1 + \mathscr{R}_p(\tau)] \, d\tau \tag{15.7-8a}$$

$$= \frac{1}{\Delta_T} \int_0^\infty \tau e^{-\tau}(1 + \mathscr{R}_{p\text{-max}} \, e^{-[0.674(\tau - 1.5)]^2}) \, d\tau \tag{15.7-8b}$$

$$= \frac{1 + 0.3298\mathscr{R}_{p\text{-max}}}{\Delta_T} \tag{15.7-8c}$$

The variance of Z is

$$\sigma_Z^2 = \frac{1}{\Delta_T} \int_0^\infty k_2^2(\tau)[1 + \mathscr{R}_p(\tau)] \, d\tau \tag{15.7-9a}$$

$$= \frac{1}{\Delta_T} \int_0^\infty \tau^2 e^{-2\tau}(1 + \mathscr{R}_{p\text{-max}} \, e^{-[0.674(\tau - 1.5)]^2}) \, d\tau \tag{15.7-9b}$$

$$= \frac{0.25 + 0.2160\mathscr{R}_{p\text{-max}}}{\Delta_T} \tag{15.7-9c}$$

With (15.7-8c) and (15.7-9c) as the parameters of the assumed Gaussian distribution of Z, P_d can be calculated as a function of system threshold. Computer calculations have been performed [2] for a double-loop integrator assuming $P_{fa} = 10^{-6}$ and $P_d = 0.5$. The results of these computations are also shown in Fig. 15.6-1.

15.8. Comparison of Single- and Double-Loop Integrators

Figure 15.6-1 shows that the maximum rf signal-to-noise ratio for a double-loop integrator is between 0.4 and 0.7 db lower than that for a single-loop integrator, for equal P_d and P_{fa}. For comparison, this figure also contains a plot of maximum signal-to-noise ratio required by an integrator with uniform weighting for a pulse train with a Gaussian envelope, assuming equal P_d and P_{fa}. The difference among all three integrators is small.

In practice, however, the difference between a single- and double-loop integrator can be significant. The optimum loop gain of a double-loop integrator is smaller than the optimum loop gain for a single-loop integrator. This is an important consideration since a feedback loop with loop gain approaching unity is difficult to stabilize. From (15.3-17), (15.3-19), (15.3-26a), and (15.3-27), the relationship between them is

$$\frac{\ln \beta_2}{\ln \beta_1} = \frac{q_2}{q_1} = 2.09 \tag{15.8-1}$$

or

$$\beta_2 = \beta_1^{2.09} \tag{15.8-2}$$

Additional flexibility is available in a double-loop integrator in that a different value of loop gain can be employed for each loop, if desired. Optimum performance is achieved in this case [1] when the geometric mean of the two loop gains is set equal to the optimum value of β_2 previously determined.

PROBLEMS

15.1. Compute the impulse response of the double-loop integrator shown in Fig. 15.2-3a and verify the result shown in Fig. 15.2-3b. Prove that the weighting function is that shown in Fig. 15.2-3c.

15.2. To verify (15.3-19) and (15.3-20), plot ρ_v of Eq. (15.3-16) as a function of θ_1 for $q_1 = -0.6838$; repeat for a value of q_1 above and below the optimum.

15.3. Evaluate the integrals in (15.3-24) and verify Eqs. (15.3-25).

15.4. Verify Eq. (15.4-23) by carrying out the indicated differentiation.

15.5. Derive Eq. (15.4-24).

15.6. Verify approximation (15.4-35).

15.7. Derive Eqs. (15.6-3), (15.6-4), and (15.6-5). Express $p(Z)$ and P_{fa} in terms of these quantities using Eqs. (15.4-28) and (15.4-36b).

15.8. Verify (15.6-12c) for mean $\bar{Z}$ by evaluating (15.6-12b). Verify (15.6-13c) for variance σ_Z^2 by evaluating (15.6-13b).

15.9. Verify Eqs. (15.7-4), (15.7-5), and (15.7-6). Express $p(Z)$ and P_{fa} in terms of these quantities using Eqs. (15.4-28) and (15.4-36b).

15.10. Verify Eqs. (15.7-8c) and (15.7-9c) by evaluating (15.7-8b) and (15.7-9b), respectively.

15.11. Plot as a function of N the increase in maximum rf signal-to-noise ratio required by a single- and double-loop integrator for a Gaussian amplitude-modulated pulse train to yield the same performance as an ideally processed constant-amplitude pulse train, using the curves in Fig. 15.6-1 for $P_d = 0.5$ and $P_{fa} = 10^{-6}$.

REFERENCES

1. Cooper, D. C., and Griffiths, J. W. R.: Video Integration in Radar and Sonar Systems, *J. Brit. IRE*, **21**: 421–433 (May, 1961).

2. Palmer, D. S., and Cooper, D. C.: An Analysis of the Performance of Weighted Integrators, *IEEE Trans. on Information Theory*, **IT-10**: (4), 296–302 (October, 1964).

3. Papoulis, A.: "The Fourier Integral and its Applications," McGraw-Hill, New York, 1962.

PART VI

SPECIAL TOPICS
IN DETECTION

In earlier chapters we considered the detection of targets for fixed integration time (number of pulses observed). Integration time is usually established by such considerations as antenna scan rate, maximum detection range, available peak and average transmitter power, and so on; these parameters, in turn, depend on the radar application. A more flexible test procedure is to vary the test length in accordance with the data received. For example, if a large signal is present, it should be possible to reduce the test period; similarly, if no signal is present, a short test should be possible.

In a sequential test, the test length is a random variable. Following each sample, one of three decisions is made: (1) terminate the test with the decision that noise alone is present; (2) terminate the test with the decision that signal plus noise is present; or (3) continue the test. The expected performance of sequential detection tests and their application to radar are discussed in Chapter 16.

The final chapter considers the optimum Bayes strategy of a surveillance radar for the detection of multiple targets. The strategy is developed by considering multiple target returns with known parameters first. The strategy is then extended to targets with unknown discrete parameters. The results obtained in this chapter relate directly to the Bayes strategy employed in Part IV of this book.

16

SEQUENTIAL
DETECTION

16.1. Introduction†

In previous chapters the total observation interval, or number of received echo pulses, is predetermined. In this chapter the number of observations required for a decision is a function of the test results obtained after each observation. Thus, the test length is a random variable; this is the principal distinguishing characteristic of sequential detection.

The theory of sequential detection is developed first from elementary concepts. A number of illustrative examples are then treated. The chapter concludes with a brief generalization to the more comprehensive risk formulation.

†Most of the material in this chapter is based on the work of Abraham Wald [1].

547

16.2. Notion of a Sequential Test

A sequential procedure for testing a simple or composite hypothesis H can be described as follows. A rule is prescribed for making one of three decisions at each stage of an experiment: accept hypothesis H, reject hypothesis H, or continue the experiment. Thus, the experiment is either terminated with the acceptance or rejection of H, or continued with an additional observation. The test length m is a random variable, since it depends on the outcome of the previous observations.

For each m, let Γ_m denote the totality of all possible samples $(v_1, v_2, \ldots, v_m)$. As shown in Fig. 16.2-1, Γ_m can be represented by an m-dimensional sample space. A decision rule is selected that subdivides this sample space into three mutually exclusive regions. After the first observation $v_1 (m = 1)$, hypothesis H is accepted if v_1 is in Γ_1'; it is rejected if v_1 is in Γ_1''; if v_1 is in region Γ_1''', the experiment is continued by making an additional observation v_2. In this case, if (v_1, v_2) falls in Γ_2', hypothesis H is accepted; if (v_1, v_2) is in region Γ_2'', the hypothesis is rejected; and so forth. The test terminates only when the sample point falls in either Γ_m' or Γ_m''.

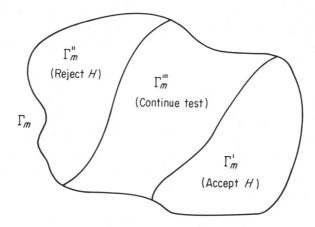

Fig. 16.2-1. Decision regions for sequential detection.

Consider a test for the presence of a signal of unknown strength in noise. For large values of signal-to-noise ratio the test can be expected to terminate with the rejection of null hypothesis H_0 after relatively few samples since our *degree of preference* for this choice is high. If signal-to-noise ratio is very small, it is reasonable to expect the null hypothesis H_0 to be selected fairly often, since our preference for noise alone is high. There will be an intermediate range of signal-to-noise values at the boundary between decision spaces where the *degree of preference* for either choice is weak. This is called the *zone of indifference*. In this region either choice could be made with only a small effect on the overall risk or penalty. In this situation, it is desirable to continue the test by making additional observations until the test is terminated with a decision in more preferred regions of sample space Γ_m. The zone of indifference in the sample space is related to practical considerations governing the costs of incorrect decisions.

To develop an optimum test, one possible approach is to select loss assignments and a criterion such as Bayes that minimizes average loss; this approach is discussed in a later section. In the following paragraphs, minimization of the average test length is employed as the optimization criterion. A test is desired that minimizes the conditional expectation of test length n with respect to the decision rule D, given the set of signal parameters $\boldsymbol{\theta}$. It has been shown by Wald and Wolfowitz [2] that regardless of error probabilities, a priori probabilities, and assigned costs, no test produces smaller values of conditional average sample numbers $E(n\,|\,0)$ and $E(n\,|\,\boldsymbol{\theta})$ than a sequential probability ratio test; it is shown later that this test also minimizes the conditional and average losses.

A sequential probability ratio test can be described as follows. Let $p_m(\mathbf{v}\,|\,\boldsymbol{\theta})$ denote the conditional probability density function of the m data samples $v_1, v_2, \ldots, v_m$ given the signal parameters $\theta_1, \theta_2, \ldots, \theta_j$; and let $p_m(\mathbf{v}\,|\,0)$ denote the probability density function of the m observed samples given that signal is absent. Then,

$$\ell_m(\mathbf{v}\,|\,\boldsymbol{\theta}) = \frac{p_m(\mathbf{v}\,|\,\boldsymbol{\theta})}{p_m(\mathbf{v}\,|\,0)} \tag{16.2-1}$$

defines a conditional probability ratio (likelihood ratio) for testing the null hypothesis H_0 against hypothesis H_1 that signal plus noise is present. Next, two positive constants $\mathscr{A}$ and $\mathscr{B}$ ($\mathscr{B} < \mathscr{A}$) are selected such that, at each stage of the experiment, if

$$\mathscr{B} < \ell_m(\mathbf{v}\,|\,\boldsymbol{\theta}) < \mathscr{A} \tag{16.2-2}$$

the experiment is continued with an additional observation; if

$$\ell_m(\mathbf{v}\,|\,\boldsymbol{\theta}) \leqslant \mathscr{B} \tag{16.2-3}$$

the test terminates with the acceptance of hypothesis H_0 (noise alone); similarly, if

$$\ell_m(\mathbf{v}\,|\,\boldsymbol{\theta}) \geqslant \mathscr{A} \tag{16.2-4}$$

the test terminates with the acceptance of hypothesis H_1. The sequential probability ratio test can be summarized as follows:

$$\text{if } \mathscr{B} < \ell_m(\mathbf{v}\,|\,\boldsymbol{\theta}) < \mathscr{A}, \qquad \text{continue test } (\mathbf{v} \text{ in } \Gamma_m''') \tag{16.2-5a}$$

$$\text{if } \mathscr{B}\, p_m(\mathbf{v}\,|\,0) \geqslant p_m(\mathbf{v}\,|\,\boldsymbol{\theta}), \qquad \text{accept } H_0 \ (\mathbf{v} \text{ in } \Gamma_m') \tag{16.2-5b}$$

$$\text{if } p_m(\mathbf{v}\,|\,\boldsymbol{\theta}) \geqslant \mathscr{A}\, p_m(\mathbf{v}\,|\,0), \qquad \text{accept } H_1 \ (\mathbf{v} \text{ in } \Gamma_m'') \tag{16.2-5c}$$

With the decision rules above, it is a simple matter to relate threshold values $\mathscr{A}$ and $\mathscr{B}$ to probability of false alarm α and probability of missed signal $\beta(\boldsymbol{\theta})$. Integrating Eqs. (16.2-5b) and (16.2-5c) over the regions Γ'_m and Γ''_m, respectively, yields

$$\mathscr{B}(1 - \alpha) \geqslant \beta(\boldsymbol{\theta}) \tag{16.2-6}$$

$$1 - \beta(\boldsymbol{\theta}) \geqslant \mathscr{A}\alpha \tag{16.2-7}$$

Inequalities (16.2-6) and (16.2-7) can be used to establish thresholds $\mathscr{A}$ and $\mathscr{B}$, given error probabilities α and $\beta(\boldsymbol{\theta})$, except for one difficulty. Since the likelihood ratio varies discretely as a function of m, an exact equality in Eqs. (16.2-6) and (16.2-7) may never occur. This is called the "excess over boundaries" problem and is discussed at length by Wald [1]. It is usually assumed that the boundaries are not exceeded by an appreciable amount, especially when m tends to be large. When the excess is neglected, it follows from (16.2-6) and (16.2-7) that

$$\mathscr{A} \cong \frac{1 - \beta(\boldsymbol{\theta})}{\alpha} \tag{16.2-8}$$

$$\mathscr{B} \cong \frac{\beta(\boldsymbol{\theta})}{1 - \alpha} \tag{16.2-9}$$

Note the simplicity with which both $\mathscr{A}$ and $\mathscr{B}$ are related to the error probabilities. Solving (16.2-8) and (16.2-9) for α and β yields

$$\alpha \cong \frac{1 - \mathscr{B}}{\mathscr{A} - \mathscr{B}} \tag{16.2-10}$$

$$\beta(\boldsymbol{\theta}) \cong \frac{\mathscr{B}(\mathscr{A} - 1)}{\mathscr{A} - \mathscr{B}} \tag{16.2-11}$$

16.3. The Operating Characteristic Function (OCF)

In this chapter only those tests are considered which eventually terminate. Wald [1] proves that the probability is unity that a sequential test will terminate if observations v_i are independent; in addition, he proves also that a sequential test will terminate with unity probability for a large class of distributions when observations v_i are not independent.

The operating characteristic function (OCF), denoted by $\mathscr{L}(\boldsymbol{\theta})$, is required for loss computations and for evaluation of the average sample number (ASN). $\mathscr{L}(\boldsymbol{\theta})$ is the conditional probability of accepting hypothesis H_0 at the end of a test, given parameters $\boldsymbol{\theta}$. It follows from this definition that

$$\mathscr{L}(0) = 1 - \alpha \tag{16.3-1}$$

$$\mathscr{L}(\boldsymbol{\theta}) = \beta(\boldsymbol{\theta}) \tag{16.3-2}$$

and, since only tests that terminate are considered, the probability that H_1

is accepted at the end of a test is

$$1 - \mathscr{L}(0) = \alpha \tag{16.3-3}$$

$$1 - \mathscr{L}(\boldsymbol{\theta}) = P_d(\boldsymbol{\theta}) \tag{16.3-4}$$

If signal-to-noise ratio χ is the only parameter involved, the OCF can be plotted as shown in Fig. 16.3-1. In this case an ideal OCF would have an impulse of unit area at $\chi = 0$, and would be zero for all $\chi > 0$.

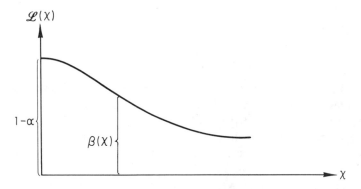

Fig. 16.3-1. The operating characteristic function (OCF).

In general a sequential ratio test is designed for a specific set of parameters denoted by *design* parameter point $\boldsymbol{\theta}_d$ in parameter space $\boldsymbol{\theta}$. However, the entire operating characteristic $\mathscr{L}(\boldsymbol{\theta})$ is required in order to compute the average sample number for all values of $\boldsymbol{\theta}$. Wald [1] developed a method for computing $\mathscr{L}(\boldsymbol{\theta})$ by use of a parametric equation. The derivation is approximate, since the "excess over boundaries" is neglected; this assumption results in negligible error for large m. Consider the expression

$$\left[\frac{p_m(\mathbf{v} \mid \boldsymbol{\theta}_d)}{p_m(\mathbf{v} \mid 0)} \right]^h \tag{16.3-5}$$

where $h = h(\boldsymbol{\theta}, \boldsymbol{\theta}_d)$ is a real number such that the expected value of (16.3-5) is equal to unity with $h \neq 0$—that is,

$$\int_{-\infty}^{\infty} \left[\frac{p_m(\mathbf{v} \mid \boldsymbol{\theta}_d)}{p_m(\mathbf{v} \mid 0)} \right]^h \cdot p_m(\mathbf{v} \mid \boldsymbol{\theta}) \, dv = 1 \tag{16.3-6}$$

It follows that the integrand in (16.3-6), which we denote by

$$f_m^*(\mathbf{v}) = \left[\frac{p_m(\mathbf{v} \mid \boldsymbol{\theta}_d)}{p_m(\mathbf{v} \mid 0)} \right]^h \cdot p_m(\mathbf{v} \mid \boldsymbol{\theta}) \tag{16.3-7}$$

is a distribution of $\mathbf{v}$. Consider first the case where $h > 0$. Let H denote the hypothesis that $p_m(\mathbf{v} \mid \boldsymbol{\theta})$ is the distribution of $\mathbf{v}$, and H^* the hypothesis that $f_m^*(\mathbf{v})$ is the distribution of $\mathbf{v}$. Consider a sequential probability ratio test with the following rules: (a) continue taking observations when

$$\mathscr{B}^h < \frac{f_m^*(\mathbf{v})}{p_m(\mathbf{v} \mid \boldsymbol{\theta})} < \mathscr{A}^h \qquad (16.3\text{-}8)$$

(b) accept hypothesis H^* when the ratio is equal to or greater than $\mathscr{A}^h$; (c) accept hypothesis H when the ratio is equal to or less than $\mathscr{B}^h$. From (16.3-7),

$$\frac{f_m^*(\mathbf{v})}{p_m(\mathbf{v} \mid \boldsymbol{\theta})} = \left[\frac{p_m(\mathbf{v} \mid \boldsymbol{\theta}_d)}{p_m(\mathbf{v} \mid 0)} \right]^h \qquad (16.3\text{-}9)$$

so that (16.3-8) can be rewritten as

$$\mathscr{B} < \frac{p_m(\mathbf{v} \mid \boldsymbol{\theta}_d)}{p_m(\mathbf{v} \mid 0)} < \mathscr{A} \qquad (16.3\text{-}10)$$

The ratios for acceptance of H or H^* can be similarly modified.

If the test between H^* and H results in the acceptance of H^*, (16.3-10) implies the acceptance of H_1; likewise the acceptance of H corresponds to the acceptance of H_0. It follows that $\mathscr{L}(0)$, the probability of accepting H_0, given $\boldsymbol{\theta} = 0$, is the same as $\mathscr{L}(\boldsymbol{\theta})$, the probability of accepting H when $f_m^*(\mathbf{v})$ is the true distribution. To calculate $\mathscr{L}(\boldsymbol{\theta})$ let α' and β' represent the error probabilities for test H^* versus H. Then, as in (16.2-8) through (16.2-11),

$$\mathscr{A}^h = \frac{1 - \beta'}{\alpha'} \qquad (16.3\text{-}11)$$

$$\mathscr{B}^h = \frac{\beta'}{1 - \alpha'} \qquad (16.3\text{-}12)$$

and

$$\alpha' = \frac{1 - \mathscr{B}^h}{\mathscr{A}^h - \mathscr{B}^h} \qquad (16.3\text{-}13)$$

$$\beta' = \frac{\mathscr{B}^h(\mathscr{A}^h - 1)}{\mathscr{A}^h - \mathscr{B}^h} \qquad (16.3\text{-}14)$$

From Eqs. (16.3-1) and (16.3-8),

$$\mathscr{L}(\boldsymbol{\theta}) = 1 - \alpha' \qquad (16.3\text{-}15)$$

and from (16.3-13),

$$\mathscr{L}(\boldsymbol{\theta}) = \frac{\mathscr{A}^h - 1}{\mathscr{A}^h - \mathscr{B}^h} = \frac{\left(\dfrac{1 - \beta'}{\alpha'} \right) - 1}{\left(\dfrac{1 - \beta'}{\alpha'} \right) - \left(\dfrac{\beta'}{1 - \alpha'} \right)} \qquad (16.3\text{-}16)$$

Equation (16.3-16) is the desired operating-characteristic function subject to condition (16.3-6), which h must satisfy. The case for $h < 0$ can be treated in a similar manner and also results in Eq. (16.3-16). It is of interest to note that from (16.3-6) $h(0, \boldsymbol{\theta}_d) = 1$ and $h(\boldsymbol{\theta}_d, \boldsymbol{\theta}_d) = -1$. The solution for h and its usefulness will become apparent from the examples in the next few sections.

16.4. The Average Sample Number (ASN)

An indirect procedure is employed to evaluate the average sample number. Let the logarithm of likelihood ratio $\ell_m(\mathbf{v} \mid \boldsymbol{\theta})$, defined in Eq. (16.2-1), be denoted by Z_m.† For $\boldsymbol{\theta} = \boldsymbol{\theta}_d$,

$$Z_m = \log \left[\frac{p_m(\mathbf{v} \mid \boldsymbol{\theta}_d)}{p_m(\mathbf{v} \mid 0)} \right] \tag{16.4-1}$$

The test procedure in terms of Z_m is: (a) continue taking samples when $\log \mathscr{B} < Z_m < \log \mathscr{A}$; (b) accept H_0 when $Z_m \leqslant \log \mathscr{B}$; and (c) accept H_1 when $Z_m \geqslant \log \mathscr{A}$.

The average value of Z_m at test termination is approximately given by (neglecting excess over boundaries) $\log \mathscr{B}$ times the probability of accepting H_0 plus $\log \mathscr{A}$ times the probability of not accepting H_0. Thus, for a terminated test of length n, given $\boldsymbol{\theta}$,

$$E(Z_n \mid \boldsymbol{\theta}) = \mathscr{L}(\boldsymbol{\theta}) \log \mathscr{B} + [1 - \mathscr{L}(\boldsymbol{\theta})] \log \mathscr{A} \tag{16.4-2}$$

For independent observations,

$$Z_n = \sum_{i=1}^{n} z_i$$

where z_i denotes the logarithm of the likelihood ratio of the ith sample:

$$z_i = \log \left[\frac{p(v_i \mid \boldsymbol{\theta}_d)}{p(v_i \mid 0)} \right] \tag{16.4-3}$$

Let $\bar{n}_\theta = E(n \mid \boldsymbol{\theta})$ denote the average number of observations required to terminate the test (average sample number). Then, for independent observations, and assuming $E(z_i \mid \boldsymbol{\theta}) = E(z \mid \boldsymbol{\theta})$ for all i,

$$E(Z_n \mid \boldsymbol{\theta}) = E \left(\sum_{i=1}^{n} z_{i \mid \theta} \right) = \bar{n}_\theta E(z \mid \boldsymbol{\theta}) \tag{16.4-4}$$

From (16.4-2) and (16.4-4), the average sample number (ASN) is given by,

$$\bar{n}_\theta = \frac{\mathscr{L}(\boldsymbol{\theta}) \log \mathscr{B} + [1 - \mathscr{L}(\boldsymbol{\theta})] \log \mathscr{A}}{E(z \mid \boldsymbol{\theta})} \tag{16.4-5}$$

When the observations are correlated, it is possible to obtain a relation for $\bar{n}_\theta$ from Eq. (16.4-2), provided $\bar{Z}_n$ is expressible as a linear function of $\bar{n}_\theta$[3].

The following test for the presence of a positive constant voltage in noise illustrates the advantages of sequential detection. Let Gaussian random variable v_i be independent with mean value $\bar{v} = \boldsymbol{\theta}_d$ for all i when signal is

---

†Z_m is a test statistic in a sequential test based on a set of design parameters $\boldsymbol{\theta}_d$.

present and $\bar{v}_i = 0$ for signal absent; the variance of v_i equal to σ^2 for all i. Then, for signal present,

$$p(v_i \mid \bar{v}, \sigma^2) = \frac{1}{\sqrt{2\pi\sigma^2}} \exp\left[-\frac{(v_i - \bar{v})^2}{2\sigma^2}\right] \qquad (16.4\text{-}6)$$

and

$$z_i = \frac{v_i \bar{v}}{\sigma^2} - \frac{(\bar{v})^2}{2\sigma^2} \qquad (16.4\text{-}7)$$

Hence,

$$E(z \mid 0, \sigma^2) = -\frac{(\bar{v})^2}{2\sigma^2} \qquad (16.4\text{-}8a)$$

$$E(z \mid \bar{v}, \sigma^2) = \frac{(\bar{v})^2}{2\sigma^2} \qquad (16.4\text{-}8b)$$

so that

$$E(n \mid 0) = \frac{(1 - \alpha) \log \mathcal{B} + \alpha \log \mathcal{A}}{-(\bar{v})^2/2\sigma^2} \qquad (16.4\text{-}9)$$

and

$$E(n \mid \bar{v}) = \frac{\beta \log \mathcal{B} + (1 - \beta) \log \mathcal{A}}{(\bar{v})^2/2\sigma^2} \qquad (16.4\text{-}10)$$

To obtain Eqs. (16.4-9) and (16.4-10), Eqs. (16.3-1) and (16.3-2) were substituted into (16.4-5). It can be shown [1] that the assumptions leading to the neglect of excess over boundaries in Eqs. (16.2-8) and (16.2-9) are conservative; that is, the true average sample numbers will never be greater than (16.4-9) and (16.4-10).

Average sequential test length $\bar{n}_\theta$ for the example above is compared next with length N of a fixed-duration, most powerful test which has the same probabilities of error α and β. For $\bar{v} > 0$, it was shown in Chapter 8 that the most powerful test is a likelihood-ratio test and the test statistic is sum random variable

$$y = \sum_{i-1}^{N} v_i$$

The conditional probability density function of y is given by

$$p(y \mid N, \sigma^2, \bar{v}) = \frac{1}{\sqrt{2\pi N\sigma^2}} \exp\left[-\frac{(y - N\bar{v})^2}{2N\sigma^2}\right] \qquad (16.4\text{-}11)$$

The probability of false alarm α can be written as

$$\alpha = \int_T^\infty p(y \mid N, \sigma^2, 0)\, dy = \Phi\left(\frac{T}{\sigma\sqrt{N}}\right) \qquad (16.4\text{-}12)$$

where $\Phi(\ )$ is defined in Eq. (9.6-2). The probability of detection is given by

$$P_d = 1 - \beta = \int_T^\infty p(y \mid N, \sigma^2, \bar{v})\, dy = \Phi\left(\frac{T - N\bar{v}}{\sigma\sqrt{N}}\right) \qquad (16.4\text{-}13)$$

Equations (16.4-12) and (16.4-13) can be rewritten as [using definitions (9.6-2) and (9.6-3)]

$$T = \sigma\sqrt{N}\ \Phi^{-1}(\alpha) \qquad (16.4\text{-}14)$$

$$T = \sigma\sqrt{N}\ \Phi^{-1}(1 - \beta) + N\bar{v} \qquad (16.4\text{-}15)$$

from which

$$N = \frac{[\Phi^{-1}(\alpha) - \Phi^{-1}(1 - \beta)]^2}{(\bar{v})^2/\sigma^2} \qquad (16.4\text{-}16)$$

The ratio of ASN to fixed test length N is given by

$$\frac{E(n\,|\,0)}{N} = -2\frac{[(1 - \alpha)\log \mathscr{B} + \alpha \log \mathscr{A}]}{[\Phi^{-1}(\alpha) - \Phi^{-1}(1 - \beta)]^2} \qquad (16.4\text{-}17)$$

$$\frac{E(n\,|\,\bar{v})}{N} = 2\frac{[\beta \log \mathscr{B} + (1 - \beta)\log \mathscr{A}]}{[\Phi^{-1}(\alpha) - \Phi^{-1}(1 - \beta)]^2} \qquad (16.4\text{-}18)$$

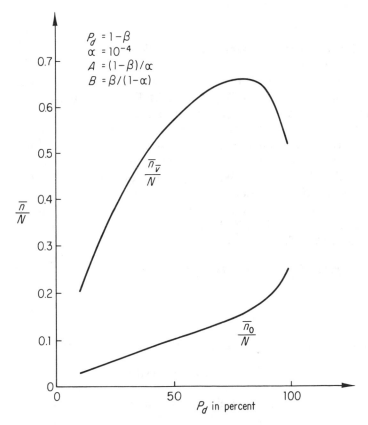

Fig. 16.4-1. Ratio of average sample number (ASN) to fixed-sample test length versus probability of detection.

For small $\alpha \, (< 10^{-4})$ and $0.1 < \beta < 0.9$, Eqs. (16.4-17) and (16.4-18) reduce to

$$\frac{\bar{n}_0}{N} \simeq - \frac{2 \log \beta}{[\Phi^{-1}(\alpha) - \Phi^{-1}(1 - \beta)]^2} \tag{16.4-19}$$

$$\frac{\bar{n}_{\bar{v}}}{N} \simeq \frac{2[\beta \log \beta + (1 - \beta) \log \mathscr{A}]}{[\Phi^{-1}(\alpha) - \Phi^{-1}(1 - \beta)]^2} \tag{16.4-20}$$

Equations (16.4-19) and (16.4-20) are plotted in Fig. 16.4-1 (for $\alpha = 10^{-4}$). The average sequential test length is smaller than the fixed sample test length in all cases: when signal is present, the average sequential test length in the region of $P_d = 50$ per cent ($\beta = 0.5$) is approximately 3/5 the length of a fixed sample test; for no signal the average sequential test length is about 1/10 as long. Since a search radar normally operates under *no-signal* conditions most of the time, there is a large potential reduction in test length possible with sequential detection; this is discussed in later sections of this chapter.

16.5. Coherent Detection in White Gaussian Noise

In this section the detection of a long coherent uniform pulse train in stationary white Gaussian noise is considered.† Except for pulse train amplitude, all signal parameters including rf phase are assumed known. This somewhat artificial case is not too different in performance from the coherent detection of a signal train with unknown initial phase (see Chapter 9). Let received waveform sample values at times t_i, corresponding to the ith observation, be given by $v(t_i) = v_i$. When signal is present, each pulse in the train has unknown signal amplitude A at the sampling instants. The noise is assumed to be Gaussian with zero mean and variance σ^2. In addition, at the sampling instants the noise variates are assumed to be uncorrelated—that is, $\overline{n_i n_j} = 0$ for $i \neq j$. Then

$$p_m(\mathbf{v} \,|\, A) = \prod_{i=1}^{m} p(v_i \,|\, A) = [p(v_i \,|\, A)]^m \tag{16.5-1a}$$

$$= (2\pi\sigma^2)^{-m/2} \exp \left[-\frac{\sum\limits_{i=1}^{m} (v_i - A)^2}{2\sigma^2} \right] \tag{16.5-1b}$$

It follows from (16.5-1a) that Eq. (16.3-6) can be rewritten as

$$\left[\int_{-\infty}^{\infty} \left[\frac{p(v_i \,|\, A_d)}{p(v_i \,|\, 0)} \right]^h \cdot p(v_i \,|\, A) \, dv_i \right]^m = 1 \tag{16.5-2}$$

where A_d is the signal amplitude for which the test has been designed.

---

†The pulse-train duration is assumed to be as long as required for a sequential test. The question of truncation is considered in Section 16.12.

Substituting Eq. (16.5-1b), with $m = 1$, into Eq. (16.5-2) permits h to be evaluated. Thus,

$$\frac{1}{\sqrt{2\pi\sigma^2}} \int_{-\infty}^{\infty} \left[\exp\left(-\frac{A_d^2 - 2A_d v_i}{2\sigma^2} \right) \right]^h \exp\left[-\frac{(v_i - A)^2}{2\sigma^2} \right] dv_i = 1 \qquad (16.5\text{-}3)$$

After integration (16.5-3) becomes

$$\exp\left[-\frac{h(h-1)A_d^2 + 2AA_d h}{2\sigma^2} \right] = 1 \qquad (16.5\text{-}4)$$

Hence, h must satisfy

$$(h^2 - h)A_d^2 + 2AA_d h = 0 \qquad (16.5\text{-}5)$$

which, in addition to $h = 0$, has the solution:

$$h = 1 - \frac{2A}{A_d} \qquad (16.5\text{-}6)$$

With (16.5-6) substituted into (16.3-16) the operating-characteristic function (OCF) is given by

$$\mathscr{L}(A) = \frac{\left(\dfrac{1 - \beta}{\alpha} \right)^{1 - 2A/A_d} - 1}{\left(\dfrac{1 - \beta}{\alpha} \right)^{1 - 2A/A_d} - \left(\dfrac{\beta}{1 - \alpha} \right)^{1 - 2A/A_d}} \qquad (16.5\text{-}7)$$

Note that for a coherent detector the OCF in (16.5-7) is independent of m, the number of observations. Equation (16.5-7) is plotted in Fig. 16.5-1 for typical values of α and β.

To find the average sample number (ASN), Eq. (16.5-1) is substituted

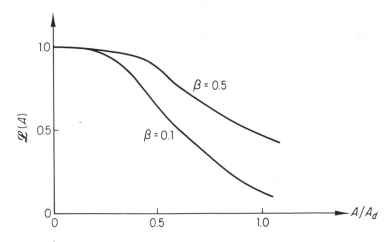

Fig. 16.5-1. OCF of coherent detector.

into Eq. (16.4-3), which yields

$$z_i = \frac{v_i A_d}{\sigma^2} - \frac{A_d^2}{2\sigma^2} = \frac{v_i A_d}{\sigma^2} - \frac{\chi_d}{2} \qquad (16.5\text{-}8)$$

where $\chi_d = A_d^2/\sigma^2$ denotes the design (peak) signal-to-noise ratio. It follows from (16.5-8) that

$$E(z_i \,|\, A) = \frac{A A_d}{\sigma^2} - \frac{\chi_d}{2} \qquad (16.5\text{-}9)$$

and

$$E(z_i \,|\, 0) = -\frac{\chi_d}{2} \qquad (16.5\text{-}10)$$

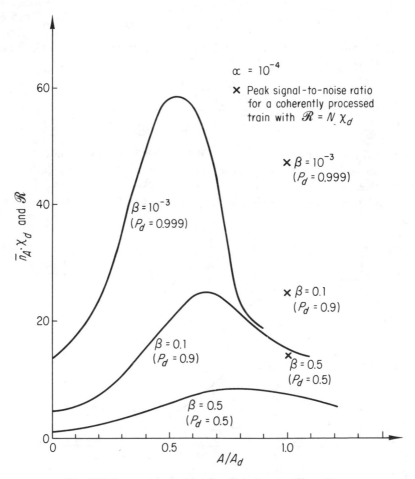

Fig. 16.5-2. $\bar{n}_A \cdot \chi_d$ and $\mathscr{R}$ versus A/A_d for a coherent detector.

With Eqs. (16.5-7), (16.5-9) and (16.5-10), the average sample number in Eq. (16.4-5) becomes (for small α)

$$\bar{n}_A = \frac{\mathscr{L}(A) \log \beta + [1 - \mathscr{L}(A)] \log \dfrac{1 - \beta}{\alpha}}{-\dfrac{\chi_d}{2}\left(1 - 2\dfrac{A}{A_d}\right)} \tag{16.5-11}$$

Note that the ASN is inversely proportional to design signal-to-noise ratio.

Equation (16.5-11) is plotted in Fig. 16.5-2 for $\alpha = 10^{-4}$ and several values of β. Note that the ASN peaks for signals of roughly half the design signal amplitude; in this region there is no pronounced tendency to cross the upper or lower thresholds. Observe also that the average sample number increases when the probability of detection is increased.

The cross marks on the figure denote the expected performance of a coherent fixed-sample test of length N with peak signal-to-noise ratio $\mathscr{R} = N\chi_d$ and the same error probabilities; these points are obtained from the curves in Chapter 9. Observe that average test length $\bar{n}_A$ is, in general, less than the duration of a fixed-sample test except for large P_d, where the ASN exceeds the fixed test length in the region of $A/A_d = 0.5$. In a later section the variance of n_θ is evaluated in order to reveal the tendency for excessively long test runs to occur. This tendency can be avoided by a process called truncation, which is discussed in Section 16.12.

For no signal ($A = 0$) the ratio of $\bar{n}_0/N$ is approximately $1:10$ for $P_d = 0.5$ and $1:3$ for $P_d = 0.999$. In the latter case the benefit to be derived from sequential testing is not as great.

16.6. Coherent Detection in Colored Gaussian Noise

The results of Section 16.5 can be extended to stationary colored Gaussian noise where the noise samples are, in general, correlated at the sampling instants. These results are applicable to the detection of signals in certain types of clutter.

From Section 3.14

$$p_m(\mathbf{v} \mid A) = \frac{\exp\left[-\frac{1}{2}(\mathbf{v} - A\mathbf{1})^T \Lambda^{-1}(\mathbf{v} - A\mathbf{1})\right]}{(2\pi)^{m/2} |\Lambda|^{1/2}} \tag{16.6-1}$$

where Λ is the noise covariance matrix with elements $\lambda_{ij} = \overline{n_i n_j}$, Λ^{-1} and $|\Lambda|$ are respectively the inverse and determinant of the covariance matrix, and $\mathbf{1}$ is a matrix consisting of a column of 1's. Equation (16.6-1) can also be expressed as

$$p_m(\mathbf{v} \mid A) = \frac{\exp\left[-\dfrac{1}{2|\Lambda|} \sum_{i,j}^{m} |\Lambda|_{ij}(v_i - A)(v_j - A)\right]}{(2\pi)^{m/2} |\Lambda|^{1/2}} \tag{16.6-2}$$

where $|\Lambda|_{ij}$ is the cofactor of element λ_{ij} in the covariance matrix. With the normalization $a = A/\sigma$ and $u_i = v_i/\sigma$, where σ^2 is the noise variance, Eq. (16.6-2) becomes

$$p_m(\mathbf{u}\,|\,A) = \frac{\exp\left[-\frac{1}{2|\boldsymbol{\rho}|}\sum_{i,j}^{m}|\boldsymbol{\rho}|_{ij}(u_i - a)(u_j - a)\right]}{(2\pi)^{m/2}|\boldsymbol{\rho}|^{1/2}} \tag{16.6-3}$$

where the elements of $\boldsymbol{\rho}$ are given by $\rho_{ij} = E[(u_i - a)(u_j - a)]$, $i \neq j$, $\rho_{ii} = 1$ for all i, and $|\boldsymbol{\rho}|_{ij}$ is the cofactor of element ρ_{ij}. If (16.6-3) is substituted into Eq. (16.3-6), the following constraint on h is obtained:

$$\exp\left[-\sum_{i,j}^{m}\frac{|\boldsymbol{\rho}|_{ij}a_d h(a_d - 2a - ha_d)}{2|\boldsymbol{\rho}|}\right] = 1 \tag{16.6-4}$$

Solving (16.6-4) yields the same relation for h obtained previously, namely,

$$h = 1 - \frac{2a}{a_d} \tag{16.6-5}$$

Thus, for white or colored Gaussian noise, the OCF is independent of the noise correlation matrix and test length m. Since the correlation matrix is not involved, the sampling instants can be spaced arbitrarily close; in the limit the results apply to the continuous case.

Substituting (16.6-3) into Eq. (16.4-1) yields the test statistic Z_m:

$$Z_m = -\frac{1}{2}\sum_{i,j}^{m}\frac{|\boldsymbol{\rho}|_{ij}}{|\boldsymbol{\rho}|}[a_d^2 - a_d(u_i + u_j)] \tag{16.6-6}$$

from which it follows that, at termination,

$$E(Z_n\,|\,a) = -\frac{1}{2}(a_d^2 - 2aa_d)E\left(\sum_{i,j}^{n}\frac{|\boldsymbol{\rho}|_{ij}}{|\boldsymbol{\rho}|}\right) \tag{16.6-7}$$

Also, from Eq. (16.4-2),

$$E(Z_n\,|\,a) = \mathscr{L}(a)\log\mathscr{B} + [1 - \mathscr{L}(a)]\log\mathscr{A} \tag{16.6-8}$$

The average sample number $\bar{n}_a$ can be found by equating Eqs. (16.6-7) and (16.6-8) if and only if the summation of the elements of the inverse matrix appearing in (16.6-7) is a simple function of n.

To illustrate the above, consider noise with exponential autocorrelation function $\phi_{nn}(\tau) = \exp(-\alpha|\tau|)$ for $\alpha > 0$. This type of noise describes a stationary Markoff process. Such noise can be generated by passing white noise through an *RC* filter; it also corresponds to the envelope of narrow-band noise that has been filtered by an *RLC* network. If all samples are assumed to be taken at equal time intervals $\Delta t = T/m$, the correlation matrix of the noise is given by

$$\rho = \begin{bmatrix} 1 & \nu & \nu^2 & \cdots & & & \nu^{n-1} \\ \nu & 1 & \nu & \cdots & & & \nu^{n-2} \\ \nu^2 & \nu & 1 & \cdots & & & \\ \cdot & & & & & & \\ \cdot & & & & & & \nu^2 \\ \nu^{n-2} & \cdots & & & & 1 & \nu \\ \nu^{n-1} & \nu^{n-2} & & \cdots & \nu^2 & \nu & 1 \end{bmatrix} \qquad (16.6\text{-}9)$$

where $\nu = \exp(-\alpha\Delta t)$. The inverse matrix is given by [4]

$$\rho^{-1} = \frac{1}{1-\nu^2} \begin{bmatrix} 1 & -\nu & 0 & 0 & \cdots & & & & 0 \\ -\nu & 1+\nu^2 & -\nu & 0 & \cdots & & & & 0 \\ 0 & -\nu & 1+\nu^2 & -\nu & 0 & \cdots & & & \\ \cdot & & & & & & & & \\ \cdot & & & & & & & & 0 \\ 0 & \cdots & & & & 0 & -\nu & 1+\nu^2 & -\nu \\ 0 & \cdots & & & & 0 & 0 & -\nu & 1 \end{bmatrix}$$

$$(16.6\text{-}10)$$

where the elements of the inverse matrix correspond to $|\rho|_{ij}/|\rho|$ in Eqs. (16.6-6) and (16.6-7). From (16.6-10) and (16.6-7),

$$E(Z_n \mid a) = -\frac{1}{2}(a_d^2 - 2aa_d)\left(\frac{\bar{n}_a[1-\nu] + 2\nu}{1+\nu}\right) \qquad (16.6\text{-}11)$$

Equating (16.6-11) and (16.6-8) yields, for large $\bar{n}_a$,

$$\bar{n}_a \cong \frac{\mathscr{L}(a)\log\mathscr{B} + [1-\mathscr{L}(a)]\log\mathscr{A}}{-\dfrac{a_d^2}{2}\left(1-\dfrac{2a}{a_d}\right)}\cdot\frac{(1+\nu)}{(1-\nu)} \qquad (16.6\text{-}12)$$

For $\nu = 0$ (zero correlation) Eq. (16.6-12) reduces to Eq. (16.5-11) (a_d^2 is equal to peak signal-to-noise ratio χ_d). For $\nu = 1$ (unity correlation) the ASN approaches infinity; this results from the fact that successive samples yield little new information. A plot of Eq. (16.6-12) is similar to that shown in Fig. 16.5-2 with a different scale factor for the ordinate.

A solution in the continuous case can be obtained as follows. Substituting (16.6-10) into (16.6-6) yields

$$Z_m = -\frac{1}{2(1-\nu^2)}[2a_d^2 + (1+\nu^2)(m-2)a_d^2 - 2a_d u_1 - 2a_d u_m$$

$$- 2a_d(1+\nu^2)\sum_{i=2}^{m-1} u_i - 2\nu(m-1)a_d^2 + 2\nu a_d(u_1+u_m) + 4\nu a_d \sum_{i=2}^{m-1} u_i]$$

$$(16.6\text{-}13)$$

For $m \gg 1$ and with end effects neglected, Eq. (16.6-13) simplifies to

$$Z_m = -\frac{(1-\nu)}{2(1+\nu)} \sum_{i=2}^{m-1} (a_d^2 - 2a_d u_i) \qquad (16.6\text{-}14)$$

For small Δt between samples, $\nu \cong 1 - \alpha\,\Delta t$; with this approximation Eq. (16.6-14) can be rewritten as

$$Z_m \cong -\frac{\alpha}{4} \sum_{i=2}^{m} (a_d^2 - 2a_d u_i)\Delta t \qquad (16.6\text{-}15)$$

Since test duration $T = m\,\Delta t$, in the limit as $\Delta t \to 0$, the following continuous solution [3] is obtained from (16.6-15):

$$Z_T \cong -\frac{\alpha a_d^2}{4} \int_0^T \left[1 - 2\frac{u(t)}{a_d}\right] dt \qquad (16.6\text{-}16)$$

This solution applies to a constant signal $s(t) = 1$; for arbitrary $s(t)$, Z_T is given by [3]

$$Z_T \cong -\frac{a_d^2}{4\alpha} \int_0^T \left[s(t) - \frac{2u(t)}{a_d}\right]\left[\alpha^2 s(t) - \ddot{s}(t)\right] dt \qquad (16.6\text{-}17)$$

In Eqs. (16.6-16) and (16.6-17), integration period T is a random variable. The test continues until either of the thresholds $\log \mathscr{A}$ or $\log \mathscr{B}$ is crossed.

16.7. Wald's Fundamental Identity

To find the characteristic function of test length n (from which the distribution of n can be obtained), a fundamental identity due to Wald is employed. For z_i defined in Eq. (16.4-3), moment-generating function $M_{z_i}(w)$ is defined as

$$M_{z_i}(w) = E[e^{z_i w}] \qquad (16.7\text{-}1)$$

which exists and is finite for w in some subset D' of the complex plane. This function is similar to the characteristic function defined in Eq. (3.12-1a) and is useful for finding the moments of a random variable. The moment generating function of a sum of independent random variables (with identical statistics),

$$Z_l = \sum_{i=1}^{l} z_i$$

is given by

$$M_{Z_l} = E[e^{Z_l w}] = [M_{z_i}(w)]^l \qquad (16.7\text{-}2)$$

For fixed N, consider the identity

$$E[e^{Z_N w}] = E[e^{Z_n w + (Z_N - Z_n)w}] \qquad (16.7\text{-}3a)$$

$$= [M_{z_i}(w)]^N \qquad (16.7\text{-}3b)$$

where n is a random variable; Eq. (16.7-3b) follows from (16.7-1) and (16.7-2). Equations (16.7-3) can be rewritten as

$$P(n \leqslant N)E[e^{Z_n w + (Z_N - Z_n)w} \,|\, n \leqslant N] + P(n > N)E[e^{Z_N w} \,|\, n > N] = [M_{z_i}(w)]^N$$
$$(16.7\text{-}4)$$

where $P(n \leqslant N)$ is the probability that $n \leqslant N$, and the expectations are conditional based on $n \leqslant N$ or $n > N$, respectively. With (16.7-2) and noting that, in the subpopulation defined by N, random variable $(Z_N - Z_n) = z_{n+1} + \ldots + z_N$ is independent of $Z_n = z_1 + \ldots + z_n$, the following relation can be derived:

$$E[e^{Z_n w + (Z_N - Z_n)w} \,|\, n \leqslant N] = E\{(e^{Z_n w})[M_{z_i}(w)]^{N-n} \,|\, n \leqslant N\} \quad (16.7\text{-}5)$$

The moment-generating function in (16.7-5) must remain within the brackets, since the expectation with respect to random variable n has not yet been performed. Substituting (16.7-5) into (16.7-4) yields

$$P(n \leqslant N)E\{e^{Z_n w}[M_{z_i}(w)]^{N-n} \,|\, n \leqslant N\}$$
$$+ [1 - P(n \leqslant N)]E[e^{Z_N w} \,|\, n > N] = [M_{z_i}(w)]^N \quad (16.7\text{-}6)$$

Dividing both sides of (16.7-6) by $[M_{z_i}(w)]^N$ results in

$$P(n \leqslant N)E\{e^{Z_n w}[M_{z_i}(w)]^{-n} \,|\, n \leqslant N\}$$
$$+ \frac{[1 - P(n \leqslant N)]E[e^{Z_N w} \,|\, n > N]}{[M_{z_i}(w)]^N} = 1 \quad (16.7\text{-}7)$$

Let D'' be the subset of the complex plane in which $|M_{z_i}(w)| \geqslant 1$ and let D be the intersection of subsets D' and D'' ($D = D' \cap D''$). Since

$$\lim_{N \to \infty} [1 - P(n \leqslant N)] = 0$$

and since $|E[e^{Z_N w} \,|\, n > N]|$ is a bounded function of N, it follows that in subset D,

$$\lim_{N \to \infty} [1 - P(n \leqslant N)] \frac{E[e^{Z_N w} \,|\, n > N]}{[M_{z_i}(w)]^N} = 0 \quad (16.7\text{-}8)$$

and

$$\lim_{N \to \infty} P(n \leqslant N)E\{e^{Z_n w}[M_{z_i}(w)]^{-n} \,|\, n \leqslant N\} = E\{e^{Z_n w}[M_{z_i}(w)]^{-n}\} \quad (16.7\text{-}9)$$

Substituting (16.7-8) and (16.7-9) into (16.7-7) yields Wald's fundamental identity:

$$E\{e^{Z_n w}[M_{z_i}(w)]^{-n}\} = 1 \quad (16.7\text{-}10)$$

for w in subset D. The moments of n can be obtained by differentiating (16.7-10) with respect to w and setting $w = 0$. It can also be used to find the density function of n.

16.8. The Characteristic Function of n

The characteristic function of random variable n can be obtained by a technique due to Wald [1]. Consider the equation,

$$M_{z_i}(w) = E[e^{z_i w}] = 1 \tag{16.8-1}$$

One root of this equation is $w = 0$. Since

$$\ddot{M}_{z_i}(w) = \frac{d^2}{dw^2} E[e^{z_i w}] = E[z_i^2 e^{z_i w}] \tag{16.8-2}$$

$\ddot{M}_{z_i}(w)$ is positive for all real $w > -\infty$. In addition $M_{z_i}(w)$ can be shown to be infinite at $w = \pm\infty$. Hence, $M_{z_i}(w)$ must possess a minimum at some real w_m, as shown in Fig. 16.8-1a or 16.8-1b. Then, if it is required that

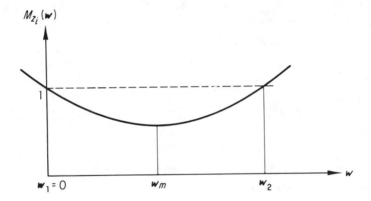

Fig. 16.8-1a. The moment-generating function of z_i for $\dot{M}_{z_i}(0)$ < 0 and $w_2 > 0$.

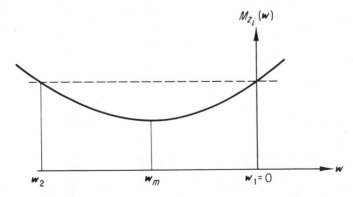

Fig. 16.8-1b. The moment-generating function of z_i for $\dot{M}_{z_i}(0)$ > 0 and $w_2 < 0$.

$\dot{M}_{z_i}(0) = E[z_i] \neq 0$, the minimum must occur at some value $w_m \neq 0$. Thus, there will be one and only one other root of Eq. (16.8-1), which is designated by w_2 in Figs. 16.8-1a and 16.8-1b.

Consider next the equation,

$$M_{z_i}(w) = E[e^{z_i w}] = e^{-j\xi} \tag{16.8-3}$$

If $M_{z_i}(w)$ is not singular at w_1 and w_2, then by analytic continuation there are at least two roots $w_1(\xi)$ and $w_2(\xi)$ of Eq. (16.8-3) such that

$$\lim_{\xi \to 0} w_1(\xi) = 0 \quad \text{and} \quad \lim_{\xi \to 0} w_2(\xi) = w_2$$

For example, if z_i is Gaussian with mean μ and variance σ^2, then,

$$M_{z_i}(w) = \frac{1}{\sqrt{2\pi\sigma^2}} \int_{-\infty}^{\infty} e^{z_i w} \cdot e^{-(z_i - \mu)^2/2\sigma^2} \, dz_i \tag{16.8-4}$$

and

$$\log M_{z_i}(w) = \mu w + \frac{\sigma^2}{2} w^2 \tag{16.8-5}$$

Combining Eqs. (16.8-3) and (16.8-5) and solving the resulting quadratic yields the following roots:

$$w_1(\xi) = -\frac{\mu}{\sigma^2} + \frac{1}{\sigma^2} \sqrt{\mu^2 - 2j\sigma^2\xi} \tag{16.8-6a}$$

$$w_2(\xi) = -\frac{\mu}{\sigma^2} - \frac{1}{\sigma^2} \sqrt{\mu^2 - 2j\sigma^2\xi} \tag{16.8-6b}$$

Wald's identity (16.7-10) can be rewritten as

$$\mathscr{L}(\theta)E_{H_0}\{e^{Z_n w}[M_{z_i}(w)]^{-n}\} + (1 - \mathscr{L}(\theta))E_{H_1}\{e^{Z_n w}[M_{z_i}(w)]^{-n}\} = 1 \tag{16.8-7}$$

where $\mathscr{L}(\theta)$ describes the OCF previously defined; $E_{H_0}\{\ \}$ denotes the conditional expected value of the bracketed quantity when H_0 is accepted, and $E_{H_1}\{\ \}$ is the conditional expectation when hypothesis H_0 is rejected. If the excess over the boundaries is neglected, $Z_n = \log \mathscr{B}$ when H_0 is accepted at termination and $Z_n = \log \mathscr{A}$ when H_1 is accepted. Hence, Eq. (16.8-7) can be rewritten as (with $\mathscr{L}(\theta) = \mathscr{L}$)

$$\mathscr{L}\mathscr{B}^w E_{H_0}\{[M_{z_i}(w)]^{-n}\} + (1 - \mathscr{L})\mathscr{A}^w E_{H_1}\{[M_{z_i}(w)]^{-n}\} = 1 \tag{16.8-8}$$

which is valid for $|M_{z_i}(w)| \geqslant 1$. Then, substituting Eq. (16.8-3) into (16.8-8) yields, for $w = w_1(\xi)$,

$$\mathscr{L}\mathscr{B}^{w_1(\xi)} E_{H_0}[e^{j\xi n}] + (1 - \mathscr{L})\mathscr{A}^{w_1(\xi)} E_{H_1}[e^{j\xi n}] = 1 \tag{16.8-9}$$

and for $w = w_2(\xi)$,

$$\mathscr{L}\mathscr{B}^{w_2(\xi)} E_{H_0}[e^{j\xi n}] + (1 - \mathscr{L})\mathscr{A}^{w_2(\xi)} E_{H_1}[e^{j\xi n}] = 1 \tag{16.8-10}$$

The solution of Eqs. (16.8-9) and (16.8-10) is given by

$$E_{H_0}[e^{j\xi n}] = \frac{\mathscr{A}^{w_2(\xi)} - \mathscr{A}^{w_1(\xi)}}{\mathscr{L}[\mathscr{B}^{w_1(\xi)}\mathscr{A}^{w_2(\xi)} - \mathscr{B}^{w_2(\xi)}\mathscr{A}^{w_1(\xi)}]} \tag{16.8-11}$$

$$E_{H_1}[e^{j\xi n}] = \frac{\mathscr{B}^{w_1(\xi)} - \mathscr{B}^{w_2(\xi)}}{(1 - \mathscr{L})[\mathscr{B}^{w_1(\xi)}\mathscr{A}^{w_2(\xi)} - \mathscr{A}^{w_1(\xi)}\mathscr{B}^{w_2(\xi)}]} \tag{16.8-12}$$

Characteristic function $C_n(\xi)$ can be written as

$$C_n(\xi) = E[e^{j\xi n}] = \mathscr{L} E_{H_0}[e^{j\xi n}] + (1 - \mathscr{L})E_{H_1}[e^{j\xi n}] \tag{16.8-13}$$

Substituting Eqs. (16.8-11) and (16.8-12) into (16.8-13) yields

$$C_n(\xi) = \frac{\mathscr{A}^{w_2(\xi)} - \mathscr{A}^{w_1(\xi)} + \mathscr{B}^{w_1(\xi)} - \mathscr{B}^{w_2(\xi)}}{\mathscr{B}^{w_1(\xi)}\mathscr{A}^{w_2(\xi)} - \mathscr{B}^{w_2(\xi)}\mathscr{A}^{w_1(\xi)}} \tag{16.8-14}$$

for all real ξ. Equation (16.8-14) is the characteristic function of random variable n.

16.9. The Distribution of n for the Coherent Detector

For the coherent detector the density function of n can be obtained from characteristic function (16.8-14) and Eqs. (16.8-6a) and (16.8-6b) for arbitrary thresholds $\mathscr{A}$ and $\mathscr{B}$. However, it is simpler to find the density functions for noise alone and signal plus noise when the error probabilities approach zero. For a coherent detector, z_i is a Gaussian variable described by Eq. (16.5-8).

Consider first the case when $\mathscr{B}$ is finite and $\mathscr{A}$ approaches infinity; this corresponds to zero false-alarm probability in the presence of noise alone. From (16.5-10) and (16.7-1), $E(z_i | 0) = \dot{M}_{z_i}(0) < 0$. From Fig. 16.8-1 this implies that the real part of $w_2(\xi)$ is positive for small ξ. Also the real part of $w_1(\xi) \to 0$ as $\xi \to 0$. Then, for small ξ, Eq. (16.8-14) is approximately given by

$$C_n(\xi | 0) \cong \frac{\mathscr{A}^{w_2(\xi)}}{\mathscr{B}^{w_1(\xi)}\mathscr{A}^{w_2(\xi)}} \tag{16.9-1a}$$

$$\cong \mathscr{B}^{-w_1(\xi)} \tag{16.9-1b}$$

Conditions $\mathscr{B} \to 0$ ($\log \mathscr{B} \to -\infty$) and $\mathscr{A}$ finite describe the case where missed detections approach zero for signal plus noise; this corresponds to signal amplitude greater than the design value. From (16.5-9) $E(z | A) = \dot{M}_{z_i}(0) > 0$ and from Fig. 16.8-1b the real part of $w_2(\xi)$ is negative for small ξ. In this case, Eq. (16.8-14) becomes (approximately)

$$C_n(\xi | \mu) \cong \frac{-\mathscr{B}^{w_2(\xi)}}{-\mathscr{B}^{w_2(\xi)}\mathscr{A}^{w_1(\xi)}} \tag{16.9-2a}$$

or

$$C_n(\xi | \mu) \cong \mathscr{A}^{-w_1(\xi)} \tag{16.9-2b}$$

The functions $w_1(\xi)$ and $w_2(\xi)$ are given for a Gaussian random variable z_i with mean μ and variance σ^2 by Eqs. (16.8-6a) and (16.8-6b). For the coherent detector the mean of z_i when signal is present is given by (16.5-9) and for noise alone by (16.5-10). It is convenient to derive these quantities by rewriting z_i in (16.5-8) as

$$z_i = y_i \chi_d - \frac{\chi_d}{2} \qquad (16.9\text{-}3)$$

where $y_i = v_i/A_d$. With definition $k = A/A_d$, the mean and variance of z_i for signal present and absent can be computed from (16.9-3); they are given by

$$E[z_i \mid 0] = -\frac{\chi_d}{2} \qquad (16.9\text{-}4)$$

$$E[z_i \mid k] = (2k - 1)\frac{\chi_d}{2} \qquad (16.9\text{-}5)$$

$$E[(z_i - \bar{z}_i)^2 \mid 0] = \chi_d \qquad (16.9\text{-}6)$$

$$E[(z_i - \bar{z}_i)^2 \mid k] = \chi_d \qquad (16.9\text{-}7)$$

Substituting Eqs. (16.9-4) and (16.9-6) into (16.8-6) yields $w_1(\xi)$ for noise alone:

$$w_1(\xi \mid 0) = \frac{1}{2} - \frac{1}{2}\sqrt{1 - \frac{8j\xi}{\chi_d}} \qquad (16.9\text{-}8)$$

With (16.9-8) Eq. (16.9-1b) becomes

$$C_n(\xi \mid 0) = \mathcal{B}^{-(1/2)+(1/2)\sqrt{1-8j\xi/\chi_d}} \qquad (16.9\text{-}9a)$$

$$= \exp\left[-\frac{\log \mathcal{B}}{2}\left(1 - \sqrt{1 - \frac{8j\xi}{\chi_d}} \right) \right] \qquad (16.9\text{-}9b)$$

For the assumption $\alpha \to 0$, it follows from Eq. (16.2-9) that $\mathcal{B}$ can be replaced by β in Eq. (16.9-9b). The probability density function of n for noise alone is obtained by evaluating the anticharacteristic function of $C_n(\xi \mid 0)$; thus,

$$p(n \mid 0) = \frac{1}{2\pi}\int_{-\infty}^{\infty} e^{(-\log \beta/2)(1-\sqrt{1-8j\xi/\chi_d})} \cdot e^{-jn\xi}\, d\xi \qquad (16.9\text{-}10)$$

With substitutions $c = -\log \beta/2$, $s = 1 - 8j\xi/\chi_d$ and $t = n\chi_d/8$, Eq. (16.9-10) can be rewritten as

$$p(n \mid 0) = \frac{\chi_d e^{c-t}}{8}\cdot\frac{1}{2\pi j}\int_{1-j\infty}^{1+j\infty} e^{-c\sqrt{s}}\cdot e^{st}\, ds \qquad (16.9\text{-}11)$$

which is a tabulated integral [5]. The result is

$$p(n \mid 0) = \frac{-\log \beta}{\sqrt{2\pi\chi_d}\; n^{3/2}}\exp\left\{ -\left[\frac{(\log \beta)^2}{2\chi_d n} + \frac{n\chi_d}{8} + \frac{\log \beta}{2} \right] \right\} \qquad (16.9\text{-}12)$$

In like manner, when signal is present [see Eq. (16.8-6a)],

$$w_1(\xi \mid k) = \left(k - \frac{1}{2}\right)\left[-1 + \sqrt{1 - \frac{8j\xi}{(2k-1)^2\chi_d}} \right] \qquad (16.9\text{-}13)$$

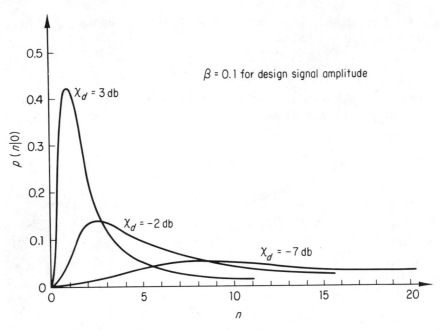

Fig. 16.9-1. Probability density of n for noise alone.

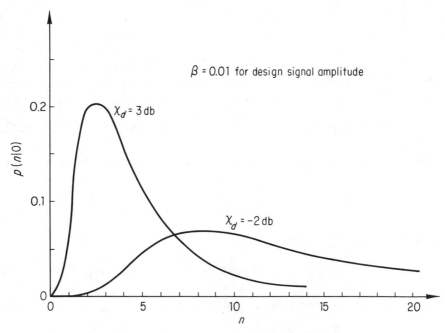

Fig. 16.9-2. Probability density of n for noise alone.

and

$$C_n(\xi \mid k) = \exp\left\{(\log \mathscr{A})\left(k - \frac{1}{2}\right)\left[1 - \sqrt{1 - \frac{8j\xi}{(2k-1)^2\chi_d}}\right]\right\} \tag{16.9-14}$$

For $k > 1$ (signal greater than the design value), $\mathscr{A} \cong 1/\alpha$; thus

$$p(n \mid k) = \frac{-\log \alpha}{\sqrt{2\pi\chi_d}\, n^{3/2}} \exp\left\{-\left[\frac{(\log \alpha)^2}{2\chi_d n} + \frac{(2k-1)^2 n\chi_d}{8} + \frac{(2k-1)\log \alpha}{2}\right]\right\} \tag{16.9-15}$$

Equations (16.9-12) and (16.9-15) give the approximate density functions of random variable n for a coherent detector for noise alone and for signal plus noise. When $k = 1$, Eqs. (16.9-12) and (16.9-15) are similar if $\log \alpha$ and $\log \beta$ are interchanged.

Equations (16.9-12) and (16.9-15) are plotted in Figs. 16.9-1 to 16.9-3. Figures 16.9-1 and 16.9-2 show that, for noise alone, the mean and variance of n increase significantly as χ_d decreases and both increase as the error probability β is reduced. For signal plus noise, the mean and variance of n are large with respect to the corresponding values for noise alone since a small error probability was selected.

The exponent in Eq. (16.9-12) has a minimum value (zero) for $n^* = -2\log \beta/\chi_d$; a Taylor series expansion about n^* is given by

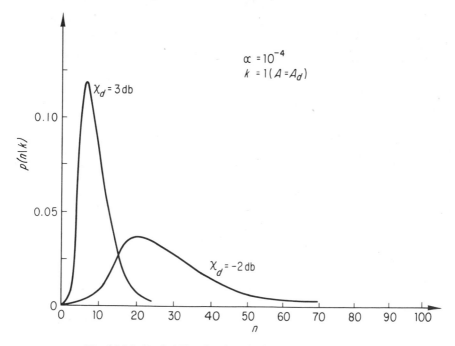

Fig. 16.9-3. Probability density of n for signal plus noise.

$$-\left[\frac{(\log \beta)^2}{2\chi_d n} + \frac{n\chi_d}{8} + \frac{\log \beta}{2}\right] = \frac{n^*\chi_d}{8}\left[\left(\frac{n-n^*}{n^*}\right)^2 - \left(\frac{n-n^*}{n^*}\right)^3 + \cdots\right]$$

$$(16.9\text{-}16)$$

For large n^*, only the quadratic term in expansion (16.9-16) is significant near the peak of density function $p(n\,|\,0)$. Further, in this region the denominator term $n^{3/2}$ in (16.9-12) can be approximated by $(n^*)^{3/2}$. Then, Eq.

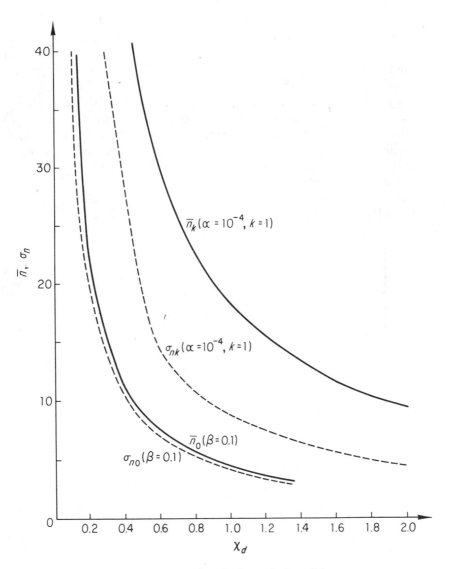

Fig. 16.9-4. ASN and σ of n for typical conditions.

(16.9-12) can be rewritten approximately as

$$p(n \,|\, 0) \cong \frac{1}{\sqrt{2\pi}} \cdot \frac{1}{\sqrt{\dfrac{8 \log \dfrac{1}{\beta}}{\chi_d^2}}} \exp\left[-\frac{\left(n - \dfrac{2 \log \dfrac{1}{\beta}}{\chi_d}\right)^2}{2\left(\dfrac{8 \log \dfrac{1}{\beta}}{\chi_d^2}\right)} \right] \qquad (16.9\text{-}17)$$

Equation (16.9-17) describes a Gaussian density function with mean

$$\bar{n}_0 \cong \frac{2 \log \dfrac{1}{\beta}}{\chi_d} \qquad (16.9\text{-}18)$$

and variance

$$\sigma_{n_0}^2 \cong \frac{8 \log \dfrac{1}{\beta}}{\chi_d^2} \qquad (16.9\text{-}19)$$

For signal plus noise, Eq. (16.9-15) can also be approximated by a Gaussian distribution with mean and variance given by (for $k = 1$)

$$\bar{n}_{(k=1)} = \frac{2 \log \dfrac{1}{\alpha}}{\chi_d} \qquad (16.9\text{-}20)$$

$$\sigma_{n_{(k=1)}}^2 = \frac{8 \log \dfrac{1}{\alpha}}{\chi_d^2} \qquad (16.9\text{-}21)$$

Expressions (16.9-18) through (16.9-21) are plotted in Fig. 16.9-4 for typical conditions: $\beta = 0.1$, $\alpha = 10^{-4}$, and $k = 1$. Equation (16.9-15) can be used for $k > 1$.

16.10. Some Remarks on the PDF of n

Expressions (16.9-12) and (16.9-15) are continuous probability density functions of n. This may seem strange, since sample number n is a discrete random variable. However, for large sample numbers and small error probabilities these density functions are good approximations to the discrete probability densities of n.

To obtain density functions (16.9-12) and (16.9-15), it was assumed that $\mathscr{B}$ was finite and $\mathscr{A}$ infinite in the first case and that $\mathscr{B}$ was zero and $\mathscr{A}$ finite in the second. For finite $\mathscr{A}$ and $\mathscr{B}$ the density function for n can be obtained from the anticharacteristic function of Eq. (16.8-14). It is shown in Wald [1] that, when z_i is Gaussian, the exact density function is given by

$$p(n \,|\, k) = \sum_{i=1}^{\infty} r_i e^{-\lambda_i} p_i(n \,|\, k) \qquad (16.10\text{-}1)$$

where λ_i and r_i are constants ($\lambda_i > 0$) and where $p_i(n \mid k)$, $i = 1, 2, \ldots$, have the form of Eq. (16.9-12) or (16.9-15).

As a final point, although the density function of n was obtained assuming Gaussian z_i, (16.9-12) and (16.9-15) are good approximations when z_i is not Gaussian if $|E(z_i)|$ and $\{E[(z_i - \bar{z}_i)^2]\}^{1/2}$ are both small with respect to $\log \mathscr{A}$ and $\log \mathscr{B}$. Consider sum random variable

$$z_i^* = \sum_{j=(i-1)r+1}^{ir} Z_j \qquad (i = 1, 2, \ldots) \qquad (16.10\text{-}2)$$

where r is a positive integer. For statistically independent Z_j with the same arbitrary density functions and subject to some weak restrictions, the probability density of sum variate z_i^* approaches a Gaussian distribution for large r by virtue of the central limit theorem. The cumulative sum

$$Z_n^* = z_1^* + z_2^* + \ldots + z_n^* \qquad (16.10\text{-}3)$$

also tends to Gaussian for large n.

16.11. ASN for Incoherent Detection of Signals of Unknown Phase

In this section it is assumed that the in-phase and quadrature components of the noise are each white Gaussian random processes of zero mean and variance σ^2. The normalized signal-plus-noise envelope r_i out of a matched filter at the expected arrival time of the ith signal is governed by the density function

$$p(r_i \mid a) = r_i \exp\left(-\frac{r_i^2 + a^2}{2}\right) I_0(ar_i), \qquad r_i > 0 \qquad (16.11\text{-}1)$$

[see Eq. (9.3-38)], where $a^2 = \mathscr{R}$ is the peak signal-to-noise ratio and is assumed to be the same for each sample.

The sequential test variate z_i is given by [see Eq. (16.4-3)]

$$z_i = \log \frac{p(r_i \mid a_d)}{p(r_i \mid 0)} = -\frac{a_d^2}{2} + \log I_0(a_d r_i) \qquad (16.11\text{-}2)$$

where a_d^2 is the design peak signal-to-noise ratio. For the weak-signal approximation $a_d \ll 1$, z_i can be expressed by a power series expansion of $\log I_0(x)$, as follows:

$$z_i = -\frac{a_d^2}{2} + \frac{a_d^2 r_i^2}{4} - \frac{a_d^4 r_i^4}{64} + O(a_d^6 r_i^6) \qquad (16.11\text{-}3)$$

The third term in Eq. (16.11-3) contributes principally to the bias of z_i; without this contribution $\bar{n}$ approaches infinity for hypothesis H_0. It can be shown that [6]

$$\overline{r_i^2} = 2\left(1 + \frac{a^2}{2}\right) \qquad (16.11\text{-}4)$$

and

$$\overline{r_i^4} = 2\left(4 + 4a^2 + \frac{a^4}{2}\right) \tag{16.11-5}$$

From (16.11-3), (16.11-4), and (16.11-5),

$$E(z_i \mid a) = \frac{a_d^2 a^2}{4} - \frac{a_d^4}{8} \tag{16.11-6}$$

The term $(a_d^4/8)$ in Eq. (16.11-6) is the bias contribution of the third term in (16.11-3). A good approximation of z_i, in the weak-signal case, is provided by retaining only the first two terms in (16.11-3) and the bias contribution of the third term, as follows:

$$z_i \cong \frac{a_d^2}{2} + \frac{a_d^2 r_i^2}{4} - \frac{a_d^4}{8} \tag{16.11-7a}$$

$$z_i \cong -\frac{a_d^2}{2}\left(1 + \frac{a_d^2}{4}\right) + \frac{a_d^2 r_i^2}{4} \tag{16.11-7b}$$

From (16.11-7b)

$$E(z_i \mid 0) = -\frac{a_d^4}{8} \tag{16.11-8}$$

and

$$E(z_i \mid a) = -\frac{a_d^4}{4}\left[\frac{a_d^2}{2} - a^2\right] \tag{16.11-9}$$

for $a_d \ll 1$, and $a \ll 1$.

The performance for arbitrary signal-to-noise ratio (a^2) is given by the OCF. From (16.3-16)

$$\mathscr{L}(a) = \frac{\mathscr{A}^h - 1}{\mathscr{A}^h - \mathscr{B}^h} \tag{16.11-10}$$

where h is now determined by [see Eqs. (16.5-2) and (16.11-1)]

$$\int_0^\infty r_i I_0(a r_i)[I_0(a_d r_i)]^h \, e^{-r_i^2/2} \, dr_i = e^{(a^2 + h a_d^2)/2} \tag{16.11-11}$$

The integral in (16.11-11) can be approximately evaluated with a power series expansion of $I_0(x)$, retaining terms only to fourth order; this yields the following approximate expression for h:

$$h \cong 1 - 2\left(\frac{a}{a_d}\right)^2 \tag{16.11-12}$$

With (16.11-10) and (16.11-12) the OCF can be plotted, as in Fig. 16.5-1. The ASN for an incoherent detector can be obtained by substituting (16.2-8), (16.2-9), and (16.11-9) into Eq. (16.4-5), which yields (for $\alpha \ll 1$, $a \ll 1$, and $a_d \ll 1$)

$$\bar{n}_a = \frac{\mathscr{L}(a) \log \beta + [1 - \mathscr{L}(a)] \log \dfrac{1 - \beta}{\alpha}}{\dfrac{a^2 a_d^2}{4} - \dfrac{a_d^4}{8}} \tag{16.11-13}$$

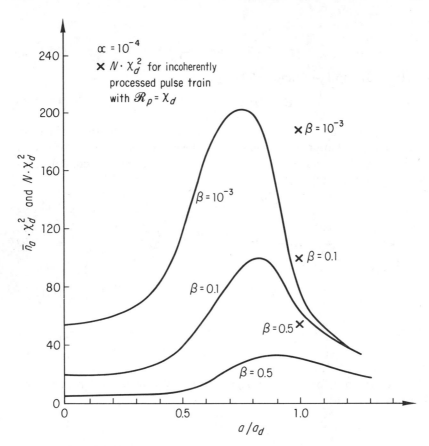

Fig. 16.11-1. $\bar{n}_a \cdot \chi_d^2$ and $N \cdot \chi_d^2$ vs. a/a_d for incoherent detector.

With (16.11-10) and (16.11-12), Eq. (16.11-13) has been evaluated for typical values of α and β and plotted in Fig. 16.11-1. Modifying Eq. (16.11-13) to the form

$$\bar{n}_a = \frac{\mathscr{L}(a) \log \beta + [1 - \mathscr{L}(a)] \log \dfrac{1 - \beta}{\alpha}}{-\dfrac{\chi_d^2}{4}\left[\dfrac{1}{2} - \left(\dfrac{a}{a_d}\right)^2\right]} \qquad (16.11\text{-}14)$$

shows that the ASN is inversely proportional to the square of design (peak) signal-to-noise ratio $\chi_d = a_d^2$.

The curves in Fig. 16.11-1 are similar to those for the coherent detector in Fig. 16.5-2 and peak at an amplitude roughly equal to two-thirds the design amplitude. The cross marks in Fig. 16.11-1 indicate $N\chi_d^2$ for fixed sample tests of the same α, β corresponding to incoherently integrated pulse trains of length N with pulse signal-to-noise ratios of $\mathscr{R}_p = \chi_d$. The plotted values were obtained from approximate expression (10.7-27) ($N \gg 1$,

$\mathscr{R}_p < 1$) with $\mathscr{R}_p = \chi_d$:

$$N\chi_d^2 = 4[\Phi^{-1}(\alpha) - \Phi^{-1}(1 - \beta)]^2 \qquad (16.11\text{-}15)$$

where $\Phi^{-1}(T)$ is defined by Eqs. (9.6-2) and (9.6-3). When there is no signal ($a = 0$), the ratio $\bar{n}_0/N$ is about $1:5$ for $P_d = 0.5$, and $P_d = 0.9$, and roughly $1:4$ for $P_d = 0.999$. Ratio $\bar{n}_0/N$ does not change as rapidly for an incoherent detector as for a coherent detector.

16.12. The Need for Truncation

Since test length n is a random variable, situations may occur where a test is excessively long, whether H_0 or H_1 is true (see, for example, Figs. 16.9-1 and 16.9-3). To avoid this situation, a sequential test can be *truncated* at a predetermined maximum permissible test length N. To insure test termination, a new rule is adopted only when $m = N$. In a test that has not terminated at $m = N$, that is,

$$\log \mathscr{B} < \sum_{i=1}^{N} z_i < \log \mathscr{A} \qquad (16.12\text{-}1)$$

a reasonable rule is to decide H_0 (no signal) when

$$\log \mathscr{B} < \sum_{i=1}^{N} z_i \leqslant 0 \qquad (16.12\text{-}2)$$

and decide H_1 (signal plus noise) when

$$0 < \sum_{i=1}^{N} z_i < \log \mathscr{A} \qquad (16.12\text{-}3)$$

When test length N is bounded in this way, error probabilities α, β differ from their design values. Upper bounds have been derived by Wald [1] for error probabilities α_T and β_T corresponding to a truncated test with a normal distribution for z_i. For selected values of α and β Wald relates the truncated test length to the test length of a fixed-sample test with the same (α, β); when the truncation point is set equal to the fixed test length, the upper bounds on α_T and β_T are about double the values for an untruncated test. When the truncation is double the length of the fixed-sample test, the upper bounds on α_T and β_T are only slightly higher than the values of α and β for an untruncated test. More accurate bounds have been obtained by Bussgang and Middleton [3]; they show that the bounds obtained by Wald are conservative.

16.13. Application of Sequential Detection to a Multiple-
Resolution-Bin Radar

In a typical radar system the antenna dwells in a particular beam position until all range (or range-doppler) cells out to the maximum unambiguous range have been searched. For a fixed-sample test of length N, the dwell time is approximately NT_r, where T_r is usually slightly greater than the maximum unambiguous range delay. When sequential detection is applied (without truncation) to the above radar, a different value of n is required for each range bin, owing to the randomness of n. As a result, the dwell time in each beam position is determined by the longest range-bin test. In such cases the advantage of a variable test length is diminished.

To investigate the penalties incurred, consider the case of k resolution bins, each with independent, zero-mean, quadrature Gaussian noise components. As before, let α and β denote the error probabilities in each resolution element. Let α_G and β_G denote respectively the overall or *global* probability of false alarm and the *global* probability of a missed detection in a particular antenna beam position. Since the probability of no global false alarms equals the joint probability of no individual false alarms, it follows for $\alpha \ll 1$ that

$$1 - \alpha_G = (1 - \alpha)^k \cong 1 - k\alpha \tag{16.13-1}$$

so that

$$\alpha_G = k\alpha \tag{16.13-2}$$

Consider a test for the presence of exactly one signal in an unknown range bin. The global probability of a missed detection is equal to the joint probability that the target is missed in the appropriate jth subtest, multiplied by the joint probability of no false alarms in all other subtests—that is,

$$\beta_G = \beta(1 - \alpha)^{k-1} = \frac{\beta(1 - \alpha)^k}{1 - \alpha} = \frac{\beta(1 - \alpha_G)}{1 - \alpha} \tag{16.13-3}$$

For $\alpha \ll 1$, it follows from (16.13-2) that $\alpha_G \ll 1$; hence the global false-dismissal probability is approximately

$$\beta_G \cong \beta \tag{16.13-4}$$

Equations (16.13-2) and (16.13-4) relate subtest error probabilities α, β to global error probabilities α_G, β_G.

The average test termination length, when noise alone is present, is given by

$$\bar{n}_0 = \alpha_G \bar{n}_{\alpha_G} + (1 - \alpha_G)\bar{n}_{1-\alpha_G} \tag{16.13-5}$$

where $\bar{n}_{\alpha_G}$ is the average test length when H_1 is falsely accepted and $\bar{n}_{1-\alpha_G}$ is the average test length when H_0 is correctly accepted. For $\alpha_G \ll 1$,

$$\bar{n}_0 \cong \bar{n}_{1-\alpha_G} \tag{16.13-6}$$

In like manner, when signal plus noise is present,

$$\bar{n}_a = \beta_G \bar{n}_{\beta_G} + (1 - \beta_G)\bar{n}_{1-\beta_G} \tag{16.13-7}$$

For $\beta_G \ll 1$ (high detection probability),

$$\bar{n}_a \cong \bar{n}_{1-\beta_G} \cong \bar{n}_{1-\beta} \tag{16.13-8}$$

Average test length $\bar{n}_{1-\beta}$ was previously computed and plotted for both coherent and incoherent detectors as a function of signal strength a.

The value $\bar{n}_{1-\alpha_G}$ is the average global test length when H_0 is correctly accepted. A test is terminated when all tests end with the correct acceptance of H_0. Denote by $p_i(n)$ the probability that H_0 is correctly accepted in the ith subtest with a test of length n; note that $p_i(n) = 1 - \alpha_i(n)$. Let $P_i(n)$ denote the cumulative probability that H_0 is correctly accepted *by or at n*; then, if $p_i(n) = p(n)$ for all i,

$$P_i(n) = P(n) = \sum_{l=1}^{n} p(l) \tag{16.13-9}$$

It follows that

$$p(n) = P(n) - P(n-1) \tag{16.13-10}$$

For k resolution bins, the probability that a global test ends with the correct acceptance of H_0, *by or at n*, is $P(n)^k$; hence the probability $p_G(n)$ that a global test ends *at n* can be expressed as

$$p_G(n) = P(n)^k - P(n-1)^k \tag{16.13-11}$$

By definition the average sample number $\bar{n}_{0_G}$ for a global test is given by

$$\bar{n}_{0_G} = \bar{n}_{1-\alpha_G} = \sum_{n=1}^{\infty} n p_G(n) \tag{16.13-12}$$

Substituting (16.13-11) into (16.13-12) yields

$$\bar{n}_{0_G} = \sum_{n=1}^{\infty} n[P(n)^k - P(n-1)^k] = \lim_{r \to \infty} \sum_{n=1}^{r} n[P(n)^k - P(n-1)^k] \tag{16.13-13}$$

Expanding the sum in (16.13-13) and combining terms results in

$$\bar{n}_{0_G} = \lim_{r \to \infty} \left[rP(r)^k - \sum_{n=1}^{r-1} P(n)^k \right] \tag{16.13-14}$$

Since $\lim_{r \to \infty} P(r)^k = 1$ for finite k, Eq. (16.13-14) can be rewritten in a form obtained by Reed and Selin [7]:

$$\bar{n}_{0_G} = \sum_{n=0}^{\infty} [1 - P(n)^k] \tag{16.13-15}$$

To compute the average global test length $\bar{n}_{0_G}$ using (16.13-15) for a coherent detector (assuming Gaussian z_i), $P(n)$ can be calculated from Eq. (16.9-12) as follows:

$$P(n) = \int_0^n p(l \mid 0) \, dl \tag{16.13-16a}$$

$$= \int_0^n \frac{-\log \beta}{\sqrt{2\pi \beta \chi_d} \, (l)^{3/2}} \, e^{-[(\log \beta)^2/2\chi_d l + l\chi_d/8]} \, dl \tag{16.13-16b}$$

and

$$P(n) = \frac{1}{\beta} \, \Phi\!\left(\frac{-\log \beta}{\sqrt{n\chi_d}} + \sqrt{\frac{n\chi_d}{4}} \right) + \Phi\!\left(\frac{-\log \beta}{\sqrt{n\chi_d}} - \sqrt{\frac{n\chi_d}{4}} \right) \tag{16.13-17}$$

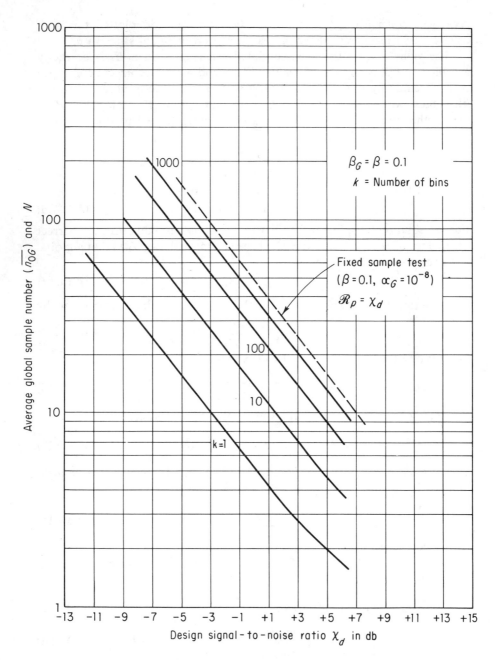

Fig. 16.13-1. Average global sample number for a coherent detector as a function of design signal-to-noise ratio with signal absent (*after Reed and Selin* [7], *courtesy of IEEE*).

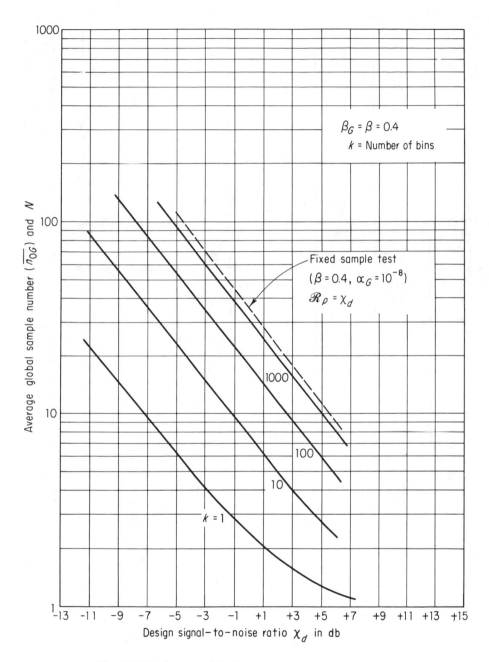

Fig. 16.13-2. Average global sample number for a coherent detector as a function of design signal-to-noise ratio with signal absent (*after Reed and Selin* [7], *courtesy of IEEE*).

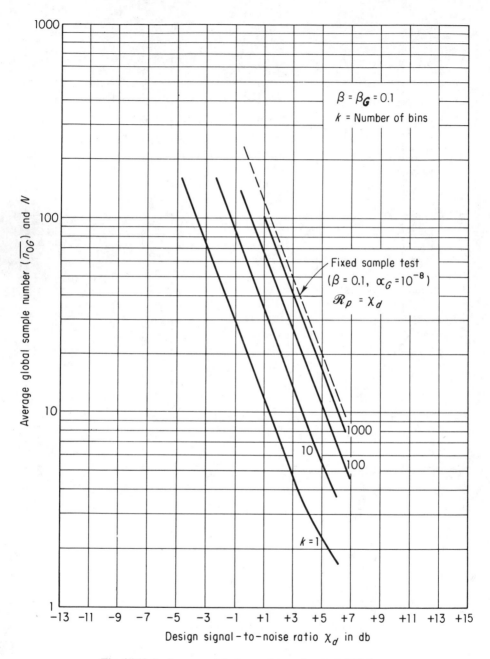

Fig. 16.13-3. Average global sample number and N for an incoherent detector as a function of design signal-to-noise ratio with signal absent.

580

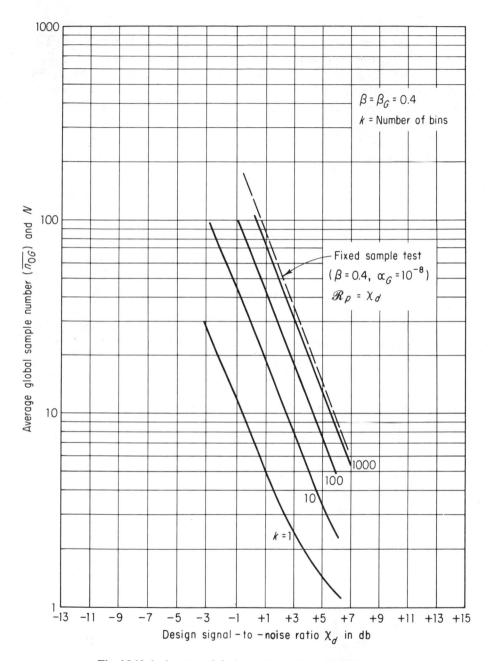

Fig. 16.13-4. Average global sample number and N for an incoherent detector as a function of design signal-to-noise ratio with signal absent.

where $\Phi(T)$ is defined by Eq. (9.6-2). Relation (16.13-17) applies when upper threshold $\mathscr{A}$ approaches infinity. When (16.13-17) is substituted into Eq. (16.13-15), $\bar{n}_{0_G}$ can be determined (see [7]). The result is shown on Figs. 16.13-1 and 16.13-2 for $\beta = 0.1$ and $\beta = 0.4$. For comparison the length of a fixed-sample test has been plotted for $k = 1000$ and global false-alarm probability $\alpha_G = 10^{-8}$; in this case there is a reduction in search time by a factor of about 1.2 when a sequential test is employed.

The previous results can be applied to an incoherent detector when n is large. The mean and variance of z_i are given by Eqs. (16.11-7b), (16.11-4), and (16.11-5):

$$E(z_i \mid 0) = -\frac{\chi_d^2}{8} \tag{16.13-18}$$

$$E[(z_i - \bar{z}_i)^2] = \frac{\chi_d^2}{4} \tag{16.13-19}$$

When (16.13-18) and (16.13-19) are compared with (16.9-4) and (16.9-6), relations (16.13-15) and (16.13-17) apply provided χ_d is replaced by $\chi_d^2/4$. The value of $\bar{n}_{0_G}$ for an incoherent detector is plotted in Figs. 16.13-3 and 16.13-4 for $\beta = 0.1$ and $\beta = 0.4$. Again, a reduction of 1.2:1 in search time (for zero signal) is obtained for $k = 1000$. This small improvement raises a question concerning the value of sequential detection for large k. For 100 resolution elements, the reduction in search time is nearly 1.8:1, as noted by Helstrom [8]. In the next section a different technique that improves on these results is considered.

The preceding results are valid for a low target-density environment (or for very small signals). For large signals, or high target densities, the ASN for signal present $\bar{n}_{a_G}$ tends to dominate the results; in this case truncation may be required. Additional results were obtained by Bussgang and Ehrman [9] in which range dependence is taken into account; see [9] for curves.

16.14. Analogue Marcus and Swerling Test

In the previous section, a test strategy was considered in which each resolution bin is individually tested for the presence of one and only one possible signal. This results in a different n for each resolution bin; with this approach the global test length is equal to the longest test length obtained.

A different strategy was considered by Marcus and Swerling [10]. At each stage of a sequential test, a binary test is performed in which H_0 is the hypothesis that noise alone is present in all resolution bins and H_1 is the hypothesis that one signal (of known amplitude) is present in the ith resolution element. The a priori probability that the signal is in any one of k resolution elements is assumed to be uniform. A likelihood-ratio test is employed at each stage until a decision is made.

Let $p_m(\mathbf{v}_j \mid 0)$ be the joint conditional probability density of the m variates $(v_{1j}, v_{2j}, \ldots, v_{mj})$ in the jth resolution element, when signal is absent. Then, for independent samples in each range bin, the joint conditional probability density of no signal in each of k resolution bins is given by

$$p_m(\mathbf{v} \mid 0) = \prod_{j=1}^{k} p_m(\mathbf{v}_j \mid 0) \tag{16.14-1}$$

When a signal (of known strength a) is present in the ith bin, with a priori probability P_i, the joint probability density is given by

$$p_m(\mathbf{v} \mid a, i, \mathbf{0} - i) = p_m(\mathbf{v}_1 \mid 0)p_m(\mathbf{v}_2 \mid 0)\ldots p_m(\mathbf{v}_i \mid a)\ldots p_m(\mathbf{v}_k \mid 0) \tag{16.14-2}$$

where $(\mathbf{0} - i)$ is the null set: $j = 1, 2, \ldots, k$ with $j \neq i$. Averaging (16.14-2) with respect to random variable i yields

$$p_m(\mathbf{v} \mid a) = \sum_{i=1}^{k} P_i p_m(\mathbf{v}_i \mid a) \prod_{\substack{j=1 \\ j \neq i}}^{k} p_m(\mathbf{v}_j \mid 0) \tag{16.14-3}$$

For uniform $P_i = 1/k$, it follows from (16.14-1) and (16.14-3) that the likelihood ratio is given by

$$\ell_m(\mathbf{v} \mid a) = \frac{p_m(\mathbf{v} \mid a)}{p_m(\mathbf{v} \mid 0)} \tag{16.14-4a}$$

$$= \frac{1}{k} \sum_{i=1}^{k} \frac{p_m(\mathbf{v}_i \mid a)}{p_m(\mathbf{v}_i \mid 0)} \tag{16.14-4b}$$

The Marcus and Swerling test strategy [10] is to continue the sequential test as long as $\ell_m(\mathbf{v} \mid a)$ falls between $\mathscr{A}$ and $\mathscr{B}$. The test terminates with either the acceptance of H_0 when $\ell_m \leqslant \mathscr{B}$ or with the acceptance of H_1 when $\ell_m \geqslant \mathscr{A}$.

For a coherent detector with z_i given by Eq. (16.9-3)

$$\ell_m(\mathbf{y} \mid a) = \frac{1}{k} \sum_{i=1}^{k} \exp\left[\sum_{j=1}^{m} \left(\chi_d y_{ij} - \frac{\chi_d}{2} \right) \right] \tag{16.14-5}$$

where $y_{ij} = v_{ij}/A_d$. When (16.14-5) is compared with thresholds $\mathscr{A}$ and $\mathscr{B}$, no simplification results by taking logarithms; further, a power series expansion of $\exp(u)$ is not useful since u is of the order of $m\chi_d$, which is not small in the cases of interest. Marcus and Swerling conducted a computer simulation to evaluate the performance of a coherent detector in a target-free environment (no signal). The results are plotted in Fig. 16.14-1. Also shown for comparison purposes are the length of a fixed-sample test with the same α, β and a curve of the performance derived for the global test procedure described in Section 16.13. Observe that for $\chi_d = 0$ db the sequential detector has an ASN that is roughly one-half the length of a fixed-sample test for $k = 1000$. The reduction in test length grows more pronounced for smaller k. Also note the significant improvement in performance of the Marcus-Swerling coherent detector (MSCD) for $\chi_d = 0$ db over a comparable test conducted in each resolution bin; in this case, the improvement increases with k.

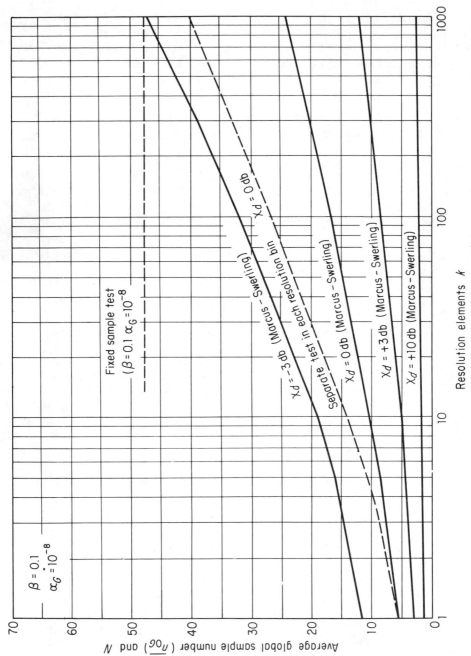

Fig. 16.14-1. ASN and N vs. number of resolution elements for a coherent detector employing a Marcus and Swerling test with signal absent (*after Marcus and Swerling [10], courtesy of IEEE*).

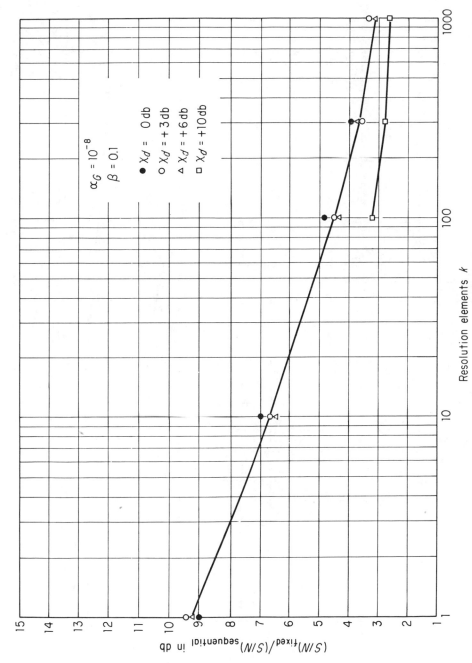

Fig. 16.14-2. Power saving of sequential coherent detector over fixed-sample detector for ASN = N (*after Marcus and Swerling* [10], *courtesy of IEEE*).

The figure shows a plot with vertical axis labeled $(S/N)_{fixed}/(S/N)_{sequential}$ in db, ranging from 1 to 15, and horizontal axis labeled "Resolution elements k" ranging from 1 to 1000. Legend: $\alpha_G = 10^{-8}$, $\beta = 0.1$, with symbols ● $X_d = 0$ db, ○ $X_d = +3$ db, △ $X_d = +6$ db, □ $X_d = +10$ db.

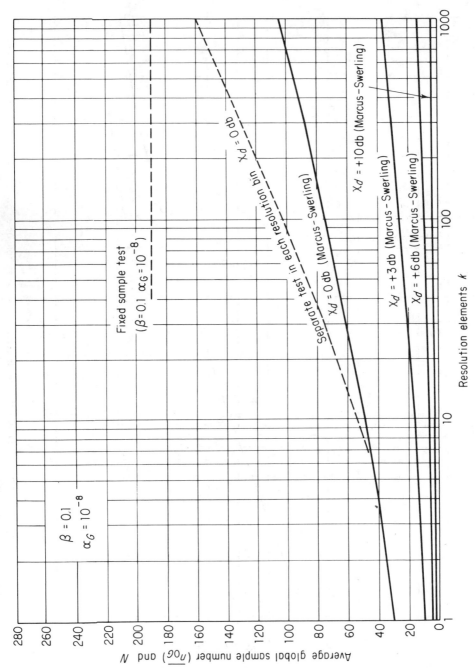

Fig. 16.14-3. ASN and N vs. number of resolution elements for incoherent detector employing a Marcus-Swerling test with signal absent (*after Marcus and Swerling* [10], *courtesy of IEEE*).

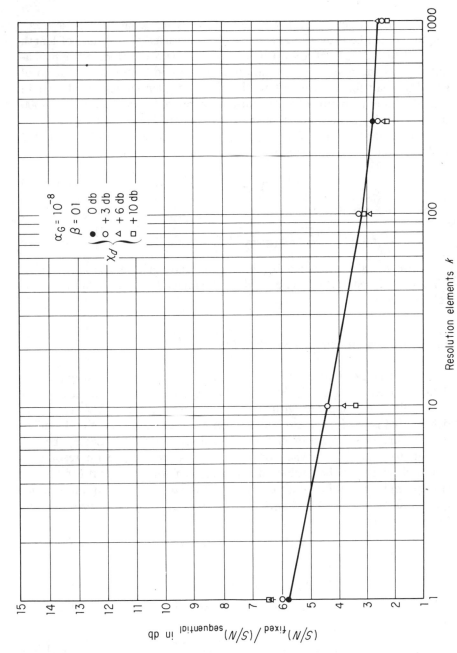

Fig. 16.14-4. Power saving of sequential incoherent detector over fixed-sample incoherent detector for ASN = N (*after Marcus and Swerling* [10], *courtesy of IEEE*).

The improvement provided by a sequential test compared to a fixed-length test can be shown in another way. For a fixed sample test of length N equal to the ASN of a sequential test, and with the same error probabilities α, β and number of resolution elements k in both tests, the ratio of signal-to-noise for a fixed test to that for a sequential test is plotted in Fig. 16.14-2 as a function of k. The ratio is expressed in decibels. At $k = 1000$ there is roughly a 3 db power saving if a sequential test is employed.

For likelihood ratio (16.14-4b) an incoherent detector can be evaluated with relation (16.11-1), assuming independent noise variates; thus

$$\ell_m(\mathbf{r} \mid a) = \frac{1}{k} \sum_{i=1}^{k} e^{-ma_d^2/2} \prod_{j=1}^{m} I_0(a_d r_{ij}) \qquad (16.14\text{-}6)$$

which can be rewritten as

$$\ell_m(\mathbf{r} \mid a) = \frac{1}{k} \exp\left(-\frac{ma_d^2}{2}\right) \sum_{i=1}^{k} \exp\left[\sum_{j=1}^{m} \log I_0(a_d r_{ij})\right] \qquad (16.14\text{-}7)$$

This case was also evaluated by simulation, and performance is shown in Figs. 16.14-3 and 16.14-4. The results are similar to those obtained for a coherent detector.

16.15. Binomial Marcus and Swerling Test

The test strategy of Marcus and Swerling has been applied to quantized data by G. M. Dillard [11]. Suppose that v_{ij}, the jth observation in resolution bin i, is compared with a reference level q; a new variable u_{ij} is defined that is one for $v_{ij} \geqslant q$ and zero for $v_{ij} < q$. A constant q and similar density functions are assumed for all resolution elements. Let P_{i0} be the probability that $v_{ij} \geqslant q$ when only noise is present in the ith resolution bin, and let P_{ia} be the probability that $v_{ij} \geqslant q$ when a signal (of known strength a) is present in the ith bin. If N_{im} is the total number of ones in the quantized sample $\mathbf{u}_i = (u_{i_1}, u_{i_2}, \ldots, u_{i_m})$, the probability density of $\mathbf{u}_i$, for noise alone, is

$$p_m(\mathbf{u}_i \mid 0) = P_{i0}^{N_{im}}(1 - P_{i0})^{m-N_{im}} \qquad (16.15\text{-}1)$$

and, for signal present,

$$p_m(\mathbf{u}_i \mid a) = P_{ia}^{N_{im}}(1 - P_{ia})^{m-N_{im}} \qquad (16.15\text{-}2)$$

With (16.15-1) and (16.15-2) the likelihood ratio given by Eq. (16.14-4b) becomes

$$\ell_m(\mathbf{u} \mid a) = \frac{1}{k} \sum_{i=1}^{k} \left(\frac{P_{ia}}{P_{i0}}\right)^{N_{im}} \left(\frac{1 - P_{ia}}{1 - P_{i0}}\right)^{m-N_{im}} \qquad (16.15\text{-}3)$$

After each observation (16.15-3) is formed and compared with $\mathscr{A}$ and $\mathscr{B}$. Hypothesis H_0 is accepted when $\ell_m(\mathbf{u} \mid a) \leqslant \mathscr{B}$ and H_1 when $\ell_m(\mathbf{u} \mid a) \geqslant \mathscr{A}$. If $\ell_m(\mathbf{u} \mid a)$ falls between $\mathscr{A}$ and $\mathscr{B}$, another observation is taken.

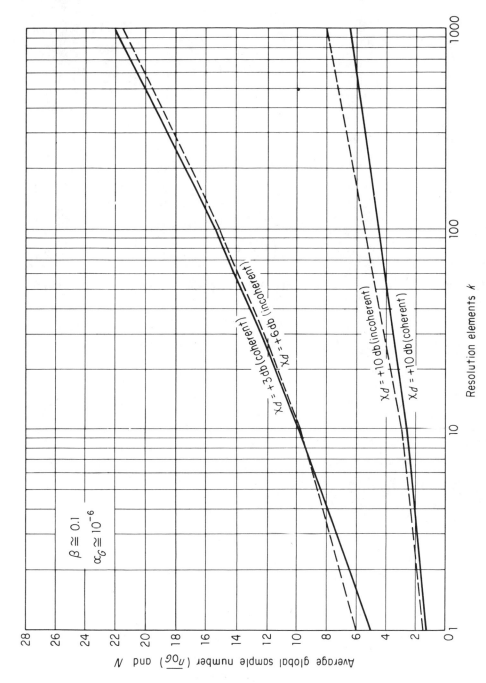

Fig. 16.15-1. ASN and N vs. number of resolution elements for coherent and incoherent detectors employing a binomial Marcus-Swerling test with signal absent (*after Dillard* [11], *courtesy of IEEE*).

From the results of Section 16.5, the expressions for P_{i0} and P_{ia} can be found. For the coherent detector,

$$P_{i0} = \Phi\left(\frac{q}{\sigma}\right) \qquad (16.15\text{-}4)$$

$$P_{ia} = \Phi\left(\frac{q}{\sigma} - a\right) \qquad (16.15\text{-}5)$$

For an incoherent detector, from Eq. (16.11-1),

$$P_{i0} = \int_{q^*}^{\infty} r_i e^{-r_i^2/2}\, dr_i = \exp\left[-\frac{(q^*)^2}{2}\right] \qquad (16.15\text{-}6)$$

where $q^* = q/\sigma$, and

$$P_{ia} = \int_{q^*}^{\infty} r_i e^{-(r_i^2 + a^2)/2} I_0(ar_i)\, dr_i \qquad (16.15\text{-}7a)$$

$$P_{ia} = 1 - \exp\left[-\frac{(q^*)^2 + a^2}{2}\right] \sum_{n=1}^{\infty} \left(\frac{q^*}{a}\right)^n I_n(aq^*) \qquad (16.15\text{-}7b)$$

where $I_n(x)$ is the modified Bessel function of the first kind and of order n.

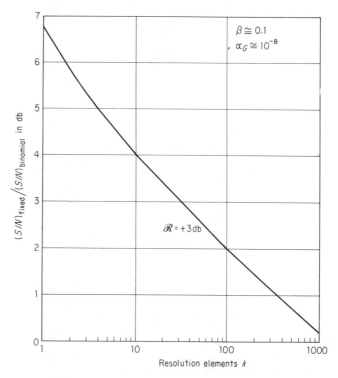

Fig. 16.15-2. Power saving of binomial coherent detector over fixed-sample coherent detector for ASN = N in a Marcus-Swerling test (*after Dillard* [11], *courtesy of IEEE*).

The integral in (16.15-7a) is called the Q function and is discussed in Section 9.3.

For any given signal level the values of P_{i0} and P_{ia} can be computed from relations (16.15-4) through (16.15-7b) as a function of q. Care must be exercised in selecting the value of q; Dillard [11] chooses a value for q that minimizes the ASN $(\bar{n}_0)$ of a sequential test in a single resolution element when only noise is present and with α, β held fixed. Wald [1] has a relation for the ASN of a binomial test which can be minimized (as discussed by Dillard). For this case a sequential test using (16.15-3) was simulated; results are plotted in Figs. 16.15-1 and 16.15-2 for a coherent and incoherent detector. Figure 16.15-2 shows a decrease in (S/N) improvement of almost 3 db for the binomial (coherent) test with respect to the analogue test when $k = 1000$. This loss can be reduced by increasing the number of quantization states.

The Marcus and Swerling test has been modified by Kendall and Reed [12] to include multiple targets. Evaluation of performance of the modified test requires simulation; to date, no results are available. (See Problem 16.8 for multiple-target strategy.)

16.16. Average Loss Formulation

The general problem considered in this chapter can be formulated in terms of average loss [1, 3]. For a fixed-sample test the average loss given by Eq. (8.5-12) can be rewritten as [with $C_{1-\beta}(\boldsymbol{\theta}) = 0$]

$$L_{fs}(D, \sigma) = \mathfrak{q}\alpha C_\alpha + \mathfrak{p}\overline{\beta_d}C_\beta \qquad (16.16\text{-}1)$$

where α is defined by Eq. (8.5-10) and $\overline{\beta_d}$ by Eq. (8.5-11) with the random parameters of the signal equal to design values $\boldsymbol{\theta}_d$. For a sequential test with random n, an average cost of experimentation is added to the above expression. Let c represent the cost of each observation; then the average loss becomes

$$L_{\text{seq}} = \mathfrak{q}[\alpha C_\alpha + c\overline{n(0)}] + \mathfrak{p}[\overline{\beta_d}C_\beta + c\overline{n(\boldsymbol{\theta}_d)}] \qquad (16.16\text{-}2)$$

where $\overline{n(0)}$ is the average test duration with noise alone and $\overline{n(\boldsymbol{\theta}_d)}$ the average duration with a signal having design parameters $\boldsymbol{\theta}_d$. Average-loss expression (16.16-2) includes the cost of both fixed-sample tests ($c = 0$) and sequential tests.

The problem can be formulated in even more general terms by analysis of loss expression (8.5-6). However, the number of integrations is a random variable, resulting in complicated expressions. This approach to sequential detection has not yet been extensively applied to the radar problem.

16.17. Two-Step Sequential Detection

A two-step approximation to a sequential detection test was first described by Finn [13]. In this test, a pulse is transmitted in each antenna beam position; if a threshold crossing occurs in any range bin (or range-doppler bin), a second pulse† of greater energy is transmitted in the same beam position. If a second threshold is exceeded in the same range cell, a target is declared present. The second transmission is sometimes called a verification pulse. This procedure is called "energy variant sequential detection" by Finn [13] and two-step detection by Brennan and Hill [14].

Following the first pulse transmission the envelope of the received signal in each range bin is compared with a threshold r_{b_1}. The false-alarm probability in each bin is given by [see Eq. (9.3-26b)]

$$P_{fa_1} = \exp\left(-\frac{r_{b_1}^2}{2}\right) \qquad (16.17\text{-}1)$$

If a signal is present, the probability of detection is given by [see Eqs. (9.3-40) and (9.3-41)]

$$P_{d_1} = Q(r_{b_1}, \sqrt{\mathscr{R}_1}) \qquad (16.17\text{-}2)$$

where the Q function is defined in (9.3-41), and $\mathscr{R}_1$ is peak signal-to-noise ratio. When threshold r_{b_1} is exceeded on the first transmission, a second pulse is transmitted in the same beam position. Only those resolution cells are examined for which threshold r_{b_1} is exceeded on the first pulse. If P_{fa_2} is the probability of a false alarm in any range bin on the second transmission, then the probability of a false alarm occurring twice in the same range bin is $P_{fa_1} \cdot P_{fa_2}$. By the methods of Chapter 13, the total false-alarm probability, assuming k resolution cells in each beam position, is given by

$$P_{fa} = 1 - (1 - P_{fa_1} \cdot P_{fa_2})^k \qquad (16.17\text{-}3)$$

Substituting (16.17-1) into (16.17-3) yields

$$P_{fa} = 1 - \left[1 - \exp\left(-\frac{r_{b_1}^2 + r_{b_2}^2}{2}\right)\right]^k \qquad (16.17\text{-}4)$$

where r_{b_2} is the threshold employed on the second sweep. For large r_{b_1} and r_{b_2} (16.17-4) can be approximated by

$$P_{fa} \cong k \exp\left(-\frac{r_{b_1}^2 + r_{b_2}^2}{2}\right) \qquad (16.17\text{-}5)$$

The probability of detecting a target after both transmissions is

$$P_d = Q(r_{b_1}, \sqrt{\mathscr{R}_1}) \cdot Q(r_{b_2}, \sqrt{\mathscr{R}_2}) \qquad (16.17\text{-}6)$$

where $\mathscr{R}_2$ is the peak signal-to-noise ratio of the second pulse (in the same range cell).

---

†To obtain higher energy on the second transmission, a pulse train can be employed; in this case, losses relating to pulse-train processing must be included.

With expressions (16.17-5) and (16.17-6), P_{fa} and P_d can be evaluated given the two thresholds and signal-to-noise ratios. Optimum performance is obtained by maximizing probability of detection P_d under the constraints of fixed false-alarm probability and fixed total average energy transmitted in each beam position. This is accomplished by optimizing the choice of threshold values and the distribution of signal energy between the first and second pulses. Brennan and Hill [14] employed a digital computer to perform the necessary calculations for $P_{fa} = 10^{-6}$. Thus, the first constraint is

$$P_{fa} = k \exp\left(-\frac{r_{b_1}^2 + r_{b_2}^2}{2}\right) = 10^{-6} \qquad (16.17\text{-}7)$$

For a low target density environment, the probability of a false alarm on the first sweep in any beam position is considerably greater than the probability of a target detection. Hence, the total average energy $\mathscr{E}$ per beam position

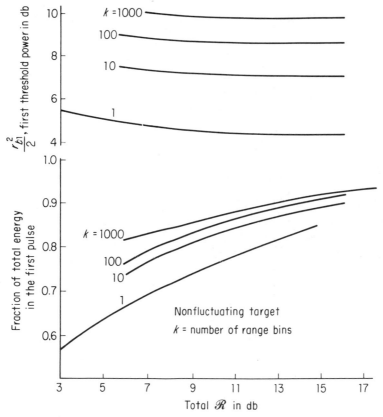

Fig. 16.17-1. Two-step optimum parameters (*after Brennan and Hill* [14], *courtesy of IEEE*).

is given by

$$\mathscr{E} \cong \mathscr{E}_1 + \mathscr{E}_2[1 - (1 - P_{fa_1})^k] \tag{16.17-8}$$

Multiplying (16.17-8) by $2/N_0$ yields,

$$\mathscr{R} \cong \mathscr{R}_1 + \mathscr{R}_2[1 - (1 - P_{fa_1})^k] \cong \mathscr{R}_1 + \mathscr{R}_2 k \exp\left(-\frac{r_{b_1}^2}{2}\right) \tag{16.17-9}$$

where $\mathscr{R}$ is the total average signal-to-noise ratio per beam position.

Optimum performance parameters can be obtained from Fig. 16.17-1. The first threshold r_{b_1} can be read from the top curves, where threshold power $(r_{b_1}^2/2)$ is plotted versus $\mathscr{R}$; the fraction of total energy on the first pulse is obtained from the lower curves. With Fig. 16.17-1 and Eq. (16.17-9), a typical result obtained for $k = 100$ and $P_{fa} = 10^{-6}$ is $\mathscr{R} = 13$ db, $\mathscr{R}_1 = 12.5$ db, and $\mathscr{R}_2 = 16.2$ db.

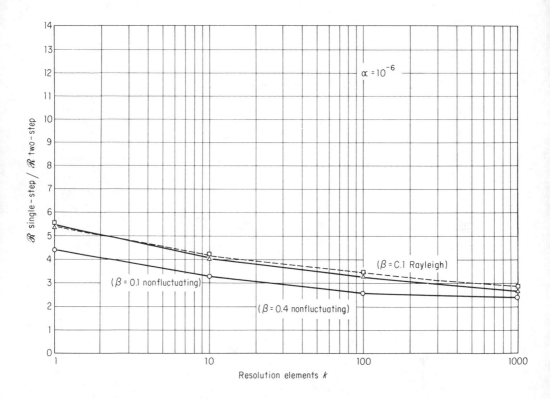

Fig. 16.17-2. Power saving of optimized two-step detector over a single-pulse system.

The power saving of an optimized two-step detector, relative to a single-pulse system with equal false-alarm and detection probabilities, is shown in Fig. 16.17-2. Results have been obtained for a nonfluctuating target for $P_d = 90$ per cent and 60 per cent, and for a Rayleigh fluctuating target for $P_d = 90$ per cent. Although these results cannot be compared accurately with earlier results, since the curves of Fig. 16.14-2 for a coherent detector must be adjusted for the difference in α and for the loss due to unknown initial phase, an approximate evaluation can be made. With some allowance for these differences, the two-step detector is relatively poor with respect to a true sequential detector for low values of k. However, for a large number of resolution bins, $k \geqslant 100$, the optimized two-step detector performs nearly as well as the sequential detector discussed in Section 16.14.

PROBLEMS

16.1. For RC noise with autocorrelation function $\phi_{nn}(\tau) = \exp\left(-\alpha\,|\tau|\right)$, $\alpha > 0$, verify the inverse matrix ρ^{-1} in (16.6-10). Show that Eq. (16.6-7) leads to Eq. (16.6-12) for large $\bar{n}_a$.

16.2. Find a linear transformation of the noise variables with covariance ρ given by (16.6-9) such that the covariance matrix of the transformed variables is diagonal. Next find Z_m given by (16.4-1) in terms of the new variables and, neglecting end effects, relate the result to Eq. (16.6-14).

16.3. In Sections 16.5 and 16.6, the signal was assumed to have a constant, but unknown, amplitude at the sampling instants. Consider a signal with Gaussian statistics, mean zero and mean-square value A^2. For Gaussian noise with mean zero and mean-square value σ^2, formulate the joint density functions of samples **v** for noise alone and for signal plus noise assuming that signal and noise have the same exponential autocorrelation function.
  (a) Find Z_m as in Section 16.6. Find the expression for Z_m for zero correlation ($\nu = 0$).
  (b) As in Section 16.6 find h for both the correlated and uncorrelated cases. Discuss the results.
  (c) For the correlated case find $\bar{Z}_n$ and $\overline{n(a)}$, where $a = A/\sigma$. Do any correlation terms appear in the result? Explain. What are the implications of this answer? (See Bussgang and Middleton [3].)

16.4. From characteristic functions (16.9-9b) and (16.9-14) find approximate expressions for $\bar{n}_0$, $\sigma^2_{n_0}$, $\bar{n}_{(k=1)}$, $\sigma^2_{(k=1)}$ using the formula

$$E[n^u] = (-j)^u \left.\frac{\partial^u C_n(\xi)}{\partial \xi^u}\right|_{\xi=0}$$

for computing the uth moment of n. Compare these expressions with Eqs. (16.9-18) to (16.9-21). Why are these results approximate?

16.5. With the results of Section 16.9 plot the pdf of $n(k)$ when $\alpha = 10^{-4}, k = 1$, for values of χ_d equal to 0 db, -6 db, -20 db, and -30 db. Repeat the plots for $n(0)$ and $\beta = 0.1$.

16.6. Carry out the integration in Eq. (16.13-16) to verify Eq. (16.13-17). [This result corresponds to $F(N)$ in reference [7].]

16.7. Using the results of Section 16.13, compute $P(n)$ and find $\bar{n}_{0_G}$ for $k = 10$ to verify the plot shown on Fig. 16.13-1.

16.8. Consider N resolution elements with a priori probability q that any given element contains a target. Let the appearance of targets of amplitude a be independent in each element.

(a) Show that

$p_m(\mathbf{v}_1, \mathbf{v}_2, \ldots, \mathbf{v}_N)$
$$= [1 - (1 - q)^N]p_m(\mathbf{v}_1, \mathbf{v}_2, \ldots, \mathbf{v}_N \,|\, \text{one or more targets present})$$
$$+ (1 - q)^N\, p_m(\mathbf{v}_1, \mathbf{v}_2, \ldots, \mathbf{v}_N \,|\, 0)$$

(b) For independent noise in each of the N elements find $p_m(\mathbf{v}_1, \mathbf{v}_2, \ldots, \mathbf{v}_N \,|\, 0)$ if $p_m(\mathbf{v}_i \,|\, 0)$ represents the pdf of vector $\mathbf{v}_i$, given that no target is present in the ith element.

(c) With the results of (b) find the marginal density function $p_m(\mathbf{v}_1, \mathbf{v}_2, \ldots, \mathbf{v}_N)$ in terms of the separate density functions. Prove that

$$p_m(\mathbf{v}_1, \mathbf{v}_2, \ldots, \mathbf{v}_N) = (1 - q)^N \prod_{i=1}^{N} p_m(\mathbf{v}_i \,|\, 0) + q(1 - q)^N p_m(\mathbf{v}_1 \,|\, a) \prod_{i=2}^{N} p_m(\mathbf{v}_i \,|\, 0)$$

$$+ \ldots + q^N \prod_{i=1}^{N} p_m(\mathbf{v}_i \,|\, a)$$

$$= \prod_{i=1}^{N} [q p_m(\mathbf{v}_i \,|\, a) + (1 - q)p_m(\mathbf{v}_i \,|\, 0)]$$

(d) With the results above show that

$p_m(\mathbf{v}_1, \ldots, \mathbf{v}_N) \,|\, \text{one or more targets are present})$

$$= \frac{\prod_{i=1}^{N} [q\, p_m(\mathbf{v}_i \,|\, a) + (1 - q)p_m(\mathbf{v}_i \,|\, 0)] - \prod_{i=1}^{N} (1 - q)p_m(\mathbf{v}_i \,|\, 0)}{1 - (1 - q)^N}$$

(e) Form the sequential probability-ratio test

$$\lambda = \frac{p_m(\mathbf{v}_1, \mathbf{v}_2, \ldots, \mathbf{v}_N \,|\, \text{one or more targets present})}{p_m(\mathbf{v}_1, \ldots, \mathbf{v}_N \,|\, 0)}$$

This ratio is compared with thresholds $\mathscr{A}$ and $\mathscr{B}$. For $Q = q/(1 - q)$, show that

$$\lambda = \frac{\prod_{i=1}^{N} (1 + Q\lambda_i) - 1}{(1 + Q)^N - 1}$$

where $\lambda_i = p_m(\mathbf{v}_i \,|\, a)/p_m(\mathbf{v}_i \,|\, 0)$ is the likelihood ratio of a single target in the ith resolution element. (The result above was obtained by Kendall and Reed [12]. The performance of this test for multiple targets has not yet been evaluated.)

(f) Show that

$$\lim_{Q \to 0} \lambda = \frac{1}{N} \sum_{i=1}^{N} \lambda_i$$

which is the Marcus and Swerling test discussed in Section 16.14.

REFERENCES

1. Wald, A.: "Sequential Analysis," Wiley, New York, 1947.

2. Wald, A., and Wolfowitz, J.: Optimum Character of the Sequential Probability Ratio Test, *Ann. Math. Stat.*, **19**: 326 (1948).

3. Bussgang, J. J., and Middleton, D.: Optimum Sequential Detection of Signals in Noise, *IRE Trans. on Information Theory*, **IT-1**: (*3*), 5–18 (December, 1955).

4. Reich, E., and Swerling, P.: The Detection of a Sine-wave in Gaussian Noise, *J. Appl. Phys.*, **24**: 289–296 (March, 1953).

5. Abramowitz, M., and Stegun, I. A.: Handbook of Mathematical Functions, U.S. Department of Commerce, National Bureau of Standards, AMS 55 (June, 1964).

6. Rice, S. O.: Mathematical Analysis of Random Noise, *Bell System Tech. J.*, **23**: 282–332 (1944); **24**: 46–156 (1945).

7. Reed, I. S., and Selin, I.: A Sequential Test for the Presence of a Signal in One of *k* Possible Positions, *IEEE Trans. on Information Theory*, **IT-9**: (*4*), 286–288 (October, 1963).

8. Helstrom, C. W.: A Range Sampled Sequential Detection System, *IRE Trans. on Information Theory*, **IT-8**: (*1*), 43–47 (January, 1962).

9. Bussgang, J. J., and Ehrman, L.: A Sequential Test for a Target in One of *k* Range Positions, *Proc. IEEE*, **53**: (*5*), 495–496 (May, 1965).

10. Marcus, M. B., and Swerling, P.: Sequential Detection in Radar with Multiple Resolution Elements, *IRE Trans. on Information Theory*, **IT-8**: (*3*), 237–245 (April, 1962).

11. Dillard, G. M.: The Binomial Marcus and Swerling Test, *IEEE Trans. on Information Theory*, **IT-11**: (*1*), 145–147 (January, 1965).

12. Kendall, W. B., and Reed, I. S.: A Sequential Test for Radar Detection of Multiple Targets, *IEEE Trans. on Information Theory*, **IT-9**: (*1*), 51–53 (January, 1963).

13. Finn, H. M.: A New Approach to Sequential Detection in Phased Array Radars, *Proc. 1963 National Winter Convention on Military Electronics* (1963).

14. Brennan, L. E., and Hill, F. S. Jr.: A Two-Step Sequential Procedure for Improving the Cumulative Probability of Detection in Radars, *IEEE Trans. on Military Electronics*, **MIL-9**: (*3 & 4*), 278–287 (July/October, 1965).

17

MULTIPLE-TARGET
DETECTION

17.1. Introduction

In this chapter Bayes decision rules are developed for optimizing the detection of multiple targets. We show that the strategy consists of an independent threshold test at each expected time of arrival for all targets in the radar field of view. This separable test strategy is shown to be optimum for the limiting cases of large and small signal-to-noise ratios; however, a different threshold level is required, in general, for large signal detection and for small signal detection. The separable test strategy is compared to the *averaging* solution, which is obtained from a Bayes strategy with cost assignments that are independent of parameters. The averaging approach is shown to be unsuitable for most radars when echo arrival times are unknown.

17.2. Approach to the Problem

Optimum detection theory for an exactly specified signal in noise is well known (see Chapter 9). The detection of a signal with unknown arrival time has been treated as an M-ary detection problem by Peterson and Birdsall [1], where the signal is assumed to occur in one of M discrete time positions. This approach was generalized by Middleton and Van Meter [2] to classes of signals. Their analysis [3] leads to averaging over the target parameter regions of uncertainty. Selin [4] employs averaging in treating the detection of a target with unknown doppler.

Multiple-target resolution has been treated by Helstrom [5] using the method of maximum likelihood; estimates of signal amplitudes are used as test statistics for the detection and resolution of multiple echoes with known arrival times and large signal-to-noise ratios. This approach, while not optimum, provides insight into the multiple-target problem. A modification of Helstrom's analysis to include unknown arrival time results in averaging. Thomas and Wolf [6] consider the detection of multiple targets for a specific set of cost assignments. They obtain a generalized likelihood-ratio test that maximizes the a posteriori probability measure over various signal groupings. The solution requires the formation of difference functions, which grow rapidly as the number of targets increases. The most direct approach to the problem of multiple-target detection (and parameter estimation) is due to Nilsson [7], who introduces a hybrid loss function that assigns penalties to both detection and estimation errors; the loss is then minimized with respect to an a posteriori probability measure. The solution requires a computer to perform a multidimensional maximization; as the number of targets grows, the solution grows in difficulty and computer capacity.

In this chapter optimum detection of multiple signals with unknown arrival times is treated from the point of view of minimizing the average loss for a cost matrix based on the general radar surveillance problem. One complete observation of the radar surveillance volume is assumed to be available for processing. Further, echo arrival times are limited to discrete positions. (The extension to continuous range delays is complex and does not add much insight.) Finally, only point targets are considered.

17.3. Detection of Two Known Targets

Consider the detection of the two nonoverlapping exactly known signals shown in Fig. 17.3-1. Signals $s_A(t)$ and $s_B(t)$ are assumed to be displaced by a multiple number of pulse widths Δ_τ; A and B represent both the re-

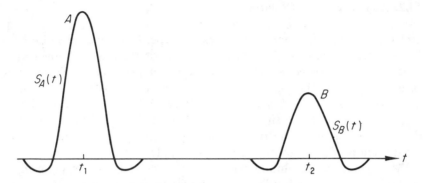

Fig. 17.3-1. Two-target problem.

spective signal amplitudes and target designations. The optimum decision rules $D(\mathbf{d}\,|\,\mathbf{v})$ that minimize the average loss $L(D, \sigma)$, defined below in Eq. (17.3-2), describe the Bayes strategy for this problem. Signal samples $s_A(t_1)$ and $s_B(t_2)$ are assumed to be statistically independent with respect to existence, and noise samples $n(t_1) = n_1$ and $n(t_2) = n_2$ are assumed to be uncorrelated and members of a stationary Gaussian random process. It is straightforward to show that

$$\overline{(v_1 - \bar{v}_1)(v_2 - \bar{v}_2)} = \overline{[s_A(t_1) + n_1 - \overline{s_A(t_1)}][s_B(t_2) + n_2 - \overline{s_B(t_2)}]} = 0 \qquad (17.3\text{-}1)$$

where v_1 and v_2 are the received data samples containing signal plus noise.

Statistical independence of the signals will generally prevail in most radar situations. Independence of the noise variates n_1 and n_2 is approximately satisfied by noise samples from a narrowband noise process that are separated in time by $1/\Delta_f$ or more, where Δ_f is the system bandwidth.

The expression for average loss is given by Eq. (8.4-7b) and is repeated here for convenience:

$$L(D, \sigma) = \int_\Omega \sigma(\mathbf{s})\, d\mathbf{s} \int_\Gamma p(\mathbf{v}\,|\,\mathbf{s})\, d\mathbf{v} \int_\Delta C(\mathbf{s}, \mathbf{d})D(\mathbf{d}\,|\,\mathbf{v})\, d\mathbf{d} \qquad (17.3\text{-}2)$$

where $\mathbf{s}$ denotes the signal components in space Ω, $\sigma(\mathbf{s})$ is the signal joint a priori probability density function, $\mathbf{v}$ are the observed samples in space Γ, $p(\mathbf{v}\,|\,\mathbf{s})$ is the conditional probability density of $\mathbf{v}$ given $\mathbf{s}$, $\mathbf{d}$ are elements in decision space Δ, $D(\mathbf{d}\,|\,\mathbf{v})$ are the decision rules, and $C(\mathbf{s}, \mathbf{d})$ is the cost function. For discrete known signals s_A and s_B, the joint a priori probability of signal is given by

$$\sigma(\mathbf{s}) = \mathfrak{p}_{00}\, \delta(s_A - 0, s_B - 0) + \mathfrak{p}_{A0}\, \delta(s_A - A, s_B - 0)$$
$$+ \mathfrak{p}_{0B}\, \delta(s_A - 0, s_B - B) + \mathfrak{p}_{AB}\, \delta(s_A - A, s_B - B) \qquad (17.3\text{-}3)$$

where $\mathfrak{p}_{00}$, $\mathfrak{p}_{A0}$, $\mathfrak{p}_{0B}$, and $\mathfrak{p}_{AB}$ are respectively the a priori probabilities of no signal, signal s_A alone, s_B alone, and both s_A and s_B, and $\delta(\ )$ denotes the Dirac delta function.

A cost matrix can be defined as shown in Table 17.3-1. The diagonal terms, which are the costs of correct decisions, have been set equal to zero. The cost of a type I error (false alarm) is denoted by C_f and the cost of a type II error (missed detection) by C_m. For a surveillance radar, the cost of a false alarm is usually small; however, the cost of a missed detection can be high. For this reason, it is assumed that $C_m \gg C_f$; in this case the cost matrix in Table 17.3-1 reduces to that shown in Table 17.3-2. From

TABLE 17.3-1

Cost Matrix for Two-Signal Detection

		Signal $\mathbf{s} = (s_A, s_B)$			
		0, 0	A, 0	0, B	A, B
	0, 0	0	C_m	C_m	$2C_m$
Decision	A, 0	C_f	0	$C_m + C_f$	C_m
$\mathbf{d} = (d_1, d_2)$	0, B	C_f	$C_m + C_f$	0	C_m
	A, B	$2C_f$	C_f	C_f	0

TABLE 17.3-2

Cost Matrix for $C_m \gg C_f$

	Signal $\mathbf{s}$			
	0, 0	A, 0	0, B	A, B
0, 0	0	C_m	C_m	$2C_m$
A, 0	C_f	0	C_m	C_m
0, B	C_f	C_m	0	C_m
A, B	$2C_f$	C_f	C_f	0

Decision $\mathbf{d}$

Eqs. (17.3-2), (17.3-3), Table 17.3-2, and the previous assumptions, the average loss can be written as†

$$
\begin{aligned}
L(D, \sigma) = &\int_{\Gamma_1}\int_{\Gamma_2} dv_1\, dv_2\, D(0, 0 \,|\, v_1, v_2)[C_m \mathfrak{p}_{A0}p(v_1 \,|\, A)p(v_2 \,|\, 0) \\
&\qquad\qquad + C_m \mathfrak{p}_{0B}p(v_1 \,|\, 0)p(v_2 \,|\, B) + 2C_m \mathfrak{p}_{AB}p(v_1 \,|\, A)p(v_2 \,|\, B)] \\
&+ \int_{\Gamma_1}\int_{\Gamma_2} dv_1\, dv_2\, D(A, 0 \,|\, v_1, v_2)[C_f \mathfrak{p}_{00}p(v_1 \,|\, 0)p(v_2 \,|\, 0) \\
&\qquad\qquad + C_m \mathfrak{p}_{0B}p(v_1 \,|\, 0)p(v_2 \,|\, B) + C_m \mathfrak{p}_{AB}p(v_1 \,|\, A)p(v_2 \,|\, B)] \\
&+ \int_{\Gamma_1}\int_{\Gamma_2} dv_1\, dv_2\, D(0, B \,|\, v_1, v_2)[C_f \mathfrak{p}_{00}p(v_1 \,|\, 0)p(v_2 \,|\, 0) \\
&\qquad\qquad + C_m \mathfrak{p}_{A0}p(v_1 \,|\, A)p(v_2 \,|\, 0) + C_m \mathfrak{p}_{AB}p(v_1 \,|\, A)p(v_2 \,|\, B)] \\
&+ \int_{\Gamma_1}\int_{\Gamma_2} dv_1\, dv_2\, D(A, B \,|\, v_1, v_2)[2C_f \mathfrak{p}_{00}p(v_1 \,|\, 0)p(v_2 \,|\, 0) \\
&\qquad\qquad + C_f \mathfrak{p}_{A0}p(v_1 \,|\, A)p(v_2 \,|\, 0) + C_f \mathfrak{p}_{0B}p(v_1 \,|\, 0)p(v_2 \,|\, B)]
\end{aligned}
$$

$$(17.3\text{-}4)$$

†This strategy is related to unconditional maximum-likelihood estimation. See Middleton [8], p. 963.

Equation (17.3-4) can be minimized—providing a minimum exists—by the following set of nonrandom Bayes decision rules:

(a) Choose decision rule $D_E(A, 0 \,|\, v_1, v_2) = 1$—that is, decide A is present and B absent, when

$$\mathfrak{p}_{A0}\ell_A(v_1) + \mathfrak{p}_{AB}\ell_A(v_1)\ell_B(v_2) > \frac{C_f\mathfrak{p}_{00}}{C_m} \qquad (17.3\text{-}5a)$$

$$\mathfrak{p}_{A0}\ell_A(v_1) > \mathfrak{p}_{0B}\ell_B(v_2) \qquad (17.3\text{-}5b)$$

$$\mathfrak{p}_{0B}\ell_B(v_2) + \mathfrak{p}_{AB}\ell_A(v_1)\ell_B(v_2) < \frac{C_f[\mathfrak{p}_{00} + \mathfrak{p}_{A0}\ell_A(v_1)]}{C_m} \qquad (17.3\text{-}5c)$$

where $\ell_A(v_1) = p(v_1\,|\,A)/p(v_1\,|\,0)$, $\ell_B(v_2) = p(v_2\,|\,B)/p(v_2\,|\,0)$. Let $D_E(A, 0\,|\,v_1, v_2) = 0$ if any of the above inequalities are not satisfied.

(b) In like manner, choose decision rule $D_E(0, B\,|\,v_1, v_2) = 1$—that is, decide B present and A absent—when

$$\mathfrak{p}_{0B}\ell_B(v_2) + \mathfrak{p}_{AB}\ell_A(v_1)\ell_B(v_2) > \frac{C_f\mathfrak{p}_{00}}{C_m} \qquad (17.3\text{-}6a)$$

$$\mathfrak{p}_{0B}\ell_B(v_2) > \mathfrak{p}_{A0}\ell_A(v_1) \qquad (17.3\text{-}6b)$$

$$\mathfrak{p}_{A0}\ell_A(v_1) + \mathfrak{p}_{AB}\ell_A(v_1)\ell_B(v_2) < \frac{C_f[\mathfrak{p}_{00} + \mathfrak{p}_{0B}\ell_B(v_2)]}{C_m} \qquad (17.3\text{-}6c)$$

and let $D_E(0, B\,|\,v_1, v_2) = 0$ if any of the inequalities are reversed.

(c) Likewise, choose decision rule $D_E(0, 0\,|\,v_1, v_2) = 1$—that is, decide both A and B are not present—when

$$\mathfrak{p}_{A0}\ell_A(v_1) + \mathfrak{p}_{0B}\ell_B(v_2) + 2\mathfrak{p}_{AB}\ell_A(v_1)\ell_B(v_2) < \frac{2C_f\mathfrak{p}_{00}}{C_m} \qquad (17.3\text{-}7a)$$

$$\mathfrak{p}_{A0}\ell_A(v_1) + \mathfrak{p}_{AB}\ell_A(v_1)\ell_B(v_2) < \frac{C_f\mathfrak{p}_{00}}{C_m} \qquad (17.3\text{-}7b)$$

$$\mathfrak{p}_{0B}\ell_B(v_2) + \mathfrak{p}_{AB}\ell_A(v_1)\ell_B(v_2) < \frac{C_f\mathfrak{p}_{00}}{C_m} \qquad (17.3\text{-}7c)$$

Let $D_E(0, 0\,|\,v_1, v_2) = 0$ if any of the inequalities are reversed.

(d) Finally, choose decision rule $D_E(A, B\,|\,v_1, v_2) = 1$—that is, both A and B present—when

$$\mathfrak{p}_{A0}\ell_A(v_1) + \mathfrak{p}_{0B}\ell_B(v_2) + 2\mathfrak{p}_{AB}\ell_A(v_1)\ell_B(v_2) > \frac{2C_f\mathfrak{p}_{00}}{C_m} \qquad (17.3\text{-}8a)$$

$$\mathfrak{p}_{A0}\ell_A(v_1) + \mathfrak{p}_{AB}\ell_A(v_1)\ell_B(v_2) > \frac{C_f[\mathfrak{p}_{00} + \mathfrak{p}_{0B}\ell_B(v_2)]}{C_m} \qquad (17.3\text{-}8b)$$

$$\mathfrak{p}_{0B}\ell_B(v_2) + \mathfrak{p}_{AB}\ell_A(v_1)\ell_B(v_2) > \frac{C_f[\mathfrak{p}_{00} + \mathfrak{p}_{A0}\ell_A(v_1)]}{C_m} \qquad (17.3\text{-}8c)$$

Let $D_E(A, B\,|\,v_1, v_2) = 0$ if any of the inequalities are reversed. At the boundaries between decision regions arbitrary selections can be made.

17.4. Interpretation of Two-Target Detection Strategy for Small Signal-to-Noise Ratios

For a general surveillance radar, the a priori probability of a target in a spatial resolution cell is normally low, so that $\mathfrak{p}_{A0}$, $\mathfrak{p}_{0B}$, and $\mathfrak{p}_{AB}$ are each very small compared to unity. Then, owing to the assumed independence of A and B,

$$\mathfrak{p}_{A0} = \mathfrak{p}_A(1 - \mathfrak{p}_B) \cong \mathfrak{p}_A \qquad (17.4\text{-}1a)$$

$$\mathfrak{p}_{0B} = \mathfrak{p}_B(1 - \mathfrak{p}_A) \cong \mathfrak{p}_B \qquad (17.4\text{-}1b)$$

$$\mathfrak{p}_{AB} = \mathfrak{p}_A \cdot \mathfrak{p}_B \qquad (17.4\text{-}1c)$$

$$\mathfrak{p}_{00} \cong 1 \qquad (17.4\text{-}1d)$$

where $\mathfrak{p}_A$ and $\mathfrak{p}_B$ are the separate a priori probabilities of the existence of A and B, respectively. It follows that $\mathfrak{p}_{AB} \ll \mathfrak{p}_{A0}$ or $\mathfrak{p}_{0B}$.

The likelihood ratios for exactly known signals with amplitudes A and B, respectively, in Gaussian noise can be written as [see Eq. (9.2-10)]

$$\ell_A(v_1) = \exp\left(-\frac{A^2}{2\sigma^2} + \frac{v_1 A}{\sigma^2}\right) \qquad (17.4\text{-}2)$$

$$\ell_B(v_2) = \exp\left(-\frac{B^2}{2\sigma^2} + \frac{v_2 B}{\sigma^2}\right) \qquad (17.4\text{-}3)$$

where σ^2 is the average noise power, and v_1 and v_2 denote the observed data at the matched filter outputs, at expected arrival times t_1 and t_2. For small signal-to-noise ratios ($A^2/\sigma^2 < 1$ and $B^2/\sigma^2 < 1$), $\ell_A(v_1)$ and $\ell_B(v_2)$ tend to unity. For this case decision rules (17.3-5) through (17.3-8) reduce to the following compact form:

(a) Choose $D_B(A, 0 \,|\, v_1, v_2) = 1$ when

$$\mathfrak{p}_A \ell_A(v_1) > \frac{C_f \mathfrak{p}_{00}}{C_m} \qquad (17.4\text{-}4a)$$

$$\mathfrak{p}_B \ell_B(v_2) < \frac{C_f \mathfrak{p}_{00}}{C_m} \qquad (17.4\text{-}4b)$$

Inequality (17.3-5b) can be neglected since (17.4-4) assures its satisfaction. If either of the above inequalities is not satisfied, let $D_B(A, 0 \,|\, v_1, v_2) = 0$.

(b) In like manner, choose $D_B(0, B \,|\, v_1, v_2) = 1$ when

$$\mathfrak{p}_B \ell_B(v_2) > \frac{C_f \mathfrak{p}_{00}}{C_m} \qquad (17.4\text{-}5a)$$

$$\mathfrak{p}_A \ell_A(v_1) < \frac{C_f \mathfrak{p}_{00}}{C_m} \qquad (17.4\text{-}5b)$$

Again the second inequality in (17.3-6) can be neglected. Let $D_B(0, B \,|\, v_1, v_2) = 0$ if the above are not true.

(c) Next choose $D_B(0, 0 \,|\, v_1, v_2) = 1$ when

$$p_A \ell_A(v_1) < \frac{C_f p_{00}}{C_m} \tag{17.4-6a}$$

$$p_B \ell_B(v_2) < \frac{C_f p_{00}}{C_m} \tag{17.4-6b}$$

Inequality (17.3-7a) is the sum of (17.3-7b) and (17.3-7c) and can be neglected since (17.4-6) assures its satisfaction. Let $D_B(0, 0 \mid v_1, v_2) = 0$ when (17.4-6) is not satisfied.

(d) Finally, choose $D_B(A, B \mid v_1, v_2) = 1$ when

$$p_A \ell_A(v_1) > \frac{\dot{C}_f p_{00}}{C_m} \tag{17.4-7a}$$

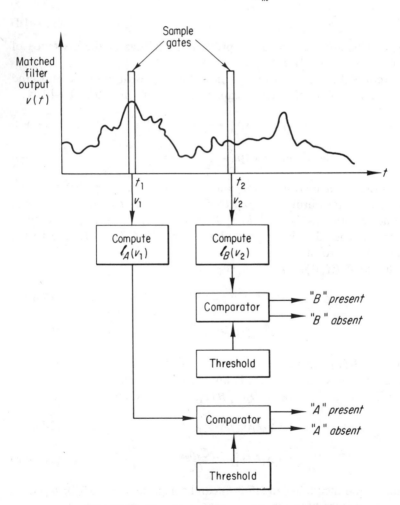

Fig. 17.4-1. Structure of Bayes detector for two known signals in noise (small S/N).

$$\mathfrak{p}_B \ell_B(v_2) > \frac{C_f \mathfrak{p}_{00}}{C_m} \qquad (17.4\text{-}7b)$$

and let $D_B(A, B \mid v_1, v_2) = 0$ when the opposite are true. [Again inequality (17.3-8a) is redundant and can be neglected.]

Rules (17.4-4) to (17.4-7) describe independent decisions at times t_1 and t_2. Thus, to decide A is present and not B, first compare $\ell_A(v_1)$ with a threshold and decide independently whether A is present; then, compare $\ell_B(v_2)$ with a different threshold and decide independently whether B is present. This can be restated as: choose $D_B(A, 0 \mid v_1, v_2) = 1$ when

$$D_B(A \mid v_1) = 1$$
$$D_B(0 \mid v_2) = 1 \qquad (17.4\text{-}8)$$

and so forth. Hence, the decision rules are separable and can be obtained by factoring the original decision rules. The structure of the Bayes decision receiver for small signal-to-noise ratio is shown in Fig. 17.4-1.

17.5. Interpretation of the Two-Target Detection Strategy for Large Signal-to-Noise Ratios

For large signal-to-noise ratios $(A^2/\sigma^2 > 1$ and $B^2/\sigma^2 > 1)$ decision rules (17.3-5) through (17.3-8) can be simplified. If only signal A is present, $\mathfrak{p}_A \ell_A(v_1)$ will be very large and $\mathfrak{p}_B \ell_B(v_2)$ will be very small most of the time. Then inequalities (17.3-5a) and (17.3-5c) reduce to

$$\ell_A(v_1) > \frac{C_f \mathfrak{p}_{00}}{\mathfrak{p}_A C_m} \qquad (17.5\text{-}1a)$$

$$\ell_B(v_2) < \frac{C_f}{\mathfrak{p}_B C_m} \qquad (17.5\text{-}1b)$$

Assume next that signal B alone is present. Then inequalities (17.3-6a) and (17.3-6c) simplify to

$$\ell_B(v_2) > \frac{C_f \mathfrak{p}_{00}}{\mathfrak{p}_B C_m} \qquad (17.5\text{-}2a)$$

$$\ell_A(v_1) < \frac{C_f}{\mathfrak{p}_A C_m} \qquad (17.5\text{-}2b)$$

A comparison of Eqs. (17.5-1a), (17.5-1b), (17.5-2a), and (17.5-2b) shows that the test for A when A alone is present requires a different threshold from the test for A when B alone is present (see Fig. 17.5-1). A similar result applies to the test for B. Equations (17.3-5b) and (17.3-6b) are not trivial in this case. For example, $\ell_A(v_1)$ can exceed the lower threshold of Fig. 17.5-1 but not the upper one. If $\ell_B(v_2)$ is just slightly below the upper threshold (and $\mathfrak{p}_A = \mathfrak{p}_B$), $\ell_B(v_2)$ may exceed $\ell_A(v_1)$, which violates inequality

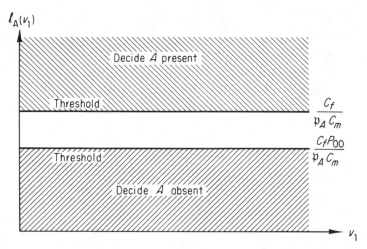

Fig. 17.5-1. Threshold test for presence of A alone (large S/N).

(17.3-5b). Thus, all three inequalities are required for a test of the presence of one target alone.

From Fig. 17.5-1 it is evident that the two thresholds differ by the factor $\mathfrak{p}_{00}$. The lower threshold corresponds to that obtained in the small signal-to-noise ratio case. Since $\mathfrak{p}_{00}$ approaches unity for a surveillance radar, the two thresholds in (17.5-1) are practically identical; for this situation the tests that yield decisions $D_B(A, 0 \mid v_1, v_2)$ and $D_B(0, B \mid v_1, v_2)$ reduce to the small signal-to-noise solution shown in Fig. 17.4-1. For large signal-to-noise ratios, inequalities (17.3-7) reduce to (17.4-6); inequalities (17.3-8) reduce to

$$\mathfrak{p}_A \ell_A(v_1) > \frac{C_f}{C_m} \tag{17.5-3a}$$

$$\mathfrak{p}_B \ell_B(v_2) > \frac{C_f}{C_m} \tag{17.5-3b}$$

For $\mathfrak{p}_{00}$ approaching unity, Eqs. (17.5-3a) and (17.5-3b) correspond to (17.4-7a) and (17.4-7b), respectively. Hence, the test strategy for large signal-to-noise ratios (and $\mathfrak{p}_{00} \cong 1$) is identical to that for small signal-to-noise ratios. However, in the region where $\mathfrak{p}_A \ell_A(v_1)$ and $\mathfrak{p}_B \ell_B(v_2)$ are of the order of unity, the decision rules are complex [see Eq. (17.3-8b), for example] and do not simplify.

17.6. Bayes Strategy for N Exactly Known Signals

For the detection of N exactly known signals, a square cost matrix similar to that in Table 17.3-2 can be formulated. The size of such a matrix grows as 2^{2N}. For example, the matrix for $N = 10$ contains about one million terms; approximately one thousand inequalities must be satisfied to obtain a set of Bayes rules. However, most of these conditions are redundant.

The N-target problem simplifies for small signal-to-noise ratios. In this case, higher-order terms of the form

$$\prod_i \mathfrak{p}_i \ell_i(v_i)$$

where $\mathfrak{p}_i$ is the a priori probability of the ith target alone, are much smaller than the first-order terms $\mathfrak{p}_j \ell_j$.† Further, $\mathfrak{p}_j \ell_j$ is much less than unity for all j, and $\mathfrak{p}_{0,0,\dots,0}$ (the a priori probability of no targets) is nearly equal to unity. Then, Eq. (17.3-2) for average loss becomes

$$L(D, \sigma) = \int_\Gamma D[0 \mid \mathbf{v}] \left[\sum_{i=1}^{N} C_m \mathfrak{p}_i \ell_i(v_i) \right] p(\mathbf{v} \mid \mathbf{0}) \, d\mathbf{v}$$

$$+ \int_\Gamma D[1, (0-1) \mid \mathbf{v}] \left[C_f \mathfrak{p}_0 + \sum_{i=2}^{N} C_m \mathfrak{p}_i \ell_i(v_i) \right] p(\mathbf{v} \mid \mathbf{0}) \, d\mathbf{v}$$

$$+ \int_\Gamma D[k, (0-k) \mid \mathbf{v}] \left[C_f \mathfrak{p}_0 + \sum_{\substack{i=1 \\ i \approx k}}^{N} C_m \mathfrak{p}_i \ell_i(v_i) \right] p(\mathbf{v} \mid \mathbf{0}) \, d\mathbf{v}$$

$$+ \int_\Gamma D[1, k, (0-1, k) \mid \mathbf{v}] \Big[2 C_f \mathfrak{p}_0 + C_f \mathfrak{p}_1 \ell_1(v_1) + C_f \mathfrak{p}_k \ell_k(v_k)$$

$$+ \sum_{\substack{i=2 \\ i \approx 1, k}}^{N} C_m \mathfrak{p}_i \ell_i(v_i) \Big] p(\mathbf{v} \mid \mathbf{0}) \, d\mathbf{v}$$

$$+ \int_\Gamma D[k, j, (0-k, j) \mid \mathbf{v}] \Big[2 C_f \mathfrak{p}_0 + C_f \mathfrak{p}_k \ell_k(v_k) + C_f \mathfrak{p}_j \ell_j(v_j)$$

$$+ \sum_{\substack{i=1 \\ i \approx j, k}}^{N} C_m \mathfrak{p}_i \ell_i(v_i) \Big] p(\mathbf{v} \mid \mathbf{0}) \, d\mathbf{v}$$

$+$ (all higher-order joint terms)

$$+ \int_\Gamma D[1, 2, \dots, N, (0-1, 2, \dots, N) \mid \mathbf{v}]$$

$$\cdot \left[N C_f \mathfrak{p}_0 + (N-1) \sum_{i=1}^{N} C_f \mathfrak{p}_i \ell_i(v_i) + \dots \right] p(\mathbf{v} \mid \mathbf{0}) \, d\mathbf{v} \qquad (17.6\text{-}1)$$

The Bayes strategy consists in choosing $D_B[1, (0-1) \mid \mathbf{v}] = 1$ when

$$\mathfrak{p}_1 \ell_1(v_1) > \frac{C_f \mathfrak{p}_0}{C_m} \qquad (17.6\text{-}2\text{a})$$

$$\mathfrak{p}_1 \ell_1(v_1) > \mathfrak{p}_k \ell_k(v_k), \qquad \text{all } k \neq 1 \qquad (17.6\text{-}2\text{b})$$

$$\mathfrak{p}_k \ell_k(v_k) < \frac{C_f \mathfrak{p}_0}{C_m} \qquad k \neq 1 \qquad (17.6\text{-}2\text{c})$$

$$\vdots$$

$$\sum_{\substack{i \\ i \approx 1}} \mathfrak{p}_i \ell_i(v_i) < \frac{(N-1) C_f \mathfrak{p}_0}{C_m} \qquad (17.6\text{-}2\text{d})$$

†The joint a priori probability of more than one target is assumed equal to the product of the individual a priori probabilities of each target alone.

Inequality (17.6-2a) follows from the first term in (17.6-1), (17.6-2b) from the third term, (17.6-2c) from the fourth term, and (17.6-2d) from the last term. Most of these conditions are redundant. Hence, the Bayes decision rule $D_B[1, (0 - 1)|\mathbf{v}] = 1$ (target 1 alone) is chosen when

$$\mathfrak{p}_1 \ell_1(v_1) > \frac{C_f \mathfrak{p}_0}{C_m} \tag{17.6-3a}$$

$$\mathfrak{p}_k \ell_k(v_k) < \frac{C_f \mathfrak{p}_0}{C_m}, \qquad k \neq 1 \tag{17.6-3b}$$

and $D_B[1, (0 - 1)|\mathbf{v}] = 0$ when any of the inequalities above is not satisfied. The test for the kth target is similar. These tests are separable, each consisting of a separate threshold test in each range bin (or, more generally, for each expected parameter set).

For large signal-to-noise ratios and $\mathfrak{p}_0$ approaching one, similar results are obtained. Thus, the optimum Bayes system for the detection of N exactly known signals, when signal-to-noise ratio is either small or large, corresponds to the simple strategy shown in Fig. 17.6-1. This strategy is subject only to very general conditions on the the cost matrix and a priori probabilities.

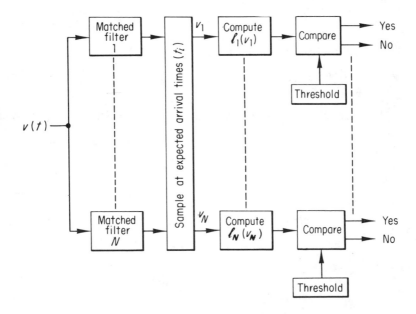

Fig. 17.6-1. Structure of Bayes receiver for detection of N expected targets in Gaussian noise.

17.7. Bayes Detection Strategy for N Targets with Unknown Times of Arrival

Consider the detection of a single target with unknown time of arrival at either of two discrete positions t_1 or t_2 shown in Fig. 17.3-1. Signals s_A and s_B now represent the same target at delays t_1 and t_2, respectively; the signal amplitudes are assumed equal.

Since the joint occurrence of both signals is impossible, $\mathfrak{p}_{AB} = 0$. Then, with the costs in Table 17.3-2, an average loss function similar to Eq. (17.3-4) can be derived, as follows:

$$L(\mathbf{D}, \sigma) = \int_{\Gamma_1} \int_{\Gamma_2} dv_1 \, dv_2 \, D(0, 0 \,|\, v_1, v_2)[C_m \mathfrak{p}_A p(v_1 \,|\, A) p(v_2 \,|\, 0)$$

$$+ \, C_m \mathfrak{p}_B p(v_1 \,|\, 0) p(v_2, B)]$$

$$+ \int_{\Gamma_1} \int_{\Gamma_2} dv_1 \, dv_2 \, D(A, 0 \,|\, v_1, v_2)[C_f \mathfrak{p}_{00} p(v_1 \,|\, 0) p(v_2 \,|\, 0)$$

$$+ \, C_m \mathfrak{p}_B p(v_1 \,|\, 0) p(v_2 \,|\, B)]$$

$$+ \int_{\Gamma_1} \int_{\Gamma_2} dv_1 \, dv_2 \, D(0, B \,|\, v_1, v_2)[C_f \mathfrak{p}_{00} p(v_1 \,|\, 0) p(v_2 \,|\, 0)$$

$$+ \, C_m \mathfrak{p}_A p(v_1 \,|\, A) p(v_2 \,|\, 0)]$$

$$+ \int_{\Gamma_1} \int_{\Gamma_2} dv_1 \, dv_2 \, D(A, B \,|\, v_1, v_2)[2C_f \mathfrak{p}_{00} p(v_1 \,|\, 0) p(v_2 \,|\, 0)$$

$$+ \, C_f \mathfrak{p}_A p(v_1 \,|\, A) p(v_2 \,|\, 0) + C_f \mathfrak{p}_B p(v_1 \,|\, 0) p(v_2 \,|\, B)]$$

$$(17.7\text{-}1)$$

Minimizing average loss in Eq. (17.7-1) yields the following Bayes decision rules (for $C_f \ll C_m$):

$$D_B(0, 0 \,|\, v_1, v_2) = 1 \quad \text{when} \quad \begin{cases} \mathfrak{p}_A \ell_a(v_1) < \dfrac{C_f \mathfrak{p}_{00}}{C_m} & (17.7\text{-}2a) \\[2ex] \mathfrak{p}_B \ell_B(v_2) < \dfrac{C_f \mathfrak{p}_{00}}{C_m} & (17.7\text{-}2b) \\[2ex] \mathfrak{p}_A \ell_A(v_1) + \mathfrak{p}_B \ell_B(v_2) < \dfrac{2C_f \mathfrak{p}_{00}}{C_m} & (17.7\text{-}2c) \end{cases}$$

$$D_B(A, 0 \,|\, v_1, v_2) = 1 \quad \text{when} \quad \begin{cases} \mathfrak{p}_A \ell_A(v_1) > \dfrac{C_f \mathfrak{p}_{00}}{C_m} & (17.7\text{-}3a) \\[2ex] \mathfrak{p}_A \ell_A(v_1) > \mathfrak{p}_B \ell_B(v_2) & (17.7\text{-}3b) \\[2ex] \mathfrak{p}_B \ell_B(v_2) < \dfrac{C_f[\mathfrak{p}_{00} + \mathfrak{p}_A \ell_A(v_1)]}{C_m} & (17.7\text{-}3c) \end{cases}$$

$$D_B(0, B \,|\, v_1, v_2) = 1 \quad \text{when} \quad \begin{cases} \mathfrak{p}_B \ell_B(v_2) > \dfrac{C_f \mathfrak{p}_{00}}{C_m} & (17.7\text{-}4a) \\[2ex] \mathfrak{p}_B \ell_B(v_2) > \mathfrak{p}_A \ell_A(v_1) & (17.7\text{-}4b) \\[2ex] \mathfrak{p}_A \ell_A(v_1) < \dfrac{C_f[\mathfrak{p}_{00} + \mathfrak{p}_B \ell_B(v_2)]}{C_m} & (17.7\text{-}4c) \end{cases}$$

$$D_B(A, B \mid v_1, v_2) = 1 \quad \text{when} \begin{cases} \mathfrak{p}_B \ell_B(v_2) > \dfrac{C_f[\mathfrak{p}_{00} + \mathfrak{p}_A \ell_A(v_1)]}{C_m} & (17.7\text{-}5a) \\[3ex] \mathfrak{p}_A \ell_A(v_1) > \dfrac{C_f[\mathfrak{p}_{00} + \mathfrak{p}_B \ell_B(v_2)]}{C_m} & (17.7\text{-}5b) \\[3ex] \mathfrak{p}_A \ell_A(v_1) + \mathfrak{p}_B \ell_B(v_2) > \dfrac{2C_f \mathfrak{p}_{00}}{C_m} & (17.7\text{-}5c) \end{cases}$$

and the corresponding D_B is set equal to zero when any of inequalities (17.7-2) through (17.7-5) is not satisfied.

As in Section 17.4, for small signal-to-noise ratios and small $\mathfrak{p}_A$ and $\mathfrak{p}_B$,† and for $\mathfrak{p}_{00}$ approaching unity, Eqs. (17.7-2) through (17.7-5) reduce to the decision rules:

$$D_B(0, 0 \mid v_1, v_2) = 1 \quad \text{when} \begin{cases} \mathfrak{p}_A \ell_A(v_1) < \dfrac{C_f \mathfrak{p}_{00}}{C_m} & (17.7\text{-}6a) \\[3ex] \mathfrak{p}_B \ell_B(v_2) < \dfrac{C_f \mathfrak{p}_{00}}{C_m} & (17.7\text{-}6b) \end{cases}$$

$$D_B(A, 0 \mid v_1, v_2) = 1 \quad \text{when} \begin{cases} \mathfrak{p}_A \ell_A(v_1) > \dfrac{C_f \mathfrak{p}_{00}}{C_m} & (17.7\text{-}7a) \\[3ex] \mathfrak{p}_B \ell_B(v_2) < \dfrac{C_f \mathfrak{p}_{00}}{C_m} & (17.7\text{-}7b) \end{cases}$$

$$D_B(0, B \mid v_1, v_2) = 1 \quad \text{when} \begin{cases} \mathfrak{p}_B \ell_B(v_2) > \dfrac{C_f \mathfrak{p}_{00}}{C_m} & (17.7\text{-}8a) \\[3ex] \mathfrak{p}_A \ell_A(v_1) < \dfrac{C_f \mathfrak{p}_{00}}{C_m} & (17.7\text{-}8b) \end{cases}$$

$$D_B(A, B \mid v_1, v_2) = 1 \quad \text{when} \begin{cases} \mathfrak{p}_B \ell_B(v_2) > \dfrac{C_f \mathfrak{p}_{00}}{C_m} & (17.7\text{-}9a) \\[3ex] \mathfrak{p}_A \ell_A(v_1) > \dfrac{C_f \mathfrak{p}_{00}}{C_m} & (17.7\text{-}9b) \end{cases}$$

and the corresponding D_B is set equal to zero when any of inequalities (17.7-6) through (17.7-9) is not satisfied; other inequalities are redundant.

Note that, according to the hypothesis, the target can appear either at position t_1 or t_2, but not at both. Hence the decision $D_B(A, B \mid v_1, v_2) = 1$ corresponds to at least one false alarm. This is not very serious, since false alarms do not influence detectability at the correct time of arrival.

From these rules it is apparent that when signal-to-noise ratio is small a test for a target of unknown time of arrival consists of a simple threshold comparison at each expected arrival time; this result does not apply, in general, for large signal-to-noise ratios ($\mathfrak{p}_i \ell_i > 1$).‡ The test implementation is shown in Fig. 17.4-1. The thresholds are determined by the assigned

---

† In general, $\mathfrak{p}_i$ will be very small when there are many range bins in a radar observation interval.

‡ Comparisons of likelihood ratios at various arrival times are required at large (S/N).

costs and the a priori probabilities. When extended to N discrete times of arrival, a test is performed at each of these times (see Fig. 17.6-1).

A comparison of Eqs. (17.7-6) through (17.7-9) with Eqs. (17.4-4) through (17.4-7) reveals that at low signal-to-noise ratios the solution for a signal with N discrete arrival times and the solution for N exactly known signals are identical in structure for the cost matrix of Table 17.3-2. For a radar with k resolution bins, it can be shown that the optimum Bayes strategy for one to N targets (for $N \ll k$) with unknown arrival times is a separate threshold test in each range bin in observation interval T; the threshold is a function of cost assignments and the a priori probabilities. With the assumptions that (a) the a priori existence probabilities $\mathfrak{p}_i = \mathfrak{p}_j$ for all i, j, (b) the joint a priori existence probability equals the product of individual a priori probabilities, and (c) each arrival time is uniformly distributed over the interval T, the threshold is approximately the same in each range bin for small signal-to-noise ratios. Under these conditions, the same elementary test is performed in each range bin for an arbitrary number of targets with unknown delays in a surveillance radar where a priori uncertainty is great.

17.8. Existence of a Single Target with Unknown Time of Arrival

The *existence* of a target with unknown arrival time at either of two discrete positions t_1 and t_2 (see Fig. 17.3-1) is established by deciding that the target is present either at t_1 or at t_2. An existence cost matrix can be obtained by modifying the cost matrix in Table 17.3-2 as shown in Table 17.8-1. Zero

TABLE 17.8-1

Pure Existence Cost Matrix

Signal **s**

$$
\begin{array}{c}
\\
\\
\text{Decision} \\
\mathbf{d}
\end{array}
\begin{array}{c}
\\
Decide \\
signal \\
present
\end{array}
\begin{array}{c}
\\
00 \\
\left\{ A0 \right. \\
\left. 0B \right.
\end{array}
\begin{array}{c}
00 \quad \overbrace{A0 \quad 0B}^{\begin{array}{c}Signal\\present\end{array}} \\
\begin{bmatrix}
0 & C_m & C_m \\
C_f & 0 & 0 \\
C_f & 0 & 0
\end{bmatrix}
\end{array}
$$

cost is assigned to a correct detection regardless of range location. The average loss equation is then given by

$$
L(D, \sigma) = \int_{\mathbf{r}_1} \int_{\mathbf{r}_2} dv_1 \, dv_2 \, D'(0, 0 \mid v_1, v_2)[C_m \mathfrak{p}_A p(v_1 \mid A) p(v_2 \mid 0)
$$

$$
+ C_m \mathfrak{p}_B p(v_1 \mid 0) p(v_2 \mid B)]
$$

$$+ \int_{\Gamma_1} \int_{\Gamma_2} dv_1\, dv_2\, D'(A, 0 \,|\, v_1, v_2)[C_f \mathfrak{p}_{00} p(v_1 \,|\, 0) p(v_2 \,|\, 0)]$$

$$+ \int_{\Gamma_1} \int_{\Gamma_2} dv_1\, dv_2\, D'(0, B \,|\, v_1, v_2)[C_f \mathfrak{p}_{00} p(v_1 \,|\, 0) p(v_2 \,|\, 0)] \qquad (17.8\text{-}1)$$

By previous methods minimizing Eq. (17.8-1) yields the Bayes decision rules:

Signal present:

$$\left.\begin{matrix} D'_B(A, 0 \,|\, v_1, v_2) = 1 \\ D'_B(0, B \,|\, v_1, v_2) = 1 \end{matrix}\right\} \text{ when } \quad \mathfrak{p}_A \ell_A(v_1) + \mathfrak{p}_B \ell_B(v_2) > \frac{C_f \mathfrak{p}_{00}}{C_m} \qquad (17.8\text{-}2a)$$

No signal:

$$D'_B(0, 0 \,|\, v_1, v_2) = 1 \quad \text{when} \quad \mathfrak{p}_A \ell_A(v_1) + \mathfrak{p}_B \ell_B(v_2) < \frac{C_f \mathfrak{p}_{00}}{C_m} \qquad (17.8\text{-}2b)$$

For small signal-to-noise ratios Eqs. (17.4-2) and (17.4-3) can be simplified. When the results are substituted into Eq. (17.8-2a), the threshold test for *signal present* becomes

$$\frac{A}{\sigma^2} (\mathfrak{p}_A v_1 + \mathfrak{p}_B v_2) > \mathfrak{p}_{00} \left(\frac{C_f + C_m}{C_m} \right) - 1 \qquad (17.8\text{-}3)$$

For the general case of N possible arrival times, threshold test (17.8-3) becomes

$$\frac{A}{\sigma^2} \sum_{j=1}^{N} \mathfrak{p}_j v_j > \mathfrak{p}_{00} \left(\frac{C_f + C_m}{C_m} \right) - 1 \quad (\text{signal present}) \qquad (17.8\text{-}4)$$

where $\mathfrak{p}_j$ is the a priori probability of a target in the jth range bin. The test described by (17.8-4) corresponds to *averaging* the output data over the region of expected arrival time. When the inequality above is not satisfied, the appropriate decision is no signal.

The a priori uncertainty of target arrival time is greatest when a uniform probability applies ($\mathfrak{p}_j = \mathfrak{p}_k$ for all j, k); in this case threshold test (17.8-4) reduces to

$$\sum_{1}^{N} v_j > \mathcal{K}$$

where the known parameters A, σ, and $\mathfrak{p}_j$ have been included in the threshold $\mathcal{K}$. The optimum Bayes strategy then consists of comparing a sum of the received data with a computed threshold. For small signal-to-noise ratios and for unknown target doppler with an existence cost matrix similar to Table 17.8-1, the optimum strategy again requires averaging.

When the region of uncertainty contains many resolution cells, the test strategy above results in the summation of many noise samples, and possibly one signal. This is a consequence of the fact that a pure existence cost matrix permits the acceptance of a noise pulse as a correct detection; that is, zero cost is assigned to deciding that the target is present at t_i when it is actually present at t_j. Thus, a noise pulse (regardless of location in the range trace) can be accepted as a correct detection of the target. This contrasts with the cost assignments of Table 17.3-2, where a detection in the wrong range bin is assessed as a missed detection.

The relative information gained in an averaging system is low, since gross errors in target location are possible. A receiver of this type is satisfactory when knowledge only of a target's presence is required. In general, pure existence unrelated to parameter estimation is not satisfactory for most radar applications.

17.9. Approximation of Bayes Receiver

For uniform a priori probability of a target in each resolution cell and for small signals, the threshold values of the Bayes receiver shown in Fig. 17.6-1 are the same. Then the Bayes receiver simplifies to that shown in Fig. 17.9-1. The receiver processing consists of a gated matched filter output and threshold testing at the expected times of arrival. (Gated matched filter output samples spaced $1/\Delta_f$ apart, where Δ_f is the signal bandwidth, are approximately independent.)

The discrete Bayes test in Fig. 17.9-1 can be approximated by a test

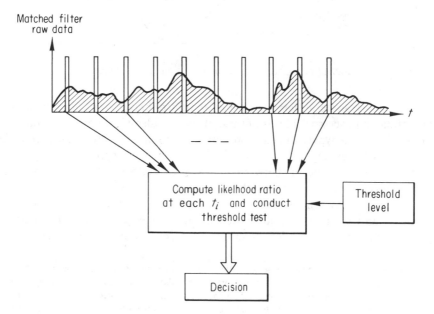

Fig. 17.9-1. Optimum Bayes search receiver.

performed continuously in real time as shown in Fig. 17.9-2. Since the target-echo initial phase is unknown in general, an envelope detector is employed following the matched filter (see Chapter 9) prior to thresholding. The detected envelope is continuously compared with a constant bias, or threshold, as shown in Fig. 17.9-2. Each threshold crossing counts as a detection.

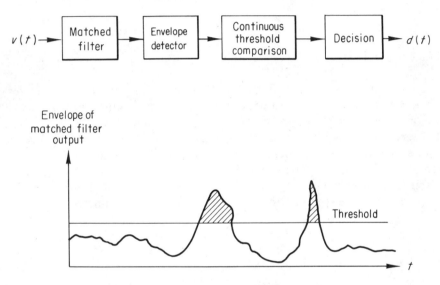

Fig. 17.9-2. Continuous approximation of Bayes receiver.

There is some difference in performance between a discrete and a continuous processor. This is illustrated by the false-alarm problem shown in Fig. 17.9-3. The waveform envelope out of the matched filter can exceed the threshold, for example, at position A. However, in a discrete sampling system a false alarm that occurs between sampling gates does not register; in a continuous system it does. One method of finding the actual false-alarm rate for a continuous system is to treat the output waveform as a stochastic process and count the number of threshold-level crossings from below.†

In Appendix B the false-alarm rate of a continuous system is computed

---

†This approach is explainable as follows. If a crossing occurs in region B of Fig. 17.9-3, the continuous system will count only one false alarm while the sampling system will count three false alarms. Since the probability of three successive false alarms, as shown in region B, is usually quite small, counting level crossings from below should give a reasonably accurate result.

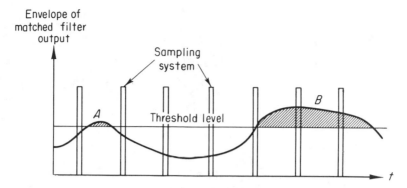

Fig. 17.9-3. False-alarm problem.

using the method above and compared with that of a discrete-sampled system. For a rectangular noise spectral density, it is shown that the mean false-alarm rate of a continuous system is approximately 3.1 times that of a sampled system for $P_{fa} = 10^{-4}$, and approximately 5.4 times the false-alarm rate of the sampled system for $P_{fa} = 10^{-12}$. These differences are not significant, especially since small-signal detectability in a continuous system is enhanced for the same reason that the false-alarm rate is greater.

17.10. Extension of Bayes Receiver to Angle and Doppler

To extend the optimum Bayes strategy to include target angle, Fig. 17.10-1 shows azimuthal beams A, B, and C being sequentially examined on successive range sweeps. Extending this approach, as indicated in Fig. 17.10-2, permits the search of a large volume surrounding a radar. The Bayes strategy corresponds to a test in each spatial resolution cell, where the cell dimensions are approximately equal to the antenna beamwidth (in two dimensions) and range resolution $\Delta_r \cong 1/\Delta_f$ in the radial direction, as shown in Fig. 17.10-2.

To extend the Bayes strategy to include unknown target doppler, note that for moderate bandwidth signals the doppler effect can be approximated by a simple frequency translation of the echo carrier frequency (see Appendix A). In analogy with the discrete time-of-arrival model, it is convenient to assume that expected doppler shifts can occur only at discrete frequencies separated by multiples of Δ_ω, where Δ_ω is equal to $2\pi\Delta_f$. Figure 17.10-3 shows two nonoverlapping doppler-shifted spectra for a simple pulsed-carrier waveform separated in doppler by $k\Delta_\omega$, where k is an integer. The Bayes strategy requires a threshold comparison at each expected time of arrival, for each expected doppler shift. This strategy can be implemented by a set of matched

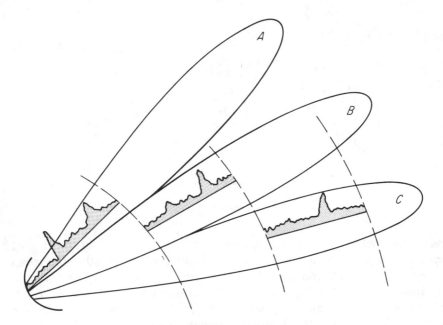

Fig. 17.10-1. Sequential search in azimuth.

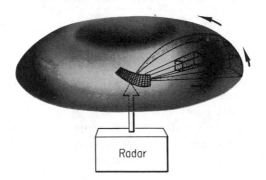

Fig. 17.10-2. Volume search of spatial resolution cells.

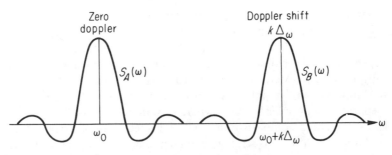

Fig. 17.10-3. Nonoverlapping doppler-shifted spectra.

filters, corresponding to the expected doppler shifts, that operate in parallel on the input data $v(t)$; each filter is followed by a sampler and threshold test. Sampled doppler-shifted matched filter outputs might appear as shown in Fig. 17.10-4. If a continuous range delay search is employed similar to that in Fig. 17.9-2, a Bayes receiver that includes doppler can be implemented as shown in Fig. 17.10-5. This corresponds closely to the range-

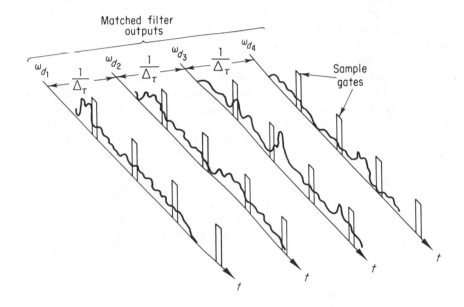

Fig. 17.10-4. Sampled doppler-shifted matched filter outputs for Bayes receiver.

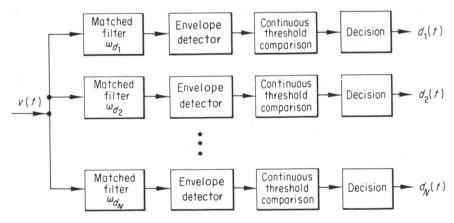

Fig. 17.10-5. Approximate Bayes receiver for range delay and doppler.

doppler processing employed in many surveillance radars, coupled with angle scanning for surveillance of a specified spatial region.

17.11. Difficulties with the Theory

The Bayes theory developed in this chapter is applicable to signals with discrete parameters and uncorrelated additive noise. In the real world, signal parameters are continuously distributed and noise is often partially correlated at sampling instants. These and other differences between theory and practice are now briefly considered.

The Bayes receiver in Fig. 17.10-5 is not optimum when phase-coded coherent radar waveforms are employed. Such waveforms typically have time-bandwidth products much greater than unity. An example of a waveform of this type is the pulsed linear fm waveform shown in Fig. 17.11-1

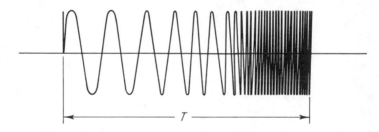

Fig. 17.11-1. Linear fm pulsed waveform.

[9]; the same waveform after matched filter processing is shown in Fig. 17.11-2. It is characterized by a large mainlobe and significant minor lobes (sidelobes) extending before and after the mainlobe. The overall duration of the output waveform is of the order of $2T$, where T is the duration of the transmitted signal. The peak-to-null width of the mainlobe is $2\pi/\Delta_\omega$, where Δ_ω is the radian signal bandwidth.

When a multiplicity of echoes is received from several closely spaced targets, the sidelobes of a large target echo limit the detectability of smaller nearby targets, and create ambiguous detections. These situations are illustrated in Fig. 17.11-3. For example, the sidelobes of the large return at A may interfere with the detectability of a small target echo at C as a result of the constructive or destructive rf addition of return C with the sidelobes of A. In addition, the sidelobes of target A may themselves exceed the threshold (see positions B), causing false indications of targets at these positions.

Note also that the threshold crossing at C occurs between sampling

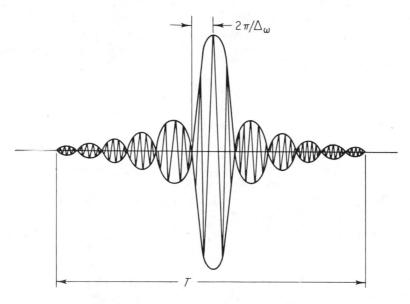

Fig. 17.11-2. Linear fm matched filter output.

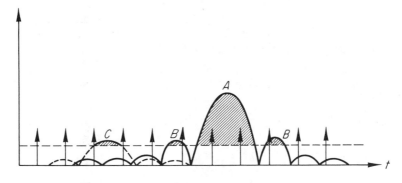

Fig. 17.11-3. Multitarget problem for both continuous and sampled radar.

gates, so that the probability of detection is different for a sampled and a continuous system; this effect is similar to the difference in false-alarm rates discussed in Section 17.9 and Appendix B. There is another difficulty associated with the large target at A: the continuous system indicates one target present while the sampling system indicates two. On the average the continuous system can be expected to produce fewer mistakes in data interpretation.

When the expected target doppler† shifts of closely spaced targets are described by continuous distributions, the Bayes optimization problem is very complex. Multiple-target resolution in range-doppler-angle involves both detection and simultaneous parameter estimation and is treated in a companion volume.

PROBLEMS

17.1. For two exactly known signals and the cost matrix in Table 17.3-1:
   (a) Let $C_m = C_f$ and develop the detection strategy for small signal-to-noise ratios; interpret the results.
   (b) If $p_{AB} \neq p_{A0} \cdot p_{0B}$, but $p_{AB} \ll p_{A0}$ and $p_{AB} \ll p_{0B}$, what changes in the detection strategy result?

17.2. Repeat problem 17.1 for large signal-to-noise ratio. Compare the detection rules for this case with previously developed rules. What effect does part 17.1(b) have on the results?

17.3. Construct a cost matrix similar to that in Table 17.3-2 for three exactly known nonoverlapping signals. Develop the Bayes detection strategy for both small and large signal-to-noise ratios.

17.4. Interpret problem 17.3 for the case of two nonoverlapping targets—one exactly known, and the other with unknown time of arrival described by the two remaining positions. How are the results affected if the a priori probabilities of existence of both targets are equal and the probability of each time of arrival for the second target is 0.5?

17.5. Verify the Bayes strategy described in the last paragraph of Section 17.7 for one to N targets with unknown arrival time by extending the theory of Section 17.6.

REFERENCES

1. Peterson, W., and Birdsall, T. G.: The Theory of Signal Detectability, Parts I and II, Tech. Report 13, University of Michigan (June, 1953). [Also with W. Fox, *IRE Trans. on Information Theory*, **PGIT-4:** 171 (1954).]

2. Middleton, D., and Van Meter, D.: On Optimum Multiple-Alternative Detection of Signals in Noise, *Trans. IRE*, **IT-1:** (2), 1–9 (September, 1955).

3. Wainstein, L. A., and Zubakov, V. P.: "Extraction of Signals from Noise," Prentice-Hall, Englewood Cliffs, N.J., 1962.

4. Selin, I.: Detection of Coherent Radar Returns of Unknown Doppler Shift, *Trans. IEEE*, **IT-11:** 396–400 (July, 1965).

†Or other parameters, except amplitude and initial phase.

5. Helstrom, C. W.: "Statistical Theory of Signal Detection," Macmillan, New York, 1960.
6. Thomas, J. B., and Wolf, J. K.: On the Statistical Detection Problem for Multiple Signals, *Trans. IRE*, **IT-8**: 274–280 (July, 1962).
7. Nilsson, N. J.: On the Optimum Range Resolution of Radar Signals in Noise, *IRE Trans. on Information Theory*, **IT-7**: (*4*), 245–253 (October, 1961).
8. Middleton, D.: "An Introduction to Statistical Communication Theory," McGraw-Hill, New York, 1960.
9. Cook, C. E., and Bernfeld, M.: "Radar Signals," p. 163, Academic Press, New York, 1967.

NARROWBAND REPRESENTATION OF ECHOES FROM MOVING TARGETS

If transmitted radar signal $s(t)$ is denoted by

$$s(t) = \text{Re}\left[\tilde{s}(t)e^{j\omega_0 t}\right] \tag{A-1}$$

where $\tilde{s}(t)$ is the complex envelope of $s(t)$, then echo $s_R(t)$ from a stationary target at range R_0 can be expressed as

$$s_R(t) = s(t - \tau_0) = \text{Re}\left[\tilde{s}(t - \tau_0)e^{j\omega(t - \tau_0)}\right] \tag{A-2}$$

where range delay τ_0 to the target is given by

$$\tau_0 = \frac{2R_0}{c} \tag{A-3}$$

and c is the velocity of propagation in the medium.

Suppose a target is moving at constant velocity v with respect to the radar. Then target range $R(t)$ at time t is given by

$$R(t) = R_0 + vt \tag{A-4}$$

where

$$R(0) = R_0 \tag{A-5}$$

By analogy with Eq. (A-2) the radar echo from a moving target with varying time delay $\tau(t)$ can be written as

$$s_R(t) = s[t - \tau(t)] = \text{Re}\,\{\bar{s}[t - \tau(t)]e^{j\omega_0[t - \tau(t)]}\} \tag{A-6}$$

It follows from equation (A-6) that a point on waveform $s_R(t)$ received at time t was transmitted at $[t - \tau(t)]$; this point was incident on the target at time $[t - \frac{1}{2}\tau(t)]$, at which time the target range was $R[t - \frac{1}{2}\tau(t)]$. The round-trip distance traveled by this point on the waveform is $2R[t - \frac{1}{2}\tau(t)]$; this distance, from the definition of $\tau(t)$, is also equal to $c\tau(t)$; that is,

$$2R[t - \tfrac{1}{2}\tau(t)] = c\tau(t) \tag{A-7}$$

Combining (A-4) and (A-7) yields

$$R_0 + v\left[t - \frac{1}{2}\tau(t)\right] = \frac{c}{2}\,\tau(t) \tag{A-8}$$

Solving (A-8) for $\tau(t)$, we have

$$\tau(t) = \frac{2R_0/c}{1 + v/c} + \frac{2vt/c}{1 + v/c} \tag{A-9}$$

With (A-9) substituted into (A-6), $s_R(t)$ becomes

$$s_R(t) = \text{Re}\left(\bar{s}\left[\left(\frac{c-v}{c+v}\right)t - \frac{2R_0/c}{1+v/c}\right]\right.$$
$$\left. \cdot \exp\left\{j\omega_0\left[\left(\frac{c-v}{c+v}\right)t - \frac{2R_0}{c+v}\right]\right\}\right) \tag{A-10a}$$

$$= \text{Re}\left\{\bar{s}\left[\left(\frac{c-v}{c+v}\right)\left(t - \frac{\tau_0}{1-v/c}\right)\right]\right.$$
$$\left. \cdot \exp\left[j\left(\frac{c-v}{c+v}\right)\omega_0\left(t - \frac{\tau_0}{1-v/c}\right)\right]\right\} \tag{A-10b}$$

$$= \text{Re}\left\{\bar{s}\left[\left(\frac{c-v}{c+v}\right)(t - \tau_0')\right]\exp\left[j\left(\frac{c-v}{c+v}\right)\omega_0(t - \tau_0')\right]\right\} \tag{A-10c}$$

where

$$\tau_0' = \frac{\tau_0}{1 - v/c} \tag{A-11}$$

Equation (A-10c) can be simplified. The factor $(c - v)/(c + v)$ that appears in the argument of complex envelope $\bar{s}(t)$ in (A-10c) represents a stretching or compression of the signal time-scale, depending on the sign of v. In most applications this effect is negligible; for a transmitted signal of duration T, the error in the argument of $s(t)$ incurred by the neglect of this factor is at most

$$\left(1 - \frac{c-v}{c+v}\right)T = \frac{2v/c}{1+v/c}T \simeq \frac{2vT}{c} \tag{A-12}$$

For a signal of bandwidth $\Delta_\omega = 2\pi\Delta_f$, $\bar{s}(t)$ does not change appreciably in an interval $1/\Delta_f$. Hence the error will be negligible if

$$\frac{2vT}{c} \ll \frac{1}{\Delta_f} \tag{A-13a}$$

or

$$\left(\frac{2v}{c}\right)T\Delta_f \ll 1 \tag{A-13b}$$

For example, a target moving 10,000 miles per hour with respect to the radar results in $v/c = 1.5 \times 10^{-5}$; inequality (A-13b) is satisfied for a waveform time-bandwidth product $T\Delta_f$ that is approximately less than 1000.

In many practical cases τ'_0 can be replaced by τ_0 in the argument of complex envelope $\tilde{s}(t)$ in Eq. (A-10c); the fractional error incurred in this case is

$$\frac{\tau'_0 - \tau_0}{\tau'_0} = \frac{\left(\dfrac{\tau_0}{1 - v/c}\right) - \tau_0}{\left(\dfrac{\tau_0}{1 - v/c}\right)} = \frac{v}{c} \tag{A-14}$$

For target velocities less than 10,000 mph, the fractional error is less than 1.5×10^{-5} (or 0.0015 per cent), which can usually be neglected. With the approximations just described to the argument of complex envelope $\tilde{s}(t)$, Eq. (A-10c) can be rewritten as

$$s_R(t) \cong \mathrm{Re}\left\{\tilde{s}(t - \tau_0)\exp\left[j\left(\frac{c-v}{c+v}\right)\omega_0(t - \tau'_0)\right]\right\} \tag{A-15}$$

Next the exponent in Eq. (A-15) is simplified. Note that

$$\frac{c-v}{c+v} = \frac{1 - v/c}{1 + v/c} \cong \left(1 - \frac{v}{c}\right)^2 \tag{A-16a}$$

$$\cong 1 - \frac{2v}{c} \tag{A-16b}$$

Define ω_d by

$$\omega_d = \frac{2v}{c}\omega_0 \tag{A-17}$$

From (A-16b) and (A-17) the exponent in Eq. (A-15) can be approximately rewritten as

$$(\omega_0 - \omega_d)(t - \tau'_0) = (\omega_0 - \omega_d)t - (\omega_0 - \omega_d)\tau'_0 \tag{A-18}$$

Approximating τ'_0 once again by τ_0 in (A-18) introduces an error in phase whose magnitude is approximately $(v/c)\omega_0\tau_0$. For a microwave radar, the phase error (modulo 2π) can be a significant fraction of 2π radians. For a 10,000 mph target and $\omega_0 = 2\pi \times 10^9$, a phase change of the order of 1 radian occurs in a period of 10 microseconds. However, since initial phase is assumed to be a random variable in most range and doppler estimation problems with a uniform probability density in the interval $(0, 2\pi)$, τ'_0 can usually be replaced by τ_0 without additional error.

With these approximations, Eq. (A-15) for the received signal becomes

$$s_R(t) \cong \mathrm{Re}\left[\tilde{s}(t - \tau_0)e^{j(\omega_0 - \omega_d)(t - \tau_0)}\right] \tag{A-19}$$

Equation (A-16) is the conventional doppler approximation for a narrowband echo from a moving target.

APPENDIX B

FALSE-ALARM RATE
OF NARROWBAND NOISE

In this appendix the false-alarm rate is evaluated for a sampled range-bin receiver and a receiver that employs a continuous threshold comparison, assuming input narrowband stationary Gaussian noise. In the latter type of receiver a false alarm occurs when envelope $V(t)$ of the narrowband noise process crosses threshold V^* from below (positive slope). The mean false-alarm rate R_{fa} is evaluated using the following formula due to Rice [1]:

$$R_{fa} = \int_0^\infty \dot{V} p(V^*, \dot{V}) \, d\dot{V} \qquad \text{(B-1)}$$

where $\dot{V}(t)$ is the derivative of $V(t)$ and $p(V^*, \dot{V})$ is the joint probability density function of V^* and $\dot{V}$.

It was shown in Section 4.11 that quadrature components $x(t)$ and $y(t)$ of a narrowband waveform $z(t)$ are each members of stationary Gaussian processes when $z(t)$ is a sample member of a stationary Gaussian process. Quadrature components $x(t)$ and $y(t)$ have the following properties:

$$\overline{x(t)} = \overline{y(t)} = 0 \qquad \text{(B-2)}$$

$$\overline{x(t) \cdot y(s)} = 0 \qquad s \neq t \qquad \text{(B-3)}$$

626

$$\sigma_1^2 = \overline{x^2(t)} = \overline{y^2(t)} = \overline{z^2(t)} \tag{B-4}$$

Variance σ_1^2 is related to autocorrelation functions $\phi_{xx}(\tau)$ and $\phi_{\cdots}(\tau)$ as follows:

$$\sigma_1^2 = \phi_{xx}(0) = \phi_{yy}(0) \tag{B-5}$$

Random waveforms $\dot{x}(t)$ and $\dot{y}(t)$ can be shown† to be members of stationary Gaussian processes with mean and variance, respectively, given by

$$\overline{\dot{x}(t)} = \overline{\dot{y}(t)} = 0 \tag{B-6}$$

$$\sigma_2^2 = \overline{[\dot{x}(t)]^2} = \overline{[\dot{y}(t)]^2} \tag{B-7}$$

Variance σ_2^2 is related to autocorrelation function $\phi_{xx}(\tau)$ by (see [2], Chap. 8)

$$\overline{[\dot{x}(t)]^2} = \overline{\frac{dx(t)}{dt} \cdot \frac{dx(s)}{ds}}\bigg|_{t=s} = \frac{\partial^2}{\partial t\, \partial s} \overline{x(t)\cdot x(s)}\bigg|_{t=s} \tag{B-8a}$$

$$= \frac{\partial^2}{\partial t\, \partial s}\, \phi_{xx}(t - s)\bigg|_{t=s} \tag{B-8b}$$

$$= -\ddot{\phi}_{xx}(0) \tag{B-8c}$$

Combining (B-7) and (B-8c) yields

$$\sigma_2^2 = |\ddot{\phi}_{xx}(0)| \tag{B-9}$$

Further, it can be shown that $x(t)$, $\dot{x}(t)$, $y(t)$, and $\dot{y}(t)$ are uncorrelated at the same instant of time [2].

The properties above yield the joint density function $p(x, y, \dot{x}, \dot{y})$:

$$p(x, y, \dot{x}, \dot{y}) = \frac{1}{(2\pi)^2 \phi_0 |\ddot{\phi}_0|} \exp\left(-\frac{x^2 + y^2}{2\phi_0} - \frac{\dot{x}^2 + \dot{y}^2}{2|\ddot{\phi}_0|}\right) \tag{B-10}$$

where ϕ_0 denotes $\phi_{xx}(0)$ and $|\ddot{\phi}_0|$ denotes $|\ddot{\phi}_{xx}(0)|$.

A transformation to polar coordinates $V, \dot{V}, \theta, \dot{\theta}$ is obtained with the relations

$$x = V \cos\theta \tag{B-11a}$$

$$y = V \sin\theta \tag{B-11b}$$

$$\dot{x} = \dot{V}\cos\theta - V\dot{\theta}\sin\theta \tag{B-11c}$$

$$\dot{y} = \dot{V}\sin\theta + V\dot{\theta}\cos\theta \tag{B-11d}$$

The Jacobian of the transformation is given by

$$J = \begin{vmatrix} \cos\theta & 0 & -V\sin\theta & 0 \\ \sin\theta & 0 & V\cos\theta & 0 \\ -\dot{\theta}\sin\theta & \cos\theta & -\dot{V}\sin\theta - V\dot{\theta}\cos\theta & -V\sin\theta \\ \dot{\theta}\cos\theta & \sin\theta & \dot{V}\cos\theta - V\dot{\theta}\sin\theta & V\cos\theta \end{vmatrix} \tag{B-12a}$$

$$J = -V^2 \tag{B-12b}$$

†See Middleton [2], Chap. 8.

With the method described in Section 3.10 and Eqs. (B-10) through (B-12), $p(V, \dot{V}, \theta, \dot{\theta})$ can be found:

$$p(V, \dot{V}, \theta, \dot{\theta}) = \frac{V^2}{(2\pi)^2 \phi_0 |\ddot{\phi}_0|} \exp\left(-\frac{V^2}{2\phi_0} - \frac{\dot{V}^2 + V^2 \dot{\theta}^2}{2|\ddot{\phi}_0|}\right) \qquad \text{(B-13)}$$

The marginal probability density function $p(V, \dot{V})$ is obtained by integrating (B-13) with respect to θ over the interval $(0, 2\pi)$ and the result with respect to $\dot{\theta}$ over the interval $(-\infty, \infty)$ to yield

$$p(V, \dot{V}) = \frac{y}{\sqrt{2\pi |\ddot{\phi}_0|} \, \phi_0} \exp\left(-\frac{V^2}{2\phi_0} - \frac{\dot{V}^2}{2|\ddot{\phi}_0|}\right) \qquad \text{(B-14)}$$

Substituting Eq. (B-14) into (B-1) results in

$$R_{fa} = \frac{V^*}{\phi_0} \sqrt{\frac{|\ddot{\phi}_0|}{2\pi}} \exp\left[-\frac{(V^*)^2}{2\phi_0}\right] \qquad \text{(B-15)}$$

The false-alarm rate of a sampled range-bin receiver, denoted for convenience by R'_{fa}, is given by†

$$R'_{fa} = mP_{fa} \qquad \text{(B-16)}$$

where m is the number of range bins searched per second and P_{fa} is the false-alarm probability in each bin. By definition, P_{fa} is the probability that $V(t)$ exceeds V^* at a sampling instant. From transformations (B-11a) and (B-11b) and the properties of x and y, it is straightforward to show that

$$p(V) = V e^{-V^2/2\phi_0} \qquad \text{(B-17)}$$

Hence

$$P_{fa} = \int_{V^*}^{\infty} V e^{-V^2/2\phi_0} \, dV = e^{-V^{*2}/2\phi_0} \qquad \text{(B-18)}$$

From Eqs. (B-15) through (B-18),

$$\frac{R_{fa}}{R'_{fa}} = \frac{1}{m} \sqrt{\frac{|\ddot{\phi}_0|}{\pi\phi_0} \ln \frac{1}{P_{fa}}} \qquad \text{(B-19)}$$

To estimate the magnitude of ratio R_{fa}/R'_{fa}, let the power spectral density of quadrature components $x(t)$ and $y(t)$ be rectangular with bandwidth Δ_ω as shown in Fig. B-1. From Wiener-Khintchine relation (4.6-2),

$$\phi_0 = \phi_{xx}(0) = \frac{1}{2\pi} \int_{-\infty}^{\infty} G_{xx}(\omega) \, d\omega = \frac{A\Delta_\omega}{2\pi} \qquad \text{(B-20)}$$

and

$$|\ddot{\phi}_0| = |\ddot{\phi}_{xx}(0)| = \left|\frac{1}{2\pi} \int_{-\infty}^{\infty} -\omega^2 G_{xx}(\omega) \, d\omega\right| = \frac{A\Delta_\omega^3}{24\pi} \qquad \text{(B-21)}$$

For a rectangular spectral density, a range-delay cell is approximately $2\pi/\Delta_\omega$ seconds wide; hence, the number of range bins searched per second is

†This quantity is the reciprocal of false-alarm time, i.e., $T_{fa} = 1/R'_{fa}$.

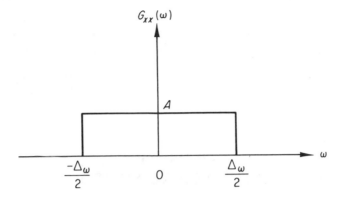

Fig. B-1. Spectral density of quadrature noise components.

$$m \simeq \frac{\Delta_\omega}{2\pi} \tag{B-22}$$

Substituting (B-20) through (B-22) into (B-19) yields

$$\frac{R_{fa}}{R'_{fa}} = \sqrt{\frac{\pi}{3} \ln \frac{1}{P_{fa}}} = 1.55 \sqrt{\log_{10} \frac{1}{P_{fa}}} \tag{B-23}$$

R_{fa} is approximately 3.1 times R'_{fa} for $P_{fa} = 10^{-4}$ and about 5.4 times R'_{fa} for $P_{fa} = 10^{-12}$. These results indicate that the false-alarm rates are not significantly different. Although false-alarm rate R_{fa} is somewhat higher, there is a compensating improvement in the detectability of small signals in a receiver employing continuous threshold comparison. It can be concluded, therefore, that a sampled range-bin receiver and a receiver that employs a continuous threshold comparison are similar in performance.

REFERENCES

1. Rice, S. O.: Mathematical Analysis of Random Noise, *Bell System Tech. J.*, **23**: 282–332 (July, 1944); **24**: 46–156 (January, 1945).
2. Middleton, D.: "Introduction to Statistical Communication Theory," McGraw-Hill, New York, 1960.

GLOSSARY
OF PRINCIPAL SYMBOLS

The principal symbols and conventions used in this book are defined below. Some symbols represent more than one quantity; the different meanings are clarified in the text. The following conventions are adopted throughout the book:

1. Boldface type is used for vectors and matrices, e.g., $\mathbf{\Lambda}$.
2. The symbol $|\quad|$ is used for "magnitude of" and for "determinant of," e.g., $|\mathbf{\Lambda}|$.
3. A superscript T signifies "transpose of," e.g., $\mathbf{\Lambda}^T$.
4. A superscript asterisk generally denotes the complex conjugate of a complex quantity, e.g., $f^*(t)$.
5. A tilde signifies the complex envelope of a complex waveform, e.g., $\tilde{f}(t)$.
6. A caret signifies the Hilbert transform of a quantity, e.g., $\hat{f}(t)$, and an estimate of a random variable.
7. An intervening dot ($\cdot$) signifies a dot product or simple product.
8. An intervening star signifies the convolution of two quantities, e.g., $f * g$.

630

9. A vinculum generally denotes a statistical average, as does the expectation operator $E[\quad]$, e.g., $\bar{f}$ and $E[f]$; the pointed brackets $\langle\quad\rangle$ generally represent a time average, e.g., $\langle f \rangle$.

10. A derivative is usually identified by a dot above the quantity. Higher-order derivatives are indicated by multiple dots or by a parenthesized quantity above the function, which shows the order of the derivative, e.g., $\overset{(m)}{f}$.

11. Σ, $\prod$, and $\int$ are the usual summation, product, and integral signs.

12. Var signifies the "variance of," e.g., var x.

SYMBOLS

A	= an amplitude coefficient, an event in probability space
$\mathscr{A}$	= a threshold level for a sequential detection test
$A(\omega)$	= a filter amplitude spectrum
A_0	= a constant
A^0	= a constant
ASN	= average sample number
a	= a constant, integral limit, a normalized amplitude
a_k	= a coefficient, a Fourier cosine coefficient
a_0	= a constant
$\mathbf{a}^{(m)}$	= an array of terms derived from a noise probability density function
$a(t)$	= amplitude modulation of an rf waveform
$a_n(t)$	= amplitude modulation of noise
$a_s(t)$	= amplitude modulation of a signal
$a_v(t)$	= amplitude modulation of a data waveform
arg	= argument of
α_k	= a Fourier coefficient for a narrowband waveform, and a normalized kth central moment
α	= a constant, false-alarm probability (type I error probability), an arbitrary phase angle, a parameter in the generalized Q function, a coefficient in the exponential RC kernel, a variable
α_G	= false-alarm probability for a global test
α_{NP}	= type I error probability for Neyman-Pearson strategy
α_{tr}	= false-alarm probability for a truncated test
B	= an event in probability space, signal amplitude, a ratio, a limit in Fresnel integrals
$\mathscr{B}$	= a threshold level for a sequential test
B^0	= a constant
b	= a constant, an integral limit, a linear FM sweep rate
b_k	= a Fourier sine coefficient, a coefficient in an orthogonal expansion of the RC kernel

$\mathbf{b}^{(m)}$	= a derived array of terms
β	= a constant, false-dismissal probability (type II error probability), an exponent in the RC kernel, a parameter in the generalized Q function, a variable, solid angle of radar beam
$\bar{\beta}$	= type II error probability averaged with respect to signal distribution
β_G	= type II error probability for a global test
β_i	= integrator loop gain
β_k	= a Fourier coefficient for a narrowband waveform
β_{tr}	= type II error probability for a truncated test
$\beta(\omega)$	= a filter phase characteristic
$\beta(\boldsymbol{\theta})$	= type II error probability with functional dependence on parameters $\boldsymbol{\theta}$
$\beta_{NP}(\boldsymbol{\theta})$	= false-dismissal probability for Neyman-Pearson strategy with functional dependence on parameters $\boldsymbol{\theta}$
C	= a capacitance
$C_X(\xi)$	= characteristic function of random variable X
$C(\xi_1, \xi_2, \ldots, \xi_n)$	= joint characteristic function of $X_1, X_2, \ldots, X_n$
$\overset{(n)}{C}_X(0)$	= nth derivative of characteristic function evaluated at $\xi = 0$
$C_Y(\xi : a)$	= characteristic function of Y as a function of parameter a
$_N C_K$	= binomial coefficients
C_1	= integration contour in complex plane
$C^{(i)}$	= cosine Fresnel integral
C_{ij}	= elements of cost matrix $\mathbf{C}$
$C(\mathbf{s}, \mathbf{d})$	= simple cost or loss function
C_α	= cost of type I error (false alarm)
C_β	= cost of type II error (missed detection)
$C_{1-\alpha}$	= cost of correct decision when noise alone is present
$C_{1-\beta}$	= cost of correct signal detection
C_f	= cost of type I error (false alarm)
C_m	= cost of type II error, cost associated with detection in the wrong resolution bin
$C_{ij}(\boldsymbol{\theta})$	= elements of a cost matrix as functions of parameters $\boldsymbol{\theta}$
c	= a constant, velocity of light, cost of experimentation
c_k	= a constant, a complex Fourier coefficient, a coefficient of an orthogonal expansion
$\bar{\chi}$	= average signal-to-noise ratio
χ_c	= colored signal-to-noise ratio
χ_d	= design signal-to-noise ratio
χ_{if}	= intermediate-frequency signal-to-noise ratio
χ_L	= signal-to-noise ratio out of linear detector

$\chi_{MF}(t)$	= instantaneous signal-to-noise ratio out of matched filter
$\chi_m(t_m)$	= maximum signal-to-noise ratio out of an optimum filter
χ_{MF-1}	= normalized signal-to-noise ratio out of a matched filter
χ_{MF-max}	= maximum value of $\chi_{MF}(t)$
χ_0	= a specific signal-to-noise ratio, a single sample video signal-to-noise ratio
$\chi_0(t)$	= instantaneous signal-to-noise ratio out of linear filter
χ_p	= if signal-to-noise ratio for a single pulse
$\bar{\chi}_p$	= (averaged) peak signal-to-noise ratio, average pulse signal-to-noise ratio
χ_{REF}	= reference signal-to-noise ratio
χ_{rf}	= a radio-frequency signal-to-noise ratio
χ_{SQ}	= signal-to-noise ratio out of square-law detector
$\chi(\tau, \omega_d)$	= the radar ambiguity function
$\mathscr{D}$	= d/dt, differential operator
$D(\omega^2)$	= a polynomial in ω^2
$D(\mathbf{d}\,|\,\mathbf{v})$	= a decision rule, e.g., decide $\mathbf{d}$ given data $\mathbf{v}$
D_B	= a Bayes decision rule
D_{NP}	= a Neyman-Pearson decision rule
$D[k; (\mathbf{0} - k)|\,\mathbf{v}]$	= a decision rule that target k is present, and no others, given $\mathbf{v}$
$D[k,j; (0-k,j)|\,\mathbf{v}] =$	a decision rule that targets k and j are present, and no others, given $\mathbf{v}$
$D_1(j\omega)$	= a factor of polynomial $D(\omega^2)$
$\mathbf{d}$	= a set of decisions
d_i	= a decision
d	= a constant, a density
$\partial/\partial v$	= partial derivative
Δ_τ	= pulse width
Δ_ω	= radian frequency bandwidth
Δ_{rms}	= rms radian frequency bandwidth
$\Delta\omega$	= radian frequency interval
Δ	= decision space, a distance
Δt	= time increment
$\delta(t)$	= Dirac delta function
δ_{ij}	= Kronecker delta
$\overset{(i)}{\delta}$	= ith derivative of delta function
$\Delta_{\tau-3db}$	= 3 db pulse width
E	= signal energy
E_P	= pulse energy

$E[\ \ ]$	= expectation operator, mathematical "expectation of"
E_0	= normalized energy
$\mathscr{E}$	= normalized energy
$\mathscr{E}_p$	= normalized pulse energy
$E_\theta[P_d]$	= probability of detection averaged with respect to random parameters θ
$\mathscr{E}_s$	= known signal energy
$E_\phi[\ell(\mathbf{v}\mid\phi)]$	= likelihood ratio averaged with respect to ϕ [also $\overline{\ell(\mathbf{v}\mid\phi)_\phi}$]
$\mathscr{E}_{P-T}$	= transmit energy
e	= base of natural system of logarithms
ϵ	= a constant, an increment in time, probability of rejecting H when it is true, the level of a test
$\in$	= "belongs to or in" in set terminology.
$\tilde{e}(t)$	= complex waveform out of filter $h(t)$
$e_I(t)$	= real part of complex filter output $\tilde{e}(t)$
$e_Q(t)$	= imaginary part of the complex filter output $\tilde{e}(t)$
$\tilde{e}_{\mathrm{MF}}(t)$	= complex matched filter output
$e_{\mathrm{IMF}}(t)\ \}$ $e_{\mathrm{QMF}}(t)\ \}$	= real and imaginary components of matched filter complex output $\tilde{e}_{\mathrm{MF}}(t)$
$\tilde{e}_{\mathrm{opt}}(0)$	= complex waveform out of optimum filter evaluated at $t=0$
env (x)	= envelope of x
$F(\omega)$	= Fourier transform of $f(t)$
F_n	= noise factor
$F_p(\omega)$	= Fourier transform of a single pulse
$F_1(a, b)$	= a function of a and b
F_i	= integral function
${}_1F_1(a, c\,; z)$	= confluent hypergeometric function
f	= frequency in cycles per second (Hz)
$\mathbf{f}$	= a vector
f_c	= cutoff frequency
f_d	= doppler frequency
f_r	= pulse repetition frequency (prf)
f_s	= sampling frequency
$f(t), f(x)$	= a time function, a mapping function
$\tilde{f}(t)$	= complex envelope of $f(t)$
$f_I(t)$	= real part of $\tilde{f}(t)$
$f_k(t)$	= generalized orthogonal function
$f_p(t)$	= an rf pulse
$f_Q(t)$	= imaginary part of $\tilde{f}(t)$
$f(X)$	= a function of X
$\mathscr{F}[x]$	= Fourier transform of x

$\mathscr{F}^{-1}[X]$	= inverse Fourier transform of X
$G(\tau)$	= signal function at matched filter output
$G(\omega)$	= Fourier transform of $g(t)$
$G_T^{(x)}(\omega)$	= power spectral density of $2T$ sample of $x(t)$
$G^{(x)}(\omega)$	= limit of $G_T^{(x)}(\omega)$ as $T \to \infty$
$G_{nn}(\omega)$	= power spectral density of noise process $\{n(t)\}$
$G_{xx}(\omega)$	= power spectral density of process $\{x(t)\}$
$G_{xy}(\omega)$	= cross power spectral density of $\{x(t)\}$ and $\{y(t)\}$
G_R	= receive antenna gain
G_T	= transmit antenna gain
g	= a constant
$\tilde{g}(t)$	= complex envelope of $g(t)$
$g_I(t)$	= real part of $\tilde{g}(t)$
$g_Q(t)$	= imaginary part of $\tilde{g}(t)$
$g(t), g(x)$	= a time function, a mapping function
$\mathbf{g}$	= a vector
$g_T(t)$	= rectangular gating pulse
$\Gamma(\nu)$	= gamma function
Γ	= observation space
Γ_m	= observation space of m dimensions
γ	= a limit in the Fresnel integral, constant, an element of the correlation matrix $\boldsymbol{\rho}$ for RC noise
$\gamma(\nu, z)$	= a function of gamma functions
$\gamma(t)$	= complex received waveform
$H(\omega)$	= Fourier transform of $h(t)$
$\mathscr{H}(x_i)$	= a priori uncertainty of X_i occurring
$\mathscr{H}(X_i \mid Y_k)$	= a posteriori uncertainty of X_i given message Y_k
H_i	= hypotheses, hermite polynomials
$H_{\mathrm{MF}}(\omega)$	= Fourier transform of matched filter impulse response $h_{\mathrm{MF}}(t)$
$H_W(\omega)$	= prewhitening filter response
$H_i(x)$	= a hermite polynomial
$H(\tau)$	= noise function at matched filter output
h	= a constant, a parameter
h_i	= complex roots of a polynomial
$h(t)$	= impulse response of a linear time-invariant filter
$h_{\mathrm{opt}}(t)$	= general optimum filter impulse response
$h_{\mathrm{MF}}(t)$	= matched filter impulse response
$h_x(t)$	= filter impulse response in orthogonal expansion
$\tilde{h}_{\mathrm{opt}}(t)$	= complex optimum filter impulse response
$\tilde{h}_{\mathrm{opt}\text{-}k}$	= complex coefficients of orthogonal expansion of $\tilde{h}_{\mathrm{opt}}(t)$
$h_{ss}(t)$	= optimum filter solution for infinite observation interval
if	= intermediate frequency

$\mathscr{I}(X_i \mid Y_K)$	= information gain of X_i given message Y_K
Im ( )	= imaginary part of
i	= index of summation
$\mathbf{I}$	= identity matrix
$I_0(x)$	= modified Bessel function of order zero and argument x
$I_\gamma(x)$	= modified Bessel function of order γ
$I(\mu, s)$	= incomplete gamma function
J	= Jacobian
j	= $\sqrt{-1}$, index of summation
K	= degrees Kelvin
$K(t, s)$	= kernel of integral equation
$k_i(\tau)$	= a weighting function
k	= index of summation, a constant, a scale factor, a function of detection range, number of resolution bins
k_B	= Boltzmann's constant
k_i	= a constant, a weighting function
k_o	= optimum value of k
$\mathscr{K}$	= a threshold
$\mathscr{K}(\boldsymbol{\theta})$	= a decision surface separating regions of space Γ
$\mathscr{K}_{-1}, \mathscr{K}_{-2}, \mathscr{K}_{-3}$	= modified threshold values
L	= attenuation loss, integration loss, expected loss
$L(T_0)$	= average loss
L_2	= Lebesgue class of integrable square functions
$L(D, \sigma)$	= average loss rating
$L_B(D_B, \sigma)$	= Bayes average loss rating
$L_c(D \mid \mathbf{s})$	= conditional loss rating
L_{fS}	= fixed-sample loss
L_{SEQ}	= average loss for a sequential test
$\mathscr{L}(\boldsymbol{\theta})$	= operating characteristic function
l	= index of summation, a coefficient
l.i.m.	= limit in the mean
ln	= natural logarithm
log	= logarithm to base 10
$\ell(\mathbf{v})$	= likelihood ratio for data $\mathbf{v}$
ℓ_0	= a threshold
$\ell(v \mid A)$	= likelihood ratio at location of signal A
$\ell(\mathbf{v} \mid \boldsymbol{\theta})$	= conditional likelihood ratio of data $\mathbf{v}$ given parameters $\boldsymbol{\theta}$. Various forms are used with specific parameters, such as $\ell(\mathbf{v} \mid \boldsymbol{\phi}, a)$, etc.
$\overline{\ell(\mathbf{v} \mid \boldsymbol{\theta})}_\theta$	= conditional likelihood ratio averaged with respect to $\boldsymbol{\theta}$
Λ	= covariance matrix
Λ^{-1}	= inverse of covariance matrix

$\|\Lambda\|_{ik}$	= cofactor of element i-k in Λ
$\|\Lambda\|$	= determinant of covariance matrix
λ_k	= an eigenvalue, a coefficient
λ	= Lagrangian multiplier, wavelength, a cost ratio, mean number of false alarms per second
λ_{ik}	= elements of covariance matrix Λ
M	= dimension of sample space, a decision-space plane, a constant, substitution variable, number of resolution cells, second threshold, number of exactly known signals, number of cumulative observations
$M_{v,\mu}(z)$	= a function related to the confluent hypergeometric function
$M_z(v)$	= moment-generating function
m	= index of summation, mean value, number of samples in sequential testing before termination, given number, number of resolution bins, number of decisions per second
mag	= magnitude of
m_{ik}	= joint moment of X^i and Y^k
m_n	= nth moment of X
μ_n	= nth central moment
μ_{ik}	= joint central moment of $(X - \bar{X})^i$ and $(Y - \bar{Y})^k$
$N_0/2$	= white-noise spectral density in watts per Hz
N	= a number, number of hits, number of observation intervals
$\mathbf{N}$	= a point in multidimensional noise space
$N(\omega^2)$	= a polynomial in ω^2
N_{fa}	= number of effective range cells in false-alarm time
N_{im}	= total number of ones in an observed sample
$N_1(j\omega)$	= a factor of polynomial $N(\omega^2)$
n	= number of trials, number of discrete events in sample space, given number, number of observations
$\mathbf{n}$	= a point in noise space
$n(t)$	= a noise waveform
$\tilde{n}(t)$	= complex envelope of $n(t)$
n_i	= noise samples
$n_i(t)$	= input noise
$n_o(t)$	= output noise
$n_o'(t)$	= output noise from a suboptimum filter
$\bar{n}_\theta$	= average sample number $= E[n\|\boldsymbol{\theta}]$
$n_I(t)$	= real part of $\tilde{n}(t)$
$n_Q(t)$	= imaginary part of $\tilde{n}(t)$
n_k	= noise coefficients of an orthogonal expansion

$\bar{n}_{\alpha_G}$	= average test length when H_1 is falsely accepted
n_0	= nominal value of n
$\bar{n}_0$	= average sample number for signal absent
$n_W(t)$	= a white-noise sample waveform
$\nu_A(n)$	= number of occurrences of A in n trials
ν_i	= a variate related to envelope ($\frac{1}{2} \sum r_i^2$)
$\nu_1(\xi)$	= a root of an equation
$\nu_2(\xi)$	= a root of an equation
$\nu(t)$	= complex noise function
$O(x^4)$	= order of x^4
Ω	= signal space, solid angle of search frame
Ω_i	= normalized envelope of ith waveform
ω	= angular frequency in radians per second
ω_0	= carrier frequency in radians per second
ω_c	= cutoff frequency in radians per second
ω_d	= doppler frequency in radians per second
ω_s	= sampling frequency in radians per second
$P(A)$	= probability of event A
$P(A, B)$	= joint probability of A and B
$P(B \mid A)$	= conditional probability of B given A
P_d	= probability of detection
P_{cd}	= cumulative probability of detection
$P_d(R_i)$	= probability of detection as a function of range R_i
$P(H_1), P(H_0)$	= a priori probability that hypothesis H_1 or H_0 is true
$P_{i\text{-}0}$	= probability that H is accepted, with only noise present, in the ith resolution bin
$P_{i\text{-}a}$	= probability that H is accepted, with signal present, in the ith resolution bin
$P(H_i \mid E_j)$	= probability of hypothesis H_i given event E_j
$P(0 \mid \mathbf{y})$	= probability of no signal given $\mathbf{y}$
$P(s \mid \mathbf{y})$	= probability of signal present given $\mathbf{y}$
$P(n)$	= cumulative probability that H_0 is accepted by or at n
$P(0), \mathfrak{p}_0$	= a priori probability of no signal
P_R	= received power
$P(s), \mathfrak{p}_s$	= a priori probability of signal
P_T	= transmitted power
P_{te}	= total probability of error
$P(X_i)$	= a priori probability that message contains signal X_i
$P(x)$	= probability distribution function
$P(x, y)$	= joint probability distribution function
$\Pr(X \text{ in } S_i)$	= probability that X lies in subset S_i
$\Pr(X \leqslant x)$	= probability distribution function
P_{fa}	= false-alarm probability

$\left.\begin{array}{l} P(T \leqslant T_0 \mid H_i) \\ P(T > T_0 \mid H_i) \end{array}\right\}$	= the probabilities that measured time T is less than or exceeds threshold T_0 given that hypothesis H_i is true
$\mathfrak{p}$	= a priori probability of signal present
p	= a variable, probability that a waveform sample exceeds first threshold
$p_A(K)$	= continuous distribution that approximates discrete $p(K)$
pdf	= probability density function
$p(\mathbf{d} \mid \mathbf{s})$	= conditional pdf of decisions $\mathbf{d}$ given signal $\mathbf{s}$
p_i	= weighting of delta probability density function at $X = x_i$, a priori probability of ith target alone
$\mathfrak{p}_i$	= a priori probability of ith target alone
$p_G(n)$	= probability that global test ends at n
$p_i(n)$	= probability that H_0 is correctly accepted in ith subtest at n
$p(k)$	= probability that k out of N independent samples exceed first threshold
p_n	= probability that envelope exceeds first threshold with noise alone
$p_n(\mathbf{n})$	= pdf of noise $\mathbf{n}$
$p_n(\mathbf{v} \mid \mathbf{s})$	= conditional pdf of data $\mathbf{v}$ given $\mathbf{s}$
$\mathfrak{p}_0$	= a priori probability of no targets
$\mathfrak{p}_{00}, \mathfrak{p}_{A0}, \mathfrak{p}_{0B}$	= a priori probability of no signals, of signal A alone, of B alone, etc.
$p(r_i; a)$	= pdf of r_i as a function of parameter a
$p(v_i; a)$	= pdf of v_i as a function of a
p_s	= probability that envelope exceeds first threshold with signal present
$p(\mathbf{n})$	= joint probability density function of noise $\mathbf{n}$
$p(\mathbf{v} \mid \mathbf{s})$	= conditional pdf of $\mathbf{v}$ given $\mathbf{s}$
$\overline{p(\mathbf{v} \mid \mathbf{s})}_\mathbf{s}$	= conditional pdf of $\mathbf{v}$ averaged over $\mathbf{s}$; the marginal pdf of $\mathbf{v}$
$p(\mathbf{v} \mid \boldsymbol{\theta})$	= conditional pdf of data $\mathbf{v}$ given parameters $\boldsymbol{\theta}$; various forms are encountered, $p(\mathbf{v} \mid \boldsymbol{\phi}, a)$, etc.
$\overline{p(\mathbf{v} \mid \boldsymbol{\theta})}_\theta$	= conditional pdf of $\mathbf{v}$ averaged over $\boldsymbol{\theta}$
$p(\mathbf{X})$	= joint pdf of components $X_1, X_2, \ldots, X_n$
$p(x)$	= probability density function
$p(x, y)$	= joint probability density function
$p(x \mid y)$	= conditional pdf of x given y
$p(y \mid u_x)$	= conditional pdf of y given message u_x, likelihood function of y
$p(y \mid x)$	= conditional pdf of y given x
$\overline{p(Y)}_\theta$	= pdf of sum variable Y averaged over random parameters $\boldsymbol{\theta}$

$\Phi(x)$	= defined integral of normal function
$\Phi^{-1}(x)$	= inverse of $\Phi(x)$
ϕ	= a dummy variable, initial phase, spatial angle
$\phi_{xx}(t_1, t_2)$	= autocorrelation function of random process $\{x(t)\}$
$\phi_{xy}(t_1, t_2)$	= cross-correlation function of processes $\{x(t)\}$ and $\{y(t)\}$
$\tilde{\phi}_{xx}(t_1, t_2)$	= complex autocorrelation function
$\phi(\tau)$	= autocorrelation function of a stationary process
$\phi(0)$	= autocorrelation function evaluated at $\tau = 0$
$\ddot{\phi}(0)$	= second derivative of autocorrelation function evaluated at $\tau = 0$
$\phi(x)$	= normal function
$\overset{(i)}{\phi}(x)$	= ith derivative of normal function
$\Psi(t)$	= complex analytic signal $= s(t) + j\hat{s}(t) \cong \tilde{s}(t) \exp[j\omega_0 t]$ for a narrowband signal
$\Psi(\omega)$	= Fourier transform of $\Psi(t)$
$\Psi_k(t)$	= an eigenfunction, orthogonal function (normalized)
$Q(\alpha, \beta)$	= Q function (defined in Chapter 9)
$Q(\tau)$	= a complex video waveform
q	= a priori probability that no signal is present
q	= a reference level
q^*	= a normalized reference level
$q(x)$	= a sufficient statistic
$\tilde{q}(\tau)$	= a complex video waveform
$\tilde{q}_s(\tau)$	= a complex low-frequency signal waveform
$\tilde{q}_n(\tau)$	= a complex low-frequency noise waveform
$R_{\max}$	= maximum range
R	= a resistance, range
R_{fa}	= mean false-alarm rate
$R(\omega)$	= real part of Fourier transform $F(\omega)$
$R_{xx}(\tau)$	= time autocorrelation function
$R_{xy}(\tau)$	= time cross-correlation function
Re ( )	= real part of
$\mathscr{R}_p$	= single pulse peak signal-to-noise ratio $(2E_p/N_0)$
$\mathscr{R}$	= peak signal-to-noise ratio $(2E/N_0)$
$R_{ss}(t)$	= known signal autocorrelation function
$R(t)$	= an envelope function
$R(\mathbf{v})$	= a real envelope as a function of data $\mathbf{v}$
R_N	= normalization range
R_0	= a radar range
$\bar{R}_t$	= a mean range
r	= envelope of a narrowband waveform, range variable
r_b	= a threshold

rf	= radio frequency
rms	= root mean square
$r_{ss}(t)$	= time autocorrelation function for finite-energy signals
r_i	= proportionality factor, envelope
$\mathfrak{r}$	= normalized envelope of narrowband waveform
$\mathfrak{r}_i$	= normalized envelope of ith waveform
$\|\boldsymbol{\rho}\|_{ij}$	= cofactor of i-j element in $\boldsymbol{\rho}$
ρ	= correlation coefficient, normalized covariance of X and Y, antenna radiation efficiency
ρ_a	= antenna aperture efficiency
ρ_e	= antenna efficiency factor
ρ_{ij}	= elements of matrix $\boldsymbol{\rho}$
ρ_v	= a defined voltage signal-to-noise ratio
$S(\omega)$	= Fourier transform of $s(t)$
$S_p(\omega)$	= Fourier transform of a periodic signal
$s_x(t)$	= a weighted signal
$S_x(\omega)$	= Fourier transform of $s_x(t)$
s	= a variable, subset of complex plane
$\mathbf{s}$	= points in signal space (a vector)
$s(t)$	= a real signal
$\hat{s}(t)$	= Hilbert transform of $s(t)$
$\tilde{s}(t)$	= complex envelope of signal $s(t)$
$s_A(t)$	= a known signal
$s_B(t)$	= a known signal
$S^{(i)}$	= sine Fresnel integral
$s_i(t)$	= input signal
$s_o(t)$	= output signal
$s_o'(t)$	= output signal from a suboptimum filter
$s_I(t)$	= real component of complex envelope $\tilde{s}(t)$
$\hat{s}_I(t)$	= in-phase rf component of narrowband signal $s(t)$
$s_{I\text{-}k}$	= real component of $\tilde{s}_k$
$s_{I-1}(t), s_{Q-1}(t), \ldots$	= normalized in-phase and quadrature signal components
$\mathbf{s}_i(\boldsymbol{\theta})$	= signal samples with parameters $\boldsymbol{\theta}$ in the ith observation interval
$\tilde{s}_{-1}(t)$	= a complex normalized envelope
$\tilde{s}_k$	= complex coefficient of orthogonal expansion
$s_Q(t)$	= imaginary (quadrature) component of $\tilde{s}(t)$
$s_{Q\text{-}k}$	= imaginary component of $\tilde{s}_k$
$\hat{s}_Q(t)$	= quadrature rf component of narrowband signal $s(t)$
$\mathbf{s}_x$	= signal vector
$s_x(t)$	= known message
$s_{x\text{-}k}$	= signal coefficient in an orthogonal expansion

$(S/N)_{\text{aver}}$	= average signal-to-noise ratio
$\text{Si}(x)$	= sine integral of x
$\Sigma_I(t)$	= normalized in-phase rf component of narrowband signal $s(t)$
$\Sigma_Q(t)$	= normalized quadrature rf component of narrowband signal $s(t)$
σ_T	= target cross section
σ	= standard deviation, variable of integration, target cross section
σ^2	= variance, second central moment
$\sigma_{n\text{-}k}^2$	= variance of n, given signal parameter k
σ_Y^2	= variance of Y
$\sigma(\mathbf{s})$	= a priori pdf of signal $\mathbf{s}$
$\sigma_{lf}(\boldsymbol{\theta})$	= least favorable a priori pdf of parameters $\boldsymbol{\theta}$
T	= finite observation interval, search interval duration, waveform duration, clocked time
$\mathscr{T}$	= temperature
$\mathscr{T}_a$	= effective antenna temperature
T_d	= dwell time, interpulse period
T_f	= frame time
$\mathscr{T}_i$	= input temperature
T_0	= a threshold
$T_{\max}, T_{\min}$	= maximum and minimum clocked times
T_s	= waveform duration
$\mathscr{T}_r$	= effective noise temperature of a receiver
$T_v(m, n, r)$	= incomplete Toronto function
$\mathscr{T}_{\text{eq}}$	= equivalent input noise temperature
t	= time variable
t_i	= expected arrival time
t_m	= time at which signal-to-noise ratio is a maximum
Δt	= time increment
τ_{fa}	= false-alarm time
τ	= time variable, echo range delay, time difference, crossing time
τ_d	= time delay
τ_{gr}	= group delay
τ_0	= true target echo delay
$\left.\begin{array}{l}\theta(t), \theta_s(t)\\ \theta_n(t), \theta_v(t)\end{array}\right\}$	= phase modulation of an rf waveform
$\boldsymbol{\theta}$	= vector denoting a set of parameters
$\boldsymbol{\theta}_d$	= set of design signal parameters
θ	= angle, initial rf phase
U	= a random variable

U_{ij}	= a binary random variable in ith resolution element and in jth observation
$\mathbf{U}_x$	= a derived signal matrix
u	= a dummy variable, a real number associated with random variable U
$u_x(t)$	= a signal or message
V	= a random variable, velocity
$V(t)$	= a received waveform
$v(t)$	= a real received waveform, signal plus noise data
v	= a dummy variable, velocity, a real number associated with random variable V
$\tilde{v}(t)$	= complex envelope of waveform $v(t)$
$\mathbf{v}$	= set of points in observation space
$\mathbf{v}_N$	= N-dimensional approximation of $\mathbf{v}$
$v_{Ik-1}, v_{Ik-2}, \ldots$	= normalized real coefficients of an orthogonal expansion
$v_{Qk-1}, v_{Qk-2}, \ldots$	= normalized imaginary coefficients of an orthogonal expansion
v_{ij}	= data observed in ith resolution bin at jth sample
$\mathbf{W}$	= a matrix
$W(\theta, \phi)$	= beam pattern weighting
$W(x)$	= a function
w	= a variable, transform variable
$w(\mathbf{s})$	= a priori pdf of signal $\mathbf{s}$
X	= a variable, random variable
X_S	= a standardized random variable
$X(\omega)$	= imaginary part of Fourier transform $F(\omega)$
X_i	= a random variable
$x(t)$	= a component waveform of a narrowband process, a sample member of a random process, a time waveform, a message
x_i	= real number associated with random variable X_i, the coefficients of an orthogonal expansion
$\{x(t)\}$	= a random process
$\mathbf{x}$	= a column vector, matrix, or array of elements
$\mathbf{x}_N$	= N-dimensional approximation of $\mathbf{x}$
ξ	= a variable, a complex variable
$\boldsymbol{\xi}$	= a vector
ξ_0	= saddle-point location in complex plane
Y_S	= a standardized random variable
Y	= a variable, a random variable, a normalized sum random variable
Y_e	= estimated value of Y
Y_b	= a threshold

$\mathbf{Y}$	= a derived matrix
y	= a sum random variable
$y(t)$	= component waveform of a narrowband process, a member of a random process, a time waveform, an observed waveform
y_i	= coefficients of an orthogonal expansion
$\{y(t)\}$	= a random process
$\mathbf{y}$	= a column vector or array of elements
$\mathbf{y}_N$	= N-dimensional approximation of $\mathbf{y}$
$\tilde{y}(t)$	= complex envelope of real waveform $y(t)$
$\dot{y}_I(T)$	= a defined in-phase real filter output
$\dot{y}_Q(T)$	= a defined quadrature real filter output
Z_m	= sum random variable, logarithm of the likelihood ratio of dimension m
Z_T	= limit of Z_m in the continuous case
$Z_{n\text{-}cs}$	= a cumulative sum random variable
z	= a sum random variable, a substitution variable, the logarithm of the likelihood ratio
$\{z(t)\}$	= a random process
z_i	= logarithm of likelihood ratio
$z_{i\text{-}s}$	= a sub-sum random variable

AUTHOR INDEX

Numbers in italic indicate pages on which the complete references appear.

Abramowitz, M., 508, 509, 510, 519, *520*, 567, *597*

Barker, R. H., 170, *184*
Barton, D. K., 455, 459, 460, 464, 466, 470, 471, *475*
Battin, R. H., *107*, *138*, 193, *201*
Bendat, J. S., *138*
Bennett, W. R., *138*
Bennington, H. D., 35, 36, 37, 38, *39*
Bernfeld, M., 157, 181, *184*, 618, *621*
Birdsall, T. G., 599, *620*
Blackwell, D., 253, *286*
Blake, L. V., 460, 464, *475*
Brennan, L. E., *494*, 592, 593, *597*
Brockner, C. E., 18, 20, *39*
Bussgang, J. J., 553, 562, 575, 582, 591, 595, *597*

Campbell, G. A., 94, *107*, 345, 346, *372*, 378, 404, 409, 426, *445*
Cook, C. E., 6, *38*, 155, 157, 181, *184*, 618, *621*
Cooper, D. C., 526, 529, 531, 535, 537, 539, 540, 541, 542, *543*
Courant, R., 146, 185, 194, *200*
Cramer, H., 97, *107*, *286*, 508, *520*

Davenport, W. B., *107*, 138, *138*, 146
Davies, I. L., 140, *141*, 202, 203, 205, 206, 209, 228, 229, *251*
DiFranco, J. V., 59, *73*, 158, 159, 160, 161, *184*, 309, *335*
Dillard, G. M., 588, 589, 590, 591, *597*
Diss, C. E., 35, *39*
Doob, J. L., 238, *251*
Dooley, L. G., 35, *39*
Dwork, B. M., 186, *200*

Edrington, T. S., *445*, 477, *494*
Ehrman, L., 582, *597*
Endreson, K., 515, 516, 517, 518, *520*
Everett, R. R., 35, 36, 37, 38, *39*

Fehlner, L. F., 287, 288, *288*, 348, *372*, 390, 404, 410, 427, *445*
Feller, W., *107*
Feshbach, H., 186, 193, *200*
Finn, H. M., 592, *597*
Foster, R. M., 94, *107*, 345, 346, *372*, 378, 404, 409, 426, *445*
Franklin, P., 86, 89
Fry, T. C., 508, *520*

Gabor, D., 58, 69, *73*
George, S. F., 162, 165, 166, *184*
Girshick, M. A., 253, *286*
Goldstein, A., 461, *475*
Grenander, U., 323, *335*
Griffiths, J. W. R., 526, 529, 531, 542, *543*
Guillemin, E. A., 235, *251*

Hall, W. M., 467, *475*
Hansen, V. G., 518, *520*
Harrington, J. V., 498, 502, *519*
Hartley, R. V. L., 140, *140*
Heatley, A. H., 348, 371, *372*
Hedemark, R., 515, 516, 517, 518, *520*
Helstrom, C. W., 6, *39*, 193, *201*, 229, *251*, 285, *286*, 582, *597*, 599, *620*
Hilbert, D., 146, 186, 194, *200*
Hill, F. S., Jr., 592, 593, *597*

Kailath, T., 243, 244, 245, 247, 248, 249, *251*
Kelly, E. J., 320, 323, 330, *335*
Kendall, W. B., 591, 596, *597*
Kerr, D. E., 456, *475*

Laning, J. H., *107*, *138*, 193, *201*
Lawson, J. L., 140, *140*, 154, 155, 171, *184*, 253, 272, *286*
Lehman, E. L., 281, *286*
Linder, I. W., Jr., 498, 514, *519*
Magnus, W., 393, *445*
Mallett, J. D., *494*
Marcum, J. I., 6, *38*, *39*, 287, *288*, 307, 309, 316, *335*, 344, 347, 348, 370, 371, *372*, 410, *445*, 465, *475*, 503, *520*

Marcus, M. B., 582, 583, 584, 585, 586, 587, 597
Mentzer, J. R., 456, *475*
Middleton, D., 6, *39*, *138*, 140, *141*, 186, *200*, 235, *251*, 253, 273, *286*, *335*, 553, 562, 575, 591, 595, *597*, 599, 601, *620*, *621*, 626, *629*
Morse, P. M., 186, 193, *200*

Neyman, J., 253, *285*, 282, *286*
Nilsson, N. J., 599, *621*
North, D. O., 6, *38*, 139, *140*

Oberhettinger, F., 393, *445*

Pachares, J., 348, *372*
Paley, R. E., 181, *184*
Palmer, D. S., 531, 535, 537, 539, 540, 541, *543*
Papoulis, A., 49, 50, *73*, 84, *107*, *138*, 147, 161, 532, 533, 534, *543*
Pearson, E., 253, 282, *285*, *286*
Pearson, K., 347, *372*
Peterson, W., 599, *620*
Pierce, B. O., 504, *520*
Pierce, J. R., 449, *475*
Price, R. C., 18, 20, *39*

Ragazzini, J. R., 145, 181, *184*, 192, *201*
Reed, I. S., 320, 323, 330, *335*, 577, 578, 579, 582, 591, 596, *597*
Reich, E., 561, *597*
Rice., S. O., *138*, 316, *335*, 572, *597*, 627, *629*
Ridenour, L. N., 4, 12, 13, 26, 27, 31, 32, *38*
Root, W. L., *107*, 138, *138*, 146, 320, 323, 330, *335*
Rubin, W. L., 59, *73*, 309, *335*

Schwartz, M., 440, *445*
Selin, I., 577, 578, 579, 582. 596, *597*, 599, *620*
Shannon, C. E., 140, *140*, 202, *251*
Siebert, W. M., 6, *38*
Siegel, K. M., 27, *39*
Silver, S., 452, 453, 460, 462, *475*

Skolnik, M. I., 28, *39*
Sokolnikoff, E. S., 192, *200*
Sokolnikoff, I. S., 192, *200*
Stegun, I. A., 508, 509, 510, 519, *520*, 567, *597*
Storer, J. E., 170, *184*
Swerling, P., 6, 28, *39*, 287, *288*, 374, 377, 421, 439, 440, *445*, 483, *494*, 498, 502, 514, 519, *519*, 561, 582, 583, 584, 585, 586, 587, *597*

Thomas, J. B., 599, *621*
Turin, G. L., 144, 167, *184*
Turyn, R., 170, *184*

Uhlenbeck, G. E., 140, *140*, 154, 155, 171, *184*, 253, 272, *286*
Urkowitz, H., 171, *184*

Vance, P. R., 35, *39*
Van der Ziel, A., 449, 467, *475*
Van Meter, D., 140, *141*, 253, *286*, 599, *620*
Van Vleck, J., 186, *200*, 461, *475*
Ville, J., 58, *73*

Wainstein, L. A., 6, *39*, 57, *73*, 177, *184*, 599, *620*
Wald, A., 140, *141*, 253, 273, 275, 282, *285*, *286*, 547, 549, 550, 551, 554, 564, 571, 575, 591, *597*
Watson, G. N., 311, *335*, 393, *445*
Wiener, N., 181, *184*
Wolf, J. K., 599, *621*
Wolfowitz, J., 549, *597*
Woodward, P. M., 5, *38*, 140, *141*, 202, 203, 205, 206, 209, 228, 229, *251*, *335*

Youla, D. C., 60, 68, *73*

Zadeh, L. A., 145, 181, *184*, 192, *201*
Zamanakos, A., 162, 165, 166, *184*
Zraket, C. A., 35, 36, 37, 38, *39*
Zubakov, V. P., 6, *39*, 57, *73*, 177, *184*, 599, *620*

SUBJECT INDEX

Adaptive system, 283
Alternative hypothesis, 258
Ambiguity function, 5–6, 229, 250
Amplitude-modulated pulse train, 521
  optimum receiver structure, 521–524
Analytic approximations for multi-hit
    detection:
  nonfluctuating pulse train, 364–369
  Swerling I pulse train, 390–392
  Swerling II pulse train, 406–407
  Swerling III pulse train, 421–424
  Swerling IV pulse train, 438–439
Analytic approximations for single-hit
    detection, 315–319
  exactly known signal, 315
  fluctuating signal, 319
  signal of unknown phase, 316
Analytic signal, 58–59, 69–71
  Gabor relations, 69–71
    mean-square bandwidth, 71
    normalized first moment, 69
    normalized second moment, 70
    signal energy, 69
AN/FPS-6 tracking radar, 21
AN/SPG-49 Talos missile-tracking radars,
    22
Antenna, 450
  antenna gain, 450–454
  aperture efficiency, 453
  effective cross section, 453
  gain function, 450–457
  illumination function, 453–454
  noise temperature, 458–460
  power gain, 451
  radiation efficiency, 454
  radiation efficiency factor, 451
  radiation field, 450
  radiation pattern, 451
  receiving cross section, 452
  receiving pattern, 452
  solid beamwidth, 451
Anticharacteristic function, 94
A posteriori detection, 229–233
  existence probability, 233
  marginal a posteriori probability density,
    231

A posteriori detection (*Cont.*):
  structure of the optimum receiver, 231
A posteriori distribution, 210
A posteriori receiver, 209–219
A posteriori theory, 202–249
ARIS, 30–35
  AMR (Atlantic Missile Range), 32
  CDCE (Central Data Conversion
    Equipment), 31
  data-processing system, 31
  OCC (Operations Control Center), 31–32
Atmospheric attenuation, 461–462
Autocorrelation function, 111
Average loss, 255, 262, 277
Average sample number (ASN), 553–556
  (*see also* Sequential detection)

Bayes criterion, 290, 599
Bayes decision rule, 209, 257, 265–270, 602,
    603–606
  acceptance region, 266
  Bayes receiver, 269, 608, 617
  examples, 267–270
    constant (dc) signal in Gaussian noise,
      267–269
    Gaussian signal in Gaussian noise,
      269–270
  generalized likelihood ratio, 266
  rejection region, 266
Bayes strategy, 337, 600–611 (*see also*
    Multiple-target detection)
  false-alarm probability, 337
  target detection, 337
Beacon radar equation, 472
Beam shape loss, 463–464
Binary detection, 263–265
  average loss, 265
  cost matrix, 264
  null hypothesis, 263
Binary detector, 497–514
  block diagram, 498–499
  for colored noise, 500
  description, 498–499
  optimizing thresholds, 502, 506
  performance of, 499–514
    extension to fluctuating targets, 514

Binary detector (*Cont.*):
  probability of detection, 502
  probability of false alarm, 502
  sensitivity to pulse interference, 498
Binomial distribution, 500
  Gram-Charlier approximation
    (Edgeworth series), 507
Blip-scan ratio, 476
British Home Chain, 4

Central limit theorem, 96, 104, 359
Central moments, 103
Characteristic function, 94–96, 103, 345,
    365, 377, 394, 409, 426, 562
  anticharacteristic function, 94, 379, 404,
    410, 426
  joint characteristic function, 95
  moments of a random variable, 95
  standard tables, 94
  sum random variable, 95
Characteristic function of n, 564–566
Chebyshev's inequality, 106
Chi-square distribution, 346
Clutter, 24–25, 170–171
Clutter power spectrum, 171
Coherent detection, 289–333
Coherent processing (pulse train), 338–339
Coincidence detection, 515–518
  double coincidence, 515–516
  performance in Gaussian noise, 517–518
  sensitivity to large random-pulse
    interference, 516
  triple coincidence, 515–516
Collapsing loss, 464–467
Colored noise likelihood ratio, 319–324
Colored noise with infinite observation
    interval, 196–198
  optimum filter, 197
  prewhitening filter, 196
Comparison of pulse train performance,
    441–443
Comparison of single- and double-loop
    integrators, 540, 542
Completeness, 321
Complex cost functions, 278–280
  decision rule, 279
Complex envelope, 58, 60, 222
Complex representations of narrowband
    waveforms, 57–61
  analytic signal, 58
  complex envelope, 58
  complex exponential function, 58
  normalized rms difference, 60
Complex noise process, 320
Composite alternative, 258
Conditional loss, 262–263
Conditional probability, 79

Conditional probability density, 219
Confluent hypergeometric function, 311,
    393
Convolution integral (convolution theorem),
    47, 87, 242
Correlation, 93
Correlation coefficient, 103
Correlation receiver, 203, 214, 216
Covariance, 92, 103
Covariance function, 92, 99
  correlation coefficient, 92
  covariance matrix, 99, 234, 559
  normalized covariance, 92
Cross correlation, 114, 236
Cross correlator, 293
Cumulative detection probability, 288,
    476–484
  moving targets, 481–484
  stationary targets, 477–481

Decision rule, 258
Delay-line integrator, 334, 524–525
Detection in colored Gaussian noise,
    324–332
  exactly known signal, 324–328
    receiver performance, 327
    receiver structure, 324
  signal of unknown phase, 328–332
    receiver performance, 330
    receiver structure, 328
    test strategy, 329
  signal of unknown phase and amplitude,
    332
Detection in white Gaussian noise, 291–315
  exactly known signal, 291–298
    detection characteristic, 297
    receiver performance, 294
    receiver structure, 291
    test statistic, 293
  one-dominant-plus-Rayleigh fluctuating
    signal, 313–315
    detection characteristic, 314
  Rayleigh fluctuating signal, 309–313
    detection characteristic, 312
    receiver performance, 310
    receiver structure, 309
  signal of unknown phase, 298–309
    detection characteristic, 308
    receiver implementation, 301
    receiver performance, 302
    receiver structure, 298
    test strategy, 301
Detection problem, 219
Deterministic random process, 238
Direct probability, 231
Distribution of n, 566–571
Doppler approximation, 623–625

Dot product, 56
Double-loop integrator:
  description, 526–527
  detection probability for Gaussian
    envelope, 540–542
  parameter optimization, 530–531
  performance, 540
  performance evaluation procedure,
    531–536
    accuracy check, 536–538
  weighting function of, 526–527
Double threshold detection (*see* Binary
  detector)
Dwell time, 468

Edgeworth series, 364
Effective input temperature, 449–450
Eigenfunctions, 67, 321
Eigenvalues, 67, 321
Energy spectral density, 115
Energy variant sequential detection, 592
Equivalent video signal-to-noise ratio,
  503
Ergodicity, 113–114
  conditions, 114
  definition, 114
Error probabilities, 270–271
  false-alarm probability, 271, 554, 576
  false-dismissal probability, 271, 576
  level of test, 270
  power of test, 270
ESAR array radar, 15
Excess over boundaries, 550–551
Existence, 611–613 (*see also* Multiple-
  target detection)
  average loss, 611
  Bayes strategy, 612
  cost matrix, 611
  difficulties, 618–620
Expectation (*see* Statistical averages)

False-alarm rate, 614, 626–629
False-alarm time, 465
Finite-time cross-correlation, 215
Fluctuating target, 25–29, 373–374 (*see also*
  Radar cross section *and* Swerling
  pulse trains)
  one-dominant-plus-Rayleigh, 29
  radar signatures, 29
  Rayleigh, 28
Fourier integral theorem (*see* Fourier
  transform)
Fourier inversion formula, 44
Fourier series, 49–51
  expansion of narrowband waveforms, 61
  periodic functions, 49
  sine-cosine form, 51

Fourier transform, 44–48
  of analytic signal, 69–71
  Fourier transform pairs, 46
  sufficient conditions, 44
    absolute integrability, 44
    Plancherel's theorem, 44
  table of transforms, 48
FPS-35 radar, 11
Fruit machine, 4
Fredholm integral equation:
  first kind, 146, 185, 242
  second kind, 67, 194
Frame time, 468

Gabor relations, 69–71 (*see also*
  Analytic signal)
Gamma function, 311
  incomplete, 347
Gaussian (normal) probability density,
  96–100
  central moments, 98
  characteristic function, 96
  example, 96
  multidimensional distribution, 98–100
  transformation of variables, 100
Gaussian random process, 122–124
  covariance matrix, 123
  definition, 123
  Fourier series representation, 124–127
  narrowband Gaussian random process,
    129–136
    covariance matrix of quadrature
      components, 135
    covariance of quadrature components,
      131–132
    cross correlation of quadrature
      processes, 134–135
    envelope and phase random processes,
      136
    joint density of quadrature
      components, 135
    marginal density of envelope, 132
    quadrature components, 130
    spectral density, 129
    orthonormal expansion (Karhunen-
      Loéve), 127–129
Generalized likelihood ratio, 277, 299,
  310, 339, 375, 393
Gram-Charlier series, 359, 362–364, 507
  definition, 359
  Edgeworth grouping, 364
Ground noise, 460

Hermite polynomials, 362
Hermitian symmetric, 321
Hilbert transform, 58
Hypothesis testing, 253

Ideal observer, 271–272, 277–278
Ideal receiver, 5, 209–210
  a posteriori receiver, 209
  direct probability, 209
  inverse probability, 209
Identity matrix, 107
Impulse function, 49
Incoherent pulse train, 336
Incomplete gamma function, 347
Incomplete Toronto function, 307, 309, 348
Information theory, 203–209
  average information transfer, 250
  information measure:
    continuous case, 208–209
    discrete case, 203–208
    units, 206
  information transfer, 204
  uncertainty, 205
Instantaneous fluctuation power, 120
Instantaneous power, 120, 296
Instantaneous signal power, 296, 298
Integrable square, 66
Intuitive substitute criterion, 281–282, 290,
    299, 323–324, 337
Inverse filter, 171
Inverse Fourier transform, 168
Inverse matrix, 561, 595
Inverse probability, 209, 231

Jacobian, 89, 102, 132, 627
Jamming radar equation, 473
Joint central moments, 103
Joint marginal density, 102
Joint moments, 103
Joint probability, 78
  conditional, 79
  independent events, 78
  overlapping events, 79
Joint probability density, 84
  definition, 84
  example, 84
  useful relations, 84–86
Joint probability distribution, 84–86

Karhunen-Loéve expansion, 127, 239, 319,
    321–322, 325 (*see also* Orthogonal
    function representations)
Kernel, 67
  hermitian symmetric, 67
  real symmetric, 67

Least mean-square departure, 214
Lebesgue class L^2, 66
Level of test, 280
Likelihood ratio, 232, 596
Limit in the mean, 45, 66, 321

Linear dependence, 93
Linear mean-square regression line, 93
Loss function, 261–263
  average loss, 261
  Bayes rule, 261
  conditional loss, 261

Marcus and Swerling test, 582–591 (*see also*
    Sequential detection)
  average sample number, 584, 586
  binomial Marcus and Swerling test,
    588–591
    average sample number, 589
    power saving, 590
  coherent detector, 583
  power saving, 585, 587
  test strategy, 583
Marginal probability density, 86, 102, 596
Marginal probability distribution, 86
Markoff process, 560
MARS, 30 (*see also* ARIS)
Matched filter, 146–170
  conjugate filter, 148
  example, 147
  peak signal-to-noise, 149
  physical realizability, 181
Matched filter approximations, 149–170
  digital-coded pulse with rectangular
    envelope, 167–170
    binary sequences, 170
    output signal, 169
    physical realizability, 168
    transfer function, 168
  finite pulse train comb filter, 161–167
    delay-line comb filter, 167
    North sine weighted comb filter,
      165–166
    output signal, 162
    pulse-doppler processor, 167, 183
    rectangular sine weighted comb filter,
      165
    uniform comb filter, 162
  linear fm pulse with rectangular envelope
    155–161
    group delay, 161
    output signal, 156
    phase approximation, 158
    rectangular approximation, 157
    signal spectrum, 157
    time-bandwidth product, 158
  rectangular rf pulse, 149–155
    comparison of filters, 153
    output signal, 151
    rectangular approximation, 152
Mathematical expectation, 91, 103 (*see
    also* Statistical averages)

Maximization of signal-to-noise ratio, 144–146
  solution to integral equation, 145–146
Maximum signal-to-noise ratio (colored noise), 198–199
Mean-square bandwidth, 71
Minimax criterion, 274–277, 278
  definition, 275
  example, 276
  properties, 275
Mismatch loss, 464
MIT Lincoln Laboratory Millstone Hill Radar, 19
Mixed kernel (white plus colored noise), 193–196
  autocorrelation function, 193
  complementary solution, 195
  optimum filter impulsive response, 195
  particular solution, 195
  spectral density, 193
Modified likelihood ratio (colored noise), 323
Modified Neyman-Pearson strategy, 273–274, 278, 280
Modified Rayleigh density function, 307
Moment-generating function, 562, 564
Moments of random variables, 91–92, 95
  (*see also* Statistical averages) .
Multiple-target detection, 598–611
  approach, 599
  existence of single target, 611–613
  N exactly known signals, 606–611
    Bayes strategy, 607–608
    structure of Bayes receiver 608
    unknown times of arrival, 609–611
  optimum receiver approximation, 613–618
    approximate Bayes receiver, 617
    extension to angle and doppler, 615
  two known signals, 599–606
    average loss, 600–601
    Bayes decision rules, 602, 603–606
    cost matrix, 601
    interpretation, 603–606
    structure of Bayes detector, 604
Mutually exclusive events, 77

Narrowband representation, 57–61, 623
Neyman-Pearson criterion, 272–274
  definition, 273
  Lagrange multiplier, 273
  modified Neyman-Pearson, 273
Noise factor, 449, 456–458
Noise temperature, 457–458
Noise types, 41
Nondeterministic random process, 238

Nonfluctuating incoherent pulse train, 339–359, 364–371
  envelope detector, 344
  integration loss, 359
  large-signal optimum receiver, 342–343
    block diagram, 343
  linear versus square-law, 370–371
  optimum receiver, 339–344
    analytic approximations, 364–369
    block diagram, 341
  small-signal optimum receiver, 342
    block diagram, 343
    small-signal receiver performance, 344–358
Nonrandom decision rules, 277
Normal distribution (*see* Gaussian probability density)
Normalized second moment, 199
Null hypothesis, 258

One-dominant-plus-Rayleigh probability density 313, 407
One-dominant-plus-Rayleigh target, 281, 333
Operating characteristic function, 550–553, 557
Optimum filter, 144–147, 185–199, 325–327
  output signal, 176
  peak output signal-to-noise ratio, 144, 149, 151, 156, 170, 176, 199
  colored LRC kernel, 199
  colored noise, 185–199
  colored RC noise, 186–191
  examples (RC noise), 189
  method of solution, 186
  rational kernel, 186
  time-domain implementation, 191
Optimum reception theories, 139–140
  decision theory, 140
  "ideal observer," 140
  maximization of output signal-to-noise ratio, 139
Orthogonal function representations, 66–68
  completeness, 66
  eigenfunctions, 67
  eigenvalues, 67
  Gram-Schmidt orthogonalization procedure, 68
  orthonormal set, 66
  sufficient conditions, 68

Paley-Wiener criterion, 238
Parameter estimation, 177–181, 260, 611
  doppler, 178, 615
    signal spectra, 178
  time of arrival, 178–181, 611–615

Parseval's formula, 48
Partially correlated fluctuating pulse
    trains, 439–440
Peak output signal-to-noise ratio, 144,
    149, 151, 156, 170, 176, 199
Probability density of n, 571–572 (*see also*
    Sequential detection)
Positive definite, 321
Positive definite matrix, 235
Power-aperture minimization, 484–493
Power functions, 281
Power of test, 280
Power series expansion, 311, 342, 393, 573
Power spectral density, 115–116, 119,
    121–122, 148, 629
  correspondences, 122
  cross-spectral, 119
  definition, 116
  shape, 121
    colored, 121
    white, 121
  table, 121
  Wiener-Khintchine theorem, 117–121
Predetection integration, 289, 338
Preferred Neyman-Pearson strategy, 280–
    281
Probability, 74–107
  compound event, 77
  definition, 74–76
  mutually exclusive events, 77
  probability of an event, 80
  relative frequency of occurrence, 76
  statistical regularity, 75
Probability density function, 81–86, 101
  Cauchy density function, 106
  definition, 81
  examples, 82, 83
  mass density interpretation, 82
  Rayleigh probability density, 106
  uniform probability density, 105
Probability distribution, 81–84, 101
  definition, 81
  examples, 82, 83
  properties, 81–84
Propagation loss, 462–463
  propagation loss factor, 462

Q function, 307, 309
Quadrature components, 60 (*see also*
    Complex representations of narrow-
    band waveforms)

Radar:
  classifications, 7–8
  electronic scan, 15
  recent developments, 5–7
  special-purpose radars, 29

Radar (*Cont.*):
  system integration, 30–38
  uses, 4, 29
Radar ambiguity function, 5–6, 229, 250
Radar cross section, 25–29, 455–456
  backscattering coefficient, 455
  effective echoing area, 455
  optical region, 456
  Rayleigh region, 455–456
  resonance region, 456
  simple bodies, 456
  time varying, 373–374
    pulse-to-pulse, 374
    scan-to-scan, 373–374
Radar equation, 333, 446–473
  classical, 446–448
  search, 468–470
  and signal-to-noise ratio, 448–450
Radar statistical problem, 24–29
Random channels (*see* Time varying
    channels)
Random event, 76
Randomized decision rule, 259
Random process, 108–136
  characterization, 109–111
  cross-correlation function, 112
  ensemble of functions, 110
  independent random processes, 112
Random variable, 79–80
  definition, 80
  random function, 80
Range estimation, 219–229
  envelope detector, 224
  fine structure in a posteriori density, 223
  ideal receiver, 220
  passive matched filter implementation,
    221
  range accuracy, 228
  sufficient receiver, 221
Rational kernel solution of Fredholm
    integral equation (first kind), 191–193
  complementary solution, 192
  "steady-state" solution, 192
Rayleigh target, 281
Rayleigh probability density, 133, 304,
    305, 376
Resolution cell, 15, 16, 290–291, 337,
    615–618
  definition, 15
  typical, 16
Rician probability density, 307

Saddle-point integration, 532-534
SAGE (Semiautomatic Ground
    Environment System), 29, 35–38
  computer, 36–37
  direction center, 35–36, 38

Sample space, 76–78
  compound event, 77
  definition, 76
  discrete, 76
  mutually exclusive events, 77
  sample space for a die, 77
Sampling-theorem representation of noise, 218
Sampling theorem (Shannon-Kotelnikov), 51–57, 61–66
  high-frequency band-limited signals, 52, 61–66
  high-frequency sampling in Cartesian and polar coordinates, 64
  reconstruction of signals, 54
  sampled approximation of signal, 54, 55
  vector representation, 55
  video band-limited signals, 51–57
Scanning loss, 463–464
Schwarz inequality, 186
Search problem, 8–15, 615–618
  detection, 8
  fan beam, 11–12
  pencil beam search, 14
  ring search, 15
  sequential search, 9, 616
  target existence, 10, 29
  3-D radar, 12
  V-beam, 12–13
  volumetric search, 8, 616
Search radar equation, 468–470
Search radar status, 470–472
Sequential detection, 547–595
  average loss, 591
  average sample number, 553–556
  coherent detection, 556–562
    average sample number, 559
    continuous solution, 561
    operating-characteristic function, 557
    test statistic, 560
  degree of preference, 548
  incoherent detection, 572–575
    average sample number, 573
    operating characteristic function, 573
  Marcus and Swerling test, 582–591
  multiple-resolution-bin radar, 575–582
    average global test length, 577
    average sample number, 577
    coherent detector, 578
    global probability of false alarm, 576
    global probability of missed detection, 576
    incoherent detector, 580, 582
  notion of a sequential test, 548–550
  operating characteristic function, 550–553
  probability density function of n, 567–568
  probability of detection, 554

Sequential detection (*Cont.*):
  probability of false alarm, 554
  sequential probability ratio test, 549
  two-step sequential detection, 592–595
    false-alarm probability, 592
    optimum parameters, 593
    power saving, 594–595
    probability of detection, 592
    zone of indifference, 549
Sequential search, 616
Sequential testing, 282 (*see also* Sequential detection)
Shannon-Kotelnikov theorem (*see* Sampling theorem)
Simple alternative, 258
Single-loop integrator:
  description, 525–526
  detection probability for Gaussian envelope, 538–540
  parameter optimization, 528–530
  performance, 540
  performance-evaluation procedure, 531–536
    accuracy check, 536–538
    weighting function, 525–526
Sky noise, 459–460
  atmospheric absorption noise, 459
  cosmic noise, 459
Spectral density (*see* Power spectral density)
Spectrum of analytic signal, 59
Stationarity, 113–115
  nonstationary processes, 113
  strict-sense stationarity, 113
  wide-sense stationarity, 113
Statistical averages, 89–94, 102–103
  central moments, 91, 103
  covariance, 92, 103
  ensemble average, 90
  joint central moments, 92, 103
  joint moment, 92, 103
  linear operator, 91
  mathematical expectation, 90, 102
  mean, 90, 102
  mean-square value, 91, 103
  moments of random variables, 91, 103
  root-mean-square, 91
  standard deviation, 91
  variance, 91, 103
Statistical decision theory, 203, 252–263
  cost matrix, 255
  criterion of optimality, 253
  decision rule, 253
  elements, 253
  hypotheses, 253
  loss function, 261–263
  loss matrix, 255
Statistical independence, 80, 101

Statistical inference, 253
Stirling's approximation, 347
Sufficient receiver, 214–219, 229–239, 241, 243–249
    active correlation, 216
    cross correlation, 215
    cross-correlation receiver for colored noise, 237
    matched filter receiver, 217
    passive filter realization, 243
    prewhitening filter, 237
    random channel, 243–249
    sufficient statistic for colored noise, 235, 241
Sum of random variables, 86–88
    definition, 87
    joint sample space, 87
    probability density, 87
Superposition integral, 47, 87, 242
Surveillance volume, 8–16, 599
Swerling I pulse train, 374–392
    optimum receiver performance, 377–390
    receiver structure, 374–376
        analytic approximations of performance, 390–392
        integration loss, 390
Swerling II pulse train, 392–407
    optimum receiver performance, 394–404
    receiver structure, 393–394
        analytic approximations, 406–407
        integration loss, 404–406
Swerling III pulse train, 407–424
    optimum receiver performance, 408–421
    receiver structure, 408
        analytic approximations, 421–424
        integration loss, 421
Swerling IV pulse train, 424–439
    optimum receiver performance, 426–437
    receiver structure, 425–426
        analytic approximations, 438
        integration loss, 427, 438
System loss factor, 450, 461

Talos missile tracking radars, 22
Target existence, 225, 233, 611–613
Taylor series expansion, 569–570
Time autocorrelation, 115, 148
Time average, 114
Time-varying channels, 243–249
    channel covariance matrix, 245
    deterministic channel, 247
    estimating filter, 248
    examples, 246–249
    likelihood function, 245
    optimum receiver structure, 247, 248
    prewhitening filter, 248
    random channel, 247–248
    sufficient statistic, 246

Time-varying channels (*Cont.*):
    tapped delay-line representation, 244
Time-varying filter, 183
TPS-34 radar, 14
Track problem, 16–24
    amplitude comparison, 17
    closed-loop null tracking system, 21
    conical scan, 18, 20
    curve fitting, 23
    early-late gate, 20 (*see also* Parameter estimation)
    extraction, 16
    interpolative tracking, 23
    monopulse tracking, 17–20
    null tracking, 22
    sequential lobing, 20
    target parameters, 17
    track-while-scan, 23
Transformation of random variables, 88–89
    inverse functions, 89
    Jacobian, 89
    one-to-one mapping, 88
Truncation, 559, 575
Two-step detection, 592–595 (*see also* Sequential detection)

Unbiased tests, 281
Unconditional maximum-likelihood estimation, 601
Uniformly most powerful test, 280, 310, 313
Uniform probability density, 133
Unit step function, 83
Urkowitz filter, 170–177
    comparison with rectangular filter, 176–177
    inverse filter, 171
    peak output signal-to-clutter ratio, 172
    transfer function, 173
U. S. Tactical Air Command Forward Director Post, 30
    operations shelter, 30, 32
    radar control center, 31

Vandermonde matrix, 193
Variance, 91, 103
Volume search (*see* Search problem)

Wald's fundamental identity, 562–564
Weighted integrators, 521-542
Wiener-Hopf integral equation, 186, 193
Wiener-Khintchine theorem, 117–122, 146, 234
    correspondences, 122
    definition, 117
    example, 119
    power spectral density, 119
    sufficient condition, 117
    table, 121